Guided-Wave Optics

Special Issue Editor
Boris Malomed

MDPI • Basel • Beijing • Wuhan • Barcelona • Belgrade

MDPI

Special Issue Editor
Boris Malomed
Tel Aviv University
Israel

Editorial Office
MDPI AG
St. Alban-Anlage 66
Basel, Switzerland

This edition is a reprint of the Special Issue published online in the open access journal *Applied Sciences* (ISSN 2076-3417) in 2017 (available at: http://www.mdpi.com/journal/applsci/special_issues/guided_wave_optics).

For citation purposes, cite each article independently as indicated on the article page online and as indicated below:

Author 1; Author 2. Article title. *Journal Name* **Year**, *Article number*, page range.

First Edition 2017

ISBN 978-3-03842-614-1 (Pbk)
ISBN 978-3-03842-615-8 (PDF)

Table of Contents

About the Special Issue Editor

Boris Malomed Boris Malomed was born in Minsk (Belarus, ex-USSR) in 1955. He graduated from the Department of Physics of the Belorussian State University in Minsk in 1977. He received his PhD in physics from the Moscow Physico-Technical Institute in 1981, and D.Sc. (докторская физико-математических наук, alias habilitation) from the N. N. Bogoliubov Institute for Theoretical Physics of the Ukrainian Academy of Sciences (Kiev) in 1989. He worked in Moscow till September 1991 as a senior researcher (ведущий научный сотрудник)) at the Institute for Oceanology of the Russian Academy of Sciences. He has been with the Tel Aviv University since 1991, as an associate professor in 1991–1999, and as a full professor since 1999. He has been a chaired professor (holding a personal research chair on "Optical Solitons") since 2012. He is an editor of journal "Chaos, Solitons & Fractals" since November 2017. His current h-index is 66 (Web of Science), 68 (Scopus), and 80 (Google Scholar).

His recent research work has been focused on theoretical studies of nonlinear waves and nonlinear dynamics in optics, matter waves (Bose–Einstein condensates), dissipative media (modeled by Ginzburg–Landau equations), dynamical lattices, and in related systems. The main subjects of these studies are solitons (self-trapped solitary waves) and solitary vortices, as well as multi-soliton complexes and periodic or quasi-periodic patterns, in models of conservative and dissipative nonlinear media. These localized and delocalized structures have been studied in both one- and multidimensional settings. In particular, in the latter case, a challenging problem is the stabilization of solitons and solitary vortices against the collapse (spontaneous formation of singularities) and splitting. Addressing this problem, his recent works have revealed stable localized two- and three-dimensional modes carrying intrinsic topological structures, such as multiple vortices, semi-vortices (two-component bound states of vortices and zero-vorticity solitons), and hopfions (twisted vortex rings, which carry two independent topological charges). He also continues to work in other directions which belong to the above-mentioned general area, such as solitons supported by nonlinear lattices, i.e., spatially periodic modulations of the local strength of the self-interaction of the underlying physical fields, solitons in media with nonlocal interactions (in particular, in dipolar Bose–Einstein condensates), nonlinear systems with the so-called PT symmetry, which is represented by spatially separated and mutually balanced gain and loss elements, and others.

Currently, his research work is supported, inter alia, by a major grant jointly provided by NSF (USA) and Binational (US-Israel) Science Foundation (BSF), on the topic of dynamics of matter-wave solitons, and by another major personal research grant (with Prof. Malomed as a single principal investigator) on the topic of solitons in spin-orbit-coupled systems, provided by the Israel Science Foundation. Collaborators in the NSF-BSF project are Profs. Randall Hulet (Rice University, Houston), Maxim Olshanii (University of Massachusetts, Boston), and Vladimir Yurovsky (Tel Aviv University). Prof. Malomed also maintains active research collaborations with other colleagues in several countries, including China, Japan, USA, Italy, Spain, Portugal, France, UK, Russia, Serbia, Romania, Mexico, Chile, Brazil, and Israel. He was recently appointed a senior international consultant at the Foshan University in China, and an international expert of the Guangdong province, China.

Prof. Malomed is currently an adviser to a postdoc, three PhD and two MS students. His publication list includes two books, ca. 20 review articles, and more than 1000 original papers in peer-reviewed journals. His h-index is 66, 68, and 79, as given, respectively, by the Web of Science, Scopus, and Google Scholar, with the total number of citations in excess of 33,400. This means that he belongs to a rather elite group of researchers whose h-index exceeds their age. Prof. Malomed is No. 30 in the nation-wide ranking list of Israeli scientists (in all sciences), as provided by Google Scholar. He served, in the period of 2009–2015, as a Divisional Associate Editor of Physical Review Letters. He is an Outstanding Referee of the American Physical Society and Optical Society of America.

applied sciences

MDPI

Editorial

Editorial: Guided-Wave Optics

Boris A. Malomed [1,2]

[1] Department of Physical Electronics, School of Electrical Engineering, Faculty of Engineering, Tel Aviv University, Tel Aviv 69978, Israel; malomed@post.tau.ac.il
[2] ITMO University, St. Petersburg 197101, Russia

Received: 11 September 2017; Accepted: 15 September 2017; Published: 20 September 2017

1. An Overview of the Topic and Its Ramifications

1.1. Introduction

Guided waves represent a vast class of phenomena in which the propagation of collective excitations in various media is steered in required directions by fixed (or, sometimes, reconfigurable) conduits. Arguably, the most well-known and practically important waveguides are single-mode and multi-mode optical fibers [1,2], including their more sophisticated version in the form of photonic crystal fibers [3] and hollow metallic structures transmitting microwave radiation [4]. Light pipes, in the form of hollow tubes with reflecting inner surfaces, are used in illumination techniques. On the other hand, medical stethoscopes offer a commonly known example of a practically important acoustic waveguide. New directions of studies in photonics are focused on waveguides for plasmonic waves on metallic surfaces [5–7] (which provide the possibility of using wavelengths much smaller than those corresponding to the traditional optical range, and thus offer opportunities to build much more compact photonic devices) and on the other hand, on the guided transmission of terahertz waves, which also have a great potential for applications [8].

Outside of the realm of photonics (optics and plasmonics) and acoustics, wave propagation plays a profoundly important role in many other areas; accordingly, waveguiding settings have drawn a great deal of interest in those areas as well. In particular, as concerns hydrodynamics, natural waveguides—which may be very long—exist for internal waves propagating in stratified liquids (e.g., in the ocean) [9]. Various settings in the form of waveguides for matter waves are well known in studies of Bose–Einstein condensates in ultracold bosonic gases [10,11]. In solid-state physics, guided propagation regimes for magnon waves in ferromagnetic media are a subject of theoretical and experimental studies [12]. In superconductivity, long Josephson junctions are waveguides for plasma waves [8,13]. The significance of waveguiding in plasma physics is also well-known; e.g., Ref [14–16].

Below, a very brief overview of basic theoretical models and experimental realizations of various physical implementations of the waveguiding phenomenology is given. The text is structured according to the character of the guided wave propagation: linear or nonlinear and conservative or dissipative, as well as according to the materials used in the underlying settings, natural or artificial.

This presentation definitely does not aim to include an exhaustive bibliography on this vast research area. References are given chiefly to review articles and books summarizing the known results, rather than to original papers where the results were first published. However, in some cases original papers are also cited if it is necessary in the context of the presentation.

1.2. Linear Waveguides

The basic waveguiding structure is a single-mode conduit, designed with a sufficiently small transverse size and boundary conditions at the boundary between the guiding core and surrounding cladding, which admits the propagation of a single transverse mode, while all higher-order modes get imaginary propagation constants (i.e., they cannot actually propagate). A commonly known—and arguably the most important—example is provided by single-mode optical fibers (although, strictly speaking, all such fibers are bimodal if the polarization of light is taken into consideration) [17,18]. Single-mode waveguides are crucially important components of telecommunication systems, while other applications (e.g., the delivery of powerful laser beams for material processing and the creation of complex spatiotemporal patterns) are best served by multimode conduits [19,20].

Parallel to waveguiding fibers, planar waveguides are a subject of many studies in optics. In the corresponding models, as well as in their fiber-optics counterparts, the evolution variable is the propagation distance, z; see Equation (2) below (this is a common feature of all guided-wave-propagation settings, not only in optics, but in other physical realizations of waveguides as well). Meanwhile, the transverse coordinate, x, in the *spatial domain* plays the same role as the reduced-time variable,

$$\tau \equiv t - V_{gr}^{-1}z, \tag{1}$$

where t is the time proper, and V_{gr} is the group velocity of the carrier wave in the *temporal domain* in fiber optics. The waveguiding structure in the planar waveguide is represented (roughly speaking) by a stripe with a locally increased effective refractive index.

Effective equations which model the temporal-domain propagation of optical waves in fibers and the spatial-domain propagation in planar waveguides are similar to each other, taking the form of the linear Schrödinger equation for local amplitude u of the electromagnetic wave, which is written here in terms of the spatial-domain propagation, and in the scaled form:

$$i\frac{\partial u}{\partial z} + \frac{1}{2}\frac{\partial^2 u}{\partial x^2} - U(x)u = 0. \tag{2}$$

In particular, the aforementioned stripe waveguiding channel is represented by trapping potential $U(x)$ in Equation (2), while the second derivative in Equation (2) represents the paraxial (weak) transverse diffraction in the planar waveguide. A ubiquitous form of the potential is

$$U(x) = -\epsilon \operatorname{sech}^2(x/l), \tag{3}$$

where $\epsilon > 0$ determines the effective depth of the potential well, and l determines its width. In the temporal domain, the transverse coordinate, x, is replaced by the above-mentioned temporal variable (1). and the diffraction term in Equation (2) is replaced by $-(\beta/2)\partial^2 u/\partial\tau^2$, where β is the coefficient of the group-velocity dispersion ($\beta > 0$ and $\beta < 0$ correspond to the normal and anomalous dispersion, respectively).

Further, the similarity between the wave-propagation Equation (2) in optics and the Schrödinger equation in quantum mechanics suggests a similarity between the guided transmission of waves in the guiding channel and propagation of real quantum particles in holding channel potentials [21]. The consideration of the transport of quantum particles in such channels gives rise to many intriguing peculiarities, such as the consideration of curved guiding channels. In this context, it is relevant to mention a well-known result which demonstrates a strong effect of the confinement imposed by a pipe-shaped potential on the character of the effectively one-dimensional mutual scattering of two quantum particles, which amounts to full reflection of the colliding particles [22]. This theoretical prediction had suggested the experimental realization of the concept of the *Tonks–Girardeau gas*; i.e., a gas composed of *hard-core bosons*, which bounce back from each other when they collide [23,24].

A natural generalization of single-channel waveguides is provided by a *coupler*, which may be considered as a set of two parallel waveguiding cores, coupled in the transverse direction by tunneling of guided wave fields steered by each tunnel in the longitudinal direction. The respective system of coupled equations for amplitudes u and v of electromagnetic waves in the two cores is [25] (cf. Equation (2)):

$$i\frac{\partial u}{\partial z} + \frac{1}{2}\frac{\partial^2 u}{\partial x^2} + \kappa v - U(x)u = 0.$$

$$i\frac{\partial v}{\partial z} + \frac{1}{2}\frac{\partial^2 v}{\partial x^2} + \kappa u - U(x)v = 0,$$

(4)

where κ is the coefficient of the linear inter-core coupling.

The next step is to consider *arrayed systems*, composed of many parallel guiding cores, which are also coupled in the transverse direction(s) by the tunneling of longitudinally guided wave fields (planar and bulk arrays have, respectively, one or two transverse coordinates). The simplest model of such a guiding medium is provided by the two- or three-dimensional scaled Schrödinger equation with a periodic transverse potential, which represents the (idealized) structure of the multi-core bundle:

$$i\frac{\partial u}{\partial z} + \frac{1}{2}\left(\frac{\partial^2 u}{\partial x^2} + \frac{\partial^2 u}{\partial y^2}\right) - \epsilon\left[\cos\left(\frac{2\pi x}{l}\right) + \cos\left(\frac{2\pi y}{l}\right)\right]u = 0.$$

(5)

Here l is the array's period (defined in scaled units, in which Equation (5) is written), and 2ϵ is the scaled depth of the effective trapping potential. In particular, in optics bulk arrays have been created as permanent structures by burning (also by means of an optical technology) a large number of parallel guiding cores in a bulk piece of silica [26]. As concerns planar guiding arrays, an interesting ramification of the topic is the propagation of optical waves in such arrays made with a curved shape [27]. On the other hand, a technology for the creation of reconfigurable *virtual* conduit patterns in the form of photonic lattices was elaborated for photorefractive materials [28]. The latter technology makes use of the fundamental property of the photorefractive materials, in which the propagation conditions for light with ordinary and extraordinary polarizations are linear and nonlinear, respectively. To create a photonic lattice, the experimentalist first illuminates the sample by counterpropagating pairs of mutually coherent laser beams in the ordinary polarization, which create a classical interference pattern in the photorefractive crystal, which is an effectively linear medium for these beams. Next, a probe beam is launched, with the extraordinary polarization in the transverse direction. Due to its inherent nonlinearity, the probe beam is affected by the originally created photonic lattice, as if it is a material structure that creates a spatially periodic modulation of the local refractive index in the transverse directions; i.e., essentially, another version of the multi-core guiding structure.

The propagation of light or waves of a different physical nature in arrays with weak coupling between guiding cores may be naturally approximated by the discrete Schrödinger equation. The basic realization of such a medium is represented by planar arrays of parallel optical waveguides coupled by evanescent waves penetrating dielectric barriers separating individual cores, the basic model being a scaled discrete version of Equation (2):

$$i\frac{du_n}{dz} + \frac{1}{2}\left(u_{n+1} + u_{n-1} - 2u_n\right) - U_n u_n = 0,$$

(6)

where the discrete coordinate, n, which replaces x, is the number of the guiding core in the array. The study of light propagation in various multi-core systems—which may be approximated by lattice models similar to Equation (6)—is a vast area known as *discrete optics* [29].

1.3. Nonlinear Waveguides

In many situations, tightly confined guided waves propagating in conduits with a small effective cross-sectional area acquire high amplitudes, which is a source of a great many fascinating nonlinear effects. In particular, waveguides often provide a combination of the nonlinearity, group-velocity dispersion, and low (or sometimes completely negligible) losses which are necessary ingredients for the creation of *solitons* (robust self-trapped solitary waves). The simplest and ubiquitous model of the nonlinear wave propagation is based on the nonlinear Schrödinger equation (NLSE), which in the simplest case includes a cubic term. In optics, this term represents the Kerr effect; i.e., nonlinear self-focusing (or, sometimes, self-defocusing) of light in the dielectric medium. The accordingly amended linear Schrödinger Equation (2) becomes the NLSE:

$$i\frac{\partial u}{\partial z} + \frac{1}{2}\frac{\partial^2 u}{\partial x^2} - U(x)u + \sigma|u|^2 u = 0, \tag{7}$$

where $\sigma = +1$ and -1 corresponds, respectively, to the self-focusing and defocusing nonlinearity; i.e., self-attraction and self-repulsion of light in the nonlinear medium. Equations (4) and (5) each acquire the same cubic terms as in Equation (7). In particular, the nonlinear version of Equation (4),

$$i\frac{\partial u}{\partial z} + \frac{1}{2}\frac{\partial^2 u}{\partial x^2} + \kappa v - U(x)u + \sigma|u|^2 u = 0,$$

$$i\frac{\partial v}{\partial z} + \frac{1}{2}\frac{\partial^2 v}{\partial x^2} + \kappa u - U(x)v + \sigma|v|^2 v = 0, \tag{8}$$

is the basic model of nonlinear couplers, their remarkable property being *spontaneous symmetry breaking* in the case of self-focusing in the parallel-coupled cores, $\sigma = +1$ [25,30,31].

A remarkable property of the one-dimensional NLSE in the absence of the potential ($U = 0$ in Equation (7)) is that it is an *integrable equation* for which a very large number of exact solutions—including multi-soliton states—can be produced by means of a mathematical technique based on the *inverse scattering transform* [32–34]. These are bright and dark solitons in the cases of self-focusing and defocusing, respectively. In particular, the exact bright-soliton solution to Equation (7) with $\sigma = +1$ and $U = 0$ is

$$u(x,z) = \eta \exp\left(\frac{i}{2}\left(\eta^2 - c^2\right)z + icx\right)\operatorname{sech}\left(\eta(x - cz)\right), \tag{9}$$

where η and c are, respectively, the arbitrary amplitude and velocity of the soliton (in fact, in the spatial domain—in terms of which Equation (7) is written—the soliton represents a self-trapped light beam, and accordingly c is not a velocity, but rather a parameter which determines the tilt of the beam in the (x, z) plane).

The discrete Schrödinger Equation (6) also has its natural nonlinear counterpart in the form of discrete NLSE:

$$i\frac{du_n}{dz} + \frac{1}{2}\left(u_{n+1} + u_{n-1} - 2u_n\right) - U_n u_n + \sigma|u_n|^2 u_n = 0; \tag{10}$$

i.e., a discrete version of NLSE (7). The discrete NLSE gives rise to discrete solitons and their bound states, which cannot be found in an exact form, but may be efficiently produced by numerical and approximate analytical methods [35]. The propagation of nonlinear waves in discrete waveguiding arrays was the subject of numerous theoretical and experimental works [29,36].

The multidimensional extension of the NLSE also has direct realizations in optics, as well as in the mean-field model of atomic Bose–Einstein condensates (BECs) [37,38], and in many other areas. In particular, the spatial-domain light propagation in bulk media is modelled by the effectively two-dimensional version of Equation (7), with two transverse coordinates (x, y):

$$i\frac{\partial u}{\partial z} + \frac{1}{2}\left(\frac{\partial^2 u}{\partial x^2} + \frac{\partial^2 u}{\partial y^2}\right) - U(x,y)u + \sigma|u|^2 u = 0. \tag{11}$$

Unlike its one-dimensional counterpart (7), Equation (11) in the free space ($U(x, y) = 0$) is not integrable. It admits formal soliton solutions, looked for as

$$u(x,y;z) = \exp\left(ikz + iS\theta\right)U_S(r), \tag{12}$$

in terms of the polar coordinates (r, θ) in the (x, y) plane, where $k > 0$ is a real propagation constant, $S = 0, \pm 1, \pm 2, ...$, is an integer *vorticity* that may be embedded in the two-dimensional soliton (shaping it as a *vortex ring*), and $U_S(r)$ is a real radial amplitude function satisfying boundary conditions $U_S(r) \sim \exp\left(-\sqrt{2k}r\right)$ at $r \to \infty$, and $U(r) \sim r^{|S|}$ at $r \to 0$. Solitons (12) with $S = 0$ are often called *Townes solitons* [39]. However, the Townes solitons—as well as their vortex counterparts, with $S \neq 0$ in Equation (12)—are completely unstable, being vulnerable to destruction by the *critical collapse* (formation of a singularity after a finite propagation distance) in the case of $S = 0$, and by a still stronger instability which splits vortex rings with $S \neq 0$ [39].

An important example of nonintegrable one-dimensional system modelling nonlinear light propagation in optics is the system of coupled-mode equations which describe the fiber Bragg gratings (i.e., nonlinear optical fibers with a periodic lattice of local defects permanently written in their cladding, with a period equal to half the wavelength of light coupled into this waveguide). The coupled-mode equations govern the evolution of amplitudes u and v of right- and left-traveling waves, which are mutually converted (reflected) into each other by the Bragg grating [40,41]:

$$iu_t + iu_x + \kappa v + \left(\frac{1}{2}|u|^2 + |v|^2\right)u = 0,$$

$$iv_t - iv_x + \kappa u + \left(\frac{1}{2}|v|^2 + |u|^2\right)v = 0,$$

(13)

where κ is the Bragg-grating reflectivity, and the group velocity of the light waves in the fiber is scaled to be 1. This system admits exact solutions in the form of solitons, but it is not an integrable one. Such solitons—moving in the fiber Bragg grating as in the waveguide—have been created in the experiment [42]. Roughly half of the soliton family is stable, and half unstable.

The use of fiber Bragg gratings operating in the linear regime has grown into a large industry with many applications, such as sensors, dispersion compensators, optical buffers, etc. [43].

Another fundamentally important nonlinear model for the guided wave propagation is the one with the quadratic (alias second-harmonic) nonlinearity, instead of the cubic (Kerr) term in NLSE (7). The model is based on the propagation equations for complex amplitudes $u(x, z)$ and $v(x, z)$ of the fundamental and second harmonics [44,45]:

$$iu_z + \frac{1}{2}u_{xx} + vu^* = 0,$$

$$2iv_z - qv + \frac{1}{2}v_{xx} + \frac{1}{2}u^2 = 0,$$

(14)

where q is a real mismatch parameter. Although it is a nonintegrable system, Equations (14) also give rise to solitons, which are generically found in a numerical form. These solitons form a family which is chiefly stable, with a small instability area [44,45].

In BEC models, Equation (7), with evolution variable z replaced by (scaled) time, t, is called the Gross–Pitaevskii equation, in which the cubic term represents—in the mean-field approximation—an average effect of collisions between atoms [37,38]. The natural sign of the collision-induced term corresponds to self-repulsion (self-defocusing) (i.e., $\sigma = -1$ in Equation (7)), but for atomic species such as ^{7}Li, ^{39}K, and ^{85}Rb, the sign may be switched to self-attraction by means of the Feshbach resonance, which is in turn controlled by a magnetic or laser field acting on the experimental setup [46].

Theoretical and experimental work with solitons and other diverse nonlinear effects (such as the modulational instability [47] and rogue waves [48,49], shock waves, separation of immiscible components in binary systems, kinks, and domain walls [50], instantons [51], etc.) is a huge research area in many branches of physics [52], including optics [47], matter waves in atomic BECs [53], and BECs of quasi-particles (in particular, excitons-polaritons) [54], plasmas [55], ferromagnetic media [56], long Josephson junctions in superconductivity [57], acoustics [58], etc. In many cases, waveguiding settings offer media in which many species of solitons can be created and/or stabilized if the solitons do not exist (or exist but are unstable) in the respective uniform media. Characteristic examples are various methods elaborated for the stabilization of three-dimensional spatiotemporal solitons ("light bullets" [59]), which are subject to strong instabilities in both two- and three-dimensional uniform media [60–62]. It was demonstrated experimentally that both fundamental spatiotemporal solitons [63] and ones with embedded vorticity [64] can be made stable (in fact, as semi-discrete solitons) in the above-mentioned systems created as bundles of parallel waveguiding cores in bulk silica samples [26]. In fact, the commonly known stability of temporal optical solitons in nonlinear fibers [47] is also an example of the stabilization of a localized mode which is—strictly speaking—a three-dimensional one, with the self-trapping in the temporal (longitudinal) direction induced by the nonlinearity, while the transverse trapping is secured by the fiber's guiding properties, which are not essentially affected by the nonlinearity. Furthermore, the stability of matter–wave solitons in cigar-shaped trapping potentials [53] is provided by a similar mechanism, in spite of a completely different physical nature of the latter setting: the longitudinal self-trapping is induced by the self-attraction of the condensate, due to attractive interactions between atoms, while the confining potential prevents spreading of the condensate's wave function in the transverse directions. Moderate deviation from the effective one-dimensionality essentially affects the shape of the matter–wave solitons, but still relies upon the trapping potential to prevent the collapse of the three-dimensional self-attractive condensate [65].

1.4. Waveguides Built of Artificial Materials

The experimental and theoretical results outlined above were obtained in naturally existing media (and, accordingly, theoretical models of such media), or in settings produced by straightforward modifications of natural media, such as the aforementioned multi-core bundled guiding structures burnt in bulk silica [26,63,64].

Still natural—but more unusual—optical materials are photonic crystals (PhCs) [66] and quasicrystals [67,68], as well as PhC-based heterostructures and interfaces [69], and PhC fibers [70–72]; i.e., holey fibers in which inner voids form a PhC structure in the transverse plane. The difference between the traditional monolithic conduits (which guide light by means of the appropriate transverse profile of the refractive index) and PhCs is that PhCs implement the *bandgap-guidance* principle, steering the transmission of different optical modes according to the spectral bandgap structure, as induced by the underlying crystalline lattice.

Related to PhC fibers are waveguides built as large-radius hollow fibers, with a specially designed multi-layer cladding, which—by means of the Bragg-reflection mechanism (acting in the radial direction)—support the *omniguiding* regime of the transmission of light in such conduits. As a result, the omniguiding fibers (alias *Bragg fibers*) may provide a quasi-single regime of the propagation for selected modes, even if the large-area fiber is a multi-mode one. This is possible because all the modes except for the selected one will be suppressed by strong losses [73].

It is relevant to mention that another guidance mechanism is also possible which makes use of lattice structures similar to those underlying PhCs and PhC fibers; however, differently from them, these are *nonlinear lattices* [74]; i.e., spatially periodic modulations of the local nonlinearity coefficient. Naturally, such nonlinear lattices and their combinations with the usual linear lattices [75] are appropriate for steering nonlinear modes—first of all, solitons [74,75].

Furthermore, a new mechanism (thus far elaborated theoretically) for guided transmission of one- and two-dimensional spatial optical solitons, as well as their matter–wave counterparts in BEC, makes use of a purely self-defocusing nonlinearity, growing from center to periphery in the D-dimensional space faster than r^D, where r is the radial coordinate [76]. This scheme was predicted to stabilize a large number of diverse self-trapped (soliton-like) modes, both fundamental ones and complex topologically-organized objects, such as three-dimensional *hopfions* [77] (i.e., vortex rings with internal twist which carry two independent topological numbers: the vorticity and the twist).

PhCs and their various modifications may indeed be considered as natural materials because such structures are found in various animals, accounting for their coloration [78]. On the other hand, the recent progress in photonics has produced remarkable results in the form of artificially built media, which exhibit completely novel properties that are not possible in natural media; a very important example is provided by *left-handed metamaterials*, featuring negative values of the refractive index [79,80]. This property may be used for realization of fascinating applications, such as superlensing (which breaks the diffraction limit of imaging [81]), and optical cloaking (lending partial invisibility to small objects [82]). Other well-known examples of purposely designed artificial optical media with extraordinary properties include hyperbolic metamaterials, whose tensors of the dielectric permittivity and/or magnetic permeability feature principal values of opposite signs [83,84], planar metasurfaces [85,86], epsilon-near-zero materials, in which the refractive index nearly vanishes [87], photonic topological insulators [88,89] (which exemplify the area of *topological photonics* [90]), and others. The use of such media opens numerous possibilities to implement diverse optical effects, including nonlinear ones [91] and guided-wave propagation, in forms that were not known previously (for instance, in the form of the surface waveguiding in photonic topological insulators, which is immune to scattering on defects because the scattering is suppressed by the topology of the guiding system), and are unified under the name of *metaoptics* [92,93]. Another unifying concept is *nanophotonics*, the name originating from the fact that many of these materials are assembled of elements with sizes measured on the nanometer scale (which is deeply subwavelength, in terms of optics). One of the fundamentally interesting subjects of nanophotonics is trapping and transmitting light in *nanowires*. Nanowires are optical filaments (usually made of silicon) whose diameter—measured in nanometers—is much smaller than the wavelength of light, while a typical length may be a few millimeters; one of their important applications is in solar photovoltaic elements [94].

1.5. Dissipative and Parity-Time Symmetric Waveguides

The brief discussion of the waveguiding mechanisms given above did not address the presence of losses and the necessity of compensating them by gain. This assumption is valid for relatively short propagation distances, as well as in the case when the compensating gain matches the action of losses so accurately that both factors may be simultaneously neglected in the first approximation. In reality, losses are an inevitably existing gradient in plasmonics and metamaterials, as the respective waveguides are based on metallic elements, which introduce the Ohmic dissipation.

Generally speaking, if the medium is essentially lossy, the above-mentioned *index-guiding* and *bandgap-guiding* mechanisms which define the guiding channel(s), respectively, in terms of a transverse profile of the local refractive index, or the transmission-band structure induced by the PhC or PhC fiber may be replaced by a *gain-guiding scheme* in which the signal propagates in a lossy planar or bulk medium along a narrow stripe of gain locally embedded into the medium [95–97].

A recently developed topic which is closely related to the light transmission in dissipative waveguides deals with *the parity-time ($\mathcal{PT}$) symmetry*, which implies balance between symmetrically (in space) placed gain and loss elements. A paradigmatic model (it often includes nonlinearity, although the $\mathcal{PT}$ symmetry is by itself a linear property) is represented by NLSE (7), in which the potential is made *complex*, with real and imaginary parts being, respectively, spatially even and odd ones:

$$i\frac{\partial u}{\partial z} + \frac{1}{2}\frac{\partial^2 u}{\partial x^2} - [U_r(x) + iU_i(x)]\,u + \sigma|u|^2 u = 0,$$
$$U_r(-x) = U_r(x),\; U_i(-x_- = -U_i(x). \tag{15}$$

Another fundamental realization of the $\mathcal{PT}$ symmetry in optics and related fields is offered by a coupler, in which one core carries uniformly distributed gain, and the parallel-coupled one is uniformly lossy, the accordingly modified Equation (8) being

$$i\frac{\partial u}{\partial z} + \frac{1}{2}\frac{\partial^2 u}{\partial x^2} + \kappa v - U(x)u + \sigma|u|^2 u = i\gamma u,$$

$$i\frac{\partial v}{\partial z} + \frac{1}{2}\frac{\partial^2 v}{\partial x^2} + \kappa u - U(x)v + \sigma|v|^2 v = -i\gamma v,$$

(16)

where $\gamma > 0$ is the gain–loss coefficient. The $\mathcal{PT}$ symmetry has been experimentally realized in photonics, and a large number of guided-wave-propagation regimes have been investigated in such systems [98–101]. In particular, as concerns solitons, although $\mathcal{PT}$-symmetric systems belong to the class of dissipative ones, where solitons generally exist as isolated *attractors*, selected by the condition of the double balance between the dispersion (or diffraction) and nonlinearity, and between the gain and loss (the latter principle is very important for the creation of stable temporal solitons in fiber lasers [102]), in $\mathcal{PT}$-symmetric systems solitons exist in *continuous families*, similar to their counterparts in conservative models [100,101]. In addition to the interest to fundamental studies, systems with the $\mathcal{PT}$ symmetry offer promising applications, such as "light diodes", admitting unidirectional propagation of light in the waveguide, and lasers operating in the $\mathcal{PT}$-symmetric regime [103].

2. Annotation of Articles Included in the Special Issue

The present Special Issue is composed of a collection of **20** contributions, which include **5** relatively brief reviews summarizing recently obtained results in various areas of the guided-wave propagation in photonics, and **15** original papers reporting novel findings in this broad field. The contributions may be naturally grouped according to different forms and manifestations of the guided-wave propagation addressed in these works. Accordingly, the list of papers published in the Special Issue (following below) is divided into 11 topics **(A)–(K)**, and review articles are highlighted. In all cases, subjects addressed in the papers are sufficiently clearly defined by their titles.

(A) A batch of three papers may be classified as addressing problems arising in the fundamental (general) theory of the guided wave transmission in conservative (i.e., lossless) nonlinear media.

(A1) J. Fujioka, A. Gómez-Rodríguez, and Á. Espinosa-Cerón, Pulse Propagation Models with Bands of Forbidden Frequencies or Forbidden Wavenumbers: A Consequence of Abandoning the Slowly Varying Envelope Approximation and Taking into Account Higher-Order Dispersion. *Appl. Sci.* **2017**, *7*, 340.

(A2) Chan, H.N.; Chow, K.W. Rogue Wave Modes for the Coupled Nonlinear Schrödinger System with Three Components: A Computational Study, *Appl. Sci.* **2017**, *7*, 559.

(A3) Govindarajan, A.; Malomed, B.A.; Mahalingam, A.; Uthayakumar, A. Modulational Instability in Linearly Coupled Asymmetric Dual-Core Fibers. *Appl. Sci.* **2017**, *7*, 645.

(B) A related topic is the study of bright and dark soliton in various settings. This topic is represented in the Special Issue by the following four contributions, one of them being a review article:

(B1) Mai, Z.; Xu, H.; Lin, F.; Liu, Y.; Fu, S.; Li, Y. Dark Solitons and Grey Solitons in Waveguide Arrays with Long-Range Linear Coupling Effects. *Appl. Sci.* **2017**, *7*, 311.

(B2) Katsimiga, G.C.; Stockhofe, J.; Kevrekidis, P.G.; Schmelcher, P. Stability and Dynamics of Dark-Bright Soliton Bound States Away from the Integrable Limit. *Appl. Sci.* **2017**, *7*, 388.

(B3) Rodriguez, P.; Jimenez, J.; Guillet, T.; Ackemann, T. Polarization Properties of Laser Solitons. *Appl. Sci.* **2017**, *7*, 442.

(B4) Mitschke, R.F.; Mahnke, C.; Hause, A. Soliton Content of Fiber-Optic Light Pulses. *Appl. Sci.* **2017**, *7*, 635.

(C) Specific aspects of transmission in optical waveguides are considered in the following three papers (the first two address problems of direct relevance to practical applications):

(C1) Lamy, M.; Finot, C.; Fatome, J.; Arocas, J.; Weeber, J.C.; Hammani, K. Demonstration of High-Speed Optical Transmission at 2 µm in Titanium Dioxide Waveguides. *Appl. Sci.* **2017**, *7*, 63.

(C2) Memon, F.A.; Morichetti, F.; Melloni, A. Waveguiding Light into Silicon Oxycarbide. *Appl. Sci.* **2017**, *7*, 561.

(C3) Morales, J.D.H.; Rodríguez-Lara, B.M. Photon Propagation through Linearly Active Dimers. *Appl. Sci.* **2017**, *7*, 587.

(D) Different aspects of the transmission of light in waveguides based on fiber Bragg gratings is considered in two papers:

(D1) Yang, S.-C.; He, Y.-J.; Wun, Y.-J. Designing a Novel High-Performance FBG-OADM Based on Finite Element and Eigenmode Expansion Methods. *Appl. Sci.* **2017**, *7*, 44.

(D2) Review: Liu, Y.; Fu, S.; Malomed, B.A.; Khoo, I.C.; Zhou, J. Ultrafast Optical Signal Processing with Bragg Structures. *Appl. Sci.* **2017**, *7*, 556.

(E) A specific phenomenon of bound states existing in the continuous spectrum of a waveguide built as an array of dielectric spheres is summarized in the following Review article:

Bulgakov, E.N.; Sadreev, A.F.; Maksimov, D.N. Light Trapping above the Light Cone in One-Dimensional Arrays of Dielectric Spheres. *Appl. Sci.* **2017**, *7*, 147.

(F) A topic of the propagation of self-accelerating beams in the form of Airy waves is overviewed in a Brief Review:

Zhang, Y.; Zhong, H.; Belić, M.R.; Zhang, Y. Guided Self-Accelerating Airy Beams-A Mini-Review. *Appl. Sci.* **2017**, *7*, 34.

(G) A specific aspect of the light propagation in metamaterials is considered in:

Mazzone, V.; Gongora, J.S.T.; Fratalocchi, A. Near-Field Coupling and Mode Competition in Multiple Anapole Systems. *Appl. Sci.* **2017**, *7*, 542.

(H) Some fundamental aspects of the light transmission in dissipative waveguides are addressed in the following paper:

Descalzi, O.; Cartes, C. Stochastic and Higher-Order Effects on Exploding Pulses. *Appl. Sci.* **2017**, *7*, 887.

(I) Theoretical studies of the propagation of light in $\mathcal{PT}$-symmetric nonlinear waveguides are represented by an original paper,

D'Ambroise, J.; Kevrekidis, P.G. Existence, Stability and Dynamics of Nonlinear Modes in a 2D Partially $\mathcal{PT}$ Symmetric Potential. *Appl. Sci.* **2017**, *7*, 223.

(J) The propagation of plasmonic waves is addressed in the following two experimental works, with direct implications for applications:

(J1) Moon, K.; Lee, T.; Lee, Y.J.; Kwon, S. A Metal-Insulator-Metal Deep Subwavelength Cavity Based on Cutoff Frequency Modulation. *Appl. Sci.* **2017**, *7*, 86.

(J2) Iwanaga, M. Perfect Light Absorbers Made of Tungsten-Ceramic Membranes. *Appl. Sci.* **2017**, *7*, 458.

(K) Specific aspects of the general topic of fiber lasers, which are significant to fundamental and applied studies alike, are the subject of a Review article:

de Araújo, C.B.; Gomes, A.S.L.; Raposo, E.P. Lévy Statistics and the Glassy Behavior of Light in Random Fiber Lasers. *Appl. Sci.* **2017**, *7*, 644.

Conflicts of Interest: The author declares no conflicts of interest.

References

1. Crisp, J.; Elliot, B. *Introduction to Fiber Optics*, 3rd ed.; Elsevier: Oxford, UK, 2005.
2. Agrawal, G.P. *Fiber-Optic Communication Systems*; John Wiley & Sons: Hoboken, NJ, USA, 2010.
3. Poli, F.; Cucinotta, A. *Photonic Crystal Fibers: Properties and Applications*; Springer: Dordrecht, The Netherlands, 2007.
4. Yao, J. Microwave Photonics. *J. Lightwave Technol.* **2009**, *27*, 314–335.
5. Hutter, E.; Fendler, J.H. Exploitation of localized surface plasmon resonance. *Adv. Mater.* **2004**, *16*, 1685–1706.

6. Maier, S.A.; Atwater, H.A. Plasmonics: Localization and guiding of electromagnetic energy in metal/dielectric structures. *J. Appl. Phys.* **2005**, *98*, 011101.
7. Gramotnev, D.K.; Bozhevolnyi, S.I. Plasmonics beyond the diffraction limit. *Nat. Photonics* **2010**, *4*, 83–91.
8. Savel'ev, S.; Yampol'skii, V.A.; Rakhmanov, A.L.; Nori, F. Terahertz Josephson plasma waves in layered superconductors: Spectrum, generation, nonlinear and quantum phenomena. *Rep. Prog. Phys.* **2010**, *73*, 026501.
9. Grimshaw, R.; Pelinovsky, E.; Talipova, T. Modelling internal solitary waves in the coastal ocean. *Surveys Geophys.* **2007**, *28*, 273–298.
10. Bongs, K.; Burger, S.; Dettmer, S.; Hellweg, D.; Arlt, J.; Ertmer, W.; Sengstock, K. Waveguide for Bose-Einstein condensates. *Phys. Rev. A* **2001**, *63*, 031602.
11. Lesanovsky, I.; von Klitzing, W. Time-averaged adiabatic potentials: Versatile matter-wave guides and atom traps. *Phys. Rev. Lett.* **2007**, *99*, 083001.
12. Khitun, A.; Bao, M.; Wang, K.L. Magnonic logic circuits. *J. Phys. D Appl. Phys.* **2010**, *43*, 264005.
13. Rigetti, C.; Gambetta, J.M.; Poletto, S.; Plourde, B.L.T.; Chow, J.M.; Corcoles, A.D.; Smolin, J.A.; Merkel, S.T.; Rozen, J.B.; Keefe, G.A.; et al. Superconducting qubit in a waveguide cavity with a coherence time approaching 0.1 ms. *Phys. Rev. B* **2012**, *68*, 100506.
14. Borisov, A.S.B.; Borovskiy, A.V.; Shriryaev, O.B.; Korovkin, W.; Prokhorov, A.M.; Solem, J.C.; Luk, S.T.; Boyer, K.; Rhodes, C.K. Relativistic and charge-displacement self-channeling of intense ultrashort laser-pulses in plasmas. *Phys. Rev. A* **1992**, *45*, 5830–5845.
15. Milchberg, H.M.; Durfee, C.G.; McIlrath, T.J. High-order frequency conversion in the plasma waveguide. *Phys. Rev. Lett.* **1995**, *75*, 2494–2497.
16. Spence, D.J.; Hooker, S.M. Investigation of a hydrogen plasma waveguide. *Phys. Rev. E* **2001**, *63*, 015401.
17. Snyder, A.W.; Love, J.D. *Optical Waveguide Theory*; Chapman & Hall: London, UK, 1995.
18. Cronin, N.J. *Microwave and Optical Waveguides*; Institute of Physics Publishing: London, UK, 1995.
19. Zervas, M.N.; Codemard, C.A. High power fiber lasers: A review. *IEEE J. Sel. Top. Quantum Electron.* **2014**, *20*, 0904123.
20. Wright, L.G.; Christodoulides, D.N.; Wise, F. Controllable spatiotemporal nonlinear effects in multimode fibres. *Nat. Photonics* **2015**, *9*, 306.
21. Exner, P.; Kovařík, H. *Quantum Waveguides*; Springer: Cham, Vietnam, 2015.
22. Olshanii, M. Atomic scattering in the presence of an external confinement and a gas of impenetrable bosons. *Phys. Rev. Lett.* **1998**, *81*, 938–941.
23. Paredes, B.; Widera, A.; Murg, V.; Mandel, O.; Folling, S.; Cirac, I.; Shlyapnikov, G.V.; Hansch, T.W.; Bloch, I. Tonks-Girardeau gas of ultracold atoms in an optical lattice. *Nature* **2004**, *429*, 277–281.
24. Bloch, I.; Dalibard, J.; Zwerger, W. Many-body physics with ultracold gases. *Rev. Mod. Phys.* **2008**, *80*, 885–964.
25. Snyder, A.W.; Mitchell, D.J.; Poladian, L.; Rowland, D.R.; Chen, Y. Physics of nonlinear fiber couplers. *J. Opt. Soc. Am. B* **1991**, *8*, 2101–2118.
26. Szameit, A.; Nolte, S. Discrete optics in femtosecond-laser-written photonic structures. *J. Phys. B At. Mol. Opt. Phys.* **2010**, *43*, 163001.
27. Garanovich, I.L.; Longhi, S.; Sukhorukov, A.A.; Kivshar, Y.S. Light propagation and localization in modulated photonic lattices and waveguides. *Phys. Rep.* **2012**, *518*, 1–79.
28. Fleischer, J.W.; Bartal, G.; Cohen, O.; Schwartz, T.; Manela, O.; Freedman, B.; Segev, M.; Buljan, H.; Efremidis, N.K. Spatial photonics in nonlinear waveguide arrays. *Opt. Express* **2005**, *13*, 1780–1796.
29. Lederer, F.; Stegeman, G.I.; Christodoulides, D.N.; Assanto, G.; Segev, M.; Silberberg, Y. Discrete solitons in optics. *Phys. Rep.* **2008**, *463*, 1–126.
30. Malomed, B.A. Variational methods in nonlinear fiber optics and related fields. *Prog. Opt.* **2002**, *43*, 71–193.
31. Malomed, B.A. (Ed.) *Spontaneous Symmetry Breaking, Self-Trapping, and Josephson Oscillations*; Springer-Verlag: Berlin, Germany, 2013.
32. Zakharov, V.E.; Manakov, S.V.; Novikov, S.P.; Pitaevskii, L.P. *Theory of Solitons*; Nauka Publishers: Moscow, Russia, 1980. (In Russian); English Translation: New York, NY, USA, 1984.
33. Newell, A.C. *Solitons and Inverse Scattering Transform*; SIAM: Philadelphia, PA, USA, 1987.
34. Ablowitz, M.J.; Segur, H. *Solitons and Inverse Scattering Transform*; SIAM: Philadelphia, PA, USA, 2000.

35. Kevrekidis, P.G. *The Discrete Nonlinear Schrödinger Equation: Mathematical Analysis, Numerical Computations, and Physical Perspectives*; Springer: Berlin, Germany, 2009.
36. Hennig, D.; Tsironis, G.P. Wave transmission in nonlinear lattices. *Phys. Rep.* **1999**, *307*, 333–432.
37. Pitaevskii, L.P.; Stringari, S. *Bose-Einstein Condensation*; Clarendon: Oxford, UK, 2003.
38. Smith, H.; Pethick, C.J. *Bose-Einstein Condensation in Dilute Gases*, 2nd ed.; Cambridge University Press: Cambridge, UK, 2008.
39. Fibich, G. *The Nonlinear Schrödinger Equation: Singular Solutions and Optical Collapse*; Springer: Cham, Vietnam, 2015.
40. De Sterke, C.M.; Sipe, J.E. Gap solitons. *Prog. Opt.* **1994**, *33*, 203–260.
41. Aceves, A.B. Optical gap solitons: Past, present, and future; theory and experiments. *Chaos* **2000**, *10*, 584–589.
42. Eggleton, B.J.; Slusher, R.E.; de Sterke, C.M.; Krug, P.A.; Sipe, J.E. Bragg grating solitons. *Phys. Rev. Lett.* **1996**, *76*, 1627–1630.
43. Kashyap, R. *Fiber Bragg Gratings*, 2nd ed.; Academic Press: Burlington, MA, USA, 2009.
44. Etrich, C.; Lederer, F.; Malomed, B.A.; Peschel, T.; Peschel, U. Optical solitons in media with a quadratic nonlinearity. *Prog. Opt.* **2000**, *41*, 483–568.
45. Buryak, A.V.; di Trapani, P.; Skryabin, D.V.; Trillo, S. Optical solitons due to quadratic nonlinearities: From basic physics to futuristic applications. *Phys. Rep.* **2002**, *370*, 63–235.
46. Chin, C.; Grimm, R.; Julienne, P.; Tiesinga, E. Feshbach resonances in ultracold gases. *Rev. Mod. Phys.* **2010**, *82*, 1225–1285.
47. Kivshar, Y.S.; Agrawal, G.P. *Optical Solitons: From Fibers to Photonic Crystals*; Academic Press: San Diego, CA, USA, 2003.
48. Kharif, C.; Pelinovsky, E. Physical mechanisms of the rogue wave phenomenon. *Eur. J. Mech. B Fluids* **2003**, *22*, 603–634.
49. Onorato, M.; Residori, S.; Bortolozzo, U.; Montina, A.; Arecchi, F.T. Rogue waves and their generating mechanisms in different physical contexts. *Phys. Rep.* **2013**, *528*, 47–89.
50. Vachaspati, T. *Kinks and Domain Walls*; Cambridge University Press: Cambridge, UK, 2006.
51. Rajaraman, R. *Solitons and Instantons*; North Holland: Amsterdam, The Netherlands, 1982.
52. Dauxois, T.; Peyrard, M. *Physics of Solitons*; Cambridge University Press: Cambridge, UK, 2006.
53. Strecker, K.E.; Partridge, G.B.; Truscott, A.G.; Hulet, A.G. Bright matter wave solitons in Bose-Einstein condensates. *New J. Phys.* **2003**, *5*, 73.1–73.8.
54. Cerda-Méndez, E.A.; Sarkar, D.; Krizhanovskii, D.N.; Gavrilov, S.S.; Biermann, K.; Skolnick, M.S.; Santos, P.V. Exciton-polariton gap solitons in two-dimensional lattices. *Phys. Rev. Lett.* **2013**, *111*, 146401.
55. Petviashvili, V.I.; Pokhotelov, O.A. *Solitary Waves in Plasmas and Atmosphere*; Routledge: London, UK, 1997.
56. Bar'yakhtar, V.G.; Chetkin, M.V.; Ivanov, B.A.; Gadetskii, S.N. *Dynamics of Topological Magnetic Solitons*; Springer-Verlag: Berlin, Germany, 1994.
57. Ustinov, A.V. Solitons in Josephson junctions. *Phys. D* **1998**, *123*, 315–329.
58. Naugolnykh, K.; Ostrovsky, L. *Nonlinear Wave Processes in Acoustics*; Cambridge University Press: Cambridge, UK, 1998.
59. Silberberg, Y. Collapse of optical pulses. *Opt. Lett.* **1990**, *15*, 1282–1284.
60. Malomed, B.A.; Mihalache, D.; Wise, F.; Torner, L. Spatiotemporal Optical Solitons. *J. Opt. B Quantum Semicl. Opt.* **2005**, *7*, R53–R72.
61. Malomed, B.A.; Mihalache, D.; Wise, F.; Torner, L. Viewpoint: On multidimensional solitons and their legacy in contemporary Atomic, Molecular and Optical physics. *J. Phys. B At. Mol. Opt. Phys.* **2016**, *49*, 170502.
62. Malomed, B.A. Multidimensional solitons: Well-established results and novel findings. *Eur. Phys. J. Spec. Top.* **2016**, *25*, 2507–2532.
63. Minardi, S.; Eilenberger, F.; Kartashov, Y.V.; Szameit, A.; Röpke, J.; Kobelke, J.; Schuster, K.; Bartelt, H.; Nolte, S.; Torner, L.; et al. Three-dimensional light bullets in arrays of waveguides. *Phys. Rev. Lett.* **2010**, *105*, 263901.
64. Eilenberger, F.; Prater, K.; Minardi, S.; Geiss, R.; Röpke, U.; Kobelke, J.; Schuster, K.; Bartelt, H.; Nolte, S.; Tünnermann, A.; et al. Observation of discrete, vortex light bullets. *Phys. Rev. X* **2013**, *3*, 041031.
65. Cuevas, J.; Kevrekidis, P.G.; Malomed, B.A.; Dyke, P.; Hulet, R. Interactions of solitons with a Gaussian barrier: Splitting and recombination in quasi-1D and 3D. *New J. Phys.* **2013**, *15*, 063006.

66. Joannopoulos, J.D.; Johnson, S.G.; Winn, J.N.; Meade, R.D. *Photonic Crystals: Molding the Flow of Light*, 2nd ed.; Princeton University Press: Princeton, NJ, USA, 2008.
67. Poddubny, A.N.; Ivchenko, E.L. Photonic quasicrystalline and aperiodic structures. *Physica E* **2010**, *42*, 1871–1895.
68. Vardeny, Z.V.; Nahata, A.; Agrawal, A. Optics of photonic quasicrystals. *Nat. Photonics* **2013**, *7*, 177–187.
69. Istrate, E.; Sargent, E.H. Photonic crystal heterostructures and interfaces. *Rev. Mod. Phys.* **2006**, *78*, 455–481.
70. Benabid, F.; Roberts, P.J. Linear and nonlinear optical properties of hollow core photonic crystal fiber. *J. Mod. Opt.* **2011**, *58*, 87–124.
71. Russell, P.S.J.; Hölzer, P.; Chang, W.; Travers, A.A.A.C. Hollow-core photonic crystal fibres for gas-based nonlinear optics. *Nat. Photonics* **2014**, *8*, 278–286.
72. Saleh, M.F.; Biancalana, F. Soliton dynamics in gas-filled hollow-core photonic crystal fibers. *J. Opt.* **2016**, *18*, 013002.
73. Johnson, S.G.; Ibanescu, M.; Skorobogatiy, M.; Weisberg, O.; Engeness, T.; Soljačić, M.; Jacobs, S.A.; Joannopoulos, J.D.; Fink, Y. Low-loss asymptotically single-mode propagation in large-core OmniGuide fibers. *Opt. Exp.* **2001**, *9*, 748–779.
74. Kartashov, Y.V.; Malomed, B.A.; Torner, L. Solitons in nonlinear lattices. *Rev. Mod. Phys.* **2011**, *83*, 247–306.
75. Sakaguchi, H.; Malomed, B.A. Solitons in combined linear and nonlinear lattice potentials. *Phys. Rev. A* **2010**, *81*, 013624.
76. Borovkova, O.V.; Kartashov, Y.V.; Torner, L.; Malomed, B.A. Bright solitons from defocusing nonlinearities. *Phys. Rev. E* **2011**, *84*, 035602.
77. Kartashov, Y.V.; Malomed, B.A.; Shnir, Y.; Torner, L. Twisted toroidal vortex-solitons in inhomogeneous media with repulsive nonlinearity. *Phys. Rev. Lett.* **2014**, *113*, 264101.
78. Kinoshita, S.; Yoshioka, S.; Miyazaki, J. Physics of structural colors. *Rep. Progr. Phys.* **2008**, *17*, 076401.
79. Shalaev, V.M. Optical negative-index metamaterials. *Nat. Photonics* **2007**, *1*, 41–48.
80. Soukoulis, C.M.; Wegener, M. Past achievements and future challenges in the development of three-dimensional photonic metamaterials. *Nat. Photonics* **2011**, *5*, 523–530.
81. Jacob, Z.; Alekseyev, L.V.; Narimanov, E. Optical hyperlens: Far-field imaging beyond the diffraction limit. *Opt. Exp.* **2006**, *14*, 8247–8256.
82. Cai, W.; Chettiar, U.K.; Kildishev, A.V.; Shalaev, V.M. Optical cloaking with metamaterials. *Nat. Photonics* **2007**, *1*, 224–227.
83. Drachev, V.P.; Podolskiy, V.A.; Kildishev, A.V. Hyperbolic metamaterials: New physics behind a classical problem. *Opt. Exp.* **2013**, *21*, 15048–15064.
84. Poddubny, A.; Iorsh, I.; Belov, P.; Kivshar, Y. Hyperbolic metamaterials. *Nat. Photonics* **2013**, *7*, 958–967.
85. Kildishev, A.V.; Boltasseva, A.; Shalaev, V.M. Planar Photonics with Metasurfaces. *Science* **2013**, *339*, 1232009.
86. Chen, H.-T.; Taylor, A.J.; Yu, N. A review of metasurfaces: physics and applications. *Rep. Prog. Phys.* **2016**, *79*, 076401.
87. Liberal, I.; Engheta, N. Near-zero refractive index photonics. *Nat. Photonics* **2017**, *11*, 149–159.
88. Rechtsman, M.C.; Zeuner, J.M.; Plotnik, Y.; Lumer, Y.; Podolsky, D.; Dreisow, F.; Nolte, S.; Segev, M.; Szameit, A. Photonic Floquet topological insulators. *Nature* **2013**, *496*, 196–200.
89. Kartashov, Y.V.; Skryabin, D.V. Modulational instability and solitary waves in polariton topological insulators. *Optica* **2016**, *3*, 1228–1236.
90. Lu, L.; Joannopoulos, J.D.; Soljačić, M. Topological photonics. *Nat. Photonics* **2014**, *8*, 821–829.
91. Maimistov, A.I.; Gabitov, I.R. Nonlinear optical effects in artificial materials. *Eur. Phys. J. Spec. Top.* **2007**, *147*, 265–286.
92. Baeva, A.; Prasad, P.N.; Grenb, H.À.; Samoć, M.; Wegener, A. Metaphotonics: An emerging field with opportunities and challenges. *Phys. Rep.* **2015**, *594*, 1–60.
93. Urbas, A.M.; Jacob, Z.; Dal Negro, L.; Engheta, N.; Boaedman, A.D.; Egan, P.; Khanikaec, A.B.; Menon, V.; Ferrera, M.; Kinsey, N.; et al. Roadmap on optical metamaterials. *J. Opt.* **2016**, *18*, 093005.
94. Garnett, E.; Yang, P. Light trapping in silicon nanowire solar cells. *Nano Lett.* **2010**, *10*, 1082–1087.
95. Kartashov, Y.V.; Konotop, V.V.; Vysloukh, V.A.; Zezyulin, D.A. Guided modes and symmetry breaking supported by localized gain. In *Spontaneous Symmetry Breaking, Sepf-Trapping, and Josephson Oscillations*; Malomed, B.A., Ed.; Springer-Verlag: Berlin, Germany, 2013; pp. 149–166.
96. Malomed, B.A. Spatial solitons supported by localized gain. *J. Opt. Soc. Am. B* **2014**, *31*, 2460–2475.

97. Lobanov, V.E.; Borovkova, O.V.; Malomed, B.A. Dissipative quadratic solitons supported by a localized gain. *Phys. Rev. A* **2014**, *90*, 053820.
98. Makris, K.G.; El-Ganainy, R.; Christodoulides, D.N.; Musslimani, Z.H. $\mathcal{PT}$-Symmetric Periodic optical potentials. *Int. J. Theor. Phys.* **2011**, *50*, 1019–1041.
99. Bender, C.M.; DeKieviet, M.; Klevansky, S.P. $\mathcal{PT}$ quantum mechanics. *Phil. Trans. R. Soc. A Math. Phys. Eng. Sci.* **2013**, *371*, 20120523.
100. Sukhorukov, S.V.S.A.A.; Huang, J.; Dmitriev, S.V.; Lee, C.; Kivshar, Y.S. Nonlinear switching and solitons in $\mathcal{PT}$-symmetric photonic systems. *Laser Photonics Rev.* **2016**, *10*, 177.
101. Konotop, V.V.; Yang, J.; Zezyulin, D.A. Nonlinear waves in $\mathcal{PT}$-symmetric systems. *Rev. Mod. Phys.* **2016**, *88*, 035002.
102. Grelu, P.; Akhmediev, N. Dissipative solitons for mode-locked lasers. *Nat. Photonics* **2012**, *6*, 84–92.
103. Huang, Y.; Shen, Y.; Min, C.; Fan, S.H.; Veronić, G. Unidirectional reflectionless light propagation at exceptional points. *Nanophotonics* **2017**, *6*, 977–996.

applied
sciences

MDPI

Article

Stochastic and Higher-Order Effects on Exploding Pulses

Orazio Descalzi * and Carlos Cartes

Complex Systems Group, Facultad de Ingeniería y Ciencias Aplicadas, Universidad de los Andes,
Av. Mons. Álvaro del Portillo 12.455, Las Condes, Santiago 7620001, Chile; ccartes@gmail.com
* Correspondence: odescalzi@miuandes.cl

Received: 27 June 2017; Accepted: 22 July 2017; Published: 30 August 2017

Abstract: The influence of additive noise, multiplicative noise, and higher-order effects on exploding solitons in the framework of the prototype complex cubic-quintic Ginzburg-Landau equation is studied. Transitions from explosions to filling-in to the noisy spatially homogeneous finite amplitude solution, collapse (zero solution), and periodic exploding dissipative solitons are reported.

Keywords: exploding solitons; Ginzburg-Landau equation; mode-locked fiber lasers

1. Introduction

Soliton explosions, fascinating nonlinear phenomena in dissipative systems, have been observed in at least three key experiments. As has been reported by Cundiff et al. [1], a mode-locked laser using a Ti:Sapphire crystal can produce intermittent explosions. More recently, a different medium for explosions was reported by Broderick et al., namely, a passively mode-locked fibre laser [2]. In 2016, Liu et al. showed that in an ultrafast fiber laser, the exploding behavior could operate in a sustained but periodic mode called "successive soliton explosions" [3].

Almost all parts of these exploding objects are unstable, but nevertheless they remain localized. Localized structures in systems far from equilibrium are the result of a delicate balance between injection and dissipation of energy, nonlinearity and dispersion (compare [4] for a recent exposition of the subject). This fact leads to a generalization of the well known conservative soliton to a dissipative soliton DS (Akhmediev et al. [5]). Experimental observation of DSs, apart from explosions, shows a wide spectrum in nature including binary fluid convection, granular systems, chemical surface reactions, nonlinear optics and starch suspensions [6–15].

Explosions, being chaotic phenomena, are not identical, resulting in a random distribution of times between explosions. Real systems, where explosions were observed, are not continuous, however, explosive behavior was predicted theoretically in the complex cubic-quintic Ginzburg-Landau equation, whose parameters vary continuously [16,17].

The complex cubic-quintic Ginzburg-Landau equation (CQGLE), a prototype envelope equation, derived near the onset of a subcritical instability (inverted Hopf bifurcacion), was first introduced by Brand et al. when modeling binary mixtures [18,19]. However, Thual and Fauve were the first to report explicitly the existence of stable pulse solutions in the CQGLE [20]. In optics, this equation describes laser systems [21–23], soliton transmission lines [24], and nonlinear cavities with an external pump [25]. A natural parameter to be varied in this equation is the distance from linear onset. For a large range of parameters, the following sequence was found: stationary pulses, pulses with one and two frequencies, and finally exploding pulses [26]. This fact revealed a quasiperiodic route to chaos for spatially localized solutions [27]. In addition, we have studied the effect of small and large additive noise on the formation of localized patterns [28,29]. We concluded that weak additive noise is enough to induce explosions while the interaction of localization and noise can lead to noisy localized structures.

Recently, we have reported that multiplicative noise can reduce and even suppress the existence of explosions in localized solutions [30].

In this article, we study the influence of additive noise, multiplicative noise, and higher-order effects on exploding solitons in the framework of the prototype envelope complex cubic-quintic Ginzburg-Landau equation. Transitions from explosions to filling-in to the noisy spatially homogeneous finite amplitude solution, collapse (zero solution), and periodic exploding dissipative solitons are reported.

2. Influence of Additive and Multiplicative Noise on Exploding Dissipative Solitons

Sources of noise are always present in physical systems, which can be external or internal (as examples, we can mention thermal noise, fluctuations in pressure, temperature or electrical signals, etc.). Ordinary or partial differential equations take account of the behavior of systems under the influence of fluctuations [31–33]. Internal noise becomes typically additive [34] while external noise becomes typically multiplicative or both [35].

2.1. Stochastic Equations

The stochastic complex cubic-quintic complex Ginzburg-Landau equation (SCQGLE) under the influence of additive noise reads

$$\partial_t A = \mu A + \beta |A|^2 A + \gamma |A|^4 A + D \partial_{xx} A + \eta \, \xi. \tag{1}$$

Here, $A(x,t)$ is the complex envelope, $\beta = \beta_r + \mathrm{i}\,\beta_i$; $\gamma = \gamma_r + \mathrm{i}\,\gamma_i$; $D = D_r + \mathrm{i}\,D_i$; μ is the distance from linear onset (real), and η the noise strength. $\beta_r > 0$ and $\gamma_r < 0$ in order to guarantee that we are in the presence of an inverted Hopf bifurcation saturating to quintic order. The complex white noise $\xi(x,t)$ satisfies $\langle \xi \rangle = 0$ (zero mean), $\langle \xi(x,t)\,\xi(x',t') \rangle = 0$ and $\langle \xi(x,t)\,\xi^*(x',t') \rangle = 2\delta\,(x - x')\,\delta\,(t - t')$ (delta correlated in space and time). Discretization in space and time should not affect quantitative noise features, so that $\xi(x,t)$ becomes replaced by $(\chi_r + \mathrm{i}\chi_i)/\sqrt{dx\,dt}$, where χ_r and χ_i are uncorrelated random numbers (in space and time) obeying a normal distribution with zero mean and unit variance.

The SCQGLE with multiplicative noise that we investigate here is of the form

$$\partial_t A = (\mu + \eta\,\zeta) A + \beta |A|^2 A + \gamma |A|^4 A + D \partial_{xx} A, \tag{2}$$

where the white noise $\zeta(t)$ satisfies $\langle \zeta \rangle = 0$ (zero mean), and $\langle \zeta(t)\,\zeta(t') \rangle = \delta\,(t - t')$ (delta-correlated in time but homogeneous in space), so that $\zeta(x,t)$ is replaced by $\chi_r/\sqrt{dt}$, where χ_r corresponds to uncorrelated random numbers obeying a normal distribution with zero mean and unit variance. Additive noise implies perturbations on short length and time scales. However, homogeneous multiplicative noise leads to a homogeneous enhancement and suppression of the modulus as a function of time. As an example, we can mention electroconvection in nematic liquid crystals by superposing noise on the driving voltage [36].

2.2. Numerical Method

The parameters we used are $\beta = 1 + 0.8\,\mathrm{i}$, $\gamma = -0.1 - 0.6\,\mathrm{i}$, $D = 0.125 + 0.5\,\mathrm{i}$. The complex cubic-quintic Ginzburg-Landau equation has seven parameters. After scaling t, x and A, we can fix $\beta_r = 1$, $\gamma_r = -0.1$ and $D_r = 0.125$. Nevertheless, explosions occur as a function of μ only in a subset of the space (β_i, γ_i, D_i). According to our previous experience in explosions [26], we choose $\beta_i = 0.8$, $\gamma_i = -0.6$ and $D_i = 0.5$.

We considered periodic boundary conditions and by varying the box size we made sure that our box is sufficiently large so that none of the results presented in the following are sensitively dependent on the box size.

The bifurcation parameter μ is varied from -0.26 until -0.16. This range includes values of μ where deterministically we find stationary solutions and oscillatory solutions (with one and two frequencies) and exploding solitons [28]. Initial conditions are deterministic solutions of the CQGLE.

To numerically solve Equations (1) and (2), we used a pseudo—spectral split—step method, where the derivatives are computed in Fourier space using a fast Fourier transformation and the non-linear part is integrated in time by a fourth order Runge-Kutta method. To perform these operations, we write our equations in the following way

$$\partial_t A = M[A]A + D\partial_{xx}A, \tag{3}$$

where $M[A]A$ represents the part of the equation containing the linear and non-linear terms with the exception of the double derivative in x. If we split Equation (3) into two operators, we get $\partial_t A_L = D\partial_{xx}A_L$ and $\partial_t A_M = M[A]A_M$.

We integrate the differential part over a small interval Δt, by performing a Fourier transformation $\partial_t \hat{A}_L = -Dk^2\hat{A}_L$, where k^2 is the squared wave vector. After integrating the differential and nonlinear part separately, we obtain

$$\begin{aligned} \hat{A}_L(t + \Delta t) &= e^{-\Delta t Dk^2}\hat{A}_L(t), \\ A_M(t + \Delta t) &= e^{\Delta t M[A]}A_M(t). \end{aligned} \tag{4}$$

The second integration is performed by a fourth order Runge-Kutta method which gives us a more numerically stable result. Finally, we get the solution for A integrated over a small time step Δt

$$A(t + \Delta t) = e^{\Delta t M[A]}e^{\Delta t D\partial_{xx}}A(t). \tag{5}$$

This numerical method was implemented using a Python code with Fortran subroutines. For the numerical simulations, we used $N = 1024$ Fourier modes over a discretized grid of length $L = 50$ and a time step of size $\Delta t = 0.005$.

2.3. Results

In [28], we studied the effect of weak additive noise on the spatially localized pulses (either stationary or oscillating with one and two frequencies) concluding that small additive noise is enough to induce exploding dissipative solitons, which are mostly chaotic.

Here, we report (Figure 1a) three types of pulses emerging as a function of μ, including the range where deterministically exploding solitons exist, for varying two decades of η (logarithmic scale). For $\mu \lesssim 0.22$ and small values of η, we find (in agreement with [28]) noisy non-explosive localized solutions, either chaotic or non-chaotic. Large enough values of noise always induce explosions. Non-chaotic explosions are observed at the border separating non-explosive states from exploding solitons. For sufficiently large noise strength η, we observe a transition to filling-in, that is, a noisy spatially homogeneous finite amplitude solution.

The phase diagram shown in Figure 1a is rather insensitive to the maximum run time T, as we can notice from Figure 1b. There, the filling-in time for the transition from an exploding dissipative soliton to filling-in under the influence of additive noise for $\mu = -0.25$ is shown. The average time scale T for filling-in is plotted as a function of the noise strength η. Black solid circles represent an average over 50 realizations shown as open squares ($\square$). Inspecting this figure, we see that to decrease T by about one decade, one needs to increase η by 0.01.

For $\mu = -0.25$ and $\eta = 0$, we have a stationary solution. When a small noise ($\eta \sim 0.003$) is added to Equation (1), it acts mainly as a perturbation on short length and time scales giving the state a noisy appearance (Figure 2a). When η is increased to ~ 0.03, a perturbation starts growing in the wings. Once this peak has grown, it interacts with the main pulse forming a wide chaotic localized one (this instant is shown as a snapshot in Figure 2b). After this, rapidly the system collapses to a state, similar to the original starting peak. This is what we call *explosion*. For large enough noise ($\eta \gtrsim 0.06$), the whole system jumps to a noisy spatially homogeneous finite amplitude solution (Figure 2c).

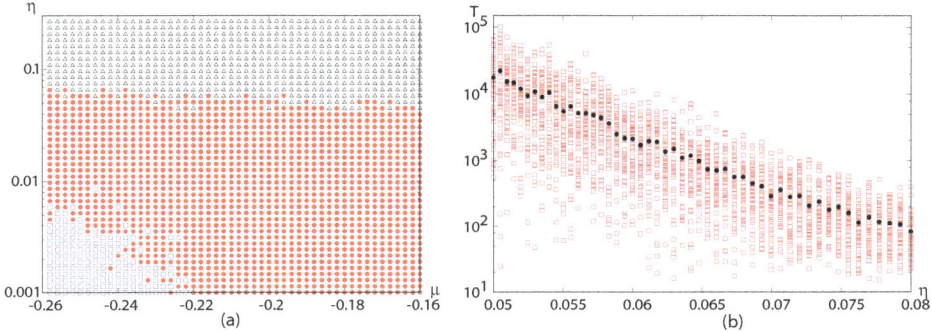

Figure 1. Phase diagram: Effect of additive noise on exploding solitons for a fixed time scale $T = 10^4$. (**a**) Black triangles ($\triangle$) denote filling-in to the spatially homogeneous finite amplitude solution under the influence of additive noise. Red solid circles ($\bullet$) stand for exploding dissipative solutions and blue squares ($\square$) for noisy non-explosive localized solutions; (**b**) Filling-in time for the transition from an exploding dissipative soliton to filling-in under the influence of additive noise for $\mu = -0.25$. The average time scale T for filling-in is shown as a function of the noise strength η. Black solid circles ($\bullet$) represent an average of over 50 realizations shown as open squares ($\square$).

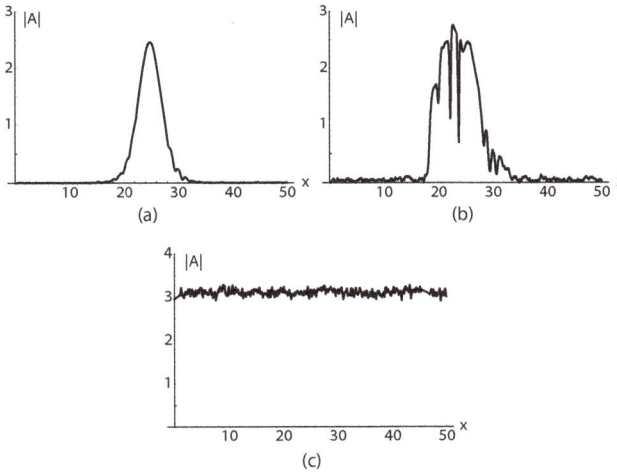

Figure 2. Three snapshots for $\mu = -0.25$ and additive noise for different values of the noise strength η. (**a**) Noisy non-explosive localized solution, $\eta = 0.003$; (**b**) Exploding dissipative solution, $\eta = 0.03$; (**c**) Noisy spatially homogeneous finite amplitude solution, $\eta = 0.2$.

For spatially homogeneous multiplicative noise, where we observe a collective enhancement and suppression (in space) of the amplitude, we report in Figure 3a three types of patterns, as a function of the bifurcation parameter μ, and η the noise strength: exploding dissipative solitons (red solid circles ($\bullet$)), oscillating localized states (blue squares ($\square$)). Either they are not explosive or their frequency of explosions is undetectable for $T = 10^4$. In the range of μ shown in Figure 3, values of $\eta \gtrsim 0.3$ induce collapse (black triangles ($\triangle$)). In a previous article [30], we reported that multiplicative noise can lead to a reduction of the number of explosions and even to the collapse of dissipative solitons.

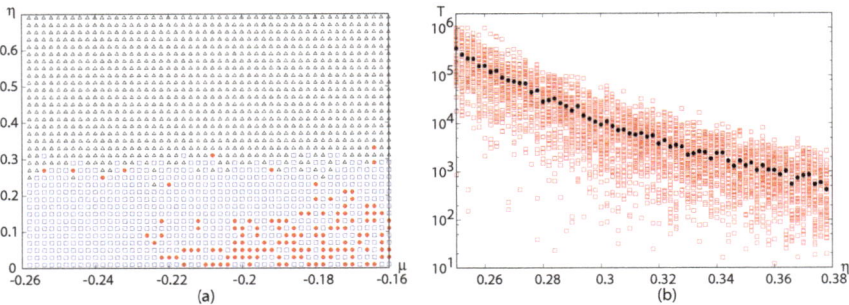

Figure 3. Phase diagram: Influence of multiplicative noise on exploding solitons for a fixed time scale $T = 10^4$. (**a**) Black triangles ($\triangle$) denote collapsed states. Red solid circles (•) stand for exploding dissipative solutions ($\square$) for oscillating localized solutions. (**b**) Collapse time for exploding dissipative solitons under the influence of multiplicative noise for $\mu = -0.18$. The average time scale T for collapse is shown as a function of the noise strength η. Black solid circles (•) represent an average over 50 realizations shown as open squares ($\square$).

In Figure 3b, collapse time for exploding dissipative solitons under the influence of multiplicative noise is shown for $\mu = -0.18$. The average time scale T for collapse is plotted as a function of the noise strength η. Black solid circles represent an average of over 50 realizations shown as open squares ($\square$). One can see from the plot that decreasing η at the border in 0.03 means to increase T in one decade.

3. Exploding Dissipative Solitons and Higher-Order Effects

The inclusion of higher-order terms to the complex cubic-quintic Ginzburg-Landau equation has a clear physical motivation, namely, modeling the propagation of short pulses along a mode-locked fiber laser. According to Agrawal [37], for short pulses ($T_0 \lesssim 1$ ps), where T_0 is the width of the pulse, one should include the higher-order effects. Therefore, the generalized pulse-propagation equation reads

$$\partial_{\bar{z}}\tilde{A} - \frac{g(\bar{z},P)}{2}\tilde{A} - i\sum_{k\geq 2}\frac{\beta_k}{k!}(i\partial_T)^k\tilde{A} = i\left(\gamma(\omega_0) + i\gamma_1\,\partial_T\right)\left(\tilde{A}(\bar{z},T)\int_0^\infty R(T')|\tilde{A}(\bar{z},T-T')|^2 dT'\right), \quad (6)$$

where $\tilde{A}(\bar{z},T)$ is the envelope (slowly varying function of $\bar{z}$ and T) of the complex electrical field in a comoving frame, g the gain or loss of energy, P the pumping power, β_k stands for the dispersion coefficients, $\gamma_1 \approx \gamma/\omega_0$, $\gamma = \omega_0 n_2/cA_{\text{eff}}$, where A_{eff} is the effective mode area, n_2 is called the nonlinear Kerr parameter, and ω_0 the carrier frequency. The integral in this equation accounts for the energy transfer resulting from intrapulse Raman scattering.

3.1. Complex Ginzburg-Landau Equation and Short Pulses

For short pulses but wide enough ($T_0 \sim 0.1$ ps), expanding $|\tilde{A}(\bar{z},T-T')|^2$ in a Taylor-series up to first order in T', and considering up to the third-order dispersion β_3, we can deduce from Equation (6) the following non-integrable quation [38]

$$\partial_{\bar{z}}\tilde{A} - \frac{g(\bar{z},P)}{2}\tilde{A} + i\frac{\beta_2}{2}\partial_T^2\tilde{A} - \frac{\beta_3}{6}\partial_T^3\tilde{A} = i\gamma|\tilde{A}|^2\tilde{A} - \frac{1}{\omega_0}\partial_T(|\tilde{A}|^2\tilde{A}) - iT_R\,\tilde{A}\,\partial_T(|\tilde{A}|^2), \quad (7)$$

where $T_R \equiv \int_0^\infty tR(t)dt$ is the first moment of the Raman response function and $\int_0^\infty R(t)dt = 1$.

One can note that $|\tilde{A}|^2$ has dimensions of power. Therefore, one can define the dimensionless variable A as $\tilde{A} = A\sqrt{|\beta_2|}/(T_0\sqrt{\gamma})$. In the same way, using the dispersion length, the dimensionless variable z: $\tilde{z} = zT_0^2/|\beta_2|$, and the dimensionless variable τ: $T = T_0\tau$. Thus, introducing the variables A, z and τ in Equation (7), we can obtain a dimensionless complex CQGLE including three higher-order effects: third-order dispersion, self-steepening, and intrapulse Raman scattering.

$$i\partial_z A + \frac{1}{2}\partial_\tau^2 A + |A|^2 A - \nu|A|^4 A = i\delta A + i\epsilon|A|^2 A + i\mu|A|^4 A + i\beta\partial_\tau^2 A, \tag{8}$$
$$+ \; i\delta_3\partial_\tau^3 A - is\partial_\tau(|A|^2 A) + \tau_R A\partial_\tau(|A|^2),$$

The left side of Equation (8) is nothing but the cubic-quintic non-linear Schrödinger equation, which is conservative. On the right side, the three first terms are related to g, taking account of the linear gain and loss of energy (δ), and for the non-linear gain or absorption of energy (ϵ, μ). The term associated to β plays the role of spectral filtering. The last three terms on the right side are precisely the higher-order effects, which are conservative and whose coefficients δ_3, s, and τ_R are defined as follows:

$$\delta_3 \equiv \frac{\beta_3}{6T_0|\beta_2|}; \; s \equiv \frac{1}{\omega_0 T_0}; \; \tau_R \equiv \frac{T_R}{T_0}. \tag{9}$$

3.2. Results

Equation (8) without considering higher-order effects is the optical version for the CQGLE studied in Section 2. Coefficients can easily be converted from Equations (1)–(8): $\delta = \mu$; $\epsilon = \beta_r$; $1 = \beta_i$; $\mu = \gamma_r$, $\nu = -\gamma_i$; $\beta = D_r$; and $\frac{D}{2} = D_i$. In optics, to use ϵ as a control parameter is meaningful because ϵ is related to the pumping power.

As in Section 2, to solve Equation (8), we used a pseudo-spectral split-step numerical method along to $N = 8192$ Fourier modes, $dt = 0.01$ and $dz = 0.004$. Our parameters are: $\delta = -0.1$, $\beta = 0.125$, $\mu = -0.1$, $\nu = 0.6$ and $\epsilon \sim 1.0$.

For ϵ around 1.02 (and $\delta = s = \tau_R = 0$), the energy $Q(z) = \int_0^T |\psi|^2 d\tau$ exhibits different maxima Q_{max}, which can be plotted in a logistic map (see Figure 4a), giving the aspect of a complex picture, natural consequence of explosive chaotic behavior. While in Figure 5a, one can observe the evolution of the amplitude $|A|$ in a $\tau - z$ plot for $\epsilon = 1.02$. There, one can notice the random distribution of locations in z between explosions (the analogue of the random distribution of times between explosions in Equation (1)). The distribution of locations in z obeys a narrow distribution centered around a mean value.

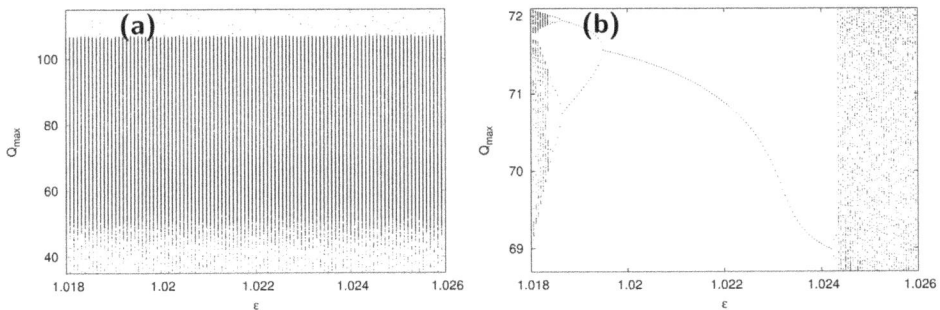

Figure 4. (**a**) Logistic map for Q_{max} without including higher-order effects. Complex picture, natural consequence of explosive chaotic behavior; (**b**) Logistic map for Q_{max} including higher-order effects. One can observe a window without chaos around $\epsilon = 1.022$.

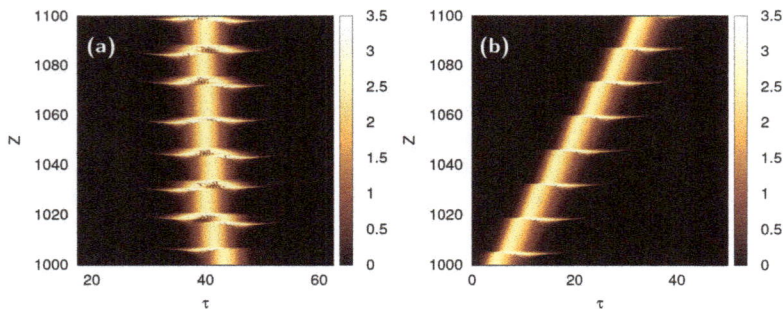

Figure 5. $\tau - z$ plot of $|A|$ following the dimensionless complex CQGLE (8) for $\epsilon = 1.02$, (**a**) without including higher-order effects, showing chaotic explosions, and (**b**) including higher-order effects, showing periodic explosions.

Considering $T_R \sim 3$ fs [37] and $T_0 \sim 0.1$ ps, we estimate $\tau \sim 0.03$, and noticing that $\omega_0 T_0 \sim N \sim 100$ (number of cycles tangled by $|A|$), s can also be estimated as $s \sim 0.01$. Now, following the same above procedure for the CQGLE, but including higher-order nonlinear and dispersive effects ($\delta_3 = 0.016, s = 0.009, \tau_R = 0.032$), we notice the appearance of windows corresponding to non-chaotic behaviors (see Figure 4b). Around $\epsilon = 1.022$, one can observe period-halving bifurcations leading to order, and period-doubling bifurcations leading to chaos. Around $\epsilon = 1.019$ and $\epsilon = 1.024$, the limit of the ratio of distances between consecutive bifurcation intervals tends to 4.6 (close to the Feigenbaum constant).

Figure 5b shows the evolution of the amplitude $|A|$ in a $\tau - z$ plot, for $\epsilon = 1.02$, considering higher-order nonlinear and dispersive effects. In contrast to Figure 5a, we see that there is a fixed distance in z between explosions and explosions repeat exactly after a period. We are in the presence of periodic explosions.

4. Conclusions and Discussion

In summary, we have studied the influence of additive noise, multiplicative noise, and higher-order effects on exploding solitons in the framework of the prototype complex cubic-quintic Ginzburg-Landau equation.

For the stochastic CQGLE with enough large additive noise, we report a transition from explosions to filling-in, that is, a noisy spatially finite solution. Under the influence of large multiplicative noise, homogeneous in space, a collapse occurs in the zero amplitude solution.

We show that the phase diagrams for different outcomes under the influence of additive or multiplicative noise are rather insensitive to the choice of the run time. The transitions to filling-in and collapse follow an exponential law for the transition time as a function of the noise strength. This is a typical behavior for a transition between different potential barriers triggered by noise.

For short pulses, we deduced a dimensionless complex CQGLE including three higher-order effects: third-order dispersion, self-steepening, and intrapulse Raman scattering. Periodic exploding dissipative solitons are reported. We notice that a long time ago, but in the context of envelope equations, the effect of nonlinear gradient terms on localized solutions was studied by Deissler and Brand [39].

Acknowledgments: We wish to thank the support of FONDECYT through Project No. 1170728.

Author Contributions: Orazio Descalzi deduced the equations and has partially written the article. Carlos Cartes has carried out the numerical simulations, data analysis and has partially written the article.

References

1. Cundiff, S.; Soto-Crespo, J.; Akhmediev, N. Experimental Evidence for Soliton Explosions. *Phys. Rev. Lett.* **2002**, *88*, 073903.
2. Runge, A.F.; Broderick, N.G.; Erkintalo, M. Observation of soliton explosions in a passively mode-locked fiber laser. *Optica* **2015**, *2*, 36–39.
3. Liu, M.; Luo, A.P.; Yan, Y.R.; Hu, S.; Liu, Y.C.; Cui, H.; Luo, Z.C.; Xu, W.C. Successive soliton explosions in an ultrafast fiber laser. *Opt. Lett.* **2016**, *41*, 1181–1184.
4. Descalzi, O.; Rosso, O.A.; Larrondo, H.A. Localized Structures in Physics and Chemistry. *Eur. Phys. J. Spec. Top.* **2014**, *223*, 1–7.
5. Akhmediev, N.; Ankiewicz, A. *Dissipative Solitons*; Lecture Notes in Physics; Springer: Berlin/Heidelberg, Germany, 2005.
6. Kolodner, P.; Bensimon, D.; Surko, C. Traveling-wave convection in an annulus. *Phys. Rev. Lett.* **1988**, *60*, 1723.
7. Niemela, J.J.; Ahlers, G.; Cannell, D.S. Localized traveling-wave states in binary-fluid convection. *Phys. Rev. Lett.* **1990**, *64*, 1365.
8. Kolodner, P. Collisions between pulses of traveling-wave convection. *Phys. Rev. A* **1991**, *44*, 6466–6479.
9. Rotermund, H.; Jakubith, S.; Von Oertzen, A.; Ertl, G. Solitons in a surface reaction. *Phys. Rev. Lett.* **1991**, *66*, 3083.
10. Umbanhowar, P.B.; Melo, F.; Swinney, H.L. Localized excitations in a vertically vibrated granular layer. *Nature* **1996**, *382*, 793.
11. Taranenko, V.; Staliunas, K.; Weiss, C. Spatial soliton laser: Localized structures in a laser with a saturable absorber in a self-imaging resonator. *Phys. Rev. A* **1997**, *56*, 1582.
12. Lioubashevski, O.; Hamiel, Y.; Agnon, A.; Reches, Z.; Fineberg, J. Oscillons and propagating solitary waves in a vertically vibrated colloidal suspension. *Phys. Rev. Lett.* **1999**, *83*, 3190.
13. Ultanir, E.A.; Stegeman, G.I.; Michaelis, D.; Lange, C.H.; Lederer, F. Stable dissipative solitons in semiconductor optical amplifiers. *Phys. Rev. Lett.* **2003**, *90*, 253903.
14. Merkt, F.S.; Deegan, R.D.; Goldman, D.I.; Rericha, E.C.; Swinney, H.L. Persistent holes in a fluid. *Phys. Rev. Lett.* **2004**, *92*, 184501.
15. Ebata, H.; Sano, M. Self-replicating holes in a vertically vibrated dense suspension. *Phys. Rev. Lett.* **2011**, *107*, 088301.
16. Soto-Crespo, J.M.; Akhmediev, N.; Ankiewicz, A. Pulsating, creeping, and erupting solitons in dissipative systems. *Phys. Rev. Lett.* **2000**, *85*, 2937–2940.
17. Akhmediev, N.; Soto-Crespo, J.; Town, G. Pulsating solitons, chaotic solitons, period doubling, and pulse coexistence in mode-locked lasers: Complex Ginzburg-Landau equation approach. *Phys. Rev. E* **2001**, *63*, 056602.
18. Brand, H.R.; Lomdahl, P.S.; Newell, A.C. Evolution of the order parameter in situations with broken rotational symmetry. *Phys. Lett. A* **1986**, *118*, 67–73.
19. Brand, H.R.; Lomdahl, P.S.; Newell, A.C. Benjamin-Feir turbulence in convective binary fluid mixtures. *Physica D* **1986**, *23*, 345–361.
20. Thual, O.; Fauve, S. Localized structures generated by subcritical instabilities. *J. Phys. France* **1988**, *49*, 1829–1833.
21. Haus, H.A. Theory of mode locking with a fast saturable absorber. *J. Appl. Phys.* **1975**, *46*, 3049–3058.
22. Belanger, P. Coupled-cavity mode locking: A nonlinear model. *JOSA B* **1991**, *8*, 2077–2081.
23. Weiss, C. Spatio-temporal structures. Part II. Vortices and defects in lasers. *Phys. Rep.* **1992**, *219*, 311–338.
24. Mollenauer, L.F.; Gordon, J.P.; Evangelides, S.G. The sliding-frequency guiding filter: An improved form of soliton jitter control. *Opt. Lett.* **1992**, *17*, 1575–1577.
25. Firth, W.; Scroggie, A. Optical bullet holes: Robust controllable localized states of a nonlinear cavity. *Phys. Rev. Lett.* **1996**, *76*, 1623.
26. Descalzi, O.; Brand, H.R. Transition from modulated to exploding dissipative solitons: Hysteresis, dynamics, and analytic aspects. *Phys. Rev. E* **2010**, *82*, 026203.
27. Descalzi, O.; Cartes, C.; Cisternas, J.; Brand, H.R. Exploding dissipative solitons: The analog of the Ruelle-Takens route for spatially localized solutions. *Phys. Rev. E* **2011**, *83*, 056214.
28. Cartes, C.; Descalzi, O.; Brand, H.R. Noise can induce explosions for dissipative solitons. *Phys. Rev. E* **2012**, *85*, 015205.

29. Descalzi, O.; Cartes, C.; Brand, H.R. Noisy localized structures induced by large noise. *Phys. Rev. E* **2015**, *91*, 020901.
30. Descalzi, O.; Cartes, C.; Brand, H.R. Multiplicative noise can lead to the collapse of dissipative solitons. *Phys. Rev. E* **2016**, *94*, 012219.
31. Van Kampen, N. *Stochastic Processes in Physics and Chemistry*; North-Holland Personal Library; Elsevier Science: Amsterdam, The Netherlands, 2011.
32. Risken, H.; Haken, H. *The Fokker-Planck Equation: Methods of Solution and Applications*, 2nd ed.; Springer: Berlin/Heidelberg, Germany, 1989.
33. Tsimring, L.S. Noise in biology. *Rep. Prog. Phys.* **2014**, *77*, 026601.
34. Graham, R. Hydrodynamic fluctuations near the convection instability. *Phys. Rev. A* **1974**, *10*, 1762.
35. Schenzle, A.; Brand, H.R. Multiplicative stochastic processes in statistical physics. *Phys. Rev. A* **1979**, *20*, 1628.
36. Brand, H.R.; Kai, S.; Wakabayashi, S. External noise can suppress the onset of spatial turbulence. *Phys. Rev. Lett.* **1985**, *54*, 555.
37. Agrawal, G. *Nonlinear Fiber Optics*; Optics and Photonics Series; Academic Press: Cambridge, MA, USA, 2013.
38. Cartes, C.; Descalzi, O. Periodic exploding dissipative solitons. *Phys. Rev. A* **2016**, *93*, 031801.
39. Deissler, R.; Brand, H.R. The effect of nonlinear gradient terms on localized states near a weakly inverted bifurcation. *Phys. Lett. A* **1990**, *146*, 252.

applied
sciences

MDPI

Article

Designing a Novel High-Performance FBG-OADM Based on Finite Element and Eigenmode Expansion Methods

Sheng-Chih Yang [1], Yue-Jing He [2,*] and Yi-Jyun Wun [1]

[1] Department of Computer Science and Information Engineering, National Chin-Yi University of Technology, Taichung 41170, Taiwan; scyang@ncut.edu.tw (S.-C.Y.); sss60712@gmail.com (Y.-J.W.)
[2] Department of Electronic Engineering, National Chin-Yi University of Technology, Taichung 41170, Taiwan
* Correspondence: yuejing@ncut.edu.tw; Tel.: +886-4-2392-4505 (ext. 7359)

Academic Editor: Boris Malomed
Received: 6 December 2016; Accepted: 28 December 2016; Published: 30 December 2016

Abstract: This study designed a novel high-performance fiber Bragg grating (FBG) optical add/drop multiplexers (OADMs) by referring to current numerical simulation methods. The proposed FBG-OADM comprises two single-mode fibers placed side by side. Both optical fibers contained an FBG featuring identical parameters and the same geometric structure. Furthermore, it fulfills the full width at half maximum (FWHM) requirement for dense wavelength-division multiplexers (DWDMs) according to the International Telecommunication Union (i.e., FWHM < 0.4 nm). Of all related numerical calculation methods, the combination of the finite element method (FEM) and eigenmode expansion method (EEM), as a focus in this study, is the only one suitable for researching and designing large-scale components. To enhance the accuracy and computational performance, this study used numerical methods—namely, the object meshing method, the boundary meshing method, the perfectly matched layer, and the perfectly reflecting boundary—to simulate the proposed FBG-OADM. The simulation results showed that the novel FBG-OADM exhibited a −3 dB bandwidth of 0.0375 nm. In addition, analysis of the spectrum revealed that the drop port achieved the power output of 0 dB at an operating wavelength of 1550 nm.

Keywords: fiber bragg grating; optical add-drop multiplexer; finite element method; eigenmode expansion method; perfectly matched layer; perfectly reflection boundary; object meshing method; boundary meshing method

---

1. Introduction

Fiber gratings refer to periodic structures that change according to the refractive index (RI) of the core of photosensitive optical fibers. According to the periodic length of fiber gratings, they can be categorized as short- or long-period fiber gratings. Short-period fiber gratings, also known as fiber Bragg gratings (FBGs), were first proposed by Hill et al. at the Canadian Communication Center in 1978 [1–3]. Of all related methods currently used to produce FBGs, the interferometric and phase mask methods are the most common ones. The interferometric method involves irradiating a photosensitive optical fiber with two interleaved beams of UV light; the UV light wavelength or angle between the interleaved beams is adjusted to change the RI of the optical fiber, thereby producing the desired fiber grating. The phase mask method involves irradiating UV light onto a phase mask, which creates constructive and destructive interference in the core of the photosensitive optical fiber. Changes in energy intensity then cause the RI of the optical fiber to display a periodic distribution [4–6].

Optical add/drop multiplexers (OADMs) are key components for creating wavelength-division multiplexers (WDMs) in fiber-optic communication networks. The main function of an OADM is to drop or add a client signal in the fiber-optic communication network. Specifically, OADMs employ wavelength division multiplexing techniques in the frequency domain to emulate the time division multiplexing capabilities of the traditional synchronous digital hierarchy standard in the time domain. In the past several decades, numerous types of OADM have been proposed, many of which are widely used in various fiber-optic communication systems [7–12]. Of all of the currently available OADMs, the most common one is composed of an FBG and two optical circulators. However, this particular OADM is disadvantageous because of its large size and high-cost, complex production process.

To overcome the disadvantages of current OADMs, the present study referenced optical coupling theory to develop a novel high-performance FBG-OADM model comprising two single mode fibers (SMFs), which are placed side by side so that they work in the same manner as a 2 × 2 optical fiber coupler, and two FBGs, which have identical parameters so that the they work in the same manner as a mode coupler. In contrast to traditional OADMs, the proposed FBG-OADM is advantageous because it is a miniature-sized all-optical fiber-based multiplexer that can be fabricated through a low-cost and simple manufacturing process [12]. In researching and designing the proposed FBG-OADM, this study combined the finite element method (FEM) and eigenmode expansion method (EEM) to perform numerical simulations for calculation and analysis.

The remainder of this paper is organized as follows: Section 2 details the geometric structure, parameters, and working principle of the proposed FBG-OADM. Optical coupling theory is used to explain how the FBG-OADM achieves outstanding bandwidth performance (−3 dB). In addition, to minimize discrepancies between simulations and actual performance results, a perfectly matched layer (PML) and perfectly reflecting boundary (PRB) were integrated into the FBG-OADM design. Section 3 explains how the FEM technique was used to solve and analyze the FBG-OADM mode. To obtain suitable mesh-cutting resolutions, the object meshing method (OMM) and boundary meshing method (BMM)—two methods that are currently used in the FEM—were adopted to determine the optimal tradeoff between computational performance and cost. Mathematically, solutions derived from partial differential equations (PDEs) with boundary conditions must be pairwise orthogonal; in other words, their orthogonal value must be equal to zero. However, in numerical simulations, because of memory and computational time constraints, achieving pairwise orthogonality between modes is impossible. In other words, errors are inevitable. Therefore, in this study, the acceptable maximum error value was set at −40 dB as the review standard. All subsequently employed mesh-cutting resolutions must, therefore, generate modes with an orthogonal value of less than −40 dB. For mesh-cutting resolutions that failed to meet this standard, the resolution was increased; subsequently the solutions to the modes were recalculated and the orthogonal values were reevaluated [12]. Section 4 introduces the roles and functions of EEM in the numerical simulations and expounds why the combined FEM-EEM approach is superior to traditional numerical simulation methods for designing and analyzing large optical components containing periodic structures. Mathematically, EEM embodies an operation concept similar to that of the Fourier series expansion method; that is, if too few expansion functions are used, discrepancies will occur between the expansion and expanded functions. In numerical operations, it is impossible to include all modes; consequently, errors are unavoidable. To overcome this problem, similar to the FEM approach adopted in this study, the acceptable maximum error (i.e., power loss) for modal expansions was set to −40 dB. Section 5 shows how the FEM and EEM were combined to research and design the proposed FBG-OADM. The design process involved the following procedure: (1) solve the FBG-OADM modes, (2) plot 2D power distribution maps of the modes, (3) review the orthogonal values of the modes, (4) identify the optimal design parameters, (5) inspect the overall power loss, (6) plot the spectrum for the drop port, (7) examine heterodyne and homodyne crosstalk, and (8) calculate the −3 dB bandwidth. Finally, Section 6 summarizes the study findings. This study developed a novel, high-performance FBG-OADM that was designed through the combined FEM-EEM approach. To minimize the number of errors, a PML and PRB were simulated through various numerical methods (i.e., the OMM and BMM) and a review standard was formulated to assess the orthogonal values and power loss. The numerical simulations showed that the proposed FBG-OADM overcomes the disadvantages of current OADMs. Moreover, the proposed FBG-OADM possesses nearly zero heterodyne and homodyne crosstalk and meets the full width at half maximum (FWHM) requirement for dense wavelength-division multiplexers (DWDMs) according to the International Telecommunication Union (ITU; FWHM < 0.4 nm).

2. Novel FBG-OADM

Figure 1 shows the geometric structure of the proposed FBG-OADM [12]. The structure comprises two SMFs placed side by side (Fibers 1 and 2), a 1 μm thick PML (purple lines), a PRB (yellow lines), a noncoupling area (L_1 and L_3); a mode-power coupling area (L_2); an HE_{11} core mode (red lines); and a cladding mode (black lines).

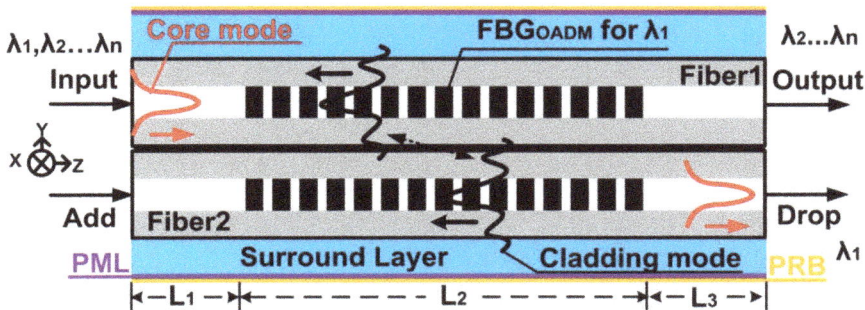

Figure 1. Side view (*X-Z* plane) of the proposed FBG-OADM.

When the drop wavelength of the FBG-OADM is set to λ_1 and N core mode signals with different wavelengths (i.e., $\lambda_1, \lambda_2, \ldots \lambda_n$) are inputted via the input port, core mode signals of all wavelengths (except for that with a wavelength of λ_1) are unaffected by the FBG and are outputted directly via the output port. However, the core mode signal with a wavelength of λ_1 is subjected to perturbation from the FBG and is coupled to the cladding mode propagated along the $-Z$ direction. Mode theory posits that the power distribution in the core mode exists in the cladding layer in the form of exponential decay (red lines). In other words, the core mode in the figure cannot detect the presence of the surround layer. Thus, the core mode cannot couple Fiber 1 to Fiber 2. Conversely, because the power in the cladding mode can be extended to the surround layer (black lines), the cladding mode can couple Fiber 1 to Fiber 2. Subsequently, the cladding mode in Fiber 2 is subjected to perturbation from the FBG, prompting the coupled core mode to be propagated along the Z direction and outputted via the drop port, thus completing the signal drop process. Regarding the signal add process, because the proposed FBG-OADM has a symmetric structural design, signal adding works according to the same principle as signal dropping; that is, when a signal is inputted via the add port of Fiber 2, it will be outputted via the output port of Fiber 1, thus completing the signal add process [12].

Compared with a previous long-period fiber grating (LPG) OADM proposed by the author of the present study (in which the LPG-OADM comprised two LPGs and a 2 × 2 optical fiber coupler) [7], the present OADM uses FBGs instead of LPGs. Roughly speaking, the overall spectrum of the two OADMs is the product of the spectrums of the two LPGs (or two FBGs) and 2 × 2 optical fiber coupler, as shown in Table 1. According to the table, the newly-proposed FBG-OADM features a narrower spectrum, indicating that it can achieve exceptional bandwidth performance (−3 dB).

Table 1. Comparison of spectrums between the LPG-OADM and FBG-OADM.

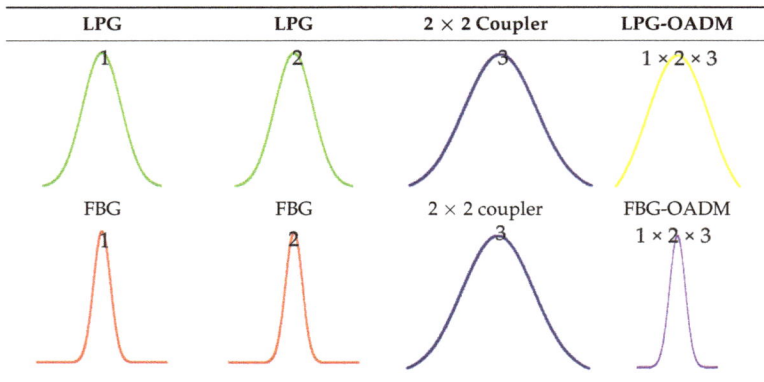

In real-world environments, signals in an optical fiber structure may take the form of one of two mode types: discrete guided modes and continuous radiation modes. However, because continuous radiation modes cannot be measured in real-world environments, they are considered a type of power loss. Moreover, the limitations imposed on numerical simulations by memory and computation time constraints render it impossible to contain continuous radiation modes. To overcome this drawback and, thus, improve the accuracy of simulation results, this study employed a PRB to convert continuous radiation modes into discrete radiation modes. In addition, to enable the converted discrete radiation modes to achieve expected power loss, a PML was incorporated into the surround layer, enabling the propagation constant of the discrete radiation modes to be $-(\alpha + i\beta)$, where α is the attenuation constant and β is the phase constant [13,14].

SMFs were employed as the optical fiber technology in this study. SMFs are a type of fiber in which the solution for the core and the nearly infinite number of solutions for the cladding layer can be identified once the size and constituent materials of the fibers are determined (using PDEs). From a physics perspective, all solutions are referred to as guided modes and each solution corresponds to an equivalent RI. Thus, the equivalent RI (i.e., n_{neff}^{core}; $n_2 < n_{neff}^{core} < n_1$) of the core mode as well as nearly infinite number of equivalent RIs (i.e., $n_{neff}^{cladding}$; $n_3 < n_{neff}^{cladding} < n_2$) of the cladding modes can be determined, where n_1 is the RI of the core layer, n_2 is the RI of the cladding layer, n_3 is the RI of the surround layer, n_{neff}^{core} is the effective RI of the core mode, and $n_{neff}^{cladding}$ is the effective RI of the cladding mode.

FBGs involve irradiating UV light signals on phase masks with a specific cycle to cause periodic variation in the RI of the fiber core. These variations in the core RI are called fiber gratings. A uniform fiber grating can be expressed using the following mathematical equation [12,15]:

$$n_1(z) = n_1 + \delta n \left[1 + \cos\left(\frac{2\pi}{\Lambda_{FBG}} z \right) \right] \tag{1}$$

where δn is the peak induced-index change, n_1 is the RI of the core, and Λ_{FBG} is the FBG period. Mathematically, before light signals are affected by fiber gratings, all guided modes in the optical fiber display pairwise orthogonality. In order words, powers between the modes do not couple to, or swap with, each other during propagation. However, as light signals become affected by fiber gratings, powers between the modes undergo perturbation, causing coupling and crosstalk between the modes.

3. Finite Element Method

The calculation principles underlying the FEM [12,16–18] are briefly explained here. The FEM can be applied to solve PDEs and it can be combined with numerical simulation methods for evaluating boundary conditions. The FEM is based on the variational principle, region segmentation, and interpolation functions. In the FEM, the variational principle is used to convert original problems (i.e., PDEs and boundary conditions) into functional functions to find the extrema. Next, geometric regions are divided, and hypothetically known element nodes are substituted into interpolation functions to describe the unknown functions in the elements. Then, multivariate linear equations are substituted into the functional functions, and boundary conditions are incorporated to identify the interpolation functions of all elements. When solutions to the interpolation functions are obtained, unknown nodes and the solutions of an entire region can be derived. The FEM procedure is summarized as follows: (1) convert the problems of PDEs and boundary conditions into quadratic functional functions to identify the extrema; (2) divide the geometric regions into subblocks, which are referred to as elements (e.g., triangular or quadrilateral elements); (3) assuming that the nodes in a triangular element are known, substitute the known nodes into the interpolation functions to create polynomial linear equations and describe the unknown variables in the elements; and (4) substitute the polynomial linear equations into the extrema-based functional functions while incorporating the boundary conditions to identify the solutions to the interpolation functions. By obtaining the unknown nodes using the interpolation function, calculate the solutions of the entire region.

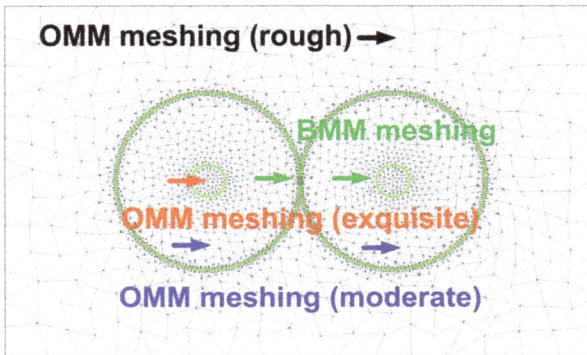

Figure 2. BMM- and OMM-derived meshes for the FBG-OADM simulation.

This study used FEM to solve the guided modes in an optical fiber structure. As shown in Figure 2, the OMM was used to obtain the standard mesh size, which was based on the geometric size of the objects [12]. Then, the BMM was used to obtain boundary meshes with a higher resolution. Since the numerical simulations were limited by memory and computation ability constraints, achieving pairwise orthogonality between modes was impossible. Thus, errors were unavoidable. Therefore, the acceptable maximum error value (i.e., the orthogonal value) was set at −40 dB as the review standard. In other words, all mesh-cutting resolutions were required to produce modes with an orthogonal value of less than −40 dB [9,12,19]. Mesh sizes that satisfied the review standard were obtained through adjusting the mesh-cutting resolutions and testing the orthogonal values. The orthogonal value equation is shown as follows [15,19]:

$$\int_{A\infty} E_{t\nu} \times H_{t\mu} \cdot \hat{z} dA = \int_{A\infty} E_{t\mu} \times H_{t\nu} \cdot \hat{z} dA = 0 \; for \; \nu \neq \mu \qquad (2)$$

The mesh size used in this study was 1:5:17:43 (i.e., the object boundary/small object/medium object/large object ratio).

4. Eigenmode Expansion Method

The primary function of the EEM was to transmit the power of simulated guided modes. The RI of FBG-OADM underwent cyclical changes and each change in RI during a cycle was called a segment. The first segment of the FBG-OADM was first dropped with each division referred to as a "block". Each block was assumed to be a uniform waveguide. Next, the FEM was used to solve all guided modes in each block, and the EEM was then employed to transmit the power between the block boundaries. Since changes in the RIs of FBG-OADM were cyclical, the power transmission in the segments during the simulations resembled the power transmission situation of the overall FBG-OADM [7,13,14].

The aforementioned information shows that, in contrast to using traditional numerical simulation techniques, this study performed FBG-OADM-based numerical simulations by using the combined FEM–EEM approach, which can complete numerical simulations within a shorter timeframe. Although the finite-difference time-domain (FDTD) method has been used in previous research [20], it requires considerable computation time and memory capacity during simulations—especially for complex structures, such as the FBG-OADM. In summary, the combined FEM–EEM approach is more efficient than conventional numerical simulation techniques for designing devices with a cyclic structure (e.g., FBG-OADM, LPG-OADM, and resonance sensors with a long-period fiber grating).

Mathematically, EEM embodies an operation concept similar to that of the Fourier series expansion method; that is, if too few expansion functions are used, discrepancies will occur between the expansion and expanded functions. In numerical operations, because including all modes is impossible, errors are unavoidable. To overcome this problem, similar to the FEM approach adopted in this study, the acceptable maximum error (i.e., power loss) for modal expansions was set to -40 dB [7,13,14]. In other words, when the power loss failed to reach the review standard, the number of modes was increased and simulations were repeated until the power loss satisfied the review standard.

5. Result and Analysis of the FBG-OADM

FBG-OADM simulations, designs, and research were made using the combined FEM–EEM approach. The parameters in Figure 3 are explained as follows: a_1 is the core radius; a_2 is the cladding layer radius; n_1 is the RI of the core layer; n_2 is the RI of the cladding layer; n_3 is the RI of the surround layer; δn is the peak induced-index change by the UV light; d_{pml} is the thickness of the PML; L_1 and L_3 are the noncoupling areas of the FBG-OADM; L_2 is the mode–power coupling area; d_{WPC} and d_{HPC} are the width and height between the cladding layer and the PML, respectively; W and H are the width and height of the overall structure, respectively. The parameters were set as follows: $a_1 = 2.25$ μm, $a_2 = 12.25$ μm, $n_1 = 1.454$, $n_2 = 1.43$, $n_3 = 1.415$, $\delta n = 1.454 \times 10^{-3}$, $d_{pml} = 1$ μm, $L_1 = 1$ μm, $L_3 = 1$ μm, $L_2 = \Lambda_{FBG} \times N_p$, $d_{WPC} = 13.7$ μm, $d_{HPC} = 11.25$ μm, $W = 78.4$ μm, and $H = 49$ μm, where N_p is the number of period in the FBG-OADM. Moreover, the distance between the two optical fibers was set to zero and the operating wavelength $\lambda = 1550$ nm. Figure 4 shows the side view (i.e., X-Z plane) of the FBG-OADM during the numerical simulations.

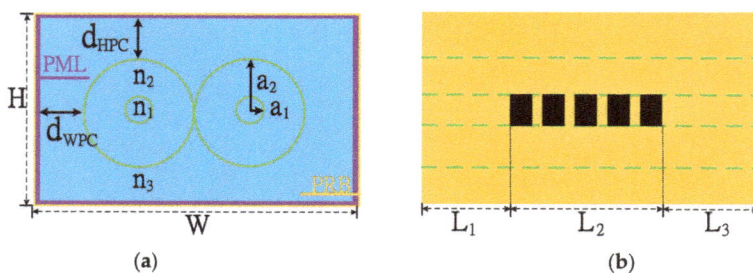

Figure 3. Geometric structure of the FBG-OADM: (**a**) sectional view from the *X-Y* plane; and (**b**) side view from the *Y-Z* plane.

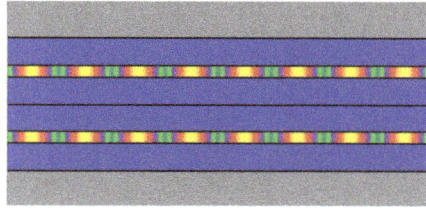

Figure 4. Side view (*X-Z* plane) of the FBG-OADM in the numerical simulation.

After repeatedly adjusting the number of modes, recalculating the solutions, and reexamining the power loss, this study found that the number of guided modes that met the power loss review standard was 80. Figures 5–8 show the power distribution maps of modes when v = 9, 22, 56, and 79, respectively; the corresponding equivalent RIs were $n_{neff}^{v=9}$ = 1.429881, $n_{neff}^{v=22}$ = 1.427474, $n_{neff}^{v=56}$ = 1.422949, and $n_{neff}^{v=79}$ = 1.420274, respectively.

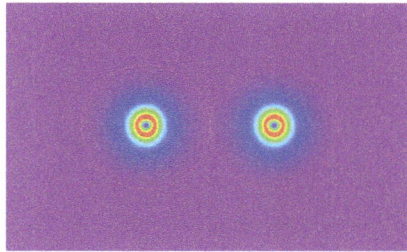

Figure 5. 2D power distribution plot for guided-mode v = 9 ($n_{neff}^{v=9}$ = 1.429881).

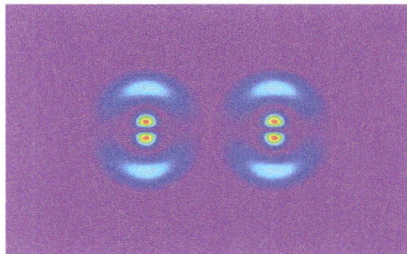

Figure 6. 2D power distribution plot for guided-mode v = 22 ($n_{neff}^{v=22}$ = 1.427474).

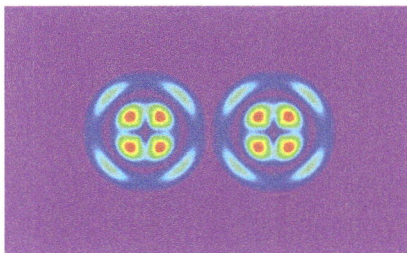

Figure 7. 2D power distribution plot for guided-mode v = 56 ($n_{neff}^{v=56}$ = 1.422949).

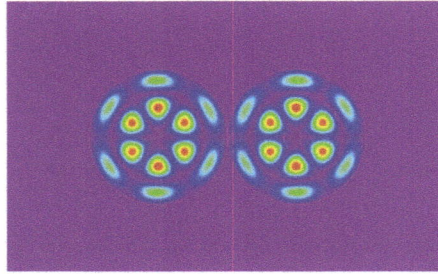

Figure 8. 2D power distribution plot for guided-mode $v = 79$ ($n_{neff}^{v=79} = 1.420274$).

According to optical coupling theory, the only guided modes that are subjected to the influence of the 2 × 2 optical fiber coupler and could subsequently transmit and couple power from Fiber 1 to Fiber 2 are the cladding modes. In addition, because of the perturbation effect of the FBG, core modes couple only to cladding modes with a powered core. In other words, of the four guided modes shown in Figures 5–8, only the one in Figure 5 ($v = 9$) achieved acceptable FBG-OADM drop and add functionality.

Mathematically, all guided modes must display pairwise orthogonality. However, in numerical simulations, because of memory and computation time constraints, it is impossible to achieve pairwise orthogonality between modes. Therefore, in this study, the OMM and BMM were combined to determine the appropriate segmentation size and formulate a review standard for the orthogonal value. Figure 9 shows the orthogonal values between the 80 modes, which confirms that the results satisfied the review standard.

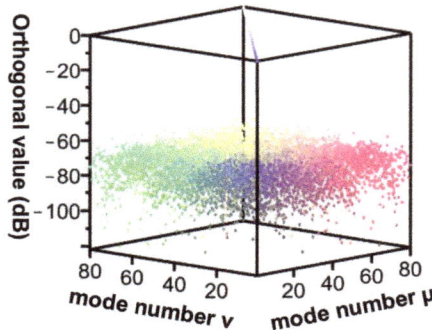

Figure 9. Relationships between the orthogonal values of 80 guided modes for the proposed FBG-OADM.

The FBG-OADM was designed after determining the optimal mesh resolution and number of modes. The ratio of the drop power to input power for the FBG-OADM must equal 1. In other words, its drop power should be 0 dB. Thus, Λ_{FBG} and N_P values that satisfied this condition had to be identified to design an FBG-OADM that achieved zero insertion loss and minimal heterodyne and homodyne crosstalk. Figure 10 shows the scanning diagram of the optimal Λ_{FBG}. According to the power transmission curve in Figure 10, the power transmission approached 0 dB when $\Lambda_{FBG} = 0.5396$ μm. Figure 11 illustrates the scanning diagram of the optimal FBG-OADM length, showing that the power transmission approached 0 dB when $L_2 = 0.9908$ cm. Next, from Equation (3), N_P (i.e., the number of period in the FBG-OADM) was calculated and found to be 18,361.

$$L_2 = \Lambda_{FBG} \times N_P \tag{3}$$

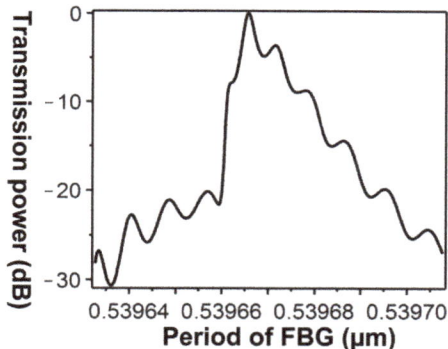

Figure 10. Scanning diagram of the optimal Λ_{FBG}.

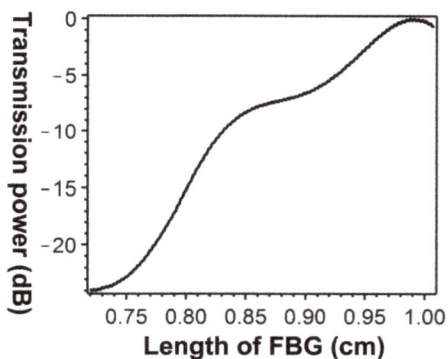

Figure 11. Scanning diagram of the optimal FBG-OADM length.

According to the EEM descriptions in Section 4, a small amount of power loss can be observed in all modes during transmissions; to address this, the power loss review standard was established. Figure 12 details the correlations between the power loss and length of the FBG-OADM, showing that the power loss in the power transmission over the entire FBG-OADM met the review standard for all segment cycles (N_p).

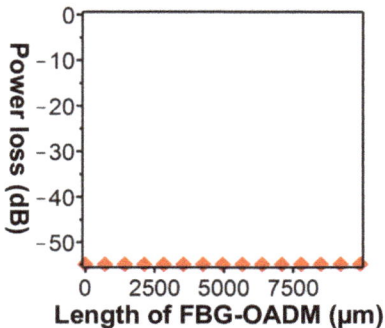

Figure 12. Diagram of the relationship between the transmission distance and power loss when the core mode was completely dropped.

To verify that the optimal parameter values were $\Lambda_{FBG} = 0.5396$ µm and $N_P = 18{,}361$, the power transmission along the *X-Z* plane was examined with light inputted via the input port of the FBG-OADM, as shown in Figure 13. The figure shows that the input power was successfully coupled to the drop port of Fiber 2. To further investigate the coupling performance, a 2D power distribution map of the input port, $z = 4955.3991$ µm, and drop port were collected, as shown in Figures 14–16. Figure 14 shows the power distribution map of the input port. Figure 15 displays the 2D power distribution during coupling, showing the perturbation effect of the FBG inducing the core mode to couple to the cladding mode, as well as the cladding mode under the effect of the 2×2 optical fiber coupler, which facilitated cross-coupling between the two optical fibers. Figure 16 is the power distribution map of the drop port, showing that the cladding mode was subjected to the perturbation effect of the FBG, causing cladding mode to couple to the core mode. Thus, the numerical simulation results in Figures 13–16 confirm the accuracy of the proposed FBG-OADM.

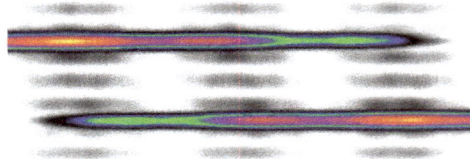

Figure 13. Diagram of the power propagation in the *X-Z* plane when the core mode was completely dropped.

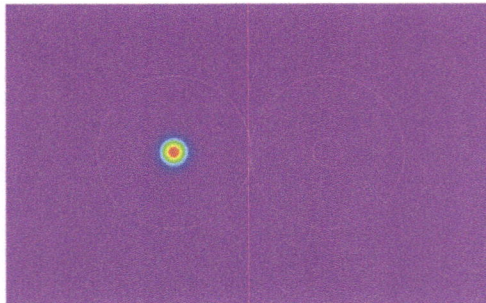

Figure 14. 2D power distribution map of the input port ($z = 0.0000$ µm).

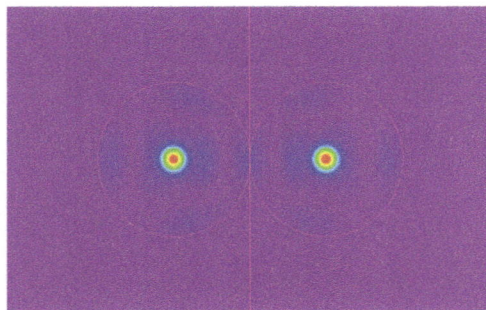

Figure 15. 2D power distribution map ($z = 4955.3991$ µm).

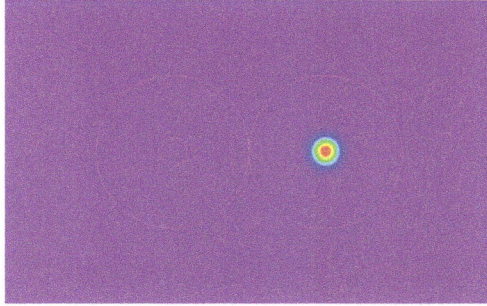

Figure 16. 2D power distribution map for the output port (z = 9910.7982 μm).

To investigate whether the −3 dB bandwidth of the proposed FBG-OADM satisfied the ITU guidelines, the spectrum of the drop port was plotted, as shown in Figure 17. The figure shows that the FBG-OADM exhibited transmission power of 0 dB in the drop port. In other words, power inputted via the input port was outputted via the drop port with zero power loss, showing that heterodyne and homodyne crosstalk were unlikely to occur in the FBG-OADM. When an OADM drops a signal, power that is not outputted via the drop port of Fiber 2 is restored in the output port in the form of residual power, which creates heterodyne crosstalk with other wavelengths. Similarly, when an OADM adds a signal, power remaining in the output port creates homodyne crosstalk. Figure 18 shows the −3 dB spectrum of the drop port, indicating that the FBG-OADM demonstrated considerably high bandwidth efficiency. In addition, the FBG-OADM exhibited an FWHM of ≤0.0375 nm (i.e., ≤0.0000375 μm), which is markedly lower than the ITU requirement for DWDMs (FWHM < 0.4 nm). In this simulation, four Intel Xeon CPUs (E7530@1.87 GHz) and 128 GB memory were used and the computation time was approximately 118 h.

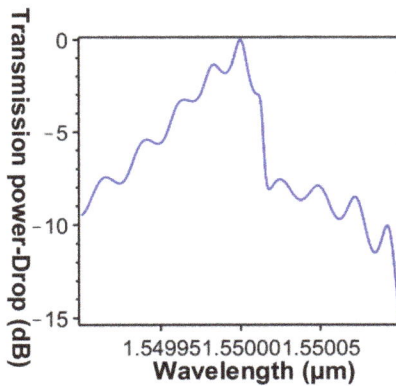

Figure 17. Spectrum of the FBG-OADM when the core mode was completely dropped.

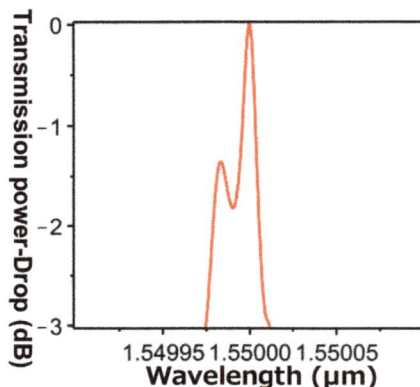

Figure 18. Spectrum of the FBG-OADM for the −3 dB bandwidth (FWHM).

6. Conclusions

The emergence of optical fiber technology in recent years has considerably improved network stability and bandwidth development. OADMs are crucial components for connecting computers in fiber-optic networks. In the past several decades, many types of OADMs have been introduced and used in network systems. Of all the OADMs currently available, the most common one is composed of an FBG and two optical circulators. However, this particular OADM is disadvantageous because of its large size and high-cost, complex production process. The proposed high-performance FBG-OADM model, which was developed on the basis of optical coupling theory, overcomes these shortcomings. The model comprises two SMFs, which are placed side by side so that they work in the same manner as a 2 × 2 optical fiber coupler, and two FBGs, which have identical parameters so that they work in the same manner as a mode coupler. To analyze the performance of the FBG-OADM, simulations were performed to predict the performance of the device in a real-world environment. The simulated system comprised an internal part (for simulating the FBG-OADM) and an external part (for simulating the radiation mode absorption). The combination of the FEM and the EEM provided a rigorous yet simple process for designing OADMs and for testing models in an environment that is comparable to the real world. Next, the results were produced in the form of graphs. Concerning the design structure, the FEM was used to obtain solutions for the guided modes in the FBG-OADM and to analyze the modes. Next, the EEM was employed to simulate the guided modes for power transmission simulations. In cyclic components such as the FBG-OADM, because guided modes are identical during a cycle, using the FEM to obtain solutions for the guided modes and using the EEM to simulate power transmission inference during each cycle yielded results representing the simulation results of the overall FBG-OADM. This considerably reduced the memory and computation time requirements for the simulations. When optical signals are inputted into an internal simulation structure, they create both discrete guided modes and continuous radiation modes. However, because continuous radiation modes are considered power loss in real-world environments, they cannot be measured. Therefore, this study designed an external simulation, in which two numerical methods were used to simulate the PRB and PML in order to reduce discrepancies between the simulations and actual performance results. The PRB was used to convert continuous radiation modes into discrete radiation modes, and the PML was used to absorb the power of the discrete radiation modes.

The numerical simulations revealed the optimal FBG-OADM cycle (i.e., Λ_{FBG} = 0.5396 μm) as well as the optimal number of period (i.e., N_P = 18,361). These two parameters were used to develop the high-performance FBG-OADM. The findings are summarized as follows: (1) the orthogonal values between all guided modes satisfied the preset orthogonal value review standard; (2) power loss during transmission in the FEM simulations met the preset power loss review standard; (3) the 2D power distribution maps of the input and drop ports showed that the FBG-OADM accurately transmitted and dropped signals; (4) the frequency distribution of the drop port showed that at an operating wavelength (λ) of 1550 nm, the power approximated 0 dB, indicating that homodyne and heterodyne crosstalk did not occur; and (5) the FBG-OADM exhibited an FWHM of 0.0375 nm, which is lower than the ITU requirement for DWDMs. These simulation results confirm the feasibility and accuracy of the proposed FBG-OADM, the advantages of which are that it is a miniature-sized all-optical fiber-based multiplexer with a low FWHM and that it can be fabricated through a low-cost and simple manufacturing process.

Acknowledgments: The author gratefully acknowledges the support provided for this study by the Ministry of Science and Technology (MOST 104-2221-E-167-013-MY2) of Taiwan.

Author Contributions: Sheng-Chih Yang proposed the research topic. Yue-Jing He performed the simulation and wrote the manuscript. Yi-Jyun Wun wrote the manuscript.

Conflicts of Interest: The authors declare no conflict of interest.

References

1. Hill, K.O.; Meltz, G. Fiber Bragg grating technology fundamentals and overview. *J. Lightwave Technol.* **1997**, *15*, 1263–1276. [CrossRef]
2. Kawasaki, B.S.; Hill, K.O.; Johnson, D.C.; Fujii, Y. Narrow-band Bragg reflectors in optical fibers. *Opt. Lett.* **1978**, *3*, 66–68. [CrossRef] [PubMed]
3. Hill, K.O.; Fujii, Y.; Johnson, D.C.; Kawasaki, B.S. Photosensitivity in optical fiber waveguides: Application to reflection filter fabrication. *Appl. Phys. Lett.* **1978**, *32*, 647–649. [CrossRef]
4. Osuch, T.; Markowski, K.; Gasior, P.; Jedrzejewski, K. Quasi-uniform fiber Bragg gratings. *J. Lightwave Technol.* **2015**, *33*, 4849–4856. [CrossRef]
5. Gagne, M.; Kashyap, R. New nanosecond Q-switched Nd:VO$_4$ laser fifth harmonic for fast hydrogen-free fiber Bragg gratings fabrication. *Opt. Commun.* **2010**, *283*, 5028–5032. [CrossRef]
6. Ozcan, A.; Digonnet, M.J.F.; Kino, G.S. Characterization of fiber Bragg gratings using spectral interferometry based on minimum-phase functions. *J. Lightwave Technol.* **2006**, *24*, 1739–1757. [CrossRef]
7. He, Y.J.; Chen, X.Y. Designing LPG-OADM based on a finite element method and an eigenmode expansion method. *IEEE Trans. Nanotechnol.* **2013**, *12*, 460–471. [CrossRef]
8. Chiaroni, D.; Yin, X.; Qiu, X.Z.; Gillis, J.; Put, J.; Bauwelinck, J.; Lanteri, D.; Blache, F.; Achouche, M.; Gripp, J. Successful experimental validation of an integrated burst mode receiver designed for 10G-GPON systems in a packet-OADM metro network. In Proceedings of the Optical Fiber Communication Conference, Los Angeles, CA, USA, 4–8 March 2012; Optical Society of America: Washington, DC, USA, 2012; pp. 1–3.
9. Munoz, R.; Martinez, R.; Pinart, C.; Sorribes, J.; Junyent, G. Experimental in-fiber GMPLS fault management for 1:1 OUPSR R-OADM networks. In Proceedings of the 31st Eruopean Conference on Optical Communication, Glasgow, UK, 25–29 September 2005; pp. 223–224.
10. Chang, C.H.; Liang, T.C.; Huang, C.Y. DWDM self-healing access ring network with cost-saving, crosstalk-free and bidirectional OADM in single fiber. *Opt. Commun.* **2009**, *282*, 4518–4523. [CrossRef]
11. Bhatia, K.S.; Kamal, T.S. Modeling and simulative performance analysis of OADM for hybrid multiplexed optical-OFDM system. *Opt. Int. J. Light Electron Opt.* **2013**, *124*, 1907–1911. [CrossRef]
12. He, Y.J. Analyzing guided-modes for novel and high performance FBG-OADM using improved FEM. In Proceedings of the 2016 International Symposium on Computer, Consumer and Control, Xi'an, China, 4–6 July 2016; pp. 1043–1046.
13. He, Y.J. Highly sensitive biochemical sensor comprising rectangular nano-metal arrays. *Sens. Actuators B* **2015**, *220*, 107–114. [CrossRef]
14. He, Y.J. High-performance localized surface plasmon resonance fiber sensor based on nano-metal-gear array. *Sens. Actuators B* **2014**, *193*, 778–787. [CrossRef]
15. He, Y.J.; Chen, X.Y. Optical characteristic research on fiber Bragg gratings utilizing finite element and eigenmode expansion Method. *Sensors* **2014**, *14*, 10876–10894. [CrossRef] [PubMed]
16. Ou, Y.X.; Pardo, D.; Chen, Y.T. Fourier finite element modeling of light emission in waveguides: 2.5-dimensional FEM approach. *Opt. Express* **2015**, *23*, 30259–30269. [CrossRef] [PubMed]

17. Liu, J.; Zou, J.; Tian, J.H.; Yuan, J.S. Analysis of electric field, ion flow density, and corona loss of same-tower double-circuit HVDC lines using improved FEM. *IEEE Trans. Power Deliv.* **2009**, *24*, 482–483.
18. Xingqi, D.; Tongy, A. A new FEM approach for open boundary Laplace's problem. *IEEE Trans. Microw. Theory Tech.* **1996**, *44*, 157–160. [CrossRef]
19. He, Y.J. Novel D-shape LSPR fiber sensor based on nano-metal strips. *Opt. Express* **2013**, *21*, 23498–23510. [CrossRef] [PubMed]
20. Tsarev, A.V. Simulation by BPM and FDTD of new thin heterogeneous optical waveguides on SOI for reconfigurable optical add/drop multiplexers. In Proceedings of the 2008 IEEE Region 8 International Conference on Computational Technologies in Electrical and Electronics Engineering, Novosibirsk, Russia, 21–25 July 2008; pp. 366–368.

applied
sciences

MDPI

Article

A Metal-Insulator-Metal Deep Subwavelength Cavity Based on Cutoff Frequency Modulation

Kihwan Moon, Tae-Woo Lee, Young Jin Lee and Soon-Hong Kwon *

Department of Physics, Chung-Ang University, Seoul 06974, Korea; sinbadra@gmail.com (K.M.);
ekqnsgl@gmail.com (T.-W.L.); youngjin.lee.91@gmail.com (Y.J.L.)
* Correspondence: soonhong.kwon@gmail.com or shkwon@cau.ac.kr; Tel.: +82-2-820-5844

Academic Editor: Boris Malomed
Received: 16 November 2016; Accepted: 10 January 2017; Published: 17 January 2017

Abstract: We propose a plasmonic cavity using the cutoff frequency of a metal-insulator-metal (MIM) first-order waveguide mode, which has a deep subwavelength physical size of $240 \times 210 \times 10$ (nm^3) $= 0.00013 \lambda_0{}^3$. The cutoff frequency is a unique property of the first-order waveguide mode and provides an effective mode gap mirror. The cutoff frequency has strong dependence on a variety of parameters including the waveguide width, insulator thickness, and insulator index. We suggest new plasmon cavities using three types of cutoff frequency modulations. The light can be confined in the cavity photonically, which is based on the spatial change of the cutoff frequency. Furthermore, we analyze cavity loss by investigating the metallic absorption, radiation, and waveguide coupling loss; the radiation loss of the higher-order cavity mode can be suppressed by multipole cancellation.

Keywords: MIM; deep subwavelength; cutoff; first order waveguide mode

1. Introduction

Surface plasmon polaritons (SPPs), electron oscillations upon coupling with photons, appear at dielectric-metal interfaces by coupling with photons [1]. Recently, many researchers have investigated the miniaturization of photonic devices by using SPPs because of their ability to manipulate photons in subwavelength-sized cavities and waveguides beyond the diffraction limit, which is a fundamental size limit of dielectric photonic devices [2–5]. In particular, plasmon cavities are exploited in deep subwavelength volume lasers [4,6], switches [7,8], index sensors [9,10], and plasmonic optical filters [11–13].

Metal-insulator-metal waveguides can strongly confine SPPs in thin insulators (even in several nanometer-scale dielectric gaps [14–16]), where it has been reported that visible light is confined in an ultrasmall dielectric layer. For this reason, MIM waveguide–based cavities were proposed to realize compact devices with various structures, such as disks [17,18], rings [19,20], and blocks [21–23]. As the dielectric layer between metals becomes thinner, the propagating SPP mode has a larger k (wavevector), thereby miniaturizing the physical size of the photonic device [13,21,24]. In MIM-based devices, there are two main optical losses: the metal absorption loss and the radiation loss. Although suppression of the absorption loss has been widely studied by introducing high-index dielectric layers inside of low-index dielectric layers [25], radiation loss into free space has not been investigated, despite the fact that it represents a large portion of the total loss.

In this study, we propose MIM cavities made by using a cutoff frequency mechanism that only appears in the dispersion relation of the first-order waveguide mode [13]. The cutoff frequency is strongly dependent on the effective size of the waveguide mode and, therefore, it can be modulated by varying the waveguide width, dielectric thickness, and index of the dielectric. We investigated the optical properties of cavities consisting of two mirror waveguides and a sandwiched waveguide by modulating the cutoff frequency with three different techniques.

The mode gap, which is due to the cutoff frequency difference, prevents radiation loss along the waveguide direction. Alternatively, the higher-order cavity mode observed in a longer cavity has more intensity nodes than the fundamental cavity mode, thereby suppressing radiation loss in the direction orthogonal to the waveguide by multipole cancellation [26].

2. Dispersion Properties and Cutoff Frequency

Figure 1a shows a schematic diagram of a MIM waveguide consisting of two silver strips with a sandwiched low-index dielectric layer. Each strip has a thickness (h) of 100 nm and a width (w) of 240 nm. The dielectric layer has a thickness (t) of 10 nm. The refractive index of the dielectric layer is set to 1.5. In this MIM waveguide, the waveguide modes can be classified into fundamental and first-order waveguide modes according to the mirror symmetry of the dominant electric field (E_z) profiles for the plane with $y = 0$ (dotted lines of Figure 1b,c). Figure 1b,c show the top and side views for the E_z profiles of the fundamental and first-order waveguide modes, respectively, with a wavelength of 1550 nm (λ_0). In both modes, the electric fields of the waveguide modes are strongly localized in the deep subwavelength cross-section of the dielectric layer, 240 (w) $\times$ 10 (t) (nm^2) = 0.00099 ($\lambda_0{}^2$), by index guiding in the y-direction and plasmonic coupling between the silver strips.

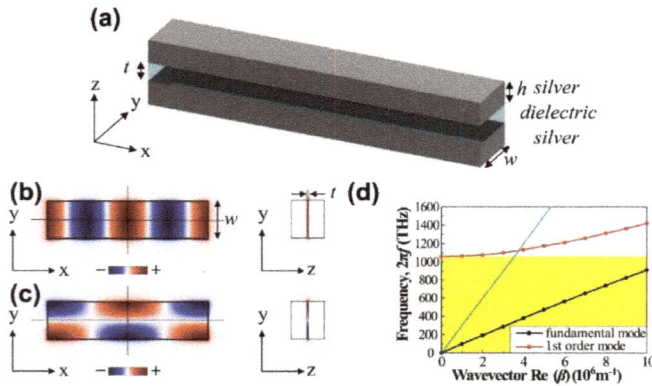

Figure 1. (**a**) Schematic of a metal-insulator-metal (MIM) waveguide consisting of two silver strips and a low-index dielectric layer. The variables w, h, and t represent the waveguide width, the thickness of the silver strip, and the thickness of the dielectric layer, respectively; (**b,c**) Top and side views of the E_z electric field profiles of the fundamental and first-order waveguide modes with a wavelength of 1550 nm, w = 240 nm, h = 100 nm, and t = 10 nm. The top view (left) is obtained in the center of the dielectric layer and the side view (right) is obtained along the solid line of the top view; (**d**) Dispersion curves of fundamental (black) and first-order (red) waveguide modes, respectively. The yellow box indicates the mode gap region for the odd waveguide mode. Light line is indicated by blue line.

Fundamental and first-order MIM waveguide modes have distinct dispersion properties, as shown in Figure 1d. The dispersion curve of the fundamental waveguide mode (black) shows a linear dependence between the frequency and the wavevector. The frequency linearly increases from the zero frequency as the wavevector increases. Alternatively, the dispersion curve of the first-order waveguide mode (red) has a nonzero lowest frequency at the zero wavevector; this is referred to as the cutoff frequency. Below the cutoff frequency, the first-order waveguide mode cannot exist in this MIM waveguide. Thus, the waveguide operates as a mirror for the first-order waveguide mode at frequencies below the cutoff frequency. The frequency region can be considered to be the mode gap of the first-order waveguide mode, which is indicated by the yellow box in Figure 1d. Therefore, the cutoff frequency mechanism can be used to make an effective mirror in the MIM-based cavity.

The cutoff frequency strongly depends on the waveguide width (w) as well as the thickness (t) and refractive index (n) of the dielectric material. This is the case because the effective index of the first-order waveguide mode is sensitive to changes in these structural parameters. By spatially modulating the mode gap with these structural parameters, a photonic well can be formed for the first-order waveguide mode, localizing photons inside of the well. In this paper, we propose three types of deep subwavelength-sized cavities using the difference in cutoff frequency in the first-order waveguide mode.

3. Results

3.1. Waveguide Width-Modulated Cavity

In this paragraph we introduce a MIM first-order mode cavity with width modulation by using the strong width-dependence of the cutoff frequency. Figure 2a shows a schematic of a cavity consisting of a broad waveguide (w = 240 nm) and two narrow mirror waveguides (w = 120 nm), where the waveguides consist of two silver strips and a low-index (n = 1.5) dielectric layer (t = 10 nm). Here, the mirror waveguide length (L_m) is set to 1000 nm.

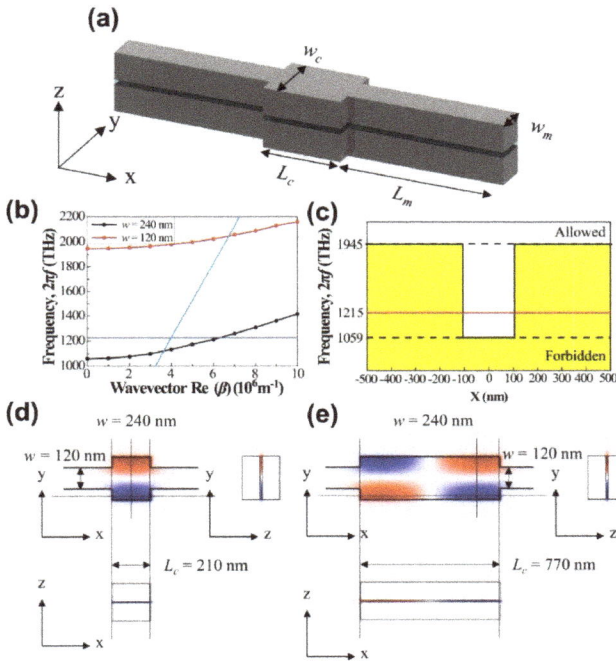

Figure 2. (**a**) Schematic of a cavity consisting of a broad waveguide (w = 240 nm) and two narrow mirror waveguides (w = 120 nm). Here, h = 100 nm, t = 10 nm, L_m = 1000 nm, and the cavity length is L_c; (**b**) Dispersion curves of first-order waveguide modes for waveguide widths of w = 240 nm (black) and w = 120 nm (red); their cutoff frequencies (cutoff wavelengths) are $2\pi f$ = 1059 THz (1779 nm) and 1945 THz (968 nm), respectively. The horizontal blue line indicates a frequency (wavelength) of 1215 THz (1550 nm). The light line is indicated by the blue line; (**c**) Cutoff frequency of the first-order waveguide mode along the *x*-axis for the cavity in (**d**). The allowed and forbidden regions are indicated by white and yellow colors, respectively. The allowed frequency region for the first-order waveguide mode is spatially changed by modulating the waveguide width; (**d**,**e**) Top and side views of the electric field profiles of the fundamental (L_c = 210 nm) and second-order (L_c = 770 nm) cavity modes, respectively. The solid lines indicate the cross-sections of the side views.

The dispersion curves of the first-order waveguide modes for broad (w = 240 nm, black) and narrow (w = 120 nm, red) waveguides are plotted in Figure 2b. Since the first-order mode in the narrower waveguide experienced more air outside of the MIM waveguide, the dispersion curve moves upward due to the smaller effective index. In particular, the cutoff frequencies (cutoff wavelength) of the broad/narrow waveguides are $2\pi f$ = 1059 THz (1779 nm) and 1945 THz (968 nm), respectively. Propagation of first-order waveguide modes with a frequency between the two cutoff frequencies is allowed for the broad waveguide (w = 240 nm); however, this is forbidden for the narrow waveguide (w = 120 nm). For example, light with a wavelength of 1550 nm (1215 THz), indicated by the horizontal blue line in Figure 2b, is only allowed in the broad waveguide region of Figure 2a. Indeed, the cavity formed a photonic well by modulating the cutoff frequency with the waveguide width, as shown in Figure 2c. The allowed frequency region from 1059 THz (1779 nm) to 1945 THz (968 nm) was localized in the broad waveguide with a length of L_c.

The resonant wavelength of the cavity mode can be controlled by changing the length (L_c) of the cavity region. In addition, the cavity mode is classified as a fundamental and higher-order cavity mode, depending on the number of intensity nodes along the y-direction. Figure 2d,e show the electric field profiles of the fundamental and second-order cavity modes with the same resonant wavelengths (1552 nm) where the dominant electric field is orthogonal to the two metal surfaces. Because both modes are based on the first-order waveguide mode, there is a common intensity node along the x-axis. However, the fundamental cavity mode in the short cavity (L_c = 210 nm) has one intensity antinode along the y-direction. Alternatively, in the long cavity (L_c = 770 nm), the second-order cavity mode has two intensity antinodes along the y-direction.

3.2. Refractive Index-Modulated Cavity

The effective waveguide width increases for dielectric materials (between the silver strips) with higher indices, increasing the cutoff frequency of the first-order waveguide mode. Therefore, a first-order mode cavity can be formed by introducing index modulation, as shown in Figure 3a. The cavity consisted of a cavity region with a low-index dielectric layer (n = 1.5) and two mirror waveguides with an air gap (n = 1.0). The waveguide width of 240 nm was kept constant. Here, the height (h) of the silver strips, the distance (t) of the silver strips, and the mirror waveguide length (L_m) were set to 100 nm, 10 nm, and 1000 nm, respectively. Figure 3b shows the dispersion curves of the first-order waveguide modes for dielectric materials with different refractive indices of n = 1.5 (black) and n = 1.0 (red). The cutoff frequency increased from 1059 THz (1779 nm) to 1493 THz (1261 nm) when the dielectric index decreased from 1.5 to 1.0. A photonic well, such as the one shown in Figure 2c, can be generated by the difference in cutoff frequencies at the boundaries, where the refractive index of the dielectric layer changes.

Figure 3c,d show the electric field profiles of the fundamental and second-order cavity modes for resonance wavelengths of 1554 nm and 1550 nm for L_c = 300 nm and L_c = 850 nm, respectively. The mode shapes are similar to those of the width-modulated cavity (Figure 2d,e).

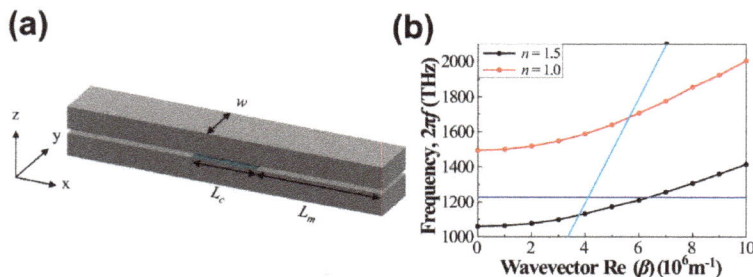

Figure 3. *Cont.*

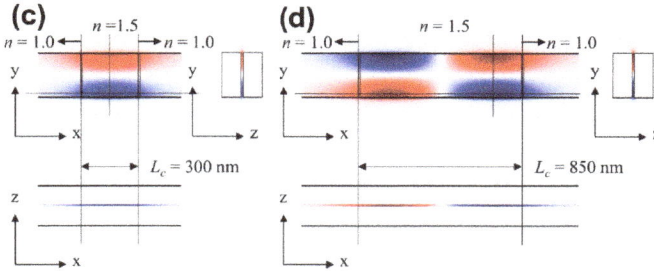

Figure 3. (**a**) Schematic of a cavity consisting of a cavity region (n = 1.5, green color) and two mirror waveguides (n = 1.0). Here, w = 240 nm, h = 100 nm, t = 10 nm, L_m = 1000 nm, and the cavity length is L_c; (**b**) Dispersion curves of first-order waveguide modes for dielectric materials (between the silver strips) with different refractive indices of n = 1.5 (blue) and n = 1.0 (red). The cutoff frequencies (wavelengths) are $2\pi f$ = 1059 THz (1779 nm) and 1493 THz (1261 nm), respectively. The horizontal blue line indicates a frequency of 1215 THz (1550 nm). The light line is indicated by the blue line; (**c**,**d**) Top and side views of the electric field profiles of fundamental (L_c = 300 nm) and second-order (L_c = 850 nm) cavity modes.

3.3. Gap Size-Modulated Cavity

In the MIM waveguide, the effective index of the waveguide mode increased for thinner dielectric layers between the metal layers [24]. Therefore, as the thickness (t) of the dielectric layer increased, the dispersion curve moved upward due to the smaller effective index (caused by weaker plasmonic coupling). Indeed, the cutoff frequency increased from 1059 THz (1779 nm) to 1945 THz (1414 nm), as shown in the dispersion curves of Figure 4b for t = 10 nm (black) and t = 30 nm (red). Here, the waveguide width (w) and mirror length (L_m) were 240 nm and 1000 nm, respectively. The heights of the cavity and mirror waveguide silver strips were 100 nm and 80 nm, respectively. Because the height of the silver was much larger than the skin depth, changes in the height did not affect the properties of the cavity. Figure 4a shows the schematic of a cavity consisting of a cavity waveguide (t = 10 nm) and two mirror waveguides (t = 30 nm), where the thickness of the dielectric layer was modulated from 10 nm to 30 nm.

Similarly to the waveguide width- or dielectric index-modulated cavities (Figures 2 and 3, respectively), in the dielectric thickness-modulated cavities, the fundamental and second-order cavity modes were observed for L_c = 220 nm (Figure 4c) and L_c = 780 nm (Figure 4d), respectively. Based on the cutoff frequency mirror modulation (similar to the width-modulated cavity (Figure 2) and the index-modulated cavity (Figure 4)), the fundamental and second-order cavity modes were observed at L_c = 220 nm (Figure 4c) and L_c = 780 nm (Figure 4d), respectively.

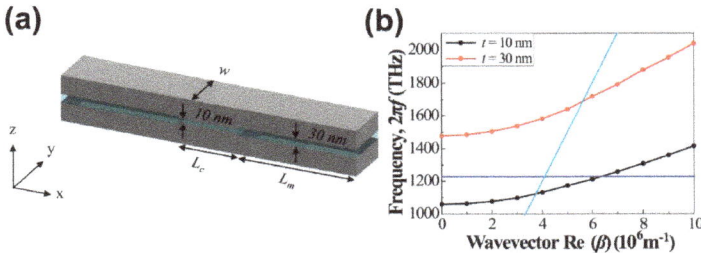

Figure 4. *Cont.*

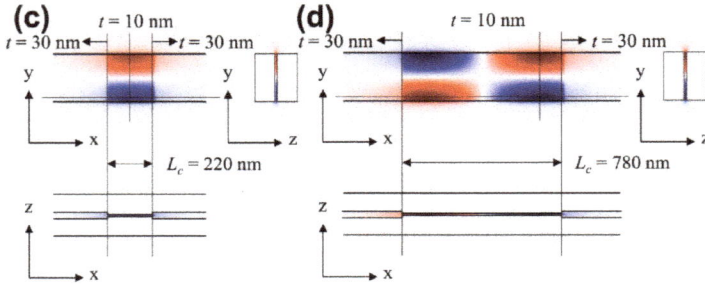

Figure 4. (**a**) Schematic of a cavity consisting of a cavity waveguide (t = 10 nm) and two mirror waveguides (t = 30 nm). The gap size between the two silver strips is modulated. In the gap, a low-index dielectric layer (n = 1.5) is assumed. Here, w = 240 nm, L_m = 1000 nm, and the heights of the cavity and mirror waveguide silver strips are 100 nm and 80 nm, respectively. The cavity waveguide length is L_c; (**b**) Dispersion curves of the first-order waveguide modes for t = 10 nm (blue) and t = 30 nm (red). The cutoff frequencies (wavelengths) are $2\pi f$ = 1059 THz (1779 nm) and 1945 THz (1414 nm). The light line is indicated by the blue line; (**c,d**) Top and side views of the electric field profiles of the fundamental (L_c = 220 nm) and second-order (L_c = 780 nm) cavity modes.

3.4. Loss Analysis for the Three Cavities

To understand the loss mechanisms of the MIM cavity modes, we investigated each type of loss (i.e., radiation loss, mirror loss, and metallic absorption loss) separately in terms of the quality (Q) factors. The Q factor is defined by $2\pi f \times$ (stored energy in the cavity/power loss) [13]. The radiation loss corresponds to radiation from the cavity's sides to free space due to the imperfect horizontal modal confinement of index-guiding. Mirror loss originates from energy loss tunneling through the cutoff frequency mirrors of the narrow waveguides, which have a finite length. In addition, metallic absorption loss results from intrinsic ohmic loss in the silver strips. We were able to obtain Q_{tot}, Q_{rad}, Q_m, and Q_{abs} by directly calculating the respective losses because the Q factors were inversely proportional to each loss, the total cavity loss, radiation loss, mirror loss, and absorption loss.

We calculated each Q factor as follows. Q_{total} was obtained from the time decay of the total energy in the cavity. Q_{abs} could be obtained by directly calculating the absorbed energy in the metal. Next, we calculated $Q_{optical}$ with the equation: $1/Q_{total} = 1/Q_{optical} + 1/Q_{abs}$. Radiation loss and mirror loss, which consists of the total optical loss, were estimated by calculating the sums of Poynting vectors into free space and through two mirror waveguides, respectively. Q_{rad} and Q_m were obtained by the ratio of the sums of Poynting vectors and the following equation, $1/Q_{optical} = 1/Q_{rad} + 1/Q_m$.

Table 1 shows the Q factors of the three cavities. The Q factors for the three different cutoff frequency-modulated cavities (i.e., width-/index-/gap-modulated cavities) were similar and showed only slight differences. Therefore, by analyzing one of the three cavities, the loss properties of all three cavities could be understood. For example, in the case of a width-modulated cavity, the total Q factors (Q_{total}) of the fundamental and the second-order cavity modes were 45.5 and 63.2, respectively. Q_m, which is inversely proportional to the mirror loss, was three orders of magnitude larger than the other Q factors when L_m = 1000 nm; therefore, tunneling loss through the cutoff frequency mirror was negligible. Q_{abs}, which is related to the absorption loss (i.e., the dominant loss of the cavities), was between 80 and 90, regardless of the fundamental/second-order cavity modes of the cavities. Additionally, the radiation loss was comparable with the absorption loss. For example, Q_{rad} of the width-modulated cavity was 101 for the fundamental cavity mode and 260 for the second-order cavity mode. Since radiation into free space is difficult to collect (because of its large divergence from 10 nanometer-scale light confinement), it is desirable to minimize this type of loss for practical application of the proposed MIM cavity. Because radiation from the two oppositely-phased E_z intensity antinodes of the second-order mode along the intensity node direction into free space was strongly suppressed by multipole cancellation [26], Q_{rad} of the second-order cavity mode became larger than that of the fundamental cavity mode, increasing Q_{total}. On the other hand, the calculated propagation length was 4.8 μm and the reported value was ~5 μm in a similar MIM waveguide [14], which is at least an order of magnitude larger than the proposed cavity size.

Table 1. Quality factors of the three cavities.

Cavity Type	Fundamental Mode				Second Order Mode			
	Q_{total}	Q_{rad}	Q_{abs}	Q_m	Q_{total}	Q_{rad}	Q_{abs}	Q_m
Width-modulated cavity (Figure 2)	45.5	101	82.8	9.05×10^5	63.2	260	83.6	1.81×10^6
Index-modulated cavity (Figure 3)	38.6	71.7	83.7	9.68×10^4	56.8	175	84.2	1.71×10^5
Gap-modulated cavity (Figure 4)	37.3	64.0	89.7	1.42×10^4	56.6	164	86.6	2.82×10^4

4. Conclusions

In summary, we proposed new plasmonic cavities based on modulating the cutoff frequency of MIM first-order waveguide modes. The first-order waveguide mode has a cutoff frequency that depends on the waveguide width, refractive index, and gap size of the dielectric layer. Light (with a wavelength of 1550 nm) can be confined in a cavity region with a deep subwavelength physical volume ($0.00013 \lambda_0^3 \sim 0.00055 \lambda_0^3$) in cavities that are modulated by changing the width, index, and gap size. The second-order cavity mode shows higher Q factors than the fundamental cavity mode because radiation along the intensity node is suppressed by multipole cancellation.

The proposed ultrasmall cavity is a good candidate for low-threshold lasers [27], ultrafast optical switches [28], and as the light source for quantum optics [6] due to its extremely small mode size of $1/10,000 \lambda_0^3$. In addition, its waveguide-based design allows for efficient light coupling with integrated detectors [29] or other photonic devices [8,30–32]. The light coupling can be achieved by using a tapered coupler from a dielectric waveguide [33] or a dielectric air slot waveguide [34]. The cutoff frequency mechanism can be widely applied to build efficient and strong mirror waveguides for various optical components operating at any wavelengths. Since the mirrors of the proposed cavity have adjustable reflectivity, unidirectional emission can be easily achievable. Usually, although the control of directionality is highly demanded for an efficient light source in quantum optics and photonic circuits, it is extremely difficult to control the directionality in the deep subwavelength cavity. The cavity mode can be directly excited by the odd waveguide mode from the dielectric waveguide with an external light source or internal light emitter such as quantum dots. The continuation of this study will involve modulating the nanocavity width, dielectric index, dielectric thickness, and structure to miniaturize key optical devices into deep subwavelength-scaled ones in integrated devices and nano-emitters.

Acknowledgments: This research was supported in part by the National Research Foundation of Korea (NRF) grant funded by the Korean government (MSIP) (No. NRF-2016R1C1B2007007) and in part by the Chung-Ang University Research Scholarship Grants in 2016.

Author Contributions: Soon-Hong Kwon conceived and designed the whole simulation and concept; Kihwan Moon and Tae-Woo Lee performed the simulations and analyzed the data; Soon-Hong Kwon, Kihwan Moon, and Young Jin Lee drafted and revised the manuscript.

Conflicts of Interest: The authors declare no conflicts of interest.

References

1. Barnes, W.L.; Dereux, A.; Ebbesen, T.W. Suface plasmon subwavelength optics. *Nature* **2003**, *424*, 824–830. [CrossRef] [PubMed]
2. Gramotnev, D.K.; Bozhevolnyi, S.I. Plasmonics beyond the diffraction limit. *Nat. Photonics* **2010**, *4*, 83–91. [CrossRef]

3.   Wei, H.; Li, Z.P.; Tian, X.R.; Wang, Z.X.; Cong, F.Z.; Liu, N.; Zhang, S.P.; Nordlander, P.; Halas, N.J.; Xu, H.X. Quantum dot-based local field imaging reveals plasmon-based interferometric logic in silver nanowire networks. *Nano Lett.* **2010**, *11*, 471–475. [CrossRef] [PubMed]

4.   Ma, R.M.; Oulton, R.F.; Sorger, V.J.; Bartal, G.; Zhang, X. Room-temperature sub-diffraction-limited plasmon laser by total internal reflection. *Nat. Mater.* **2011**, *10*, 110–113. [CrossRef] [PubMed]

5.   Zhang, S.P.; Xu, H.X. Optimizing substrate-mediated plasmon coupling toward high-performance plasmonic nanowire waveguides. *ACS Nano* **2012**, *6*, 8128–8135. [CrossRef] [PubMed]

6.   Oulton, R.F.; Sorger, V.J.; Zentgraf, T.; Ma, R.M. Plasmon lasers at deep subwavelength scale. *Nature* **2009**, *461*, 629–632. [CrossRef] [PubMed]

7.   Nikolajsen, T.; Leosson, K.; Bozhevolnyi, S.I. Surface plasmon based modulators and switches operating at telecom wavelengths. *Appl. Phys. Lett.* **2004**, *85*, 5833–5835. [CrossRef]

8.   Fang, Y.R.; Li, Z.P.; Huang, Y.Z.; Zhang, S.P.; Nordlander, P.; Halas, N.J.; Xu, H.X. Branched silver nanowires as controllable plasmon routers. *Nano Lett.* **2010**, *10*, 1950–1954. [CrossRef] [PubMed]

9.   Tsai, C.Y.; Lu, S.P.; Lin, J.W.; Lee, P.T. High sensitivity plasmonic index sensor using slablike gold nanoring arrays. *Appl. Phys. Lett.* **2011**, *98*, 153108. [CrossRef] [PubMed]

10.   Kwon, S.H. Deep subwavelength-scale metal-insulator-metal plasmonic disk cavities for refractive index sensor. *IEEE Photonics J.* **2013**, *5*, 4800107. [CrossRef]

11.   Volkov, V.S.; Bozhevolnyi, S.I.; Devaux, E.; Laluet, J.Y.; Ebbesen, T.W. Wavelength selective nanophotonic components utilizing channel plasmon polaritons. *Nano Lett.* **2007**, *7*, 880–884. [CrossRef] [PubMed]

12.   Rahimzadegan, A.; Granpayyeh, N.; Hosseini, S.P. Improved plasmonic filter, ultra-compact demultiplexer, and splitter. *J. Opt. Soc. Korea* **2014**, *18*, 261–273. [CrossRef]

13.   Lee, T.W.; Lee, D.E.; Kwon, S.H. Dual-function metal-insulator-metal plasmonic opticla filter. *IEEE Photonics J.* **2015**, *7*, 4800108. [CrossRef]

14.   Veronis, G.; Fan, S.H. Modes of subwavelength plasmonic slot waveguides. *J. Lightwave Technol.* **2007**, *25*, 2511–2521. [CrossRef]

15.   Pile, D.F.P.; Ogawa, T.; Gramotnev, D.K.; Matsuzaki, Y.; Vernon, K.C.; Yamaguchi, K.; Okamoto, T.; Haraguchi, M.; Fukui, M. Two-dimensionally localized modes of a nanscale gap plasmon waveguide. *Appl. Phys. Lett.* **2005**, *87*, 261114. [CrossRef]

16.   Tagliabue, G.; Poulikakos, D.; Eghlidi, H. Three-dimensional concentration of light in deeply sub-wavelength, laterally tapered gap-plasmon nanocavities. *Appl. Phys. Lett.* **2016**, *108*, 221108. [CrossRef]

17.   Kuttge, M.; de Abajo, F.J.G.; Polman, A. Ultasnall mode volume plasmonic nanodisk trsonators. *Nano Lett.* **2010**, *10*, 1537–1541. [CrossRef] [PubMed]

18.   Kwon, S.H. Deep subwavelength plasmonic whispering-gallery-mode cavity. *Opt. Express* **2012**, *20*, 24918–24924. [CrossRef] [PubMed]

19.   Han, Z.H.; Bozhevolnyi, S.I. Plasmon-induced transparency with detuned ultracompact Fabry-Perot resonators in integrated plasmonic devices. *Opt. Express* **2011**, *19*, 3251–3257. [CrossRef] [PubMed]

20.   Zand, I.; Abrichamian, M.S.; Berini, P. Highly tunable nanoscale metal-insulator-metal split ring core ring resonators (SRCRRs). *Opt. Express* **2013**, *21*, 79–86. [CrossRef] [PubMed]

21.   Kwon, S.H. Plasmonic ruler with angstrom distance resolution based on double metal block. *IEEE Photonics Technol. Lett.* **2013**, *25*, 1619–1622. [CrossRef]

22.   Lassiter, J.B.; McGuire, F.; Mock, J.J.; Ciralci, C.; Hill, R.T.; Wiley, B.J.; Chilkoti, A.; Smith, D.R. Plasmonic waveguide modes of film-coupled metallic nanocubes. *Nano Lett.* **2013**, *13*, 5866–5872. [CrossRef] [PubMed]

23.   Nielsen, M.G.; Gramotnev, D.K.; Pors, A.; Albrektsen, O.; Bozhevolnyi, S.I. Continuous layer gap plasmon resonators. *Opt. Express* **2011**, *19*, 19310–19322. [CrossRef] [PubMed]

24.   Miyazaki, H.T.; Kurokawa, Y. Contolled plasmon resonance in closed metal/insulator/metal nanocavities. *Appl. Phys. Lett.* **2006**, *89*, 211126. [CrossRef]

25.   Oulton, R.F.; Sorger, V.J.; Genov, D.A.; Pile, D.F.P.; Zhang, X. A hybrid plasmonic waveguide for subwavelength confinement and long-range propagation. *Nat. Photonics* **2008**, *2*, 496–500. [CrossRef]

26.   Johnson, S.G.; Fan, S.; Mekis, A.; Joannopoulos, J.D. Multipole-cancellation mechanism for high-Q cavities in the absence of a complete photonic and gap. *Appl. Phys. Lett.* **2001**, *78*, 3388–3390. [CrossRef]

27.   Hill, M.T.; Oei, Y.S.; Smalbrugge, B.; Zhu, Y.; De Vries, T.; Van Veldhoven, P.J.; Van Otten, F.W.M.; Eijkemans, T.J.; Turkiewicz, J.P.; De Waardt, H.; et al. Lasing in metallic-coated nanocavities. *Nat. Photonics* **2007**, *1*, 589–594. [CrossRef]

28. Lu, H.; Liu, X.M.; Wang, L.R.; Gong, Y.K.; Mao, D. Ultrafast all-optical switching in nanoplasmonic waveguide Kerr nonlinear resonator. *Opt. Express* **2011**, *19*, 2910–2915. [CrossRef] [PubMed]

29. Zhu, S.Y.; Lo, G.Q.; Kwong, D.L. Theorical investigation of silicide Schottky barrier detector integrated in horizontal metal-insulator-metal nanoplasmonic slot waveguide. *Opt. Express* **2011**, *19*, 15843–15854. [CrossRef] [PubMed]

30. Wang, W.; Yang, Q.; Fan, F.; Xu, H.; Wang, Z.L. Light Propagation in Curved Silver Nanowire Plasmonic Waveguides. *Nano Lett.* **2011**, *11*, 1603–1608. [CrossRef] [PubMed]

31. Wei, H.; Xu, H. Nanowire-based plasmonic waveguides and devices for integrated nanophotonic circuits. *Nanophotonics* **2012**, *1*, 155–169. [CrossRef]

32. Gramotnev, D.K.; Nielsen, M.G.; Tan, S.J.; Kurth, M.L.; Bozhevolnyi, S.I. Gap surface plasmon waveguides with enhanced integration and functionality. *Nano Lett.* **2012**, *12*, 359–363. [CrossRef] [PubMed]

33. Veronis, G.; Fan, S.H. Theoretical investigation of compact couplers between dielectric slab waveguides and two-dimensional metal-dielectric-metal plasmonic waveguides. *Opt. Express* **2007**, *15*, 1211–1221. [CrossRef] [PubMed]

34. Zhu, B.Q.; Tsang, H.K. High Coupling Efficiency Silicon Waveguide to Metal-Insulator-Metal Waveguide Mode Converter. *J. Lightwave Technol.* **2016**, *34*, 2467–2472. [CrossRef]

applied
sciences

MDPI

Review

Light Trapping above the Light Cone in One-Dimensional Arrays of Dielectric Spheres

Evgeny N. Bulgakov [1,2], Almas F. Sadreev [1,*] and Dmitrii N. Maksimov [1]

[1] Kirensky Institute of Physics, Federal Research Center KSC SB RAS, Krasnoyarsk 660036, Russia; ben@tnp.krasn.ru (E.N.B.); mdn@tnp.krasn.ru (D.N.M.)

[2] Deparment of Airspace Materials and Technology, Siberian State Aerospace University, Krasnoyarsk 660014, Russia

* Correspondence: almas@tnp.krasn.ru; Tel.: +7-391-249-4538

Academic Editor: Boris Malomed
Received: 30 November 2016; Accepted: 23 January 2017; Published: 8 February 2017

Abstract: We demonstrate bound states in the radiation continuum (BSC) in a linear periodic array of dielectric spheres in air above the light cone. We classify the BSCs by orbital angular momentum $m = 0, \pm 1, \pm 2$ according to the rotational symmetry of the array, Bloch wave vector β directed along the array according to the translational symmetry, and polarization. The most simple symmetry protected BSCs have $m = 0, \beta = 0$ and occur in a wide range of the radius of the spheres and dielectric constant. More sophisticated BSCs with $m \neq 0, \beta = 0$ exist only for a selected radius of spheres at fixed dielectric constant. We also find robust Bloch BSCs with $\beta \neq 0, m = 0$. All BSCs reside within the first but below the other diffraction continua. We show that the BSCs can be easily detected by bright features in scattering of different plane waves by the array as dependent on type of the BSC. The symmetry protected TE/TM BSCs can be traced by collapsing Fano resonance in cross-sections of normally incident TE/TM plane waves. When plane wave with circular polarization with frequency tuned to the bound states with OAM illuminates the array the spin angular momentum of the incident wave transfers into the orbital angular momentum of the BSC. This, in turn, gives rise to giant vortical power currents rotating around the array. Incident wave with linear polarization with frequency tuned to the Bloch bound state in the continuum induces giant laminar power currents. At last, the plane wave with linear polarization incident under tilt relative to the axis of array excites Poynting currents spiralling around the array. It is demonstrated numerically that quasi-bound leaky modes of the array can propagate both stationary waves and light pulses to a distance of 60 wavelengths at the frequencies close to the bound states in the radiation continuum. A semi-analytical estimate for decay rates of the guided waves is found to match the numerical data to a good accuracy.

Keywords: bound state in the continuum; Fano resonance; nanophotonics

1. Introduction

The scattering of electromagnetic waves by an ensemble of dielectric spheres has a long history of research beginning with Mie who presented a rigorous theory for scattering by a single dielectric sphere [1]. The overwhelming majority of papers since the pioneering papers by Ohtaka and his coauthors [2–4] considered the periodical two- and three-dimensional arrays [5–8]. Surprisingly, less interest has been payed to scattering by a linear array of dielectric nanoparticles mostly restricted to aggregates of a finite number of spheres [9–11]. Guiding of electromagnetic waves by a linear array of dielectric spheres below the diffraction limit attracted more attention. There were two types of consideration: finite arrays [12–16] and infinite arrays which were studied by means of the coupled-dipole approximation [17–23]. Only in 2013 a full-wave analysis of waves on linear arrays of dielectric spheres below the light cone was provided by Linton, Zalipaev, and Thompson [24].

It has been widely believed that only those modes whose eigenfrequencies lie below the light cone, are confined and the rest of the eigenmodes have finite life times. However recently confined electromagnetic modes were shown to exist in various periodical arrays of:

(i) long cylindrical rods [25–43]
(ii) photonic crystal slabs [44–51] and
(iii) two-dimensional arrays of spheres [40,52].

Surprisingly, less attention is paid to the one-dimensional array of dielectric nanoparticles. Similar one may expect light trapping in the one-dimensional array of nanoparticles with the bound frequencies above the light cone. The array of dielectric spheres or discs is interesting due to the rotational symmetry giving rise to that the orbital angular momentum (OAM) of light is preserved. Therefore one can expect the the BSCs with non-zero OAM. The angular momentum is composed of the spin angular momentum (SAM) and OAM describing the polarization and the phase structure distribution of EM fields, respectively [53–55]. In this review we summarize our resent results on BSCs in arrays of dielectric spheres [56] including some findings on transport properties of dielectric arrays [57] and BSCs with OAM [58,59]. For a thorough review of BSCs the reader is addressed to [60].

2. Basic Equations for EM Wave Scattering by a Linear Array of Spheres

In the present paper we consider a free-standing one-dimensional infinite array of dielectric spheres in air Figure 1. In what follows we refer all length quantities in terms of the period h of the array.

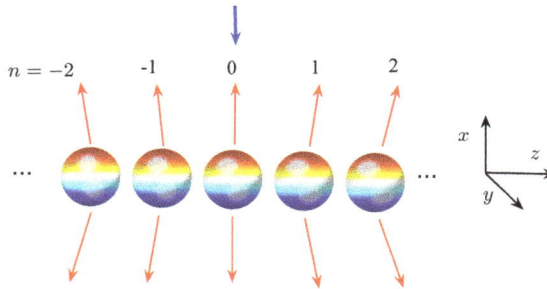

Figure 1. A periodic infinite array of dielectric spheres illuminated by a plane wave (blue arrow). The wave can be transmitted and reflected to discrete diffraction continua enumerated by integers m and n in accordance with Equations (17) and (21) shown by red arrows.

We formulate the scattering theory by a periodic array of dielectric spheres in the form similar to the approach developed for a periodic array of dielectric cylinders [37,61]

$$\widehat{S}^{-1}\Psi = \widehat{L}\Psi = \Psi_{inc}. \tag{1}$$

where the matrix $\widehat{L}$ accounts for both the scattering matrix of the isolated sphere as well as the mutual scattering events between the spheres, Ψ_{inc} is given by the incident wave, and the column Ψ consists of amplitudes a_l^m of the multipole expansion of the scattering function.

The exact expression of the matrix $\widehat{L}$ was derived by Linton et al. [24] for EM guided waves on a periodic array of dielectric spheres. For the reader's convenience we present the equations and notations from the above reference. We seek the solutions of the Maxwell equations, which obey the Bloch theorem:

$$\mathbf{E}(\mathbf{r} + \mathbf{R}_j) = e^{i\beta j}\mathbf{E}(\mathbf{r}), \mathbf{H}(\mathbf{r} + \mathbf{R}_j) = e^{i\beta j}\mathbf{H}(\mathbf{r})$$

with the Bloch wave vector β directed along the array aligned with the z-axis (see Figure 1). Here $\mathbf{R}_j = j\mathbf{e}_z$ is the position of the center of the j-th sphere and $\mathbf{e}_z$ is the unit vector along the array. Scattered electromagnetic fields are expanded in a series over vector spherical harmonics $\mathbf{M}_n^m$ and $\mathbf{N}_n^m$ [1,24].

$$\mathbf{E}(\mathbf{r}) = \sum_j e^{i\beta j} \sum_{lm} [a_l^m \mathbf{M}_l^m(\mathbf{r} - \mathbf{R}_j) + b_l^m \mathbf{N}_l^m(\mathbf{r} - \mathbf{R}_j)],$$
$$\mathbf{H}(\mathbf{r}) = -i \sum_j e^{i\beta j} \sum_{lm} [a_l^m \mathbf{N}_l^m(\mathbf{r} - \mathbf{R}_j) + b_l^m \mathbf{M}_l^m(\mathbf{r} - \mathbf{R}_j)]. \qquad (2)$$

In series (2) the first/second terms presents TE/TM spherical vector EM fields.

In absence of an incident wave Linton et al. [24] derived the homogeneous matrix equation for the amplitudes a_l^m, b_l^m.

$$Z_{TE,l}^{-1} a_l^m - \sum_v (a_v^m \mathcal{A}_{vl}^{mm} + b_v^m \mathcal{B}_{vl}^{mm}) = 0,$$
$$Z_{TM,l}^{-1} b_l^m - \sum_v (a_v^m \mathcal{B}_{vl}^{mm} + b_v^m \mathcal{A}_{vl}^{mm}) = 0, \qquad (3)$$

where summation over v begins with $max(1,m)$, and the so-called Lorenz-Mie coefficients are given by

$$Z_{TE,l} = \frac{j_l(kR)[rj_l(k_0 r)]'_{r=R} - j_l(k_0 R)[rj_l(kr)]'_{r=R}}{h_l(k_0 R)[rj_l(kr)]'_{r=R} - j_l(kR)[rh_l(k_0 r)]'_{r=R}},$$
$$Z_{TM,l} = \frac{\epsilon j_l(kR)[rj_l(k_0 r)]'_{r=R} - j_l(k_0 R)[rj_l(kr)]'_{r=R}}{h_l(k_0 R)[rj_l(kr)]'_{r=R} - \epsilon j_l(kR)[rh_l(k_0 r)]'_{r=R}}, \qquad (4)$$

where $k = \sqrt{\epsilon}k_0$ and ϵ is the dielectric constant of the spheres

$$\mathcal{A}_{lv}^{mm} = 4\pi(-1)^m i^{v-l} \sqrt{\frac{v(v+1)}{l(l+1)}} \sum_{\substack{p=|l-v|;l+v+p=even}}^{l+v} (-i)^p g_{lvp} \mathcal{G}(l,m;v,-m;p) s_p, \qquad (5)$$

$$\mathcal{B}_{lv}^{mm} = \frac{2\pi(-1)^m}{\sqrt{l(l+1)v(v+1)}} \sum_{\substack{p=|l-v|+1;l+v+p=odd}}^{l+v-1} i^{v-l-p} \sqrt{\frac{2p+1}{2p-1}} \mathcal{H}(l,m;v,-m;p) s_p. \qquad (6)$$

The coefficients

$$g_{lvp} = 1 + \frac{(l-v+p+1)(l+v-p)}{2v(2v+1)} - \frac{(v-l+p+1)(l+v+p+2)}{2(v+1)(2v+1)}, \qquad (7)$$

$$\mathcal{G}(l,m;v,\mu;p) = \frac{(-1)^{m+\mu}}{\sqrt{4\pi}} \sqrt{(2l+1)(2v+1)(2p+1)} \begin{pmatrix} l & v & p \\ m & \mu & -m-\mu \end{pmatrix} \begin{pmatrix} l & v & p \\ 0 & 0 & 0 \end{pmatrix} \qquad (8)$$

are expressed in terms of Wigner 3-j symbols,

$$\mathcal{H}(l,m;v,-m;p) = \sum_{s=-1}^{1} \mathcal{G}_s(l,m;v,-m;p) \qquad (9)$$

with

$$\mathcal{G}_0(l,m;v,-m;p) = -2m|p|\mathcal{G}(l,m;v,-m;p-1),$$
$$\mathcal{G}_{\pm 1}(l,m;v,-m;p) = \mp\sqrt{(v \pm m)(v \mp m + 1)p(p-1)}\mathcal{G}(l,m;v,-m \pm 1;p-1), \qquad (10)$$

and

$$s_p = \lambda_{p0} \sum_{j=1}^{\infty} h_p(k_0 j)(e^{i\beta j} + (-1)^p e^{-i\beta j}), \qquad (11)$$

where λ_{lm} is normalization factor given in Appendix A. The next step is to account for an incident plane wave which can be expanded over vector spherical harmonics [1,6].

$$\mathbf{E}^{\sigma}(\mathbf{r}) = \sum_{l=1}^{\infty} \sum_{-l}^{l} [q_{lm}^{\sigma} \mathbf{M}_l^m(\mathbf{r}) + p_{lm}^{\sigma} \mathbf{N}_l^m(\mathbf{r})],$$
$$\mathbf{H}^{\sigma}(\mathbf{r}) = -i \sum_{l=1}^{\infty} \sum_{-l}^{l} [p_{lm}^{\sigma} \mathbf{M}_l^m(\mathbf{r}) + q_{lm}^{\sigma} \mathbf{N}_l^m(\mathbf{r})]. \tag{12}$$

Here, index σ stands for plane TE/TM wave.

$$p_{lm}^{TE} = -F_{lm}\tau_{lm}(\alpha), \quad q_{lm}^{TE} = F_{lm}\pi_{lm}(\alpha),$$
$$p_{lm}^{TM} = -iF_{lm}\pi_{lm}(\alpha), \quad q_{lm}^{TM} = iF_{lm}\tau_{lm}(\alpha), \tag{13}$$

$k_x = -k_0 \sin\alpha, k_y = k_0 \cos\alpha,$

$$F_{lm} = (-1)^m i^l \sqrt{\frac{4\pi(2l+1)(l-m)!}{(l+m)!}},$$
$$\tau_{lm}(\alpha) = \frac{m}{\sin\alpha} P_l^m(\cos\alpha),$$
$$\pi_{lm}(\alpha) = -\frac{d}{d\alpha} P_l^m(\cos\alpha). \tag{14}$$

For a particular case of normal incidence $k_z = 0$, $\alpha = -\pi/2$ we obtain from Equation (14)

$$\tau_{lm} = -m P_l^m(0), \pi_{lm} = -\frac{d}{d\alpha} P_l^m(0). \tag{15}$$

The general equation for the amplitudes a_l^m, b_l^m which describe the scattering by a linear array of spheres takes the following form:

$$Z_{TE,l}^{-1} a_l^m - \sum_{\nu}(a_{\nu}^m \mathcal{A}_{\nu l}^{mm} + b_{\nu}^m \mathcal{B}_{\nu l}^{mm}) = q_{lm}^{\sigma},$$
$$Z_{TM,l}^{-1} b_l^m - \sum_{\nu}(a_{\nu}^m \mathcal{B}_{\nu l}^{mm} + b_{\nu}^m \mathcal{A}_{\nu l}^{mm}) = p_{lm}^{\sigma}. \tag{16}$$

Here the left hand term formulates explicitly the matrix $\widehat{L}$ in Equation (1) and the right hand term corresponds the vector of incident wave Ψ_{inc} in the space of vector spherical functions notified by two integers l, m and polarization σ.

3. The Diffraction Continua of Vector Cylindrical Modes

Thanks to the axial symmetry of the array we can exploit the vector cylindrical modes for description of the diffraction continua which are doubly degenerate in TM and TE polarizations σ. The modes can be expressed through a scalar function ψ [1].

$$\psi_{m,n}(r,\phi,z) = H_m^{(1)}(\chi_n r) e^{im\phi + ik_{z,n}z}. \tag{17}$$

Then for the TE modes we have

$$E_z = 0, \quad H_z = \psi_{m,n},$$
$$E_r = \frac{ik_0}{\chi_n^2}\frac{1}{r}\frac{\partial\psi_{m,n}}{\partial\phi}, \quad H_r = \frac{ik_z}{\chi_n^2}\frac{\partial\psi_{m,n}}{\partial r},$$
$$E_\phi = \frac{-ik_0}{\chi_n^2}\frac{\partial\psi_{m,n}}{\partial r}, \quad H_\phi = \frac{ik_z}{\chi_n^2}\frac{1}{r}\frac{\partial\psi_{m,n}}{\partial\phi}, \tag{18}$$

and for the TM modes

$$E_z = \psi_{m,n}, \quad H_z = 0,$$
$$E_r = \frac{ik_z}{\chi_n^2}\frac{\partial\psi_{m,n}}{\partial r}, \quad H_r = \frac{-ik_0}{\chi_n^2}\frac{1}{r}\frac{\partial\psi_{m,n}}{\partial\phi},$$
$$E_\phi = \frac{ik_z}{\chi_n^2}\frac{1}{r}\frac{\partial\psi_{m,n}}{\partial\phi}, \quad H_\phi = \frac{ik_0}{\chi_n^2}\frac{\partial\psi_{m,n}}{\partial r}, \tag{19}$$

where

$$\chi_n^2 = k_0^2 - k_{z,n}^2 \tag{20}$$

and

$$k_{z,n} = \beta + 2\pi n, \quad n = 0, \pm 1, \pm 2, \ldots. \tag{21}$$

In what follows we consider the BSCs in the diffraction continua specified by two quantum numbers m and n where the m is the result of the axial symmetry and n is the result of translational symmetry of the infinite linear array of the dielectric spheres. Note that each diffraction continuum is doubly degenerate relative to the polarization σ. As a result of the interplay between the frequency k_0 and the wave number $k_{z,n}$ the continua can be open (χ is real) or closed (χ is imaginary). In the present paper we restrict ourselves by the case of one, two and three open continua.

4. Classification of BSCs in the Array of Spheres

In the previous section we presented the theory for scattering of plane waves by a periodic array of dielectric spheres based on the approach by Linton et al. [24]. If there is no incident wave we have $\widehat{L}\mathbf{a} = 0$ whose solutions are bound modes of the array. There might be two kinds of the bound modes. The first type of modes have wave number $\beta > k_0$ and describe guided waves along the array. These solutions found by Linton et al. exist in some interval of the material parameters of spheres, dielectric constant ϵ or radius R, and the Bloch wave number β [24]. The second type of bound modes with $\beta < k_0$ resides above the light cone (BSCs). It is much more difficult to establish the existence of the second type of bound states because a tuning of material parameters is required. However there might exist symmetry protected BSCs which are robust with respect to the material parameters. These BSCs have been already considered in the linear array of infinitely long dielectric rods [31,33,35,37,40–45].

The axial symmetry of the array implies that the matrices $\mathcal{A}$ and $\mathcal{B}$ split into the irreducible representations of the azimuthal number m which therefore classifies the BSCs. Next, the discrete translational symmetry along the z-axis implies that the respective wave number β specifies the BSC. At last, additional optional symmetries arise due to the inversion symmetry transformation $\widehat{K}f(x,y,z) = f(x,y,-z)$ for $\beta = 0, \pi$. It follows from Equation (11) that $s_{2k+1} = 0$, and respectively from Equations (5) and (60) we obtain $\mathcal{A}_{\nu L}^{mm} = 0$ if $l + \nu$ is odd, and $\mathcal{B}_{\nu L}^{mm} = 0$ if $l + \nu$ is even. Moreover for arbitrary β: $\mathcal{B}_{\nu l}^{00} = 0$ (see Appendix A). These relations establish the selection rules for the amplitudes a_l^m, b_l^m which determine allowed BSC modes listed in the Table 1.

Table 1. Classification of the bound states in the radiation continuum (BSCs).

m	β	Type I of BSC	Type II of BSC
$\neq 0$	0	(a_{2k}^m, b_{2k+1}^m)	(a_{2k+1}^m, b_{2k}^m)
0	$\neq 0$	$(a_l^0, 0), E_z = 0$	$(0, b_l^0), H_z = 0$
0	0	$(a_{2k}^0, 0), E_z = 0$	$(0, b_{2k}^0), H_z = 0$
0	0	$(0, b_{2k+1}^0), H_z = 0$	$(a_{2k+1}^0, 0), E_z = 0$

The cartesian components of the vector spherical functions transform under the inversion of z as follows

$$M_{l,x,y}^m(\pi - \theta) = -(-1)^{l-m} M_{l,x,y}^m(\theta), \quad M_{l,z}^m(\pi - \theta) = (-1)^{l-m} M_{l,z}^m(\theta),$$
$$N_{l,x,y}^m(\pi - \theta) = (-1)^{l-m} N_{l,x,y}^m(\theta), \quad N_{l,z}^m(\pi - \theta) = -(-1)^{l-m} N_{l,z}^m(\theta). \tag{22}$$

For $\beta = 0$ we have

$$\sum_j M^m_{l,x,y}(\mathbf{r} - \mathbf{R}_j) = -(-1)^{l-m} \sum_j M^m_{l,x,y}(\widehat{K}\mathbf{r} - \mathbf{R}_j),$$

$$\sum_j M^m_{l,z}(\mathbf{r} - \mathbf{R}_j) = (-1)^{l-m} \sum_j M^m_{l,z}(\widehat{K}\mathbf{r} - \mathbf{R}_j).$$

$$\sum_j N^m_{l,x,y}(\mathbf{r} - \mathbf{R}_j) = -(-1)^{l-m} \sum_j N^m_{l,x,y}(\widehat{K}\mathbf{r} - \mathbf{R}_j),$$

$$\sum_j N^m_{l,z}(\mathbf{r} - \mathbf{R}_j) = (-1)^{l-m} \sum_j N^m_{l,z}(\widehat{K}\mathbf{r} - \mathbf{R}_j). \tag{23}$$

Then from these equations and Equation (2) one can obtain the following symmetric properties for the cartesian components of EM fields collected in Table 2.

Table 2. Symmetry properties of the eigenmodes with $\beta = 0$.

Type I	Type II
$E_{x,y}(-z) = (-1)^{m+1} E_{x,y}(z)$	$E_{x,y}(-z) = (-1)^m E_{x,y}(z)$
$E_z(-z) = (-1)^m E_z(z)$	$E_z(-z) = (-1)^{m+1} E_z(z)$
$H_{x,y}(-z) = (-1)^m H_{x,y}(z)$	$H_{x,y}(-z) = (-1)^{m+1} H_{x,y}(z)$
$H_z(-z) = (-1)^{m+1} H_z(z)$	$H_z(-z) = (-1)^m H_z(z)$

Tables 1 and 2 will be useful for the symmetry classification of the bound modes in the next sections.

5. Symmetry Protected BSCs

In this section we present numerical solutions of Equation (3) for the symmetry protected BSCs with $m = 0, \beta = 0$ embedded into the first diffraction continuum $n = 0$. They constitute the majority of the BSCs in the array. The symmetry protected BSCs are either pure TE spherical vector modes (Type I in Table 1) with $a^0_{2k} \neq 0, b^0_k = 0$ or TM spherical vector modes (Type II in Table 1) with $a^0_k = 0, b^0_{2k} \neq 0$. We show that the symmetry protected BSCs are symmetrically mismatched to the first open continuum. Below we present numerical solutions for Type I BSCs with accuracy of 10^{-4}:

$$k_0 = 4.24, \quad R = 0.3, \epsilon = 12, \quad a^0_l = \begin{pmatrix} 0 \\ 0.7563 - 0.6542i \\ 0 \end{pmatrix}, \quad b^0_l = 0, l \geq 1 \tag{24}$$

and for Type II as

$$k_0 = 4.7504, \quad R = 0.3, \epsilon = 15, \quad a^0_l = 0, \quad b^0_l = \begin{pmatrix} 0 \\ -0.6017 + 0.7988i \\ 0 \\ 0.0004 - 0.0006i \end{pmatrix}, l \geq 1, \tag{25}$$

Patterns of EM fields and EM force lines are shown in Figure 2. Hereinafter we plot only real parts of electromagnetic fields. Other patterns of the symmetry protected BSCs the reader can find in Ref. [56]. One can see from Figure 2 that in the BSC of Type I (II) electric (magnetic) force lines are parallel the sphere surface.

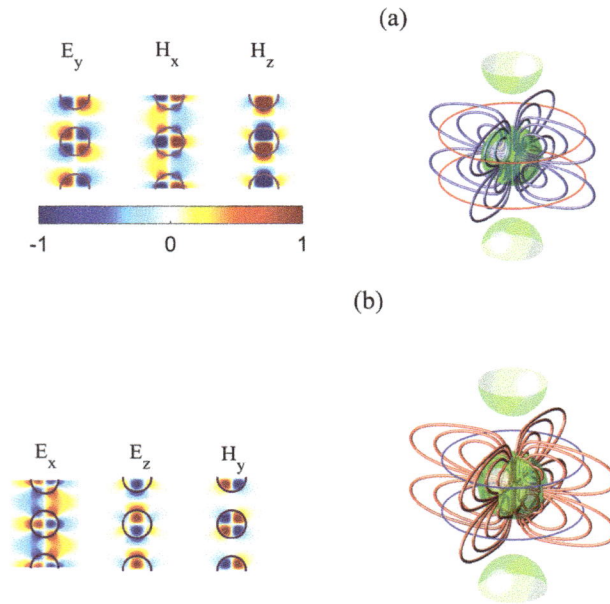

Figure 2. Patterns of the symmetry protected TE BSC (transverse electric bound state in the continuum) (24) (**a**) and TM BSC (25) (**b**). Left panels show the real parts of electromagnetic (EM) field components, right panels show the electric force lines in red and magnetic force lines in blue.

The symmetry protected Type I and Type II BSCs have qualitatively similar field structure with respect to $\mathbf{E} \leftrightarrow \mathbf{H}$ but are not degenerate because of different boundary conditions for $\mathbf{E}$ and $\mathbf{H}$ at the sphere surface. From Table 2 one can see why the eigenmodes (24) and (25) are protected by symmetry against decay into the diffraction continuum $m = 0, n = 0$. From Equations (18) and (19) we obtain that the TE/TM continuum with $k_{z,0} = 0$ ($\beta = 0$) has the only $H_z/E_z \neq 0$ independent of z. The Type I BSC has $E_z = 0$ and odd H_z so that these type of BSCs is symmetrically mismatched to both TE and TM continua. The Type II BSC has odd E_z and $H_z = 0$ to decouple from the both TE and TM continua.

Besides the fully symmetry protected BSCs from the third row in Table 1 $(a_{2k}^0, 0)$ and $(0, b_{2k}^0)$, we found a partially symmetry protected Type II BSC $(a_{2k+1}^0, 0)$ from the fourth row of Table 1:

$$k_0 = 2.934, \ R = 0.4805, \ \epsilon = 15, \ a_l^0 = \begin{pmatrix} 0.6826 + 0.0332i \\ 0 \\ -0.7291 - 0.0354i \\ 0 \\ -0.0008 \end{pmatrix}, \ b_l^0 = 0, l \geq 1, \quad (26)$$

however the Type I BSCs with $(0, b_{2k+1}^0)$ were not revealed in our computations. The BSC (26) is symmetrically mismatched relative only to the TM continuum. Zero coupling of this BSC with the TE continuum can be achieved by tuning the radius of spheres. Patterns of EM fields and EM force lines are shown in Figure 3.

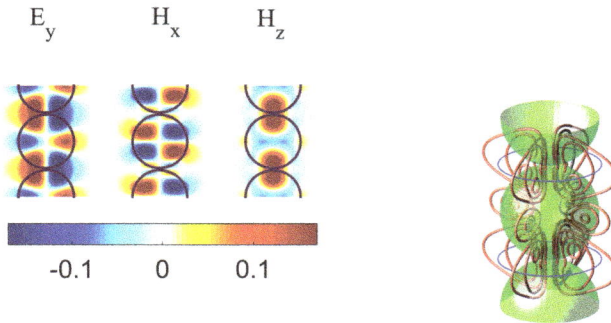

$$E_y \qquad H_x \qquad H_z$$

Figure 3. Pattern of the BSC (26) which is symmetrically protected in respect to the first TM radiation continuum and decoupled with the TE continuum by tuning of $R = 0.4805$.

5.1. $\pm m$ Degenerate BSCs with $\beta = 0$

The above described mechanism for partially symmetry protected BSCs with $m = 0$ can be exploited for even more complicated case $m \neq 0$. Their orbital angular momentum (OAM) $m \neq 0$ of these BSCs is the result of azimuthal rotation symmetry of the array and provide unique properties in the form of spinning or spiralling currents of the Poynting vector [58,59]. Besides, the system has the time reversal symmetry which implies that these BSCs are degenerate over $\pm m$. Let us start with the Type I BSC with $m = 1$ which has the odd E_z and the even H_z according to Tables 1 and 2. This BSC is symmetrically mismatched with the TM diffraction continuum $m = 1, n = 0$ which is independent of z. The coupling with the TE continuum can be cancelled by tuning the radius. The result of computation of this partially symmetry protected Type I BSC (a_{2k}, b_{2k+1}) is the following

$$m = 1, \ k_0 = 2.847, \ R = 0.3945, \ (a_l^1, \ b_l^1) = \begin{pmatrix} 0 & 0.6662 + 0.4273i \\ -0.33 + 0.5145i & 0 \\ 0 & -0.0048 - 0.0031i \\ 0 & 0 \end{pmatrix}, l \geq 1 \quad (27)$$

and shown in Figure 4a. The Type II BSC (a_{2k+1}, b_{2k}) with $m = 2$ has even E_z and odd H_z. It is symmetry protected against decay into the TE continuum with $m = 2, n = 0$ and coupling with the TM continuum is cancelled by tuning the radius with the following result

$$m = 2, \ k_0 = 3.086, \ R = 0.471, \ (a_l^2, \ b_l^2) = \begin{pmatrix} 0 & 0.6545 + 0.2013i \\ -0.2142 + 0.6964i & 0 \\ 0 & -0.0057 - 0.0018i \\ 0 & 0 \end{pmatrix}, l \geq 2. \quad (28)$$

All components of electric and magnetic fields are nonzero and localized around the array as shown in Figure 4. We show the EM field around only one sphere because the pattern is periodically repeated along the z-axis. One can see that the value of OAM m reflects in the structure of force lines in the xy-plane while the number of the amplitudes a_l^m reflects in the structure of lines along the z-axis.

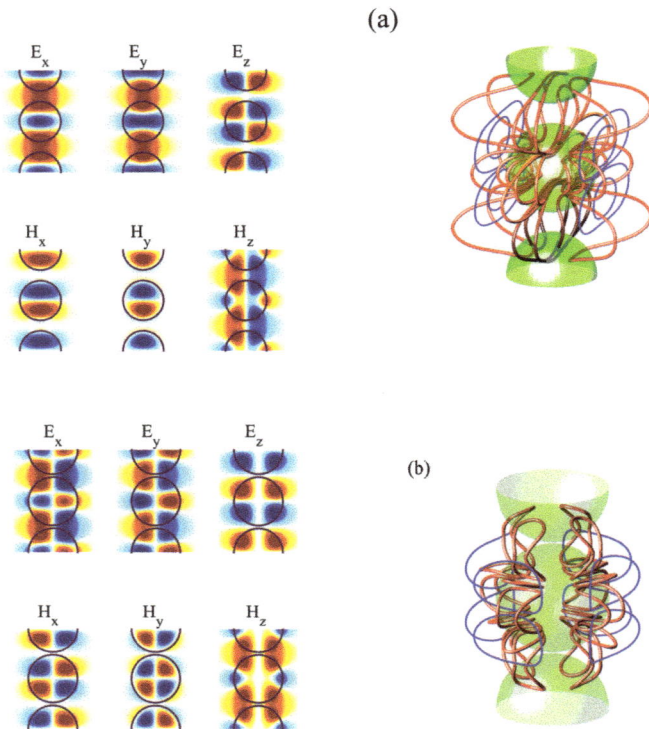

Figure 4. BSC with $\beta = 0$: (**a**) BSC (27) with orbital angular momentum (OAM) $m = 1$ and (**b**) BSC (28) with OAM $m = 2$. Electric field force lines are shown in red, magnetic field force lines are shown in blue. Both BSC have the Bloch vector $\beta = 0$.

5.2. Robust Bloch BSCs with $\beta \neq 0, m = 0$

Could the Bloch BSC occur at $\beta \neq 0$ in the continuum of free-space modes? This question was first answered positively by Porter and Evans [26] who considered acoustic trapping in an array of rods of rectangular cross-section. Marinica et al. [32] demonstrated the existence of the Bloch BSC with $\beta \neq 0$ in two parallel dielectric gratings and Ndangali and Shabanov [33] in two parallel arrays of dielectric rods. In a single array of rods positioned on the surface of bulk 2d photonic crystal multiple BSCs with $\beta \geq 0$ were considered by Chia Wei Hsu et al. [44]. The Bloch BSCs in a single array of cylindrical dielectric rods in air were also reported in Refs. [37,49,62]. Such travelling wave Bloch BSCs with the eigenfrequencies above the light cone are interesting because the array serves as a waveguide although only for fixed β (see summary of BSCs in Figure 5) in contrast to the bound states below the light cone [24].

According to Table 1 the Bloch BSCs with $\beta \neq 0, m = 0$ have only nonzero components a_l^0 or b_l^0. Let us first consider Type I BSCs with $b_l^0 = 0$ which have $E_z = 0$ and, therefore, are decoupled with the TM continuum but coupled with the TE $n = 0, m = 0$ continuum. We show numerically that this coupling can be cancelled under variation of β. The numerical results are collected in Equation (29) below with the pattern of EM fields shown in Figure 6.

$$k_0 = 3.6505, \ R = 0.4, \epsilon = 15, \ \beta = 1.2074, \ (a_l^1, b_l^1) = \begin{pmatrix} 0.1053 - 0.0638i & 0 \\ 0.1918 + 0.3161i & 0 \\ 0.6046 + 0.5572i & 0 \\ 0.7873 + 0.4777i & 0 \\ -0.0033 - 0.0054i & 0 \end{pmatrix}, \ l \geq 1. \quad (29)$$

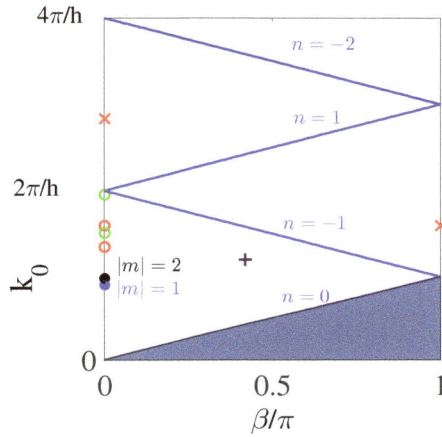

Figure 5. Summary of BSCs: the symmetry protected BSCs marked by open circles, Bloch BSC with $\beta \neq 0, m = 0$ is marked by +. Two BSCs with OAM are marked by closed circles, the partially symmetry protected BSC (9) is marked by cross. All BSC points are calculated for spheres with $\epsilon = 15$. The area filled by gray corresponds to below the light cone. Dash and dash-dot lines show thresholds where the next continua $n = \pm 1$ and $n = -2$ are opened.

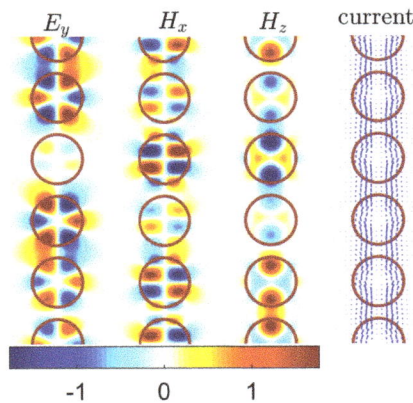

Figure 6. EM field configurations given by Equation (29) and currents of the Poynting vector of the Bloch BSC with $\beta_c = 1.2074$ for $k_{0c} = 3.6505, R_c = 0.4, \epsilon = 15$.

Although this BSC occurs at the fixed value of β there is no necessity to tune the material parameters of the spheres and therefore the BSC can be referred to as robust which is attractive from experimental viewpoint. We managed to find only Type I BSCs for $\epsilon = 15$ but none of Type II. Such a difference between the types is related to different boundary conditions for electric and magnetic fields at material interfaces. We collected all BSCs in Figure 5.

6. Light Guiding above the Light Line

The arrays of dielectric nanonparticles can serve as subwavelength waveguides to be emploied as the key components for future integrated optics [15–19,21]. They could be potentially advantageous against nanoplasmonics due to, for instance, the opportunity to control the frequencies of electric and magnetic Mie resonances by changing the geometry of high-index nanoparticles, and the absence of free carriers resulting in a high Q-factor. Arguably, the arrays of dielectric nanoparticles provide one of the most promising subwavelength set-ups for efficient light guiding [16,22] as well as more intricate effects such as resonant transmission of light [22], and optical nanoantennas [23].

So far the major theoretical tool for analyzing the infinite arrays of spherical dielectric nanoparticles has been the coupled-dipole approximation [63–65]. In that approximation guided waves in arrays of magnetodielectric spheres were first considered by Shore and Yaghjian [66,67] who derived the dispersion relation and computed the dispersion curves for dipolar waves. Recently a more tractable form of the dispersion equations was presented by the same authors [68] with the use of the polilogarithmic functions. The dipolar waves in arrays of Si dielectric nanospheres were thoroughly analyzed in [21]. It was shown that only two lowest guided modes could be fairly described by the dipole approximation which breaks down as the frequency approaches the first quadruple Mie resonance. This limits the application of the dipolar dispersion diagrams to realistic waveguides assembled of dielectric nanoparticles. As an alternative to the dipole approximation a "semiclassical" approach based on the coupling of the whispering gallery modes of individual spheres could be employed to recover the array band structure [69,70] if the wavelength is much smaller than the diameter of the spheres. The general case, however, requires a full-wave Mie scattering approach to account for all possible multipole resonances [19] involving a very complicated multiscattering picture which mathematically manifests itself in infinite multipole sums. Luckily, such an approach was recently developed by Linton, Zalipaev, and Thompson who managed to obtain a multipole dispersion relation in a closed form suitable for numerical computations [24]. The above approach was used for analyzing the spectra of dielectric arrays above the line of light in Ref. [56]. It was demonstrated that under variation of some parameter such as, for example, the radius of the spheres the leaky modes dominating the spectrum can acquire an infinite life-time. In other words, the array can support bound states in the radiation continuum (BSCs) [33,35,37,48,71]. In this letter, we will address the ability of the BSCs to propagate light along the array primarily motivated by finding new opportunities for designing subwavelength waveguides.

The dispersion diagram of an array of dielectric nanoparticles was obtained [57]. The dispersion curves were computed by solving the dispersion equations $f_{d,m}(k,\beta) = 0$, where k is the vacuum wave number $k = \omega/c$, and β is the Bloch wave number, while the subscripts d, m designate either dipole [21,68], or multipole [24] dispersion relations. For brevity we do not present the exact dispersion relations $f_{d,m}(k,\beta) = 0$. A mathematically inquisitive reader is referred to the above cited papers to examine the rather cumbersome expressions for $f_{d,m}(k,\beta)$. Here we assume that the array consists of spherical noanoparticles of radius R with dielectric constant $\epsilon = 15$ (Si) in vacuum. The centers of the nanoparticles are separated by distance a. It is worth mentioning that at a given dielectric constant the dispersion is only dependent on a single dimensionless quantity R/a. This allowed to scale the model for a microwave experiment [21]. There are three types of dipolar solutions [21], namely; longitudinal magnetic (LM), longitudinal electric (LE), and transverse electromagnetic (TEM) waves. In Figure 7 we plot the lowest frequency modes of each type in comparison against the multipole solution [24].

In all cases if the $k - \beta$ curve is above the light line $k = \beta$ the vacuum wave number becomes complex valued. The imaginary part of k is linked to the mode life-time through the following formula:

$$\tau = -[c\Im(k)]^{-1}. \tag{30}$$

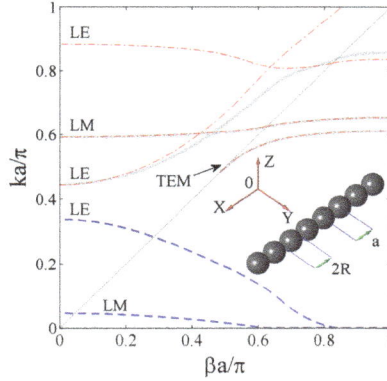

Figure 7. Dispersion diagram of an infinite array of dielectric nanospheres of radius R with dielectric constant $\epsilon = 15$, $R/a = 0.4$. The array centerline is aligned with the x-axis as shown in the south-east corner of the plot. The real parts of dipolar solutions are shown by dash-dot red lines. The thick gray lines are the real parts of the full-wave solutions; negative imaginary parts $-\Im(k)$ of the full-wave solutions are shown by blue dashed lines. The thin gray line is the light line.

Two approaches are possible for description of the leaky modes; complex frequency ω [61,72], or complex Bloch number β [21,73]. In the latter case the inverse of the imaginary part of β is the penetration depth into the array $L_\tau = [\Im(\beta)]^{-1}$. The quantities τ and L_τ are, in fact, proportional

$$L_\tau = v\tau, \tag{31}$$

where v is the group velocity $v = d\Re(\omega)/d\beta$. Here, we do not present the imaginary part of β mentioning in passing that the penetration depths for dipolar waves were analyzed in refs. [21,73]. What is important the numerical data available so far [21,56,73] indicate that all dipolar leaky modes are relatively short-lived, in particular, no dipolar BSCs were found in Ref. [56]. In compliance with Ref. [21] Figure 7 demonstrates that only two lowest eigenmodes are fairly described by the dipole approximation.

Now, let us consider the multipolar quasi-guided modes within the first radiation continuum [56]. The dispersion curves for a leaky mode for two different ratios R/a are plotted in Figure 8a. One can see that in contrast to the dipolar waves in Figure 7 now the solutions could be long-lived with the life-time Equation (30) growing up to infinity at the BSC points. It should pointed out that for both R/a of all leaky modes of the array we plot only one which has a Bloch BSC point $\Im(k) = 0$ at $\beta \neq 0$. As shown in Figure 8b the BSC exists in a wide range of parameter R/a. The magnetic and electric vectors could be found in terms of Mie coefficients a_n^m, b_n^m. For instance, outside the spheres one has for the electric vector $\mathbf{E}(\mathbf{r})$ [24]

$$\mathbf{E}(\mathbf{r}) = \sum_{j=-\infty}^{\infty} e^{iaj\beta} \sum_{n=m^*}^{\infty} \left[a_n^m \mathbf{M}_n^m (\mathbf{r} - \mathbf{r}_j) + b_n^m \mathbf{N}_n^m (\mathbf{r} - \mathbf{r}_j) \right], \tag{32}$$

where j the number of the particle in the array, m - azimuthal number, $m^* = max(1, m)$, and $\mathbf{N}_n^m(\mathbf{r}), \mathbf{M}_n^m(\mathbf{r})$ are spherical vector harmonics [1]. Only $m = 0$ Bloch BSCs were found in Ref. [56].

Our numerics indicate that for BSCs in Figure 8 the dominating term in the expansions (32) corresponds to coefficient a_3^0. In the insets in Figure 8b we plot the components of the electric and magnetic vectors of the BSC solution. One can see that the electromagnetic field is localized in the vicinity of the array. The amplitude of a wave propagating along the array attenuates exponentially according to a simple formula

$$F(x) = e^{-x/L_\tau}, \tag{33}$$

where $x = ja$ is the distance. In the vicinity of a BSC the $\omega - \beta$ dependance could be approximated as

$$\omega - \omega_0 = v_0(\beta - \beta_0) + \mathcal{O}[(\beta - \beta_0)^2], \tag{34}$$

where ω_0, β_0, v_0 are the BSC eigenfrequency, Bloch number, and group velocity, correspondingly.

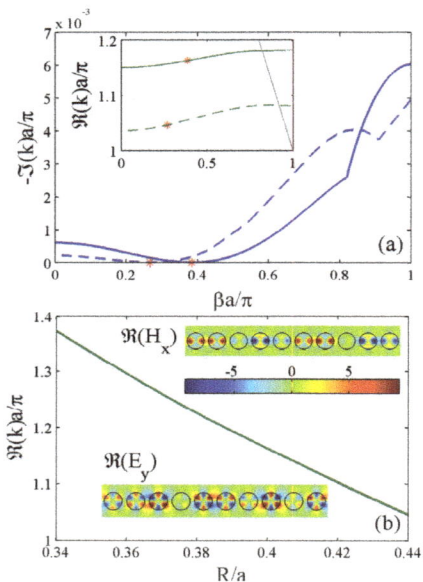

Figure 8. (**a**) Quasi-guided modes above the light line; $R/a = 0.4$—solid line, $R/a = 0.44$—dash line, $\epsilon = 15$. Imaginary part of k - the main plot, real part - the inset. The positions of the BSCs are shown by red stars. The imaginary parts are non-smooth as the real parts cross the boundary of the second radiation continuum $ka = 2\pi - \beta a$ shown by thin grey line; (**b**) Bloch BSC $\beta \neq 0$ vacuum wave number k vs. R/a, $\epsilon = 15$. The insets show the real parts of the y-component of electric vector E_y and the x-component of magnetic vector H_x in $x0y$-plane for the BSC at $R/a = 0.4$.

$$-\Im\{k\} = \alpha(\beta - \beta_0)^2 + \mathcal{O}[(\beta - \beta_0)^3]. \tag{35}$$

Combining Equations (30), (31) and (33)–(35) one obtains in the vicinity of a BSC

$$F(x) = \exp\left[-\frac{\alpha x c}{v_0^3}(\omega - \omega_0)^2\right]. \tag{36}$$

Thus, for the width of the transparency window in the frequency domain we have

$$\Delta(x) = \sqrt{\frac{v_0^3}{\alpha c}} \frac{1}{\sqrt{x}} \tag{37}$$

with α and v_0 extracted from the data in Figure 8 by a polynomial fit. Nevertheless, care is needed in applying Equation (36) as the frequency may fall out of the range where the dispersion is well approximated by the leading terms in Equations (34) and (35).

Using a full-wave multiscattering method [19] we simulated wave propagation in a finite array of 400 nanoparticles. In our numerical experiment a linearly polarized Gaussian beam [74] with the Rayleigh range $z_0 = 5a$ was focused on the first nanoparticle in the array. The wave vector of the beam was directed along the y-axis perpendicular to the array (see Figure 7), and the magnetic vector aligned with the array axis. In Figure 9a we plot the the leading Mie coefficient a_3^0 for the last nanoparticle in the array. The result shows a pronounced resonant behavior due to formation of standing waves as a consequence of the finiteness of the array. The distance between the resonances $\Delta\omega$ could be assessed as $\Delta\omega \approx \pi v_0/(aN)$ where N is the number of particles in the array. The resonant features could be averaged out by integration over small frequency intervals larger than $\Delta\omega$. The result is shown in Figure 9a in comparison against Equation (36). One can see that Equation (36) matches the numerical data to a good accuracy. The finiteness of chain also results in additional attenuation due to the radiative losses at the ends of the array. A detailed study of that effect was undertaken in Ref. [19] where it was shown that the Q factor of the finite arrays of high index nanoparticles scales as CN^3 with $0.1 < C < 10$ which makes such radiative losses negligible for $N = 400$. The discrepancy in Figure 9 is due to the higher order terms in Equations (34) and (35).

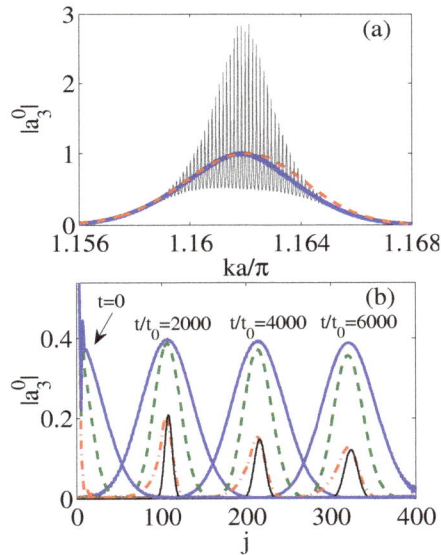

Figure 9. Light propagation in the array of 400 nanoparticles $\epsilon = 15$, $R/a = 0.4$, $v_0 = 0.054c$, $\alpha = 0.005a$. (**a**) Absolute value of the leading coefficient a_3^0 vs. wave number k of stationary wave injected into the array. The averaged data are plotted by thick blue line against the analytical result Equation (36) shown by dashed red line; (**b**) Absolute value of the leading coefficient a_3^0 vs. the particle number j for a light pulse with $k_0 = \pi 1.162/a$, $t_0 = a/c$. The pulse widths $\sigma_\omega = 0.0025/t_0$—blue solid, $\omega = 0.005/t_0$—green dashed, $\omega = 0.025/t_0$—red dash-dot lines. Thin black line shows analytical result Equation (38) for the pulse with $\sigma_\omega = 0.025/t_0$.

Finally, pulse propagation along the array was considered in the above set-up with a continuous superposition of Gaussian beams forming a Gaussian light pulse of width σ_ω in the frequency domain. The central wave number of the pulse was adjusted to the BSC wave number $k_0 = \pi 1.162/a$. At the moment $t = 0$ a light pulse was injected into the left end of the array. In Figure 9b we plot four snapshots of the leading Mie coefficient a_3^0 against the distance along the array for three different initial pulse widths σ_ω. One can clearly see in Figure 9b that the pulse propagating along the array tends to spread as the harmonics distant in the ω-space from the BSC frequency decay into the continuum. The pulse profile $f(x,t)$ could be found by Fourier-transforming the initial Gaussian pulse to the real space:

$$f(x,t) = \frac{1}{\sigma(t)} e^{-\frac{(x-v_0 t)^2}{\sigma^2(t)}} e^{i(\beta_0 x - \omega_0 t)} \tag{38}$$

with

$$\sigma^2(t) = 4a^2 \left[\left(\frac{v_0}{a\sigma_\omega} \right)^2 + \frac{c\alpha}{a^2} t \right]. \tag{39}$$

Analyzing Equation (39) for a given distance $L = v_0 t$ one can identify two possible regimes for the pulse propagation. In the "overdamped" regime the second term dominates on the left hand side of Equation (39) resulting in a noticeable spreading of the pulse in the real space. If, however, the first term dominates the pulse retains its profile during propagation time ($\sigma_\omega = 0.0025/t_0$ in Figure 9b). One finds from Equation (39) that the pulse doubles its width after travelling to the distance

$$L = \frac{3a^2}{\alpha} \left(\frac{v_0}{c} \right)^3 \left(\frac{c}{a\sigma_\omega} \right)^2. \tag{40}$$

So far the material losses due to the imaginary part of the dielectric constant were neglected. Particulary in silicon the material losses vary significantly in the optical range [75]. We ran a numerical test at 725 nm with $\Im(\epsilon) = 0.0075$ to find that the propagation distance $L \approx 100a$, $a = 421$ nm so the light can travel to approximately 60 wavelengths. It should be pointed out that in the near infrared $\lambda \approx 1000$ nm the losses can be tens of times less allowing propagation to hundreds wavelengths [21] as shown in Figure 9.

7. Emergence of the BSC in Scattering

Scattering of plane waves by periodic two-dimensional arrays of dielectric spheres originates since pioneering papers by Ohtaka et al. [3,4] (see also Ref. [5]). Scattering by aggregates of finite number of spheres was considered in the framework of multi sphere Mie scattering [6,9,11], however to our knowledge the scattering by the one-dimensional infinite array of dielectric spheres was not considered yet. The following subsections aim to present results of numerical calculation of differential and total cross-sections of the infinite array with focus to follow resonant traces of the BSCs in the cross-sections similar to the scattering by array of dielectric rods [37,44,45]. For the present case of the array of dielectric spheres we revealed different types of the BSCs. We show that excitation of corresponding quasi-BSCs needs in different ways of injection of EM waves.

The BSC has zero coupling with the continuum, i.e., the BSC has infinite quality factor [76]. Also the BSC is unique by that the solution of the scattering problem becomes ambiguous [25] and can be written as superposition of particular scattering state and the BSC [77,78].

$$\Psi = \Psi_S + \alpha \Psi_{BSC} \tag{41}$$

where α is arbitrary coefficient. This equation is well known mathematical consequence of the linear Lippmann-Schwinger Equation (1) when the inverse of the matrix $\widehat{L}$ does not exist, i.e., when $Det(\widehat{L}) = 0$. Then the BSC is the eigenmode of the matrix $\widehat{L}$

$$\widehat{L}\Psi_{BSC} = \lambda_c\Psi_{BSC} \tag{42}$$

with real eigenvalue λ_c. At the first sight the BSCs are not interesting because they are invisible for probing waves incident from the continuum. However the BSC point is isolated in the parametric space and experimentalist can approach to this point only approximately. Other words, in reality we have only the quasi-BSCs whose quality factor is restricted by set-up imperfectness. Moreover for the case of arrays of dielectric rods or nanoparticles we have always finite number of them to give rise to a leakage of trapped modes into the diffraction continua. In this case the parameter α in Equation (47) is defined by the point in the parametric space. When the point is sufficiently close to the BSC point α can achieve enormous value however as dependent on the way of approaching [78]. That observation is important for applications [26,52,78,79].

According to above equation the BSC is a null eigenvector of matrix $\widehat{L}$ with zero eigenvalue. As soon as one deviates from the BSC point in the parametric space the BSC emerges in the form of a collapsing Fano resonance. That phenomenon was observed in scattering of EM waves by arrays of rods [30,33,37,40,44,45,52,62]. The Fano resonance for the present system can be interpreted as interference of two optical paths, one through the spheres and another between the spheres. In what follows we highlight these features of the BSCs using the biorthogonal basis of eigenvectors of the non-Hermitian matrix $\widehat{L}$ [37,80].

$$\widehat{L}\mathbf{X}_f = L_f\mathbf{X}_f, \quad \widehat{L}^+\mathbf{Y}_f = L_f^*\mathbf{Y}_f, \quad \mathbf{Y}_f^+\mathbf{X}_{f'} = \delta_{ff'}. \tag{43}$$

It immediately follows that

$$\widehat{L}^{-1} = \sum_f \mathbf{X}_f \frac{1}{L_f}\mathbf{Y}_f^+. \tag{44}$$

Because of the axial symmetry matrix $\widehat{L}$ has OAM preserving block structure

$$L_{ll'}^{(m)} = \begin{pmatrix} Z_{TE,l}^{-1}\delta_{ll'} - \mathcal{A}_{ll'}^{mm} & -\mathcal{B}_{ll'}^{mm} \\ -\mathcal{B}_{ll'}^{mm} & Z_{TM,l}^{-1}\delta_{ll'} - \mathcal{A}_{ll'}^{mm} \end{pmatrix}, \tag{45}$$

where each block correspond to a specific value m.

In the nearest vicinity of the BSC point one of the complex eigenvalues L_c is close to zero. That allows us to substantially simplify Equation (44) leaving in the sum only the leading contribution related to L_c. Respectively the solution for the scattering function in Equation (1)

$$\Psi = \widehat{L}^{-1}\Psi_{inc} \tag{46}$$

is simplified as follows

$$\Psi^\sigma \approx \frac{1}{L_c}\mathbf{X}_c(\mathbf{Y}_c^+ \cdot \Psi_{inc}^\sigma), \quad \sigma = TE/TM. \tag{47}$$

This equation manifests one remarkable as well as important for applications property of the BSCs to enormously enhance the incident wave Ψ_{inc} by the factor $1/L_c$ [52,77,78].

7.1. Symmetry Protected BSCs

First the effect of enhancement of scattering function in the near zone of the infinite periodic array of dielectric spheres was shown in Ref. [56] in the vicinity to the symmetry protected BSCs with $m = 0$. As it follows from Equations(20) and (21) only one diffraction channel $n = 0$ is open for low frequencies k_0 where the majority of the BSCs occur as shown in Figure 5. Although the BSCs can not be probed directly by an incident wave they are seen as collapses of Fano resonance when the BSC point is approached in the parametric space. That phenomenon was observed for scattering of EM waves by arrays of rods [30,33,37,40,44,45,51,52,81]. In this subsection we report a similar Fano resonance collapse in the differential and total cross-sections vs. frequency when the wave number k_z tends to zero or the radius of the spheres approaches the BSC radius. The Fano resonance for the present system can be interpreted as an interference of the optical paths through and between the spheres.

Let us consider an incident plane wave with the wave vector in the x, z plane and polarizations: (a) TE polarized with electric field along the y-axis and (b) TM polarized with magnetic field along the y-axis.

For $m = 0, k_z \neq 0$ Equations (13) and (14) gives that $p_{l0}^{TE} = 0, q_{l0}^{TM} = 0$. Then taking into account that $\mathcal{B}_{vl}^{00} = 0$ (see Appendix A) we have from Equations (16) for the TE incident plane wave

$$Z_{TE,l}^{-1} a_l^0 - \sum_v a_v^0 \mathcal{A}_{vl}^{00} = q_{l0}^{TE},$$
$$Z_{TM,l}^{-1} b_l^0 - \sum_v b_v^0 \mathcal{A}_{vl}^{00} = 0. \tag{48}$$

The second equation gives $b_l^0 = 0$, and scattering of plane wave with TE polarization is given by only a_k. Then the type I BSCs is quasi BSC weakly coupled with the TE continuum for small k_z. That results in sharp resonant contribution in the cross-section $\sigma_{TE,TE}$ as shown in Figure 10a. The cross-sections $\sigma_{TE \to TM}, \sigma_{TM \to TM}$ and $\sigma_{TM \to TE}$ have no features related to these BSCs and are not shown in Figure 10a. If the plane wave falls onto the array normally $\alpha = -\pi/2$ ($k_z = 0$) we have fully invisible type I BSC that is shown by dash line in Figure 10a. Alternatively, the symmetry protected type II of the symmetry protected BSCs with the only amplitudes b_k can be observed via the cross-section $\sigma_{TM \to TM}$ as shown in Figure 10b. Thus, although the BSCs have no effect for the normal incidence they are detected by collapse of Fano resonances in total cross-sections for $k_z \to 0$.

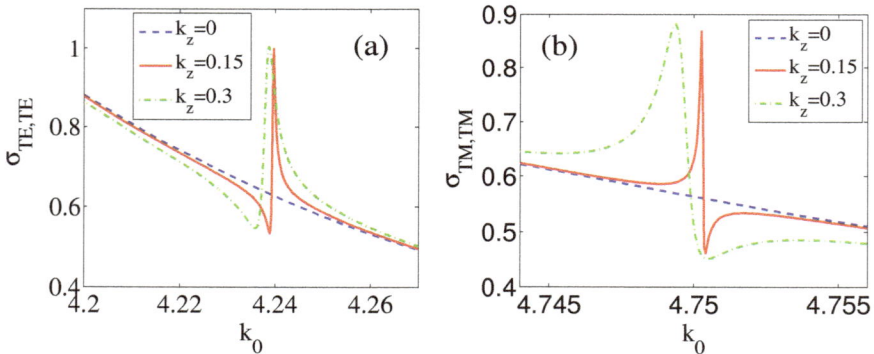

Figure 10. Total cross-section for scattering of plane wave incident by the angle ϕ onto the array. (a) Scattering of TE plane wave is strongly affected by the presence of the symmetry protected Type I BSC (24) with the eigenfrequency $k_0 = 4.24$ for $R = 0.3$, $\epsilon = 12$; (b) Scattering of TM plane wave is strongly affected by the presence of the symmetry protected Type II BSC (25) with the eigenfrequency $k_0 = 4.7504$ for $R = 0.3$, $\epsilon = 15$.

7.2. Scattering of Plane Waves in the Vicinity of the Quasi-BSCs with OAM

Next, consider the effect of the BSCs with $m = 2$ given by Equation (28) on the cross-section. We begin with the TE plane waves incident onto the array normally ($k_z = 0$). Then we have from Equations (13)–(30) that $p_{l2}^{TE} = 0, q_{l2}^{TE} \neq 0$ for odd l, and $p_{l2}^{TE} \neq 0, q_{l2}^{TE} = 0$ for even l. Therefore as Equation (16) shows there are only Type II solutions for scattered waves with the amplitudes

(a_{2k+1}, b_{2k}). Table 1 shows that they belong to the same type of BSCs with $m = 2$. Therefore in the vicinity of $R_{BSC} = 0.471$ this BSC is coupled with the TE continuum and gives the resonant contribution in the cross-section $\sigma_{TE,TE}$ that is demonstrated in Figure 11a,b. As for the scattering of the TM plane waves there are no resonant features as shown in Figure 10b by dash line. One can see in Figure 11c bright features of the differential cross-sections near the eigenfrequency of the BSC caused by the resonant contribution of the amplitude A_2 at the azimuthal angles $\phi = 0, \pm 90°, 180°$:

$$\frac{d\sigma}{d\phi} = \sigma_0 | \sum_m A_m \cos(m\phi) |^2. \tag{49}$$

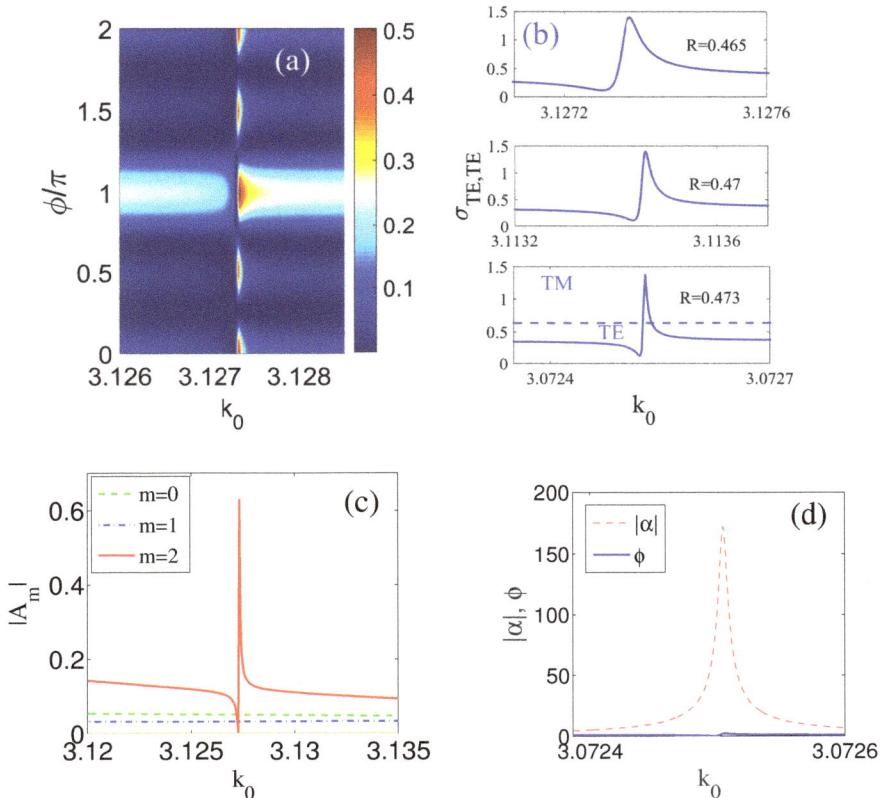

Figure 11. The effect of the BSC (28) with $m = 2, k_0 = 3.086, R = 0.471$ in: (**a**) differential cross-section vs. frequency and the azimuthal angle; (**b**) total cross-sections for different radii of the spheres close to the BSC radius (27) for plane wave illuminating the array normally; (**c**) Frequency behavior of the amplitudes A_m in the expansion (49); (**d**) Harvesting capability of the quasi BSC at $R = 0.473$. Dash red line shows the contribution of the BSC into the scattering function, blue solid line shows background ϕ.

It is clear that for the sphere radius close to $R_{BSC} = 0.471$ the BSC solution dominates in the near field zone. The solution can be presented as

$$\Psi = \alpha \Psi_{BSC} + \Phi \tag{50}$$

where α has a resonant behavior over frequency k_0 with the resonant width $\gamma \sim |R - R_{BSC}|$. Analytical expression for the resonant width can be derived following Refs. [77,78]. Thus we have slowly decaying quasi BSC modes above the light cone similar to those considered in Ref. [82]. That effect is important for concentration of light by touching spheres [52,83] notified as the harvesting capability of the system. Figure 11d illustrates the harvesting capability of the array of spheres in the vicinity of the BSC (28). Solid blue line shows the contribution of the background $\phi = ||\Phi||$ where $||\cdots||$ is the norm of vector Φ. We do not present here the scattering of plane incident normally to the array at the vicinity of the BSC with $m = \pm 1$. The results are very similar to those shown in Figure 11 except that differential cross-section has two maxima around the azimuthal angle equal $\phi = 0$ and $\phi = \pi$ while scattering near the BSC with $m = \pm 2$ shows maxima at $\phi = 0, \pi/2, \pi, 3\pi/4$.

8. Scattering of Plane Waves in the Vicinity of the Bloch BSC

In this section we consider the Bloch BSC with zero OAM whose field configuration is shown in Figure 6. Since the Bloch number $\beta_c = 1.2074$ the EM field configuration is incommensurate with the period of the array the EM field is different at each sphere. The numerical results for scattering by the array in the vicinity of this Bloch BSC are presented in Figure 12a which shows that under illumination of the array by a TE plane wave there is a resonant peak only in the total cross-section $\sigma_{TE,TE}$ [56].

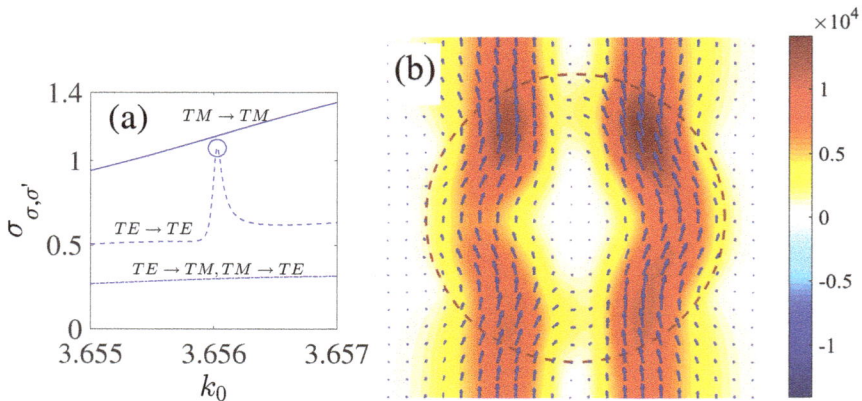

Figure 12. (a) Total cross-section for scattering of plane wave with $\beta = 1.3074$ in the vicinity of the Bloch BSC vs. the frequency; (b) The plane wave supports giant laminar power current at the point marked in the left panel by open circle. The color bar at the right indicates absolute value of the current.

If a plane wave with TE polarization, the wave vector $(k_x, 0, \beta \approx \beta_c)$ and the frequency $k_0 = k_{0c} = 3.6505$ illuminates the array, the running Bloch quasi-BSC with β is excited as shown in Figure 12b with giant laminar power flows.

9. Transfer of SAM into OAM of the BSC with $m \neq 0$

It is well known that electromagnetic (EM) fields can not only carry energy but also angular momentum. The angular momentum is composed of the spin angular momentum (SAM) and the orbital angular momentum (OAM) describing the polarization and the phase structure distribution of EM fields, respectively. The research on the OAM of EM fields has been in focus of researches since Allen et al. investigated the mechanism of the OAM in laser modes [54,84]. In contrast to SAM, which has only two possible states of left-handed and right-handed circular polarizations, the states of OAM are in principle unlimited owing to the unique characteristics of spiral flow of propagating EM waves [55]. The OAM has the potential to tremendously increase the spectral efficiency and capacity of communication systems [85]. Among numerous investigations on OAM effects, one of the subjects of intensive recent studies is the link between the near-field chirality and the far-field OAM. For different types of chiral polaritonic lenses, it was shown that the near-field chirality can lead to the tailoring optical OAM in the far-field region [86,87]. There were many proposals to generate OAM beams by use of chiral plasmonic nanostructures [86], ferrite particles [88], the monolithic integration of spiral phase plates [89], chiral polaritonic lenzes [90], and by designer metasurfaces [91], etc.

Schäferling et al. [92] have shown that chiral fields, i.e., electromagnetic fields with nonvanishing optical chirality, can occur next to symmetric nanostructures without geometrical chirality illuminated with linearly polarized light at normal incidence. Rodriguez-Fortuño et al. [93] demonstrated a planar photonic nanostructure with no chirality consisting of a silicon microdisk coupled to two waveguides. The device distinguishes the handedness of an incoming circularly polarized light beam by driving photons with opposite spins toward different waveguides. It was shown theoretically and experimentally that the fundamental resonance of a silicon microdisk resonator can inherit the angular momentum carried by anormally incident light beam and transfer it as linear momentum into one of two output waveguides. Remarkably, the microdisk is not chiral: it responds equally to the left chiral polarization and the right chiral polarization without exhibiting optical activity nor circular dichroism. Instead, it couples light to different waveguides (with opposite linear momenta) depending on the handedness of incoming light and the relative position between the microdisk and the waveguides.

The above results have a simple interpretation as an analogue with the spin-orbit interaction [94]. Rodríguez-Fortuño et al. [95] demonstrated circularly polarized dipole results in the unidirectional excitation of guided electromagnetic modes in the near field, with no preferred far-field radiation direction. In the present section we show a similar excitation of the Bloch BSC mode however with OAM in the near field transfer. Other words, we show a transfer of SAM to OAM of bound states in all-dielectric system in the near field of the array. Because of the time-reversal symmetry BSCs with OAM are degenerate with respect to the sign of m. That modifies Equation (47) as follows

$$\Psi_\sigma^m \approx \frac{1}{L_c} \sum_\pm [\mathbf{X}_c(\pm m)(Y_c(\pm m)^+ \cdot \Psi_{inc}^{\pm m,\sigma})] \tag{51}$$

where the incident wave according to Equation (16) is given

$$\Psi_{inc}^{m,\sigma} = \begin{pmatrix} sign(m)\mathbf{p}_{|m|}^\sigma \\ \mathbf{q}_{|m|}^\sigma \end{pmatrix}, m \text{ are odd}, \sigma = TE$$

$$\Psi_{inc}^{m,\sigma} = \begin{pmatrix} \mathbf{p}_{|m|}^\sigma \\ sign(m)\mathbf{q}_{|m|}^\sigma \end{pmatrix}, m \text{ are odd}, \sigma = TM$$

$$\Psi_{inc}^{m,\sigma} = \begin{pmatrix} \mathbf{p}_{|m|}^\sigma \\ sign(m)\mathbf{q}_{|m|}^\sigma \end{pmatrix}, m \text{ are even}, \sigma = TE$$

$$\Psi_{inc}^{m,\sigma} = \begin{pmatrix} sign(m)\mathbf{p}_{|m|}^\sigma \\ \mathbf{q}_{|m|}^\sigma \end{pmatrix}, m \text{ are even}, \sigma = TM. \tag{52}$$

and subvectors $\mathbf{p}^m$ and $\mathbf{q}^m$ are given by Equation (13). In particular, for the plane wave incident normally to the array $\beta = 0$ we have

$$\mathbf{p}_{|m|}^{TE} = \begin{pmatrix} 0 \\ p_{m,2}^{TE} \\ 0 \\ p_{m,4}^{TE} \\ \vdots \end{pmatrix}, \mathbf{q}_{|m|}^{TE} = \begin{pmatrix} q_{m,1}^{TE} \\ 0 \\ q_{m,3}^{TE} \\ 0 \\ \vdots \end{pmatrix}, m \text{ are odd,}$$

$$\mathbf{p}_{|m|}^{TM} = \begin{pmatrix} p_{m,1}^{TM} \\ 0 \\ p_{m,3}^{TM} \\ 0 \\ \vdots \end{pmatrix}, \mathbf{q}_{|m|}^{TM} = \begin{pmatrix} 0 \\ q_{m,2}^{TM} \\ 0 \\ q_{m,4}^{TM} \\ \vdots \end{pmatrix}, m \text{ are odd,}$$

$$\mathbf{p}_{|m|}^{TE} = \begin{pmatrix} p_{m,1}^{TE} \\ 0 \\ p_{m,3}^{TE} \\ 0 \\ \vdots \end{pmatrix}, \mathbf{q}_{|m|}^{TE} = \begin{pmatrix} 0 \\ q_{m,2}^{TE} \\ 0 \\ q_{m,4}^{TE} \\ \vdots \end{pmatrix}, m \text{ are even,}$$

$$\mathbf{p}_{|m|}^{TM} = \begin{pmatrix} 0 \\ p_{m,2}^{TM} \\ 0 \\ p_{m,4}^{TM} \\ \vdots \end{pmatrix}, \mathbf{q}_{|m|}^{TM} = \begin{pmatrix} q_{m,1}^{TM} \\ 0 \\ q_{m,3}^{TM} \\ 0 \\ \vdots \end{pmatrix}, m \text{ are even.} \tag{53}$$

By virtue of Equation (45) and $\mathcal{B}_{ll'}^{(m)} = -\mathcal{B}_{ll'}^{(-m)}$ the eigenvectors can be decomposed over the polarizations as follows

$$\mathbf{X}_c(\pm m) = \begin{pmatrix} \mathbf{x}_{TE}^m \\ \pm \mathbf{x}_{TM}^m \end{pmatrix}, \mathbf{Y}_c(\pm m) = \begin{pmatrix} \mathbf{y}_{TE}^m \\ \pm \mathbf{y}_{TM}^m \end{pmatrix}. \tag{54}$$

Then it follows from Equation (51)

$$\Psi_\sigma^m \approx \begin{cases} \frac{D_\sigma^{|m|}}{L_{c,m}}[\mathbf{X}_c(m) + (-1)^m(\mathbf{X}_c(-m)], \sigma = TE \\ \frac{D_\sigma^{|m|}}{L_{c,m}}[\mathbf{X}_c(m) + (-1)^{m+1}\mathbf{X}_c(-m)], \sigma = TM, \end{cases} \tag{55}$$

where

$$D_\sigma^{|m|} = \mathbf{y}_{TE}^+ \mathbf{p}_{|m|}^\sigma + \mathbf{y}_{TM}^+ \mathbf{q}_{|m|}^\sigma. \tag{56}$$

Assume that the elliptically polarized plane wave $\Psi_{inc}^{TE} + \alpha\Psi_{inc}^{TM}$ is incident with small β. By taking

$$\alpha = \frac{D_{TE}^{|m|}}{D_{TM}^{|m|}} \tag{57}$$

we obtain from Equation (55) that

$$\Psi_\sigma \approx F_{|m|}\mathbf{X}_c^{+m}, F_{|m|} = \frac{2D_{TE}^{|m|}}{L_{cm}}. \tag{58}$$

The scattering function has only a contribution with the positive OAM $m > 0$. Here we introduced the enhancement factor F which defines to what extent the scattering function is amplified in the near zone. Respectively for $D_{TE} = -\alpha D_{TM}$ the scattering function has only a contribution with the negative OAM $m < 0$.

One can show from Equations (53), (54) and (56) that asymptotically $D_{TM}^{|m|} \to 0$ for $\beta \to 0$. From Equation (58) it follows that the enhancement factor for scattering of plane waves in the vicinity of the BSC point is determined by the ratio D_σ / L_c. In what follows we sweep the frequency of the incident wave k_0 and the angle of incidence defined by β in the vicinity of the BSCs with OAM $m = 1$ and $m = 2$. Figure 13 illustrates the behavior of the enhancement factor in the plane of the frequency k_0 and β calculated with the use of Equation (58). Following the line with $|\alpha| = 1$ we found the maximal enhancement marked by open green circles in Figure 13 for the following parameters. (i) For the case of the BSC with $m = 1$ the optimal parameters are $k_0 = k_{0c} + 0.0025$, $\beta = 0.0052$, $\alpha = 0.63 + 0.77i$ for $R = R_c - 0.0005$; (ii) For the case of the BSC with $m = 2$: $k_0 = k_{0c} + 0.02$, $\beta = 0.031$, $\alpha = 0.31 + 0.94i$ for $R = R_c - 0.003$. Fixing these parameters except β we plot the lowest eigenvalue of the matrix $\hat{L}$ in Figure 14 and the values of $|D_\sigma|$ Equation (56) in Figure 15 versus β. From these Figures one can see that, first, the enhancement is determined by the lowest eigenvalue L_{cm} while D_{TE} is almost constant. Second, the value D_{TM} grows from zero. Therefore, to achieve enhancement one has to inject a plane wave with elliptic polarization. In what follows we take for simplicity the circular polarization $|\alpha| = 1$ of the incident wave.

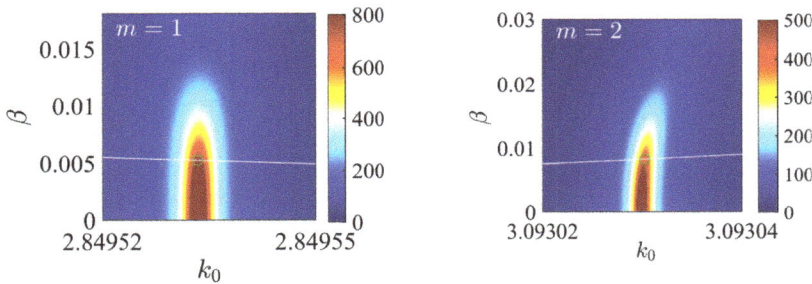

Figure 13. Enhancement factor $|F_m|$ vs. k_0 and β. White line corresponds to polarization (57) $|\alpha| = 1$. Open circles mark of maximal enhancement.

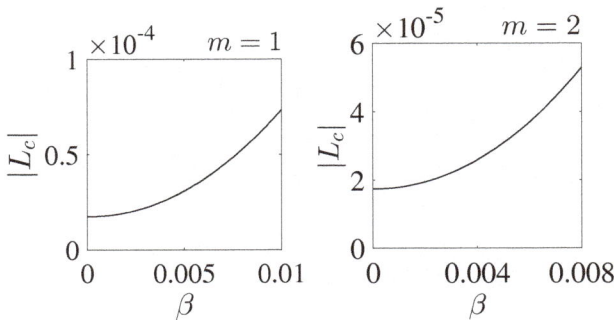

Figure 14. The lowest eigenvalue $|L_c|$ of matrix (45) in the vicinity of the BSCs with OAM.

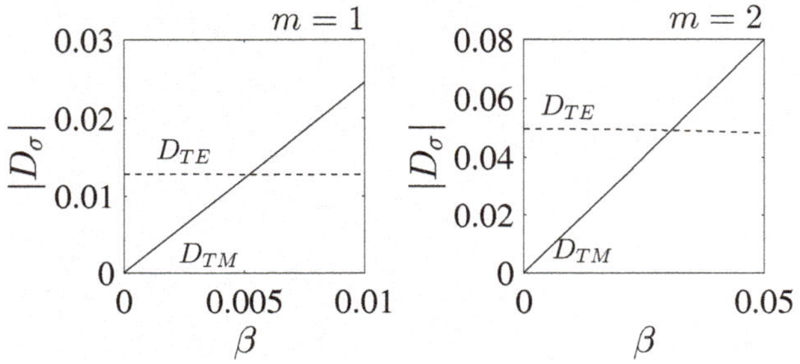

Figure 15. The values D_σ given by Equation (56) in the vicinity of the BSCs with OAM.

Because of the smallness of the eigenvalue L_{cm} in Equation (55) EM fields given by the scattering function can reach extremely high values near the spheres. Clearly this is an effect of the BSCs with infinitely high quality factor that presents a possibility to enormously enhance the incident light [40,52,78]. In Figure 16 we demonstrate that the enhancement is very sensitive to the choice of the sphere radius in the vicinity of R_c when other parameters are tuned to the BSC point.

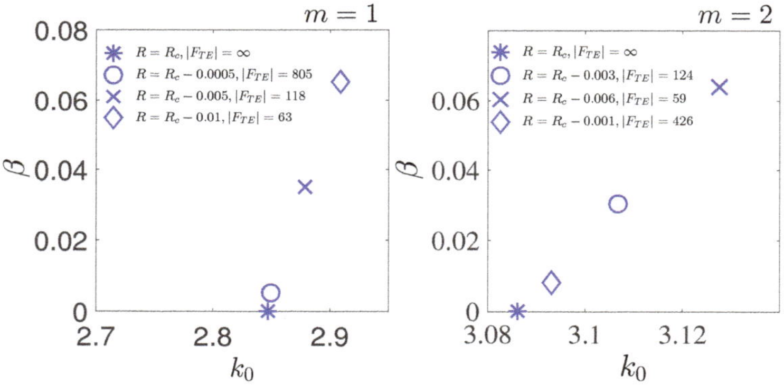

Figure 16. Values of the maximal enhancement factor $|F|$ vs. k_0 and β as dependent on radius of spheres for $|\alpha| = 1$.

Thanks to carrying OAM the BSC with $m \neq 0$ supports vortical power currents [96] as demonstrated in Figure 17. Owing to the enhancement of the scattered field in the near zone the spinning currents can reach giant values with respect to the incident power currents as demonstrated in Figure 18. All currents are measured in terms of the incident power with $\beta = 0.00517$ for the case $m = 1$ and $\beta = 0.0307$ for the case $m = 2$. The value of the current is extremely high inside the spheres but rapidly drops outside the spheres as shown in Figure 19. As soon as the polarization is linear, for example $\alpha = 0$, vortical currents around the array vanish as demonstrated in Figure 20.

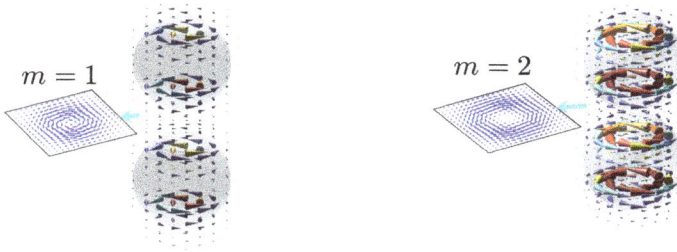

Figure 17. Pointing current circulates around the spheres when circularly polarized light is injected. Currents around other spheres are repeating periodically.

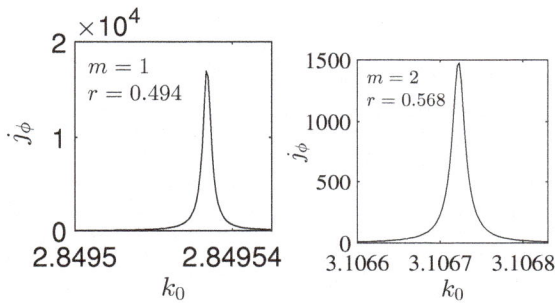

Figure 18. Value of angular component of the power current around the spheres at distance r from the center of sphere and $z = 0$.

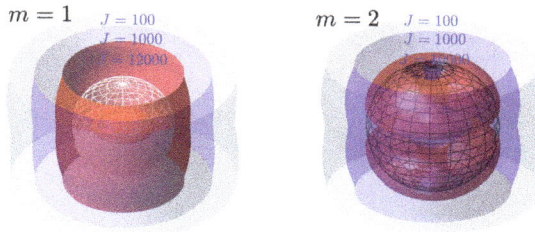

Figure 19. Iso surfaces of constant angular component of the power current.

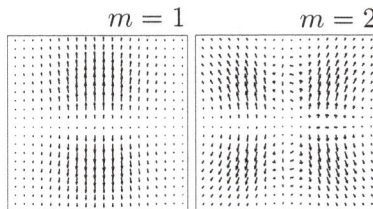

Figure 20. Power currents in the middle plane between spheres induced by linearly polarized light with $\alpha = 0$.

Figures 21 and 22 demonstrate that the orbital angular momentum of the BSCs affects the scattering of plane waves with linear polarization. The effect is a conversion of the incident polarizations $TE \to TM$ and *visa versa*. For the normally incident waves $\beta = 0$ there is no polarization conversion and no resonant peaks in the total cross-sections $TM \to TM$. Once the angle of incidence deviates from zero $\beta \neq 0$ all three total cross-sections acquire resonant response as shown in Figure 21. Note that there is polarization conversion when the frequency is far from the BSC frequencies. The absence of polarization conversion is clearly seen in the differential cross-section $TE \to TM$ as shown in Figure 22. It is also remarkable that this cross-section distinctively reflects the value of the OAM m.

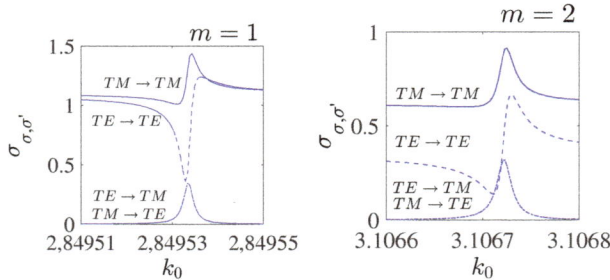

Figure 21. Total cross-section for scattering of plane wave by the array in the vicinity of the BSC with OAM at $\beta = 0.0052$ (left panel) and $\beta = 0.031$.

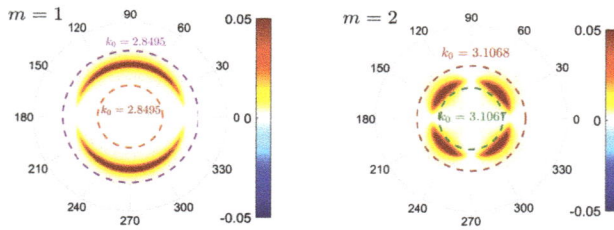

Figure 22. Differential cross-section for scattering of plane wave illuminating the array in the vicinity of the BSC with OAM.

10. Propagating Bloch BSCs with Orbital Angular Momentum

$$\vec{\psi} \approx \vec{X}_{BSC}^{m=1} B_{m=1} \oplus \vec{X}_{BSC}^{m=-1} B_{m=-1} \tag{59}$$

where

$$B_{m=\pm 1} = \frac{1}{L_c}(D^{TE} \pm \kappa D^{TM}), \quad D^{\sigma} = \vec{Y}_{BSC}^{+} \vec{\psi}^{\sigma}, \tag{60}$$

$\sigma = TE, TM$ labels the polarization of electromagnetic field. $\vec{Y}_{BSC}^{+}$ is the left eigenvector of the matrix $\vec{Y}_{BSC}^{+}\hat{L} = L_c \vec{Y}_{BSC}^{+}$. This eigenvector becomes a true BSC when $L_c = 0$. Here the amplitudes $B_{m=\pm 1}$ are the responses with OAM $m = \pm 1$ to the incident wave with linear polarization in the vicinity of the BSC point as dependent on the sign of κ. In particular the case $\kappa > 0$ is shown in Figure 23a which demonstrates a resonant enhancement in the vicinity of the BSC frequency $k_{0c} = 4.327$. What is more important Figure 23 shows that the amplitudes $B_{m=\pm 1}$ are substantially different for a plane wave with oblique incidence. The change of the sign of κ interchanges priority of $B_{m=\pm1}$ that in turn changes the direction of spiralling.

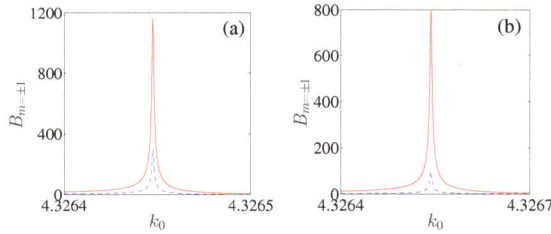

Figure 23. The enhancement factor $B_{m=1}$ (dash line) and $B_{m=-1}$ (solid line) vs. frequency for the parameters listed in Figure 1 and $\beta = \beta_c + 0.0062$ and (**a**) $\kappa = 1$ and (**b**) $\kappa = 0.5$. In the case $\kappa = -1, -0.5$ dash and solid lines are interchanged.

The spiralling currents near by the array become giant because of enhancement of the EM fields for the frequency close to the BSC frequency as shown in Figure 24. This Figure is complemented by Figure 25a,b with iso-surfaces of absolute value and the azimuthal component in Figure 25b of the Poynting vector which demonstrates the enhancement.

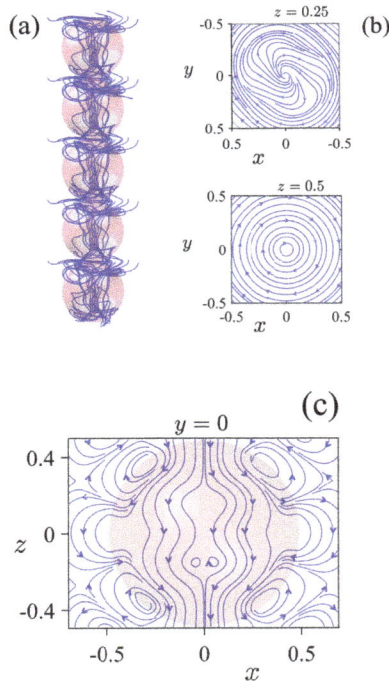

Figure 24. (**a**) Streamlines of Pointing vector; (**b**) currents in the x, y plane at selected slices along the array axis $z = 0.25$ (inside the sphere) and $z = 0.5$ (between spheres); (**c**) currents in the x, z plane at selected slices $y = 0$.

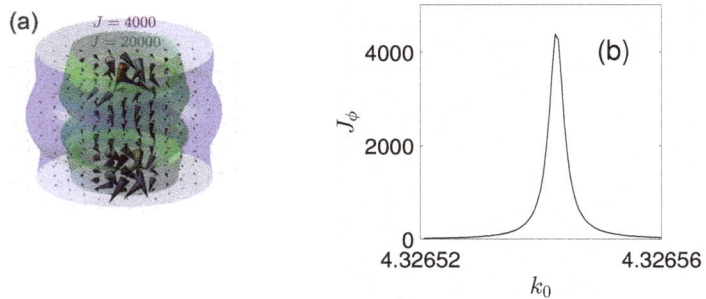

Figure 25. (a) Iso surfaces of Poynting vector at two selected values and (b) value of azimuthal component of the Poynting vector normalized to that of the incident wave around the spheres at distance $r = 0.588$ from the center of sphere and $z = 0$.

The next unique property of the BSC is related to the Fano resonance collapse [78,97] that reflects in resonant features of the cross sections of the array in the vicinity of the BSC frequency. Similar to the BSCs with $\beta = 0$ and $m \neq 0$ which were presented in Ref. [58] the Bloch BSC with OAM demonstrate resonant features shown in Figure 26.

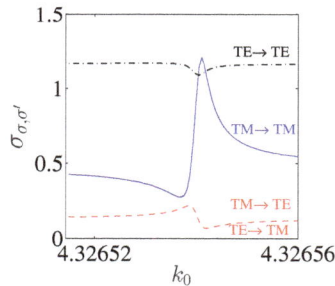

Figure 26. Total cross-section for scattering of plane wave by the array in the vicinity of the Bloch BSC with OAM at $\beta = \beta_c + 0.0062$.

11. Array with the Finite Number of Dielectric Spheres

For finite number of the spheres the translational invariance is broken. Then Equation (2) can be modified as follows [24]

$$\mathbf{E}(\mathbf{r}) = \sum_{j=1}^{N} \sum_{lm} [a_j^{lm} \mathbf{M}_l^m(\mathbf{r} - \mathbf{R}_j) + b_j^{lm} \mathbf{N}_l^m(\mathbf{r} - \mathbf{R}_j)],$$

$$\mathbf{H}(\mathbf{r}) = -i \sum_{j=1}^{N} \sum_{lm} [a_j^{lm} \mathbf{N}_l^m(\mathbf{r} - \mathbf{R}_j) + b_j^{lm} \mathbf{M}_l^m(\mathbf{r} - \mathbf{R}_j)]. \tag{61}$$

The expansion coefficients a_j^{lm}, b_j^{lm} were found numerically [11] and presented in Figure 27.

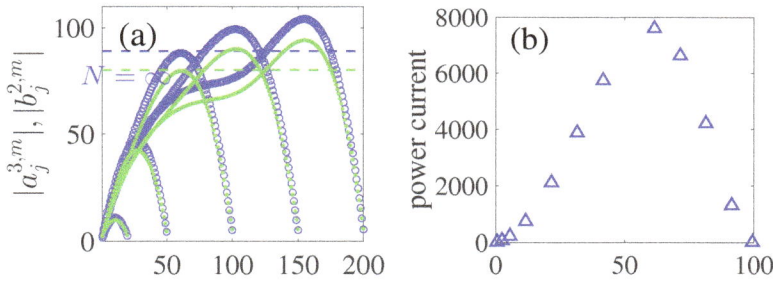

Figure 27. (a) Values of coefficients a_j^{lm} (open blue circles) and b_j^{lm} (closed green circles) in Equation (61) and (b) values of power current at $z = n + 0.5, r = 0.25$ where the current is maximal for different number of spheres with $R = 0.468$ for $k_0 = 3.10705, k_z = 0.0307, \alpha = -0.31 + 0.95i$.

At the first sight it seems that for growing N the solution for amplitudes a_j and b_j should saturate except in the vicinity of the edges of the finite array. However the EM field is a massless field which has no characteristic scale. Hence we have the behavior of the amplitudes as shown in Figure 27. Nevertheless we can see a tendency for saturation of the amplitudes to the maximal value with the growth of N, but we always observe non-negligible effect of the edges of the finite array. That affects the transfer of SAM of incident light into the giant vortical currents in the vicinity of a quasi-BSC. A similar effect which transforms the BSCs into quasi-BSCs is the volatility of the material parameters of the spheres.

12. Summary and Discussion

Recently the BSCs above the light line were shown to exist in various systems of one-dimensional arrays of dielectric rods and holes in a dielectric slab [31–33,37,38,44–46]. Similar acoustic BSCs called embedded trapped Rayleigh-Bloch surface waves were obtained in system of material rods [26,29,47,98]. One could ask why BSCs occur in periodic dielectric structures (gratings) but not in homogeneous structures like a slab or a rod which can support guided EM modes below the light line only. Let us begin with the simplest textbook system of a dielectric slab infinitely long in the x, y plane with the dielectric constant $\epsilon > 1$. The Maxwell equations can be solved by separation of variables for scalar function $\psi(x, y, z) = e^{ik_x x + ik_y y} \psi(z)$ to result in bound states below the light line $k_0^2 = k_x^2 + k_y^2$ [99] while all solutions above the light line are leaky [100]. The situation can be cardinally changed by replacing the continual translational symmetry by the discrete symmetry $\epsilon(x, y, z) = \epsilon(x + p, y, z)$ where $p = 0, \pm 1, \pm 2, \ldots$. All space variables are measured in the period length.. Then the radiation continua of plane waves $e^{ik_{x,n} x + ik_y y + ik_z z}$ are quantized $k_{x,n} = \beta + 2\pi n$, $n = 0, \pm 1, \pm 2, \ldots$ with the frequency $k_0^2 = k_{x,n}^2 + k_y^2 + k_z^2$. Here β is the Bloch wave vector along the x-axis, and the integer n refers to the diffraction continua [33]. The physical interpretation of this statement is related to that the slab with the discrete translational symmetry can be considered as a one-dimensional diffraction lattice in the x-direction. Let us take for simplicity $\beta = 0, k_y = 0$. Assume there is a bound solution with the eigenfrequency $k_{0,BSC} > 0$ which is coupled with all diffraction continua enumerated by n. Let $k_{0,BSC} < 2\pi$, i.e., the BSC resides in the first diffraction continua but below the others. Because of the symmetry or by variation of the material parameters of the modulated slab we can achieve that the coupling of the solution with the first diffraction continuum equals zero [37,38,44–46]. However the solution is coupled with evanescent continua $n = 1, 2, \ldots$ giving rise to exponential decay of the bound solution over the z-axis. The length of localization is given by $L \sim \dfrac{1}{\sqrt{4\pi^2 - k_{0,BSC}^2}}$. Therefore, the evanescent diffraction continua play a principal role in the space configuration of the BSCs. Moreover, one can see from Figure 5 that in the limit $h \to \infty$ the BSCs frequency $k_{0BSC} \to 0$ to leave no room for the BSCs with $k_{0,BSC} > 0$.

In the present paper we chose another strategy to quantize the radiation continuum. We replace the rod with continual translational symmetry by a periodic array of dielectric spheres. Because of the axial symmetry of the array aligned along the z-axis the quantized continua are specified by two integers, m and n. The first integer is the azimuthal quantum number and the second number defines discrete directions of outgoing cylindrical waves (17) given by the wave vector $k_{z,n} = \beta + 2\pi n$ in each sector m where β is the Bloch vector along the array. Bottoms of the particular continua with $m = 0$ and $n = 0, \pm 1$ and $n = -2$ are shown in Figure 5. By arguments similar to those presented above for the grated slab we obtain that the BSC with $\beta = 0$ embedded into the first radiation continuum $m = 0, n = 0$ is localized around the array.

The symmetry of the system is also important for classifications of the BSCs which are labelled by the azimuthal number of the continuum m of cylindrical vectorial waves and the Bloch wave vector β. The symmetry properties of the BSC play a very important role since it is difficult to provide a zero coupling even with the lowest continua $n = 0$ because of the degeneracy in polarization. Nevertheless the symmetry allows to decouple the BSC at least with some particular continua.

(1) The symmetry protected BSCs constitute the vast majority of BSCs which are symmetrically mismatched with the first diffraction continuum $m = 0, n = 0$ of both polarizations. The EM field configurations of such BSCs presented in Figure 2 show hybridizations of a few orbital numbers $l = 2, 4, 6, \ldots$ which specify the BSCs as multipoles of high order. Therefore the BSC solutions can not be obtained by the use of the dipole approximation [20,21]. The most remarkable property from experimental viewpoint is the robustness of the BSCs relative to choice of the material parameters of the dielectric spheres. We present in Figure 3 an example of the BSC which is symmetry protected relative to the TM diffraction continuum but a zero coupling to the TE continuum obtained through variation of the sphere radius.
(2) We demonstrated that the BSC can be accessed not only by variation of the material parameters but also by variation of Bloch wave vector β along the array axis. Patterns of the Bloch BSCs are presented in Figure 3.
(3) By tuning of the radius of the spheres we found BSCs in the next sectors of continua with $m \neq 0$. These BSCs shown in Figure 4 are remarkable by that they carry the OAM with spinning Poynting vectors.
(4) The most sophisticated Bloch BSCs with OAM demonstrate spiralling Poynting vector as shown in Figures 24 and 25.

The advantage of dielectric structures is a high quality factor and a wide range of BSC wavelengths from microns (photonics) to centimeter (microwave) as dependent on the choice of the radius of the spheres. Although the BSCs exist only in selected points in the parametric space there is a nearest vicinity of the BSC point where the BSC predominantly contributes into the cross-section and the EM field in the near field zone as seen from Figures 10 and 11. That leads to extremely efficient light harvesting capabilities [83]. The far zone EM fields can also show abundant features related to the BSCs. In particular Figure 11a demonstrates the effect of antenna when the BSC with azimuthal number $m = 2$ converts the EM energy into the perpendicular directions.

The BSCs with OAM $m = \pm 1$ emerge in the response of the array to incident plane waves with circular polarization. A transfer of the SAM of the incident plane wave into the OAM of EM field takes place for any frequency and wave vector of the incident wave as shown in Figure 13. The transfer results in the power current spinning around the array. The most remarkable is that as seen from Figures 13 and 16 in the nearest vicinity of the BSCs with $m \neq 0$ the array supports giant vortical power currents which are directly related to the extremal enhancement of the scattered field. The value of the current is also sensitive to the distance from the array. It rapidly goes down away from spheres as shown in Figure 19.

Theoretically the value of the circulating currents can grow up to infinity in the BSC point. However there is a difference between the present theory and possible experimental realization of the transfer of SAM into OAM, that is (1) a finite number of the spheres and (2) there are always some losses when the waves transport through the sample because of material for spheres. The most profound effect of finite arrays is that the BSCs become quasi-BSCs because, unlike a plasmonic sphere, finite dielectric systems can not support BSCs [101–103]. Therefore the effect of giant vortical currents around the array can be suppressed. Indeed, as our calculations show in Figure 27 for finite number N the currents decay for approaching to ends of the array. However with since $N \geq 100$ at middle of the array the BSC is restoring.

The next problem which can seriously damage the effect of giant spinning currents is the complex dielectric permittivity $\epsilon = \epsilon' + i\epsilon''$. Fortunately, for silicon dielectric particles there is a wide frequency window in the nearest infrared range where the ϵ'' is extremely small [75]. The advantage of dielectric structures is a wide range of BSC wavelengths from microns (photonics) to centimeter (microwave range) as dependent on the choice of the radius of spheres. Losses when the waves transport through the array result in the finite free path length $L = v_g/\epsilon''\omega$ where v_g is the group velocity. Therefore it is sufficient to take the number of spheres not exceeding L/h where h is the period of the array. This problem was considered in details in Ref. [57].

Moreover we have shown the propagating Bloch BSC with both $\beta \neq 0$ and OAM $m = \pm 1$ for the array of dielectric particles with the permittivity around 10. The Bloch vector β_c can be tuned by variation of the permittivity. Although we revealed only the value of OAM $m = \pm 1$ in general there is no restriction for the Bloch BSCs with OAM $|m| > 1$. To the best of our knowledge, the BSCs with OAM guided along the array has not been previously advanced as a possible fundamental effect for device applications. One of the most important application we consider that the array is capable for lasing through the BSC as it was demonstrated in Refs. [104,105]. However the principal feature of lasing by the array of dielectric spheres is that the laser beam carries the OAM without use of special chiral symmetry broken media [86,90,91].

Acknowledgments: The work was supported by Russian Science Foundation through grant 14-12-00266.

Conflicts of Interest: The authors declare no conflict of interest.

Appendix A

The value $\mathcal{B}_{l\nu}^{00}$ is expressed via

$$\mathcal{H}(l,0,\nu,0,p) = \mathcal{G}_+ + \mathcal{G}_- \tag{A1}$$

for $l + \nu + p$ odd according to Equations (10)–(60) where

$$\mathcal{G}_\pm = \mp\sqrt{\nu(\nu+1)p(p-1)}\mathcal{G}(l,0,\nu,\pm1,p-1)$$

$$\mathcal{G}(l,0,\nu,\pm1,p-1) = -\sqrt{(2l+1)(2\nu+1)(2p-1)}\begin{pmatrix} l & \nu & p-1 \\ 0 & \pm1 & \mp1 \end{pmatrix}\begin{pmatrix} l & \nu & p-1 \\ 0 & 0 & 0 \end{pmatrix} \tag{A2}$$

according to Equation (8). Using the property of 3j-symbols

$$\begin{pmatrix} j_1 & j_2 & j_3 \\ m_1 & m_2 & m_3 \end{pmatrix} = (-1)^{j_1+j_2+j_3}\begin{pmatrix} j_1 & j_2 & j_3 \\ -m_1 & -m_2 & -m_3 \end{pmatrix} \tag{A3}$$

we obtain

$$\begin{pmatrix} l & \nu & p-1 \\ 0 & 1 & -1 \end{pmatrix} = \begin{pmatrix} l & \nu & p-1 \\ 0 & -1 & 1 \end{pmatrix} \tag{A4}$$

Appl. Sci. **2017**, 7, 147

if $l + v + p - 1$ is even. Therefore we have from Equations (A1) and (A2) that $\mathcal{H}(l, 0, v, 0, p) = 0$ and respectively, $\mathcal{B}_{lv}^{00} = 0$.

References

1. Stratton, J.A. *Electromagnetic Theory*; McGraw-Hill Book Company, Inc.: NewYork, NY, USA, 1941.
2. Ohtaka, K. Energy band of photons and low-energy photon diffraction. *Phys. Rev. B* **1979**, *19*, 5057–5067.
3. Ohtaka, K. Scattering theory of low-energy photon diffraction. *J. Phys. C Solid State Phys.* **1980**, *13*, 667–680.
4. Miyazaki, H.; Ohtaka, K. Near-field images of a monolayer of periodically arrayed dielectric spheres. *Phys. Rev. B* **1998**, *58*, 6920–6937.
5. Modinos, A. Scattering of electromagnetic waves by a plane of spheres-formalism. *Phys. A Stat. Mech. Its Appl.* **1987**, *141*, 575–588.
6. Bruning, J.; Lo, Y. Multiple scattering of EM waves by spheres part I—Multipole expansion and ray-optical solutions. *IEEE Trans. Antennas Propag.* **1971**, *19*, 378–390.
7. García de Abajo, F.J. Colloquium: Light scattering by particle and hole arrays. *Rev. Mod. Phys.* **2007**, *79*, 1267–1290.
8. Wang, K.X.; Yu, Z.; Sandhu, S.; Liu, V.; Fan, S. Condition for perfect antireflection by optical resonance at material interface. *Optica* **2014**, *1*, 388, doi:10.1364/OPTICA.1.000388.
9. Fuller, K.A.; Kattawar, G.W. Consummate solution to the problem of classical electromagnetic scattering by an ensemble of spheres I: Linear chains. *Opt. Lett.* **1988**, *13*, 90–92.
10. Hamid, A.K.; Ciric, I.; Hamid, M. Iterative solution of the scattering by an arbitrary configuration of conducting or dielectric spheres. *IEE Proc. H Microw. Antennas Propag.* **1991**, *138*, 565.
11. Xu, Y.L. Electromagnetic scattering by an aggregate of spheres. *Appl. Opt.* **1995**, *34*, 4573–4588.
12. Mackowski, D.W. Calculation of total cross sections of multiple-sphere clusters. *J. Opt. Soc. Am. A* **1994**, *11*, 2851–2861.
13. Quirantes, A.; Arroyo, F.; Quirantes-Ros, J. Multiple light scattering by spherical particle systems and its dependence on concentration: A T-matrix study. *J. Colloid Interface Sci.* **2001**, *240*, 7882.
14. Luan, P.G.; Chang, K.D. Transmission characteristics of finite periodic dielectric waveguides. *Opt. Express* **2006**, *14*, 3263–3272.
15. Zhao, R.; Zhai, T.; Wang, Z.; Liu, D. Guided resonances in periodic dielectric waveguides. *J. Lightwave Technol.* **2009**, *27*, 4544–4547.
16. Du, J.; Liu, S.; Lin, Z.; Zi, J.; Chui, S.T. Guiding electromagnetic energy below the diffraction limit with dielectric particle arrays. *Phys. Rev. A* **2009**, *79*, 205436, doi:10.1103/PhysRevA.79.051801.
17. Burin, A.L.; Cao, H.; Schatz, G.C.; Ratner, M.A. High-quality optical modes in low-dimensional arrays of nanoparticles: Application to random lasers. *J. Opt. Soc. Am. B* **2004**, *21*, 121–131.
18. Gozman, M.; Polishchuk, I.; Burin, A. Light propagation in linear arrays of spherical particles. *Phys. Lett. A* **2008**, *372*, 5250–5253.
19. Blaustein, G.S.; Gozman, M.I.; Samoylova, O.; Polishchuk, I.Y.; Burin, A.L. Guiding optical modes in chains of dielectric particles. *Opt. Express* **2007**, *15*, 17380–17391.
20. Draine, B.T.; Flatau, P.J. Discrete-dipole approximation for periodic targets: Theory and tests. *J. Opt. Soc. Am. A* **2008**, *25*, 2693–2703.
21. Savelev, R.S.; Slobozhanyuk, A.P.; Miroshnichenko, A.E.; Kivshar, Y.S.; Belov, P.A. Subwavelength waveguides composed of dielectric nanoparticles. *Phys. Rev. B* **2014**, *89*, 035435.
22. Savelev, R.S.; Filonov, D.S.; Petrov, M.I.; Krasnok, A.E.; Belov, P.A.; Kivshar, Y.S. Resonant transmission of light in chains of high-index dielectric particles. *Phys. Rev. B* **2015**, *92*, 155415.
23. Li, S.V.; Baranov, D.G.; Krasnok, A.E.; Belov, P.A. All-dielectric nanoantennas for unidirectional excitation of electromagnetic guided modes. *Appl. Phys. Lett.* **2015**, *107*, 171101.
24. Linton, C.; Zalipaev, V.; Thompson, I. Electromagnetic guided waves on linear arrays of spheres. *Wave Motion* **2013**, *50*, 29–40.
25. Bonnet-Bendhia, A.S.; Starling, F. Guided waves by electromagnetic gratings and non-uniqueness examples for the diffraction problem. *Math. Methods Appl. Sci.* **1994**, *17*, 305–338.
26. Porter, R.; Evans, D.V. Rayleigh-Bloch surface waves along periodic gratings and their connection with trapped modes in waveguides. *J. Fluid Mech.* **1999**, *386*, 233–258.

27. Evans, D.V.; Porter, R. Trapping and near-trapping by arrays of cylinders in waves. *J. Eng. Math.* **1999**, *35*, 149179.

28. Cohen, O.; Freedman, B.; Fleischer, J.W.; Segev, M.; Christodoulides, D.N. Grating-mediated waveguiding. *Phys. Rev. Lett.* **2004**, *93*, 103902, doi:10.1103/PhysRevLett.93.103902.

29. Porter, R.; Evans, D. Embedded Rayleigh-Bloch surface waves along periodic rectangular arrays. *Wave Motion* **2005**, *43*, 29–50.

30. Venakides, S.; Shipman, S.P. Resonance and bound states in photonic crystal slabs. *SIAM J. Appl. Math.* **2003**, *64*, 322–342.

31. Shipman, S.P.; Venakides, S. Resonant transmission near nonrobust periodic slab modes. *Phys. Rev. E* **2005**, *71*, 026611, doi:10.1103/PhysRevE.71.026611.

32. Marinica, D.C.; Borisov, A.G.; Shabanov, S.V. Bound states in the continuum in photonics. *Phys. Rev. Lett.* **2008**, *100*, 183902, doi:10.1103/PhysRevLett.100.183902.

33. Ndangali, R.F.; Shabanov, S.V. Electromagnetic bound states in the radiation continuum for periodic double arrays of subwavelength dielectric cylinders. *J. Math. Phys.* **2010**, *51*, 102901, doi:10.1063/1.3486358.

34. Hsueh, W.; Chen, C.; Chang, C. Bound states in the continuum in quasiperiodic systems. *Phys. Lett. A* **2010**, *374*, 4804–4807.

35. Chia, W.H.; Zhen, B.; Lee, J.; Chua, S.L.; Johnson, S.G.; Joannopoulos, J.D.; Soljačić, M. Observation of trapped light within the radiation continuum. *Nature* **2013**, *499*, 188–191.

36. Weimann, S.; Xu, Y.; Keil, R.; Miroshnichenko, A.E.; Tünnermann, A.; Nolte, S.; Sukhorukov, A.A.; Szameit, A.; Kivshar, Y.S. Compact Surface Fano States Embedded in the Continuum of Waveguide Arrays. *Phys. Rev. Lett.* **2013**, *111*, 240403, doi:10.1103/PhysRevLett.111.240403.

37. Bulgakov, E.N.; Sadreev, A.F. Bloch bound states in the radiation continuum in a periodic array of dielectric rods. *Phys. Rev. A* **2014**, *90*, 053801, doi:10.1103/PhysRevA.90.053801.

38. Hu, Z.; Lu, Y.Y. Standing waves on two-dimensional periodic dielectric waveguides. *J. Opt.* **2015**, *17*, 065601, doi:10.1088/2040-8978/17/6/065601.

39. Bykov, D.A.; Doskolovich, L.L. $\omega - k_x$ Fano line shape in photonic crystal slabs. *Phys. Rev. A* **2015**, *92*, 013845, doi:10.1103/PhysRevA.92.013845.

40. Song, M.; Yu, H.; Wang, C.; Yao, N.; Pu, M.; Luo, J.; Zhang, Z.; Luo, X. Sharp Fano resonance induced by a single layer of nanorods with perturbed periodicity. *Opt. Express* **2015**, *23*, 2895–2903.

41. Zou, C.L.; Cui, J.M.; Sun, F.W.; Xiong, X.; Zou, X.B.; Han, Z.F.; Guo, G.C. Guiding light through optical bound states in the continuum for ultrahigh-Qmicroresonators. *Laser Photonics Rev.* **2014**, *9*, 114–119.

42. Wang, Z.; Zhang, H.; Ni, L.; Hu, W.; Peng, C. Analytical Perspective of Interfering Resonances in High-Index-Contrast Periodic Photonic Structures. *IEEE J. Quantum Electron.* **2016**, *52*, 6100109, doi:10.1109/JQE.2016.2568763.

43. Li, L.; Yin, H. Bound States in the Continuum in double layer structures. *Sci. Rep.* **2016**, *6*, 26988, doi:10.1038/srep26988.

44. Hsu, C.W.; Zhen, B.; Chua, S.L.; Johnson, S.G.; Joannopoulos, J.D.; Soljačić, M. Bloch surface eigenstates within the radiation continuum. *Light Sci. Appl.* **2013**, *2*, e84, doi:10.1038/lsa.2013.40.

45. Zhen, B.; Hsu, C.W.; Lu, L.; Stone, A.D.; Soljačić, M. Topological Nature of Optical Bound States in the Continuum. *Phys. Rev. Lett.* **2014**, *113*, 257401, doi:10.1103/PhysRevLett.113.257401.

46. Yang, Y.; Peng, C.; Liang, Y.; Li, Z.; Noda, S. Analytical Perspective for Bound States in the Continuum in Photonic Crystal Slabs. *Phys. Rev. Lett.* **2014**, *113*, 037401, doi:10.1103/PhysRevLett.113.037401.

47. Colquitt, D.J.; Craster, R.V.; Antonakakis, T.; Guenneau, S. Rayleigh-Bloch waves along elastic diffraction gratings. *Proc. R. Soc. A Math. Phys. Eng. Sci.* **2014**, *471*, 20140465, doi:10.1098/rspa.2014.0465.

48. Gao, X.; Hsu, C.W.; Zhen, B.; Lin, X.; Joannopoulos, J.D.; Soljačić, M.; Chen, H. Formation Mechanism of Guided Resonances and Bound States in the Continuum in Photonic Crystal Slabs. *arXiv* **2016**, arXiv:1603.02815.

49. Hung, Y.J.; Lin, I.S. Visualization of Bloch surface waves and directional propagation effects on one-dimensional photonic crystal substrate. *Opt. Express* **2016**, *24*, 16003–16009.

50. Blanchard, C.; Hugonin, J.P.; Sauvan, C. Fano resonances in photonic crystal slabs near optical bound states in the continuum. *Phys. Rev. B* **2016**, *94*, 155303, doi:10.1103/PhysRevB.94.155303.

51. Wang, Y.; Song, J.; Dong, L.; Lu, M. Optical bound states in slotted high-contrast gratings. *J. Opt. Soc. Am. B* **2016**, *33*, 2472–2479.

52. Zhang, M.; Zhang, X. Ultrasensitive optical absorption in graphene based on bound states in the continuum. *Sci. Rep.* **2015**, *5*, 8266, doi:10.1038/srep08266.
53. Padgett, M.; Courtial, J.; Allen, L. Light's Orbital Angular Momentum. *Phys. Today* **2004**, *57*, 2004.
54. Allen, L.; Padgett, M. The Poynting vector in Laguerre-Gaussian beams and the interpretation of their angular momentum density. *Opt. Commun.* **2000**, *184*, 67–71.
55. Yao, A.M.; Padgett, M.J. Orbital angular momentum: Origins, behavior and applications. *Adv. Opt. Photon.* **2011**, *3*, 161, doi:10.1364/AOP.3.000161.
56. Bulgakov, E.N.; Sadreev, A.F. Light trapping above the light cone in a one-dimensional array of dielectric spheres. *Phys. Rev. A* **2015**, *92*, 023816, doi:10.1103/PhysRevA.92.023816.
57. Bulgakov, E.; Maksimov, D.N. Light guiding above the light line in arrays of dielectric nanospheres. *Opt. Lett.* **2016**, *41*, 3888–3891.
58. Bulgakov, E.N.; Sadreev, A.F. Transfer of spin angular momentum of an incident wave into orbital angular momentum of the bound states in the continuum in an array of dielectric spheres. *Phys. Rev. A* **2016**, *94*, 033856, doi:10.1103/PhysRevA.94.033856.
59. Bulgakov, E.N.; Sadreev, A.F. Propagating Bloch waves with orbital angular momentum above the light cone in the array of dielectric spheres. *J. Opt. Soc. Am. A* **2017**, submitted.
60. Hsu, C.W.; Zhen, B.; Stone, A.D.; Joannopoulos, J.D.; Soljačić, M. Bound states in the continuum. *Nat. Rev. Mater.* **2016**, *1*, 16048, doi:10.1038/natrevmats.2016.48.
61. Bykov, D.A.; Doskolovich, L.L. Numerical methods for calculating poles of the scattering matrix with applications in grating theory. *J. Lightwave Technol.* **2013**, *31*, 793–801.
62. Yuan, L.J.; Lu, Y.Y. Propagating Bloch modes above the light line on a periodic array of cylinders. *J. Phys. B* **2016**, doi:10.1088/1361-6455/aa5480.
63. Merchiers, O.; Moreno, F.; González, F.; Saiz, J.M. Light scattering by an ensemble of interacting dipolar particles with both electric and magnetic polarizabilities. *Phys. Rev. A* **2007**, *76*, 043834, doi:10.1103/PhysRevA.76.043834.
64. Evlyukhin, A.B.; Reinhardt, C.; Seidel, A.; Luk'yanchuk, B.S.; Chichkov, B.N. Optical response features of Si-nanoparticle arrays. *Phys. Rev. B* **2010**, *82*, 045404, doi:10.1103/PhysRevB.82.045404.
65. Wheeler, M.S.; Aitchison, J.S.; Mojahedi, M. Coupled magnetic dipole resonances in sub-wavelength dielectric particle clusters. *J. Opt. Soc. Am. B* **2010**, *27*, 1083–1091.
66. Shore, R.A.; Yaghjian, A.D. *Traveling Electromagnetic Waves on Linear Periodic Arrays of Small Lossless Penetrable Spheres*; Technical Report, DTIC Document; Defense Technical Information Center: Fort Belvoir, VA, USA, 2004.
67. Shore, R.A.; Yaghjian, A.D. Travelling electromagnetic waves on linear periodic arrays of lossless spheres. *Electron. Lett.* **2005**, *41*, 578–580.
68. Shore, R.A.; Yaghjian, A.D. Complex waves on periodic arrays of lossy and lossless permeable spheres: 1. Theory. *Radio Sci.* **2012**, *47*, RS2014, doi:10.1029/2011RS004859.
69. Deych, L.; Roslyak, A. Long-living collective optical excitations in a linear chain of microspheres. *Phys. Status Solidi C* **2005**, *2*, 3908–3911.
70. Deych, L.I.; Roslyak, O. Photonic band mixing in linear chains of optically coupled microspheres. *Phys. Rev. E* **2006**, *73*, 036606, doi:10.1103/PhysRevE.73.036606.
71. Monticone, F.; Alù, A. Embedded Photonic Eigenvalues in 3D Nanostructures. *Phys. Rev. Lett.* **2014**, *112*, 213903, doi:10.1103/PhysRevLett.112.213903.
72. Tikhodeev, S.G.; Yablonskii, A.L.; Muljarov, E.A.; Gippius, N.A.; Ishihara, T. Quasiguided modes and optical properties of photonic crystal slabs. *Phys. Rev. B* **2002**, *66*, 045102.
73. Shore, R.A.; Yaghjian, A.D. Complex waves on periodic arrays of lossy and lossless permeable spheres: 2. Numerical results. *Radio Sci.* **2012**, *47*, RS2015, doi:10.1029/2011RS004860.
74. Carrasco, S.; Saleh, B.E.A.; Teich, M.C.; Fourkas, J.T. Second- and third-harmonic generation with vector Gaussian beams. *J. Opt. Soc. Am. B* **2006**, *23*, 2134–2141.
75. Vuye, G.; Fisson, S.; Nguyen Van, V.; Wang, Y.; Rivory, J.; Abelès, F. Temperature dependence of the dielectric function of silicon using in situ spectroscopic ellipsometry. *Thin Solid Films* **1993**, *233*, 166–170.
76. Bulgakov, E.N.; Rotter, I.; Sadreev, A.F. Comment on Bound-state eigenenergy outside and inside the continuum for unstable multilevel systems. *Phys. Rev. A* **2007**, *75*, 067401, doi:10.1103/PhysRevA.75.067401.

77. Bulgakov, E.N.; Pichugin, K.N.; Sadreev, A.F.; Rotter, I. Bound states in the continuum in open Aharonov-Bohm rings. *JETP Lett.* **2006**, *84*, 430–435.

78. Sadreev, A.F.; Bulgakov, E.N.; Rotter, I. Bound states in the continuum in open quantum billiards with a variable shape. *Phys. Rev. B* **2006**, *73*, 235342, doi:10.1103/PhysRevB.73.235342.

79. Yoon, J.W.; Song, S.H.; Magnusson, R. Critical field enhancement of asymptotic optical bound states in the continuum. *Sci. Rep.* **2015**, *5*, 18301, doi:10.1038/srep18301.

80. Sadreev, A.F.; Rotter, I. S-matrix theory for transmission through billiards in tight-binding approach. *J. Phys. A Math. Gen.* **2003**, *36*, 11433.

81. Bulgakov, E.N.; Sadreev, A.F. Robust bound state in the continuum in a nonlinear microcavity embedded in a photonic crystal waveguide. *Opt. Lett.* **2014**, *39*, 5212–5215.

82. Ochiai, T.; Sakoda, K. Dispersion relation and optical transmittance of a hexagonal photonic crystal slab. *Phys. Rev. B* **2001**, *63*, 125107.

83. Fernández-Domínguez, A.I.; Maier, S.A.; Pendry, J.B. Collection and Concentration of Light by Touching Spheres: A Transformation Optics Approach. *Phys. Rev. Lett.* **2010**, *105*, 266807, doi:10.1103/PhysRevLett.105.266807.

84. Allen, L.; Beijersbergen, M.W.; Spreeuw, R.J.C.; Woerdman, J.P. Orbital angular momentum of light and the transformation of Laguerre-Gaussian laser modes. *Phys. Rev. A* **1992**, *45*, 8185–8189.

85. Čelechovský, R.; Bouchal, Z. Optical implementation of the vortex information channel. *New J. Phys.* **2007**, *9*, 328, doi:10.1088/1367-2630/9/9/328.

86. Gorodetski, Y.; Drezet, A.; Genet, C.; Ebbesen, T.W. Generating Far-Field Orbital Angular Momenta from Near-Field Optical Chirality. *Phys. Rev. Lett.* **2013**, *110*, 203906, doi:10.1103/PhysRevLett.110.203906.

87. Yu, H.; Zhang, H.; Wang, Y.; Han, S.; Yang, H.; Xu, X.; Wang, Z.; Petrov, V.; Wang, J. Optical orbital angular momentum conservation during the transfer process from plasmonic vortex lens to light. *Sci. Rep.* **2013**, *3*, 3191, doi:10.1038/srep03191.

88. Berezin, M.; Kamenetskii, E.; Shavit, R. Magnetoelectric-field microwave antennas: Far-field orbital angular momenta from chiral-topology near fields. *arXiv* **2015**, arXiv:1512.01393.

89. Žukauskas, A.; Malinauskas, M.; Brasselet, E. Monolithic generators of pseudo-nondiffracting optical vortex beams at the microscale. *Appl. Phys. Lett.* **2013**, *103*, 181122, doi:10.1063/1.4828662.

90. Dall, R.; Fraser, M.D.; Desyatnikov, A.S.; Li, G.; Brodbeck, S.; Kamp, M.; Schneider, C.; Höfling, S.; Ostrovskaya, E.A. Creation of Orbital Angular Momentum States with Chiral Polaritonic Lenses. *Phys. Rev. Lett.* **2014**, *113*, 200404, doi:10.1103/PhysRevLett.113.200404.

91. Yu, N.; Capasso, F. Flat optics with designer metasurfaces. *Nat. Mater.* **2014**, *13*, 139–150.

92. Schäferling, M.; Yin, X.; Giessen, H. Formation of chiral fields in a symmetric environment. *Opt. Express* **2012**, *20*, 26326–26336.

93. Rodriguez-Fortuño, F.J.; Barber-Sanz, I.; Puerto, D.; Griol, A.; Martínez, A. Resolving Light Handedness with an on-Chip Silicon Microdisk. *ACS Photonics* **2014**, *1*, 762–767.

94. Petersen, J.; Volz, J.; Rauschenbeutel, A. Chiral nanophotonic waveguide interface based on spin-orbit interaction of light. *Science* **2014**, *346*, 67–71.

95. Rodriguez-Fortuño, F.J.; Marino, G.; Ginzburg, P.; O'Connor, D.; Martinez, A.; Wurtz, G.A.; Zayats, A.V. Near-Field Interference for the Unidirectional Excitation of Electromagnetic Guided Modes. *Science* **2013**, *340*, 328–330.

96. Bulgakov, E.N.; Sadreev, A.F. Giant optical vortex in photonic crystal waveguide with nonlinear optical cavity. *Phys. Rev. B* **2012**, *85*, 165305, doi:10.1103/PhysRevB.85.165305.

97. Kim, C.S.; Satanin, A.M.; Joe, Y.S.; Cosby, R.M. Resonant tunneling in a quantum waveguide: Effect of a finite-size attractive impurity. *Phys. Rev. B* **1999**, *60*, 10962–10970.

98. Linton, C.; McIver, P. Embedded trapped modes in water waves and acoustics. *Wave Motion* **2007**, *45*, 16–29.

99. Jackson, J.D. *Classical Electrodynamic*; Wiley: Hoboken, NJ, USA, 1999.

100. Hu, J.; Menyuk, C.R. Understanding leaky modes: Slab waveguide revisited. *Adv. Opt. Photon.* **2009**, *1*, 58–106.

101. Silveirinha, M.G. Trapping light in open plasmonic nanostructures. *Phys. Rev. A* **2014**, *89*, 023813, doi:10.1103/PhysRevA.89.023813.

102. Alù, A.; Silveirinha, M.G.; Salandrino, A.; Engheta, N. Epsilon-near-zero metamaterials and electromagnetic sources: Tailoring the radiation phase pattern. *Phys. Rev. B* **2007**, *75*, 155410, doi:10.1103/PhysRevB.75.155410.

103. Hrebikova, I.; Jelinek, L.; Silveirinha, M.G. Embedded energy state in an open semiconductor heterostructure. *Phys. Rev. B* **2015**, *92*, 155303, doi:10.1103/PhysRevB.92.155303.

104. Zhang, Z.; Li, Y.; Liu, W.; Yang, J.; Ma, Y.; Lu, H.; Sun, Y.; Jiang, H.; Chen, H. Controllable lasing behavior enabled by compound dielectric waveguide grating structures. *Opt. Express* **2016**, *24*, 19458–19466.

105. Kodigala, A.; Lepetit, T.; Gu, Q.; Bahari, B.; Fainman, Y.; Kante, B. Bound State in the Continuum Nanophotonic Laser. *Conf. Lasers Electro-Opt.* **2016**, 1–2.

![applied sciences logo] *applied sciences*

MDPI

Article

Existence, Stability and Dynamics of Nonlinear Modes in a 2D Partially $\mathcal{PT}$ Symmetric Potential

Jennie D'Ambroise [1,*] and Panayotis G. Kevrekidis [2]

[1] Department of Mathematics, Computer & Information Science, State University of New York (SUNY) College at Old Westbury, Westbury, NY 11568, USA

[2] Department of Mathematics and Statistics, University of Massachusetts, Amherst, MA 01003, USA; kevrekid@math.umass.edu

* Correspondence: dambroisej@oldwestbury.edu; Tel.: +1-516-628-5640

Academic Editors: Boris Malomed and Paolo Minzioni
Received: 3 January 2017; Accepted: 21 February 2017; Published: 27 February 2017

Abstract: It is known that multidimensional complex potentials obeying parity-time ($\mathcal{PT}$) symmetry may possess all real spectra and continuous families of solitons. Recently, it was shown that for multi-dimensional systems, these features can persist when the parity symmetry condition is relaxed so that the potential is invariant under reflection in only a single spatial direction. We examine the existence, stability and dynamical properties of localized modes within the cubic nonlinear Schrödinger equation in such a scenario of partially $\mathcal{PT}$-symmetric potential.

Keywords: nonlinear optics; solitons; $\mathcal{PT}$-symmetry

1. Introduction

The study of $\mathcal{PT}$ (parity–time) symmetric systems was initiated through the works of Bender and collaborators [1,2]. Originally, it was proposed as an alternative to the standard quantum theory, where the Hamiltonian is postulated to be Hermitian. In these works, it was instead found that Hamiltonians invariant under $\mathcal{PT}$-symmetry, which are not necessarily Hermitian, may still give rise to completely real spectra. Thus, the proposal of Bender and co-authors was that these Hamiltonians are appropriate for the description of physical settings. In the important case of Schrödinger-type Hamiltonians, which include the usual kinetic-energy operator and the potential term, $V(x)$, the $\mathcal{PT}$-invariance is consonant with complex potentials, subject to the constraint that $V^*(x) = V(-x)$.

A decade later, it was realized (and since then it has led to a decade of particularly fruitful research efforts) that this idea can find fertile ground for its experimental realization although not in quantum mechanics where it was originally conceived. In this vein, numerous experimental realizations sprang up in the areas of linear and nonlinear optics [3–8], electronic circuits [9–11], and mechanical systems [12], among others. Very recently, this now mature field of research has been summarized in two comprehensive reviews [7,8].

One of the particularly relevant playgrounds for the exploration of the implications of $\mathcal{PT}$-symmetry is that of nonlinear optics, especially because it can controllably involve the interplay of $\mathcal{PT}$-symmetry and nonlinearity. The relevant efforts have gradually progressed from the simpler setting of coupled waveguides (bearing gain and loss) to entire $\mathcal{PT}$-symmetric lattices and the identification of optical solitons in them. These developments hold considerable promise for the potential realization of more complex settings, such as the one proposed herein. In this context, the propagation of light (in systems such as optical fibers or waveguides [7,8]) is modeled by the nonlinear Schrödinger equation of the form:

$$i\Psi_z + \Psi_{xx} + \Psi_{yy} + U(x,y)\Psi + \sigma|\Psi|^2\Psi = 0. \tag{1}$$

In the optics notation that we use here, the evolution direction is denoted by z, the propagation distance. Here, we restrict our considerations to two spatial dimensions and assume that the potential $U(x,y)$ is complex valued, representing gain and loss in the optical medium, depending on the sign of the imaginary part (negative for gain, positive for loss) of the potential. In this two-dimensional setting, the condition of full $\mathcal{PT}$-symmetry in two dimensions is that $U^*(x,y) = U(-x,-y)$. Potentials with full $\mathcal{PT}$ symmetry have been shown to support continuous families of soliton solutions [13–18]. However, an important recent development was the fact that the condition of (full) $\mathcal{PT}$ symmetry can be relaxed. That is, either the condition $U^*(x,y) = U(-x,y)$ or $U^*(x,y) = U(x,-y)$ of, so-called, partial $\mathcal{PT}$ symmetry can be imposed, yet the system will still maintain all real spectra and continuous families of soliton solutions [19]. Other subsequent results include stable vortex structures in cubic nonlinear media under the influence of partially $\mathcal{PT}$-symmetric azimuthal potentials [20].

In the original contribution of [19], only the focusing nonlinearity case was considered for two select branches of solutions and the stability of these branches was presented for isolated parametric cases (of the frequency parameter of the solution). Our aim in the present work is to provide a considerably more "spherical" perspective of the problem. The corresponding physical scenario that we have in mind involves a medium featuring a cubic nonlinearity. Given that the potential of [19] was principally featuring four "nodes", we are envisioning an implementation involving a four waveguide system. This is both in the spirit of the pioneering (two waveguide) work of [3] and also in that of subsequent proposals for trimers and quadrimers (i.e., three and four waveguide systems, respectively) [21,22]. A more remote possibility is that of a continuous cubic nonlinearity medium bearing a complex index of refraction in accordance with the prescription of [19], although, to the best of our understanding, such a possibility appears less tractable on the basis of current experimental capabilities.

In what follows, we examine the bifurcation of nonlinear modes from *all three* point spectrum eigenvalues of the underlying linear Schrödinger operator of the partially $\mathcal{PT}$-symmetric potential. Upon presenting the relevant model (Section 2), we perform the relevant continuations (Section 3) unveiling the existence of nonlinear branches *both* for the focusing and for the defocusing nonlinearity case. We also provide a systematic view towards the stability of the relevant modes (Section 4), by characterizing their principal unstable eigenvalues as a function of the intrinsic frequency parameter of the solution. In Section 5, we complement our existence and stability analysis by virtue of direct numerical simulations that manifest the result of the solutions' dynamical instability when they are found to be unstable. Finally, in Section 6, we summarize our findings and present our conclusions, as well as discussing some possibilities for future studies.

2. Model, Theoretical Setup and Linear Limit

Motivated by the partially $\mathcal{PT}$-symmetric setting of [19], we consider the complex potential $U(x,y) = V(x,y) + iW(x,y)$ where

$$
\begin{aligned}
V &= \left(ae^{-(y-y_0)^2} + be^{-(y+y_0)^2}\right)\left(e^{-(x-x_0)^2} + e^{-(x+x_0)^2}\right) \\
W &= \beta\left(ce^{-(y-y_0)^2} + de^{-(y+y_0)^2}\right)\left(e^{-(x-x_0)^2} - e^{-(x+x_0)^2}\right).
\end{aligned}
\tag{2}
$$

with real constants β, $a \neq b$ and $c \neq -d$. The potential is chosen with partial $\mathcal{PT}$-symmetry so that $U^*(x,y) = U(-x,y)$. That is, the real part is even in the x-direction with $V(x,y) = V(-x,y)$ and the imaginary part is odd in the x-direction with $-W(x,y) = W(-x,y)$. The constants a,b,c,d are chosen such that there is no symmetry in the y direction.

In [19], it is shown that the spectrum of the potential U can be all real as long as $|\beta|$ is below a threshold value, after which a ($\mathcal{PT}$-) phase transition occurs; this is a standard property of $\mathcal{PT}$-symmetric potentials. For the case of $a = 3, b = c = 2, d = 1$, the spectrum is real below the threshold value of $|\beta| \approx 0.214$; we focus on $\beta = 0.1$, i.e., we operate well below this critical point. Figure 1 shows plots of the potential U. The real part of the potential is shown on the left, while the imaginary part associated with gain-loss is on the right; the gain part of the potential corresponds to $W < 0$ and occurs for $x < 0$, while the loss part with $W > 0$ occurs for $x > 0$. Figure 2 shows the spectrum of U, i.e., eigenvalues for the underlying linear Schrödinger problem $(\nabla^2 + U)\psi_0 = \mu_0 \psi_0$. The figure also shows the magnitude of the corresponding eigenvectors for the three discrete real eigenvalues μ_0. It is from these modes that we will seek bifurcations of nonlinear solutions in what follows. It is worthwhile to mention here, in comparison, e.g., with the real four-well potential of [23] that the latter possessed four localized modes, with the fourth antisymmetric, quadrupolar one being absent from the point spectrum in the case of interest herein.

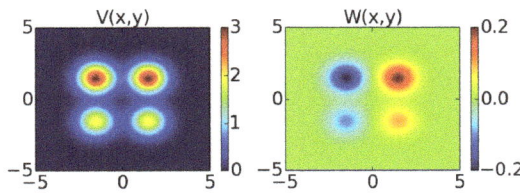

Figure 1. The plots show the spatial distribution of real (V, left panel) and imaginary (W, right panel) parts of the potential U with $x_0 = y_0 = 1.5$.

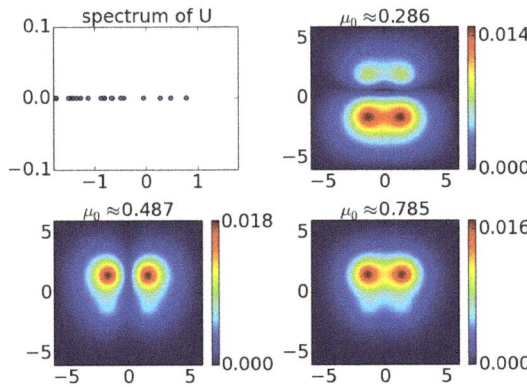

Figure 2. The top left plot shows the spectrum of Schrödinger operator associated with the potential U in the complex plane (see also the text). Plots of the magnitude of the normalized eigenvectors for the three discrete eigenvalues μ_0 are shown in the other three plots.

3. Existence: Nonlinear Modes Bifurcating from the Linear Limit

As is customary, we focus on stationary soliton solutions of (1) of the form $\Psi(x, y, z) = \psi(x, y)e^{i\mu z}$. Thus, one obtains the following stationary equation for $\psi(x, y)$.

$$\psi_{xx} + \psi_{yy} + U(x, y)\psi + \sigma|\psi|^2\psi = \mu\psi \tag{3}$$

In [19], it is discussed that a continuous family of solitons bifurcates from each of the linear solutions in the presence of nonlinearity. In order to see this, let μ_0 be a discrete simple real eigenvalue of the potential U (such as one of the positive real eigenvalues in the top left plot of Figure 2). Now, following [19], expand $\psi(x, y)$ in terms of $\epsilon = |\mu - \mu_0| << 1$ and substitute the expression

$$\psi(x, y) = \epsilon^{1/2} \left[c_0 \psi_0 + \epsilon \psi_1 + \epsilon^2 \psi_2 + \dots \right] \tag{4}$$

into Equation (3). This gives the equation for ψ_1 as

$$L \psi_1 = c_0 \left(\rho \psi_0 - \sigma |c_0|^2 |\psi_0|^2 \psi_0 \right) \tag{5}$$

where $\rho = \mathrm{sgn}(\mu - \mu_0)$ and

$$|c_0|^2 = \frac{\rho \langle \psi_0^*, \psi_0 \rangle}{\sigma \langle \psi_0^*, |\psi_0|^2 \psi_0 \rangle}. \tag{6}$$

Here, ψ_0^* plays the role of the adjoint solution to ψ_0.

Thus, in order to find solutions of (3) for $\sigma = \pm 1$, we perform a Newton continuation in the parameter μ where the initial guess for ψ is given by the first two terms of (4). The bottom left panel of Figure 3 shows how the (optical) power $P(\mu) = \int \int |\psi|^2 dx dy$ of the solution grows as a function of increasing μ for $\sigma = 1$, or as a function of decreasing μ for $\sigma = -1$ (from the linear limit). The first branch begins at the first real eigenvalue of U at $\mu_0 \approx 0.286$, the second branch at $\mu_0 \approx 0.487$, and the third branch begins at $\mu_0 \approx 0.785$. Plots of the solutions and their corresponding time evolution and stability properties are shown in the next section.

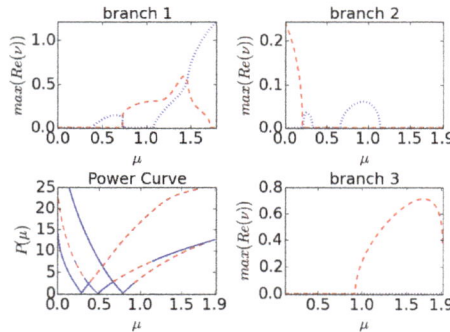

Figure 3. The bottom left plot shows the power of the solution ψ plotted in terms of the continuation parameter μ. The curves begin at the lowest power (i.e., at the linear limit) at the discrete real eigenvalues of approximately 0.286 (branch 1), 0.487 (branch 2), 0.785 (branch 3). Each power curve is drawn with its corresponding stability noted: a blue solid curve denotes a stable solution and a red dashed curve denotes an unstable solution. The other three plots track the maximum real part of eigenvalues ν as a function of the continuation parameter μ: the red dashed line represents the max real part of eigenvalues that are real (exponential instability) while the blue dotted line tracks the max real part for eigenvalues that have a nonzero imaginary part (quartets); this case corresponds to oscillatory instabilities.

As a general starting point comment for the properties of the branches, we point out that all the branches populate both the gain and the loss side. In the branch starting from $\mu_0 \approx 0.286$, all four "wells" of the potential of Figure 1 appear to be populated, with the lower intensity "nodes" being more populated and the higher intensity ones less populated. The second branch starting at $\mu_0 \approx 0.487$, as highlighted also in [19], possesses an anti-symmetric structure in x (hence the apparent vanishing of the density at the $x = 0$ line). Both in the second and in the third branch, the higher intensity nodes of the potential appear to bear a higher intensity.

4. Stability of the Nonlinear Modes: Spectral Analysis

The natural next step is to identify the stability of the solutions. This is monitored by using the linearization ansatz:

$$\Psi = e^{i\mu z}\left(\psi + \delta\left[a(x,y)e^{\nu z} + b^*(x,y)e^{\nu^* z}\right]\right) \tag{7}$$

which yields the order δ linear system

$$\begin{bmatrix} M_1 & M_2 \\ -M_2^* & -M_1^* \end{bmatrix}\begin{bmatrix} a \\ b \end{bmatrix} = -i\nu\begin{bmatrix} a \\ b \end{bmatrix} \tag{8}$$

where $M_1 = \nabla^2 + U - \mu + 2\sigma|\psi|^2$, $M_2 = \sigma\psi^2$. Thus, $\max(\mathrm{Re}(\nu)) > 0$ corresponds to instability and $\max(\mathrm{Re}(\nu)) = 0$ corresponds to (neutral) stability.

In the bottom left panel of Figure 3, the power curve is drawn with stability and instability as determined by ν noted by the solid or dashed curve, respectively. The other three plots in Figure 3 show the maximum real part of eigenvalues ν for each of the three branches: the red dashed curve is the max real part of real eigenvalue pairs, and the blue dotted curve is the max real part of eigenvalue quartets with a nonzero imaginary part. The former corresponds to exponential instabilities associated with pure growth, while the latter indicate so-called oscillatory instabilities, where growth is present concurrently with oscillations. In Figure 4, we plot some example eigenvalues in the complex plane for some sample unstable solutions. The dominant unstable eigenvalues within these can be seen to be consonant with the growth rates reported for the respective branches (and for these parameter values) in Figure 3.

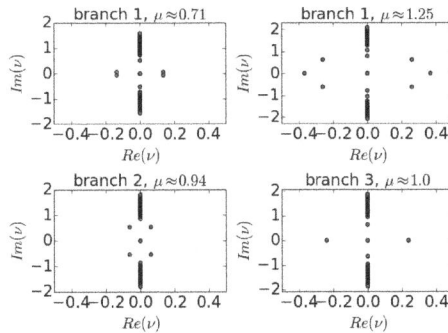

Figure 4. Eigenvalues are plotted in the complex plane $(Re(\nu), Im(\nu))$ for a few representative solutions. One can compare the maximal real part with Figure 3. For example, the top left complex plane plot here shows that for branch 1 at $\mu \approx 0.71$, the eigenvalues with the maximum real part are complex; this agrees with the top left plot of Figure 3 where at $\mu \approx 0.71$, the blue dotted curve representing complex eigenvalues is bigger. Similarly, one can check whether the other three eigenvalue plots here also agree with what is shown in Figure 3, the top right for branch 1, the bottom left for branch 2 and the bottom right for branch 3.

The overarching conclusions from this stability analysis are as follows. The lowest μ branch, being the ground state in the defocusing case, is always stable in the presence of the self-defocusing nonlinearity. For the parameters considered, generic stability is also prescribed for the third branch under self-defocusing nonlinearity. The middle branch has a narrow interval of stability and then becomes unstable, initially (as shown in the top right of Figure 3) via an oscillatory instability and then through an exponential one. In the focusing case (that was also focused on in [19] for the second and third branch), all three branches appear to be stable immediately upon their bifurcation from the linear limit, yet all three of them subsequently become unstable. Branch 1 (that was not analyzed previously) features a combination of oscillatory and exponential instabilities. Branch 2 features an oscillatory instability which, however, only arises for a finite interval of frequencies μ, and the branch restabilizes. On the other hand, for branch 3, when it becomes unstable, it is through a real pair. Branches 2 and 3 terminate in a saddle-center bifurcation near $\mu = 1.9$. The eigenvalue panels of Figure 4 confirm that the top panels of branch 1 may possess one or two concurrent types of instability (in the focusing case), branch 2 (bottom left) can only be oscillatorily unstable in the focusing case (yet as is shown in Figure 3, it can feature either type of instability in the defocusing case), while for branch 3, when it is unstable in the focusing case, it is via a real eigenvalue pair.

5. Dynamics of Unstable Solutions

Figures 5–7 show the time evolution of three unstable solutions, one on each branch, in the focusing case for the value of $\sigma = 1$. That is, they each correspond to a μ-value that is bigger than the initial discrete value μ_0 and pertain to the focusing case. The time evolution figures show a similar feature for the unstable solutions, namely that over time the magnitude of the solutions will increase on the left side of the spatial grid. This agrees with what is expected from $\mathcal{PT}$-symmetry since the left side of the spatial grid corresponds to the gain side of the potential U. Importantly, also, the nature of the instabilities varies from case to case, and is consonant with our stability expectations based on the results of the previous section.

In Figure 5, branch 1 (for the relevant value of the parameter μ) features an oscillatory instability (but with a small imaginary part). In line with this, we observe a growth that is principally exponential (cf. also the top panel for the power of the solution), yet features also some oscillation in the amplitude of the individual peaks. It should be noted here that although two peaks result in growth and two in decay (as expected by the nature of W in this case), one of them clearly dominates between the relevant amplitudes.

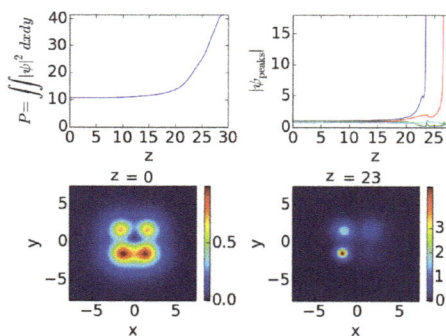

Figure 5. This figure shows the time evolution of the branch 1 solution for the value $\mu \approx 0.71$. The bottom left plot shows the magnitude of the solution $|\Psi|$ at $z = 0$. Observe that this solution has four peaks in its magnitude over the two-dimensional spatial grid. The bottom right plot shows the solution at $z = 23$. Observe that the magnitudes of the peaks on the left side have increased. The top left plot shows the time evolution of the power of the solution as a function of the evolution variable z. The top right plot here shows the evolution of the four peaks in the magnitude of the solution as a function of z (blue = bottom left peak, red = top left peak, green = bottom right peak, cyan = top right peak).

In Figure 6, it can be seen that branch 2, when unstable in the focusing case, is subject to an oscillatory instability (with a fairly significant imaginary part). Hence the growth is not pure, but is accompanied by oscillations as is clearly visible in the top left panel. In this case, among the two principal peaks of the solution of branch 2, only the left one (associated with the gain side) is populated after the evolution shown.

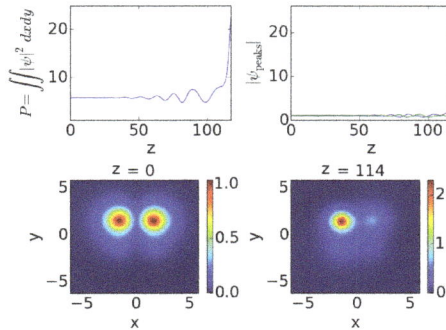

Figure 6. This figure is similar to Figure 5 (the final evolution distance however is about $z = 114$). Here, the plots correspond to the time evolution of the branch 2 solution for the value $\mu \approx 0.94$. In the top right plot, the blue curve corresponds to the left peak of the magnitude of the solution over z and the green corresponds to the right peak of the magnitude.

Lastly, in branch 3, the evolution (up to $z = 42$) manifests the existence of an exponential instability. The latter leads, once again, to the indefinite growth of the gain part of the solution, resulting in one of the associated peaks growing while the other (for $x > 0$ on the lossy side) features decay.

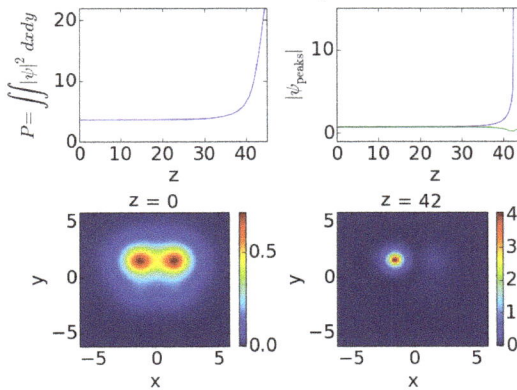

Figure 7. This figure is similar to Figure 5 (with an evolution up to distance $z = 42$). Here, the plots correspond to the evolution of the branch 3 solution for the value $\mu \approx 1.0$. In the top right plot, the blue curve corresponds to the left peak of the magnitude of the solution over z and green corresponds to the right peak of the magnitude. Clearly, once again, the gain side of the solution eventually dominates.

It is worthwhile to mention that in the case of branch 2—the only branch that was found (via our eigenvalue calculations) to be unstable in the self-defocusing case—we also attempted to perform dynamical simulations for $\sigma = -1$. Nevertheless, in all the cases considered, it was found that, fueled by the defocusing nature of the nonlinearity, a rapid spreading of the solution would take place (as z increased), resulting in the interference of the wave pattern with the domain boundaries. In Figure 8, we show an example where the maximum real part of the eigenvalues is ≈ 0.016. This weak instability is of exponential type with eigenvalues having a nonzero real part lying on the real axis similar to the eigenvalues in the bottom right panel of Figure 4 (but very close to the origin). Figure 8 shows that the solution initially seems to follow the expected pattern of dominating on the gain side (left) and there is a power growth similar to the focusing cases, but accompanied by a quick increase of the spatial extent of the solution leading to interference with the computational domain boundary.

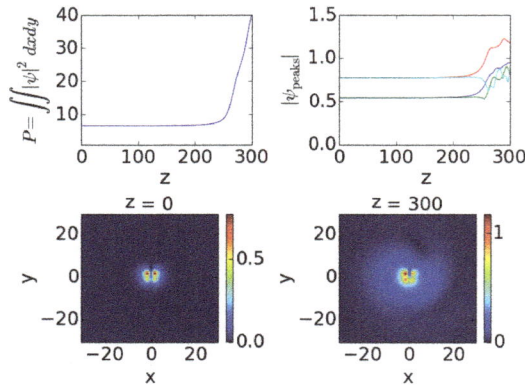

Figure 8. This figure is similar to Figure 5 (with an evolution up to distance $z = 300$). Here, the plots correspond to the evolution of the branch 2 solution for the value $\mu \approx 0.244$. In the top right plot, the colors follow the same pattern as in Figure 5 representing the four peaks. While the gain side of the solution seems to dominate, we also observe a quick increase in the spatial footprint of the solution. Shortly after the plotted time interval, the solution interferes with the boundary of the computed spatial grid.

6. Conclusions and Future Challenges

In the present work, we have revisited the partially $\mathcal{PT}$-symmetric setting originally proposed in [19] and have attempted to provide a systematic analysis of the existence, stability and evolutionary dynamics of the nonlinear modes that arise in the presence of such a potential for both self-focusing and self-defocusing nonlinearities. It was found that all three linear modes generate nonlinear counterparts. Generally, the defocusing case was found to be more robustly stable than the focusing one. In the former, two of the branches were stable for all the values of the frequency considered, while in the focusing case, all three branches developed instabilities sufficiently far from the linear limit (although all of them were spectrally stable close to it). The instabilities could be of different types, both oscillatory (as for branch 2) and exponential (as for branch 3) or even of mixed type (as for branch 1). The resulting oscillatorily or exponentially unstable dynamics, respectively, led to the gain overwhelming the dynamics and leading to indefinite growth in the one or two of the gain peaks of our four-peak potential.

Naturally, there are numerous directions that merit additional investigation. For instance, and although this would be of less direct relevance in optics, partial $\mathcal{PT}$ symmetry could be extended to three dimensions. There, it would be relevant to appreciate the differences between potentials that are partially $\mathcal{PT}$ symmetric in one direction vs. those partially $\mathcal{PT}$ symmetric in two directions. Another relevant case to explore in the context of the present mode is that where a $\mathcal{PT}$ phase transition has already occurred through the collision of the second and third linear eigenmode considered herein. Exploring the nonlinear modes and the associated stability in that case would be an interesting task in its own right. Such studies are presently under consideration and will be reported in future publications.

Acknowledgments: Panayotis G. Kevrekidis gratefully acknowledges the support of NSF-PHY-1602994, the Alexander von Humboldt Foundation, and the ERC under FP7, Marie Curie Actions, People, International Research Staff Exchange Scheme (IRSES-605096). Both authors express their gratitude to Professor Jianke Yang for initiating their interest in this direction and for numerous relevant discussions.

Author Contributions: Panayotis G. Kevrekidis and Jennie D'Ambroise designed the research; Jennie D'Ambroise carried out the numerical computations; both authors contributed substantially to the explanation of the results and to writing of the work.

Conflicts of Interest: The authors declare no conflict of interest.

References

1. Bender, C.M.; Boettcher, S. Real Spectra in Non-Hermitian Hamiltonians Having $\mathcal{PT}$ Symmetry. *Phys. Rev. Lett.* **1998**, *80*, 5243–5246.
2. Bender, C.M.; Brody, D.C.; Jones, H.F. Complex Extension of Quantum Mechanics. *Phys. Rev. Lett.* **2002**, *89*, 270401-1–270401-4.
3. Ruter, C.E.; Markris, K.G.; El-Ganainy, R.; Christodoulides, D.N.; Segev, M.; Kip, D. Observation of parity-time symmetry in optics. *Nat. Phys.* **2010**, *6*, 192–195.
4. Peng, B.; Ozdemir, S.K.; Lei, F.; Monifi, F.; Gianfreda, M.; Long, G.L.; Fan, S.; Nori, F.; Bender, C.M.; Yang, L. Parity–time-symmetric whispering-gallery microcavities. *Nat. Phys.* **2014**, *10*, 394–398.
5. Peng, B.; Ozdemir, S.K.; Rotter, S.; Yilmaz, H.; Liertzer, M.; Monifi, F.; Bender, C. M.; Nori, F.; Yang, L. Loss-induced suppression and revival of lasing. *Science* **2014**, *346*, 328–332.
6. Wimmer, M.; Regensburger A.; Miri, M.-A.; Bersch, C.; Christodoulides, D.N.; Peschel, U. Observation of optical solitons in PT-symmetric lattices. *Nat. Commun.* **2015**, *6*, 7782.
7. Suchkov, S.V.; Sukhorukov, A.A.; Huang, J.; Dmitriev, S.V.; Lee, C.; Kivshar, Y.S. Nonlinear switching and solitons in PT-symmetric photonic systems. *Laser Photonics Rev.* **2016**, *10*, 177–213.
8. Konotop, V.V.; Yang, J.; Zezyulin, D.A. Nonlinear waves in $\mathcal{PT}$-symmetric systems. *Rev. Mod. Phys.* **2016**, *88*, 035002-1–035002-65.
9. Schindler, J.; Li, A.; Zheng, M.C.; Ellis, F.M.; Kottos, T. Experimental study of active LRC circuits with $\mathcal{PT}$ symmetries. *Phys. Rev. A* **2011**, *84*, 040101-1–040101-4.
10. Schindler, J.; Lin, Z.; Lee, J.M.; Ramezani, H.; Ellis, F.M.; Kottos, T. $\mathcal{PT}$-symmetric electronics. *J. Phys. A Math. Theor.* **2012**, *45*, 444029-1–444029-17.
11. Bender, N.; Factor, S.; Bodyfelt, J.D.; Ramezani, H.; Christodoulides, D.N.; Ellis, F.M.; Kottos, T. Observation of Asymmetric Transport in Structures with Active Nonlinearities. *Phys. Rev. Lett.* **2013**, *110*, 234101-1–234101-5.
12. Bender, C.M.; Berntson, B.; Parker, D.; Samuel, E. Observation of $\mathcal{PT}$ Phase Transition in a Simple Mechanical System. *Am. J. Phys.* **2013**, *81*, 173–179.
13. Musslimani, Z.H.; Makris, K.G.; El-Ganainy, R.; Christodoulides, D.N. Optical Solitons in $\mathcal{PT}$ Periodic Potentials. *Phys. Rev. Lett.* **2008**, *100*, 030402-1–030402-4.
14. Fatkhulla, K.A.; Kartashov, Y.V.; Konotop, V.V.; Zezyulin, D.A. Solitons in $\mathcal{PT}$-symmetric nonlinear lattices. *Phys. Rev. A* **2011**, *83*, 041805(R)-1–041805(R)-4.
15. Wang, H.; Wang, J. Defect solitons in parity-time periodic potentials. *Opt. Express* **2011**, *19*, 4030–4035.
16. Lu, Z.; Zhang, Z. Defect solitons in parity-time symmetric superlattices. *Opt. Express* **2011**, *19*, 11457–11462.
17. Nixon, S.; Ge, L.; Yang, J. Stability analysis for solitons in $\mathcal{PT}$-symmetric optical lattices. *Phys. Rev. A* **2012**, *85*, 023822-1–023822-10.
18. Achilleos, V.; Kevrekidis, P.G.; Frantzeskakis, D.J.; Carretero-González, R. Dark solitons and vortices in $\mathcal{PT}$-symmetric nonlinear media: From spontaneous symmetry breaking to nonlinear $\mathcal{PT}$ phase transitions. *Phys. Rev. A* **2012**, *86*, 013808-1–013808-7.
19. Yang, J. Partially $\mathcal{PT}$ symmetric optical potentials with all-real spectra and soliton families in multidimensions. *Opt. Lett.* **2014**, *39*, 1133–1136.
20. Kartashov, Y.V.; Konotop, V.V.; Torner, L. Topological States in Partially-$\mathcal{PT}$-Symmetric Azimuthal Potentials. *Phys. Rev. Lett.* **2015** *115*, 193902.

21. Li, K.; Kevrekidis, P.G.; PT-symmetric oligomers: Analytical solutions, linear stability, and nonlinear dynamics. *Phys. Rev. E* **2011** *83*, 066608.

22. Zezyulin, D.A.; Konotop V.V.; Nonlinear Modes in Finite-Dimensional PT-Symmetric Systems. *Phys. Rev. Lett.* **2012** *108*, 213906.

23. Wang, C.; Theocharis,G.; Kevrekidis, P.G.; Whitaker, N.; Law, K.J.H.; Frantzeskakis, D.J.; Malomed, B.A. Two-dimensional paradigm for symmetry breaking: The nonlinear Schrödinger equation with a four-well potential. *Phys. Rev. E* **2009**, *80*, 046611-1–046611-9.

applied
sciences

MDPI

Article

Dark Solitons and Grey Solitons in Waveguide Arrays with Long-Range Linear Coupling Effects

Zhijie Mai [1], Haitao Xu [1], Fang Lin [1], Yan Liu [1,*], Shenhe Fu [2] and Yongyao Li [3]

[1] Department of Applied Physics, South China Agricultural University, Guangzhou 510642, China;
mhero14@163.com (Z.M.); xuhaitao@scau.edu.cn (H.X.); flin_163@163.com (F.L.)
[2] Department of Optoelectronic Engineering, Jinan University, Guangzhou 510632, China;
fushenhe@jnu.edu.cn
[3] School of Physics and Optoelectronic Engineering, Foshan University, Foshan 528000, China;
yongyaoli@gmail.com
* Correspondence: lycalm@scau.edu.cn; Tel.: +86-137-6064-7286

Academic Editors: Boris Malomed and Christophe Finot
Received: 23 December 2016; Accepted: 17 March 2017; Published: 22 March 2017

Abstract: In *J. Phys. Soc. Jpn. 83, 034404 (2014)*, we designed a scheme of waveguide arrays with long-range linear coupling effects and studied the bright solitons in this system. In this paper, we further study the dynamics of dark and grey solitons in such waveguide arrays. The numerical simulations show that the stabilities of dark solitons and grey solitons depend on the normalized decay length and the scaled input power. The width of dark solitons and the grey level of grey solitons are studied. Our results may contribute to the understanding of discrete solitons in long-range linear coupling waveguide arrays, and may have potential applications in optical communications and all-optical networks.

Keywords: discrete solitons; dark solitons; grey solitons; long-range coupling effects; waveguide arrays

1. Introduction

Nonlinear discrete systems attract considerable attention in many branches of physics, and exhibit various physical characteristics [1–3]. In the fields of optics [4–6] and Bose–Einstein condensates (BECs) [7–9], the evolution of nonlinear waves in discrete systems is a popular topic.

The basic model of a nonlinear discrete system in optics is an array of evanescently-coupled waveguides consisting of nonlinear materials. In a waveguide array, the propagation of light is primarily characterized by the coupling caused by the overlap between the fundamental modes of nearest-neighbouring waveguides. A crucial issue is to study the formation and properties of discrete solitons (DSs) in such nonlinear waveguide systems. DS formation is the result of a balance between on-site nonlinearity and the discrete diffraction induced by linear coupling among adjacent waveguides or lattice sites. DSs show strong potential for application in all-optical data processing; their most attractive feature is that DSs can enable intelligent functional operations such as routing, blocking, logic functions, and time gating in many all-optical devices [10].

An interesting extension is to investigate the formation of solitons in the presence of nonlocal effects. Trillo and colleagues studied the shock waves and dark solitons in nonlocal nonlinear media [11,12]. A next-nearest neighbour (NNN) model in which the linear coupling matrix becomes a quadruple-diagonal matrix after higher-order diffraction was studied by Kevrekidis and colleagues [13]. In 2012, Noskov and colleagues conducted significant studies of the nonlinear dipolar field in a nanoparticle train [14–16]—which can be viewed as a discrete nonlinear system—and reported that a linear coupling effect can exist among all lattice sites because of long-range dipole–dipole interactions. This system can produce all non-zero off-diagonal elements in the linear coupling matrix.

Long-range coupling in waveguide arrays is the coupling between the waveguides that were spaced with certain distance—it can affect the propagation dynamics of light field. Long-range coupling is different from that of the conventional waveguide arrays, which only consider the coupling between the adjacent waveguides (i.e., the short-range coupling). We have designed such a waveguide array with long-range coupling and studied the formation of bright solitons—see our previous paper in [17]. In this work, we further study the dynamics of dark and grey solitons in such a waveguide array, which was not considered before. The numerical results show that: regarding the dark solitons, the stability and width strongly depend on the mean power and the normalized decay length, which describes the effective length of the coupling effects in the waveguides; regarding the grey solitons, the stability and grey level are determined by the mean power and the normalized decay length.

The remainder of this paper is organized as follows. We provide a brief description of the model and basic equations in Section 2. Then, we study dark solitons in Section 3 and the grey solitons in Section 4. The paper is concluded in Section 5.

2. Model and Basic Equations

In [17], we designed a scheme for an optics experiment to apply our model of long-range linear coupling waveguide arrays. The model is built based on an AlGaAs single-mode waveguide structure. The real-scale linear coupling parameter between different waveguides can be fitted with an exponential decay, hence the long-range linear coupling effect is introduced in the system.

In our scheme, DSs can be described by the following equation, which is adapted from the discrete nonlinear Schrödinger equation [18,19]:

$$i\frac{\partial}{\partial z}u_n = \gamma|u_n|^2 u_n - \sum_m C_{mn}u_m. \tag{1}$$

Here, γ is the fixed nonlinear parameter of the system, where $\gamma = 1$ or $\gamma = -1$ indicates that the system features self-focusing or self-defocusing nonlinearity, respectively. u_n is the field amplitude of the n-th mode. Because each waveguide is identical, for simplicity, the propagation constant β is absorbed into the phase of u_n. z is the propagation distance along the waveguides, n is the number of waveguides, and the coefficient C_{mn} defines the coupling, which depends on the optical wavelength and the field overlap between m-th waveguide and n-th waveguide. Generally, Equation (1) can be expressed in matrix form as follows:

$$i\frac{\partial}{\partial z}U = (C+V)U, \tag{2}$$

where the matrix U and the elements of the matrix V are defined as $U = (u_1,\cdots,u_N)^T$ (where the superscript T indicates the transposition of the matrix and N is the number of waveguides) and $V_{mn} = \gamma|u_m|^2\delta_{mn}$ (where δ_{mn} is the Kronecker symbol), respectively.

The total power of the guide mode in the system is given by $P = \sum_n^N |u_n|^2$. In the model, we consider the matrix elements C_{mn} in Equation (1) to be given by

$$C_{mn} = \begin{cases} c_0\exp(-j/d) & (j \neq -1), \\ 0 & (j = -1), \end{cases} \tag{3}$$

where c_0 is the control parameter, $j = |m-n| - 1$, and d is the normalized decay length, which describes the effective length of the coupling effects in the waveguides. When $d \ll 1$, the system corresponds to a nearest-neighbour-coupled model. By contrast, when $d \gg 1$, the system exhibits strong coupling effects. Equation (3) forms the linear coupling matrix C, which represents the long-range linear coupling interactions among all lattice sites. In optics, when the separation between waveguides is sufficiently narrow, such higher-order cross-coupling effects can be induced [20].

The relationship between the scaled parameters in Equation (1) and the real-scale parameters are given in [17] in detail.

We assume that the soliton solutions for Equation (1) are written as

$$u_n(z) = \phi_n e^{-i\mu z}, \tag{4}$$

where ϕ_n is the stationary solution and $-\mu$ is the propagation constant, which is defined as

$$\mu = \frac{U^\dagger (C + V) U}{P}. \tag{5}$$

U is the solution of field amplitude in matrix form and $U^\dagger$ is the conjugated matrix of U. The stability of stationary solitons can be numerically determined by computing the eigenvalues for small perturbations or through direct simulations. The perturbed solution is given as $u_n = e^{-i\mu z}(\phi_n + w_n e^{i\lambda z} + v_n^* e^{-i\lambda^* z})$. Substituting this solution into Equation (1) and linearizing yields the following eigenvalue problem:

$$\begin{pmatrix} C - \mu + 2V & \Phi \\ -\Phi^* & -C + \mu - 2V \end{pmatrix} \begin{pmatrix} w \\ v \end{pmatrix} = \lambda \begin{pmatrix} w \\ v \end{pmatrix}, \tag{6}$$

where the elements of the matrix Φ are defined as $\Phi_{mn} = \gamma \phi_m^2 \delta_{mn}$.

The solution ϕ is stable if all eigenvalues λ are real.

3. Dark Solitons

In numerical simulations, we apply the imaginary time propagation (ITP) method [21] to study the fundamental solution to Equation (1) for dark solitons. We find that the stability of the dark soliton solutions changes with the scaled total power P and the normalized decay length d. When the eigenvalues λ have an imaginary part, this means that the solution of ϕ is unstable. Figure 1a,b show that the value of the imaginary part of λ (Im(λ)) varies with P and d. From Figure 1a, we can see two traits of dark solitons in this long-range linear coupling waveguide array. First, the nonlinearity strengthens when P increases, and the system requires a stronger coupling effect to achieve a stable solution for dark solitons. It can be seen that d increases as P increases. For example, when $P = 0.5$ and $d > 1.5$, the solitons are stable (Im(λ) = 0); however, when P increases to $P = 1$, the coupling effect must be enhanced to $d > 2.2$ to achieve stable solitons. Second, when the effective length d of the coupling effect is fixed, Im(λ) increases with larger P. This means that the instability is enhanced when the scaled total power P is higher. From Figure 1b, we can see when the system has either a strongly local effect ($d = 0.0001$) or a strongly coupling effect ($d = 5$); the dark soliton solution is stable regardless of the value of P. However, for $0.0001 < d < 5$, the system is subject to the combined action of the short-range and long-range effects, and the stability conditions for dark solitons are more complex. In this region of d, the tendency is that dark solitons become more unstable as P increases. Figure 1c,d show the amplitude and intensity, respectively, of the fundamental solution for dark solitons with $P = 1$ and $d = 3$. In Figure 1e, we present an example of the evolution of stable dark solitons with $P = 1$ and $d = 3$. In Figure 1f, we present an example of the evolution of unstable dark solitons with $P = 0.5$ and $d = 0.8$.

We also present the widths of dark solitons in this system for different d and P values, as shown in Figure 2. The dotted lines represent unstable dark soliton solutions (with Im(λ)≠0), and the solid lines represent stable solutions.

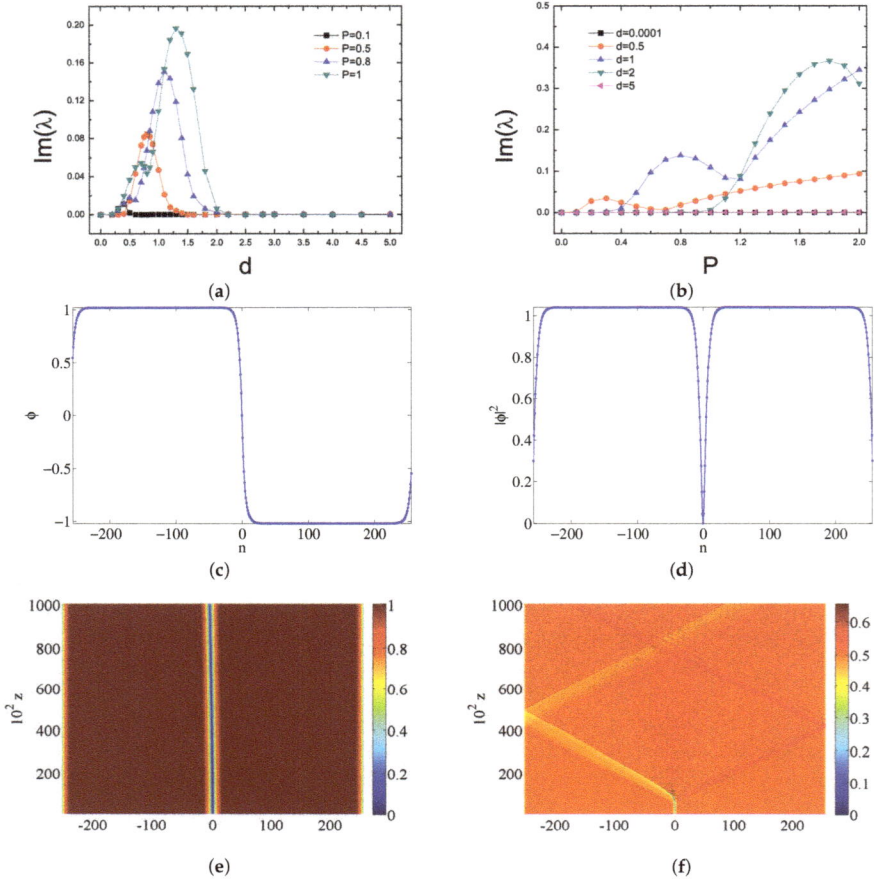

Figure 1. Dark solitons in a 513-waveguide array. (**a**) The imaginary part of the eigenvalues λ (Im(λ)) for different d values; (**b**) The imaginary part of the eigenvalues λ (Im(λ)) for different P values; (**c**) The amplitude of the fundamental solution for dark solitons ($P = 1$ and $d = 3$); (**d**) The intensity of the fundamental solution for dark solitons ($P = 1$ and $d = 3$); (**e**) The evolution of a stable solution for dark solitons ($P = 1$ and $d = 3$); (**f**) The evolution of an unstable solution for dark solitons ($P = 0.5$ and $d = 0.8$).

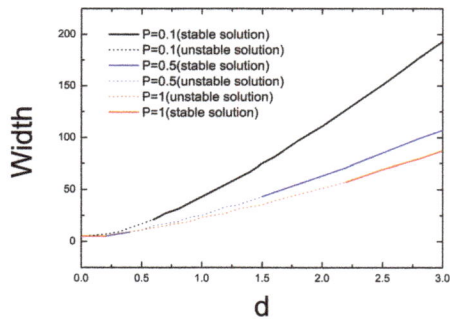

Figure 2. Widths of dark solitons for different d and P values.

4. Grey Solitons

We use the same method (ITP) to study the fundamental solution for grey solitons. Figure 3 shows the characteristics of grey solitons. Figure 3a,b show that the value of the imaginary part of λ (Im(λ)) varies with P and d. Grey solitons in this long-range linear coupling system have features similar to those of dark solitons: d increases as P increases (Figure 3a); when the effective length d of the coupling effect is fixed, Im(λ) increases with higher P (Figure 3a), and grey solitons become more unstable as P increases in the region of $0.0001 < d < 5$ (Figure 3b). However, grey solitons also show some distinct features compared with dark solitons; for example, the instability of grey solitons is stronger than that of dark solitons. First, by comparing Figure 3a,b with Figure 1a,b, we can see that Im(λ) is larger for grey solitons than for dark solitons given the same P and d. Second, the threshold in d where solitons transition from instability to stability lies at a higher value for grey solitons than for dark solitons. For example, as seen in Figure 1a, the threshold is $d = 2.2$ for $P = 1$, whereas in Figure 3a, the threshold is $d = 3$ for $P = 1$. Figure 3c,d show the amplitude and intensity, respectively, of the fundamental solution for grey solitons with $P = 1$ and $d = 3$. In Figure 3e, we present an example of the evolution of stable grey solitons with $P = 1$ and $d = 3$. In Figure 3f, we present an example of the evolution of unstable grey solitons with $P = 0.5$ and $d = 0.8$.

We also plot the grey levels for grey solitons with different d and P values, as shown in Figure 4. The grey level is defined as follows:

$$Greylevel = \frac{|max(u_n)|^2}{|min(u_n)|^2}. \tag{7}$$

The dotted lines represent unstable grey soliton solutions (with Im(λ) $\neq$ 0), and the solid lines represent stable solutions.

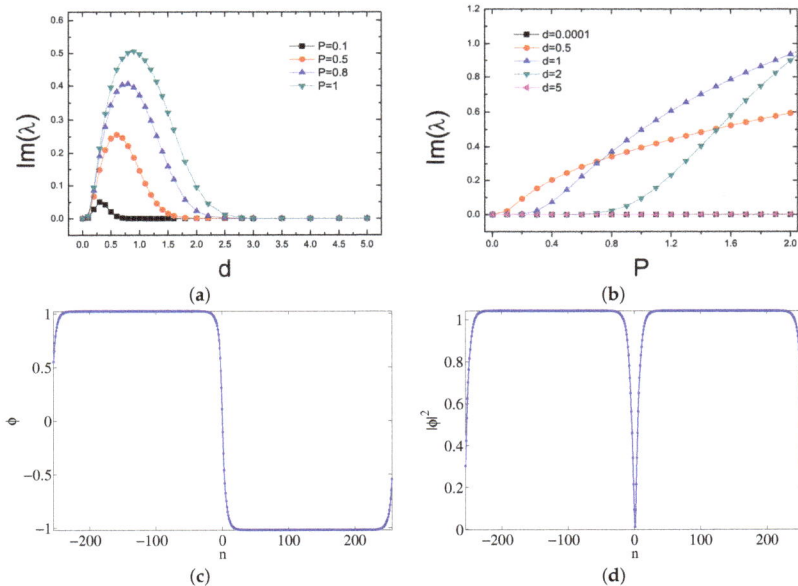

Figure 3. *Cont.*

(e)

(f)

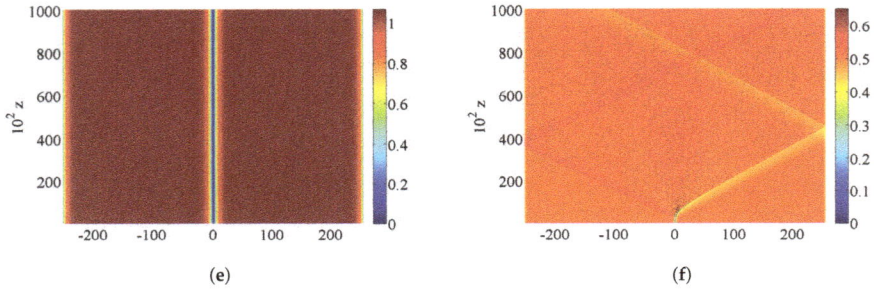

Figure 3. Grey solitons in a 512-waveguide array. (**a**) The imaginary part of the eigenvalues λ ($\text{Im}(\lambda)$) for different d values; (**b**) The imaginary part of the eigenvalues λ ($\text{Im}(\lambda)$) for different P values; (**c**) The amplitude of the fundamental solution for grey solitons ($P = 1$ and $d = 3$); (**d**) The intensity of the fundamental solution for grey solitons ($P = 1$ and $d = 3$); (**e**) The evolution of a stable solution for grey solitons ($P = 1$ and $d = 3$); (**f**) The evolution of an unstable solution for grey solitons ($P = 0.5$ and $d = 0.8$).

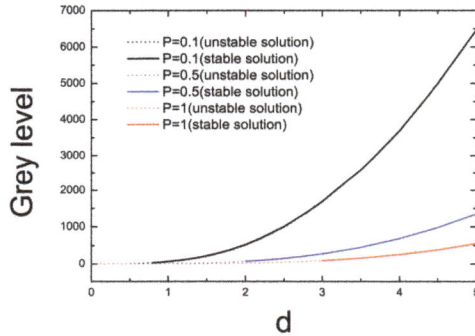

Figure 4. Grey levels of grey solitons with different d and P values.

5. Conclusions

In conclusion, we performed numerical studies of dark solitons and grey solitons in a waveguide array with long-range linear coupling effect. This system is described by the discrete nonlinear Schrödinger equation with off-diagonal elements in the linear coupling matrix filled with non-zero interaction terms. The stabilities of dark solitons and grey solitons are studied. The features of solitons such as the widths of dark solitons and the grey levels of grey solitons are comprehensively studied. Our results may fill the gap in the understanding of discrete solitons in long-range linear coupling waveguide arrays, and our design may have potential applications in optical communications and all-optical networks.

Acknowledgments: This work was supported, in part, by the National Natural Science Foundation of China through Grant Nos. 61571197, 11575063, and 61172011.

Author Contributions: Zhijie Mai, Yan Liu and Yongyao Li conceived of and designed the model. Zhijie Mai performed the numerical simulations. Haitao Xu and Fang Lin analysed the data. Shenhe Fu and Yongyao Li provided theoretical explanations. Zhijie Mai and Yan Liu wrote the paper.

Conflicts of Interest: The founding sponsors had no role in the design of the study; in the collection, analysis, or interpretation of the data; in the writing of the manuscript; or in the decision to publish the results.

References

1. Lederer, F.; Stegemanb, G.I.; Christodoulides, D.N.; Assanto, G.; Segev, M.; Silberberg, Y. Discrete solitons in optics. *Phys. Rep.* **2008**, *463*, 1–126.
2. Flach, S.; Gorbachb, A.V. Discrete breathers-Advances in theory and applications. *Phys. Rep.* **2008**, *467*, 1–116.
3. Kartashov, Y.V.; Malomed, B.A.; Torner, L. Solitons in nonlinear lattices. *Rev. Mod. Phys.* **2011**, *83*, 247–306.
4. Christodoulides, D.N.; Lederer, F.; Silberberg, Y. Discretizing light behaviour in linear and nonlinear waveguide lattices. *Nature* **2003**, *424*, 817–823.
5. Garanovich, I.L.; Longhi, S.; Sukhorukova, A.A.; Kivshar, Y.S. Light propagation and localization in modulated photonic lattices and waveguides. *Phys. Rep.* **2012**, *518*, 1–79.
6. Chen, Z.; Segev, M. Christodoulides, D.N. Optical spatial solitons: Historical overview and recent advances. *Rep. Prog. Phys.* **2012**, *75*, 086401.
7. Trombettoni, A.; Smerzi, A. Discrete solitons and breathers with dilute Bose-Einstein condensates. *Phys. Rev. Lett.* **2001**, *86*, 2353–2356.
8. Efremidis, N.K.; Christodoulides, D.N. Lattice solitons in Bose-Einstein condensates. *Phys. Rev. A* **2003**, *67*, 063608.
9. Morsch, O.; Oberthaler, M. Dynamics of Bose-Einstein condensates in optical lattices. *Rev. Mod. Phys.* **2006**, *78*, 179–215.
10. Christodoulides, D.N.; Eugenieva, E.D. Blocking and routing discrete solitons in two-dimensional networks of nonlinear waveguide arrays. *Phys. Rev. Lett.* **2001**, *87*, 160–161.
11. Ghofraniha, N.; Conti. C.; Ruocco, G.; Trillo, S. Shocks in nonlocal media. *Phys. Rev. Lett.* **2007**, *99*, 043903.
12. Armaroli, A.; Fratalocchi, A.; Trillo, S. Suppression of transverse instabilities of dark solitons and their dispersive shock waves. *Phys. Rev. A* **2012**, *80*, 72.
13. Kevrekidis, P.G.; Malomed, B.A.; Saxena, A.; Bishop, A.R.; Frantzeskakis, D.J. Higher-order lattice diffraction: Solitons in the discrete NLS equation with next-nearest-neighbor interactions. *Physica D* **2003**, *183*, 87–101.
14. Noskov, R.E.; Belov, P.A.; Kivshar, Y.S. Subwavelength modulational instability and plasmon oscillons in nanoparticle arrays. *Phys. Rev. Lett.* **2012**, *108*, 324–329.
15. Noskov, R.E.; Belov, P.A.; Kivshar, Y.S. Oscillons, soltions, and domain walls in arrays of nonlinear plasmonic nanoparticles. *Sci. Rep.* **2012**, *2*, 873.
16. Noskov, R.E.; Belov, P.A.; Kivshar, Y.S. Subwavelength plasmonic kinks in arrays of metallic nanoparticles. *Opt. Exp.* **2012**, *20*, 2733–2739.
17. Mai, Z.; Fu, S.; Wu, J.; Li, Y. Discrete solitons in waveguide arrays with long-range linearly coupled effect. *J. Phys. Soc. Jpn.* **2014**, *83*, 034404.
18. Christodoulides, D.N.; Joseph, R.I. Discrete self-focusing in nonlinear arrays of coupled waveguides. *Opt. Lett.* **1998**, *13*, 794–796;
19. Eisenberg, H.S.; Silberberg, Y.; Morandotti, R.; Boyd, A.R.; Aitchison, J.S. Discrete Spatial Optical Solitons in Waveguide Arrays. *Phys. Rev. Lett.* **1998**, *81*, 3383–3386.
20. Gordon, R. Harmonic oscillation in a spatially finite array waveguide. *Opt. Lett.* **2004**, *29*, 2752–2754.
21. Chiofalo, M.L.; Succi, S.; Tosi, M.P. Ground state of trapped interacting bose-einstein condensates by an explicit imaginary-time algorithm. *Phys. Rev. E* **2000**, *62*, 7438–7444.

![applied sciences logo] *applied sciences*

MDPI

Article

Pulse Propagation Models with Bands of Forbidden Frequencies or Forbidden Wavenumbers: A Consequence of Abandoning the Slowly Varying Envelope Approximation and Taking into Account Higher-Order Dispersion

Jorge Fujioka [1,*], Alfredo Gómez-Rodríguez [2] and Áurea Espinosa-Cerón [3]

[1] Instituto de Física, Dpto. de Sistemas Complejos, Universidad Nacional Autónoma de México, Apdo. Postal 20-364, 01000 México D.F., Mexico

[2] Instituto de Física, Dpto. de Materia Condensada, Universidad Nacional Autónoma de México, Apdo. Postal 20-364, 01000 México D.F., Mexico; alfredo@fisica.unam.mx

[3] Facultad de Ciencias, Universidad Nacional Autónoma de México, Apdo. Postal 20-364, 01000 México D.F., Mexico; ecasmir@gmail.com

* Correspondence: fujioka@fisica.unam.mx; Tel.: +52-55-5622-5078

Academic Editor: Boris Malomed

Received: 23 December 2016; Accepted: 26 March 2017; Published: 30 March 2017

Abstract: We study linear and nonlinear pulse propagation models whose linear dispersion relations present bands of forbidden frequencies or forbidden wavenumbers. These bands are due to the interplay between higher-order dispersion and one of the terms (a second-order derivative with respect to the propagation direction) which appears when we abandon the slowly varying envelope approximation. We show that as a consequence of these forbidden bands, narrow pulses radiate in a novel and peculiar way. We also show that the nonlinear equations studied in this paper have exact soliton-like solutions of different forms, some of them being *embedded* solitons. The solutions obtained (of the linear as well as the nonlinear equations) are interesting since several arguments suggest that the Cauchy problems for these equations are ill-posed, and therefore the specification of the initial conditions is a delicate issue. It is also shown that some of these equations are related to *elliptic curves*, thus suggesting that these equations might be related to other fields where these curves appear, such as the theory of *modular forms* and Weierstrass $\wp$ functions, or the design of cryptographic protocols.

Keywords: optical solitons; embedded solitons; soliton radiation; nonlinear Schrödinger equation; elliptic curves; forbidden frequencies; spectral gap

1. Introduction

The nonlinear Schrödinger (NLS) equation:

$$i\frac{\partial u}{\partial z} + \varepsilon_2 \frac{\partial^2 u}{\partial t^2} + \gamma_1 |u|^2 u = 0 \tag{1}$$

plays a central role in the study of light pulses propagating in optical fibers. In this equation z represents the distance along an optical fiber, t is the so-called *retarded time*, $u(z,t)$ is the envelope of the electric field of a laser beam, ε_2 is a real constant whose value depends on the laser's frequency, and γ_1 depends on the characteristics of the fiber (and also on the frequency of the light). It is worth mentioning that Equation (1) is sometimes referred to as the *temporal* NLS equation, to distinguish it from the *spatial* NLS equation, whose physical meaning is different, even though it has the same form as Equation (1), but with a spatial variable x (corresponding to a transversal coordinate) instead of the retarded time t.

In spite of its great importance, it should be remembered that Equation (1) is an *approximate* equation. In order to arrive at this equation several terms have been neglected. Some of these neglected terms are the following:

(a) $-iu_{ttt}$ and/or u_{4t}: these higher-order derivatives are necessary to describe sub-picosecond pulses [1–12]; in particular, conditions for including u_{4t} and discarding $-iu_{ttt}$ are discussed in [7,9],

(b) $|u|^4u$: this higher-order nonlinearity is used when we want to describe the propagation of pulses when the light intensity approaches the values which produce the "saturation" of the refractive index [13–20],

(c) $i(|u|^2u)_t$: this term is necessary to describe the self-steepening of the optical pulses [21–23],

(d) $u(|u|^2)_t$: this term is associated with the effect of Raman scattering [24–27].

The effects of including these terms in generalized NLS equations have been thoroughly studied in the literature. Moreover, an additional approximation introduced in the deduction of the NLS equation is the slowly-varying-envelope approximation (SVEA). This approximation is adequate when the complex amplitude of an optical pulse varies slowly in space and time. However, when we deal with optical pulses whose widths only contain a few cycles of the carrier wave, it is necessary to improve the SVEA. Several alternatives have been proposed to improve this approximation. As early as 1985, Christodoulides and Joseph studied an extended NLS equation which contained higher-order dispersion and the additional terms u_{zz}, $i(|u|^2u)_\tau$ and $(|u|^2u)_{\tau\tau}$, where τ was the standard (laboratory) time [28], and they found an interesting exact soliton solution by an algebraic *tour de force*. In more recent times completely different models have been proposed to describe ultra-short optical pulses, and the famous Kortweg-de Vries (KdV), modified KdV (mKdV) and sine-Gordon (sG) equations have been found useful for this purpose, as well as a combination of the last two of these equations [29,30]. Another two equations which have been useful to describe ultra-short spatiotemporal pulses are a two-dimensional sG equation [31] and a cubic generalized Kadomtsev-Petviashvili (cgKP) equation [32,33]. It is worth observing that in all these models (KdV, mKdV, sG, mKdV-sG, 2D-sG and cgKP) the temporal variable which appears in the equations is a *delayed time* (not the laboratory time), similar to the retarded time that appears in the standard NLS equation.

In the context of *spatial solitons* the *paraxial* approximation plays a role analogous to the SVEA in temporal solitons, and the effects of nonparaxiality have also been studied. In particular, the following non-paraxial extensions of the spatial NLS equation have been studied [34–36]:

$$iu_z + \alpha u_{zz} + \frac{1}{2}u_{xx} + |u|^2u = 0 \tag{2}$$

$$iu_z + \alpha u_{zz} + \varepsilon_2 u_{xx} - \gamma_1|u|^2u + \gamma_2|u|^4u = 0 \tag{3}$$

The study of Equation (2) addresses a basic question which might be of interest to any reader interested in optical solitons: what is the effect of introducing the non-paraxial term u_{zz} in the standard NLS equation? On the other hand, the study of Equation (3) considers an equally interesting question, albeit a slightly more specialized one: what is the effect of taking into account *simultaneously* nonparaxiality and higher-order nonlinearities? This question suggests that in the field of *temporal solitons* it would be interesting to investigate the effect of taking into account *simultaneously* non-SVEA terms, higher-order dispersion and higher-order nonlinearities such as $|u|^4u$. A study of this type was carried out in Ref. [28], where a model with these characteristics was proposed (except that the nonlinearity $|u|^4u$ was not considered). Complex models of this type, which include many different terms, may provide adequate descriptions for the behavior of very short pulses. However, in such models it might be difficult to appreciate the individual effect of each of the terms which have been taken into account.

In the present communication we are interested in studying approximate linear and nonlinear pulse propagation models which incorporate higher-order dispersion, higher-order nonlinearities and one of the terms that appear when we drop the SVEA (a second-order derivative with respect to the propagation direction). When we abandon the SVEA, the terms u_{zz}, $i(|u|^2 u)_\tau$ and $(|u|^2 u)_{\tau\tau}$ must be introduced in the NLS equation (τ being the standard laboratory time). However, if we use the retarded time to describe the propagation of the optical pulses, in addition to the term u_{zz}, it is also necessary to introduce a mixed derivative of the form u_{zt} in the resulting equation [37]. To study the complete equation containing these two derivatives (u_{zz} and u_{zt}), the nonlinear terms $i(|u|^2 u)_t$ and $(|u|^2 u)_{tt}$, an additional nonlinearity of the form $|u|^4 u$, and higher-order dispersive terms ($-iu_{ttt}$ and/or u_{4t}) might be interesting and important, but it is not the objective of the present work. In this communication we are only interested in studying simplified models where the terms u_{zt}, $i(|u|^2 u)_t$ and $(|u|^2 u)_{tt}$ have been discarded. It is clear that these models do not pretend to provide a quantitative accurate description of a particular real system, but we will see that the study of these simplified models reveals that the *simultaneous* presence of the non-SVEA term u_{zz} and higher-order dispersive terms such as $-iu_{ttt}$ or u_{4t} may produce interesting results. In particular, we will show that the combined effects of these terms generate bands of forbidden frequencies, which may be related to the gaps which have been experimentally observed in the spectral profiles of very narrow pulses propagating in photonic crystal fibers [38]. Moreover, we will see that just the mathematical structure of the equations considered in this paper is interesting by itself, since we will show that these equations are related to *elliptic curves*, thus suggesting that a relationship may exist between the study of optical solitons and other fields where these curves play an important role, such as the abstruse theory of modular forms and Weierstrass $\wp$ functions, or the design of cryptographic protocols.

We will begin this communication by studying the linear equations:

$$iu_z + c_0 u_{zz} + c_2 u_{tt} - ic_3 u_{ttt} = 0 \tag{4}$$

$$iu_z + c_0 u_{zz} + c_2 u_{tt} + c_4 u_{4t} = 0 \tag{5}$$

and afterwards we will focus our attention on the nonlinear equations:

$$iu_z + c_0 u_{zz} + c_2 u_{tt} + c_4 u_{4t} + \gamma_1 |u|^2 u - \gamma_2 |u|^4 u = 0 \tag{6}$$

$$iu_z + c_0 u_{zz} + c_2 u_{tt} - ic_3 u_{ttt} + c_4 u_{4t} + \gamma_1 |u|^2 u - \gamma_2 |u|^4 u = 0 \tag{7}$$

$$iu_z + c_0 u_{zz} + c_2 u_{tt} + c_4 u_{4t} + \gamma_1 |u|^2 u = 0 \tag{8}$$

The study of the linear Equations (4) and (5) will reveal how the interplay between the non-SVE term u_{zz} and the higher-order dispersive terms generates bands of forbidden frequencies or forbidden wavenumbers. Then, we will show that the nonlinear Equations (6)–(8) have exact solitons of different types, some of them being *embedded solitons*.

The structure of this paper is the following. In Section 2 we will show that the linear dispersion relations of Equations (4) and (5) have highly unusual forms. We shall see that the dispersion relation of Equation (4) is an *elliptic curve*, while that of Equation (5) presents bands of forbidden frequencies, or forbidden wavenumbers. As the occurrence of elliptic curves in this context opens the possibility of relating Equation (4)—or its solutions—to other areas where elliptic curves play an important role, in Appendix A we describe in more detail the basic characteristics of these curves, and in Appendix B we briefly discuss the relationship of these curves with modular forms and Fermat´s last theorem. In Section 3 we study the linear Equations (4) and (5). We will begin this section by paying attention to the fact that several results indicate that the Cauchy problems for these equations are *ill-posed*, and consequently the specification of initial conditions is a delicate issue. Then we present different solutions of these equations. These solutions will show that narrow pulses that evolve according to Equations (4) and (5) emit radiation in a completely novel and unexpected way, never observed in other models of optical pulses. In Section 4 exact soliton solutions of Equations (6)–(8) are presented. It is shown that Equation (6) has two different types of solitons (*embedded solitons* and solitons with a nonlinear frequency shift), Equation (7) has *moving* solitons of different heights and different velocities, and the solitons of Equation (8) are given by *squared* hyperbolic secants, thus proving that the interplay between the non-SVEA term u_{zz} and higher-order dispersive and nonlinear terms permits the existence of different types of solitons. Finally Section 5 contains the conclusions of the paper.

2. Dispersion Relations and Elliptic Curves

2.1. Dispersion Relation of Equation (4)

Let us begin by paying attention to the form of the dispersion relation of Equation (4). If we substitute the tentative solution:

$$u(z,t) = \varepsilon \, exp(i(kz - \omega t)) \tag{9}$$

in Equation (4), the following dispersion relation is easily found:

$$c_0 k^2 + k = c_3 \omega^3 - c_2 \omega^2 \tag{10}$$

and this equation can be rescaled by defining $\tilde{k} = c_0^{1/2}k$ and $\tilde{\omega} = c_3^{1/3}\omega$, thus obtaining (after supressing the tildes):

$$k^2 + a_3 k = \omega^3 + a_2 \omega^2 \tag{11}$$

where $a_3 = c_0^{-1/2}$ and $a_2 = -c_2 c_3^{-2/3}$. This equation describes an *elliptic curve* (if $a_2 \neq 0$ and $a_3 \neq 0$), which is an interesting result since elliptic curves have important roles in areas which might seem completely unrelated to the study of light pulses in optical fibers, such as the abstruse theory of *modular forms* [39], or the development of *cryptographic protocols* [40]. For the sake of simplicity in the following we will also say that Equation (10) itself is an elliptic curve, although, being rigorous, we should say that Equation (10) is an elliptic curve *up to scaling*. In other words, it is an elliptic curve distorted by a non-uniform scaling (i.e., a scale transformation which applies different scale factors in the k and ω axes).

Now let us pay more attention to the shape of the curves described by the dispersion relation (10). From Equation (10) it follows that the wavenumber k is given by the following function of the frequency:

$$k(\omega) = \frac{-1 \pm \sqrt{1 - 4c_0(c_2\omega^2 - c_3\omega^3)}}{2c_0} \tag{12}$$

and the form of this function is quite different from the dispersion relations found in other optical systems. If we choose, for example, the coefficients $c_0 = 1/30$, $c_2 = 1/2$ and $c_3 = 1/15$, the form of the curve $k(\omega)$ is shown in Figure 1.

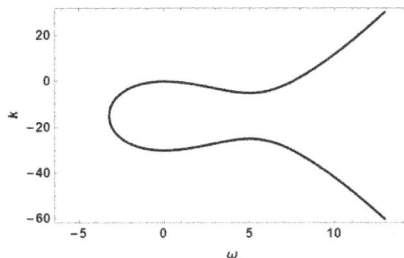

Figure 1. Dispersion relation of Equation (4), given by Equation (12), with $c_0 = 1/30$, $c_2 = 1/2$ and $c_3 = 1/15$.

We can see that the curve is symmetrical with respect to a horizontal line, but completely asymmetrical with respect to the *k*-axis, which is an unusual behavior. However, the behavior of $k(\omega)$ can be more bizarre. If we choose the coefficients $c_0 = 1/40$, $c_2 = 1/2$ and $c_3 = 1/50$, $k(\omega)$ has the form shown in Figure 2.

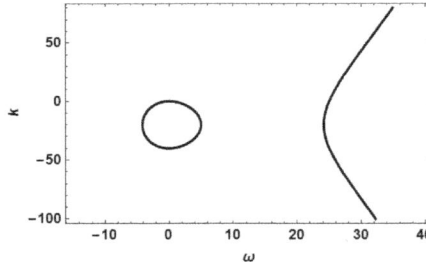

Figure 2. Dispersion relation of Equation (4), given by Equation (12), with $c_0 = 1/40$, $c_2 = 1/2$ and $c_3 = 1/50$.

We can see that in this case the dispersion relation exhibits two bands of forbidden frequencies, thus implying that small-amplitude linear waves cannot propagate with frequencies in the ranges:

$$-\infty < \omega < \omega_1 \text{ and } \omega_2 < \omega < \omega_3, \tag{13}$$

where:

$$\omega_1 = 10\left(1 - \sqrt{2}\right), \ \omega_2 = 5 \text{ and } \omega_3 = 10\left(1 + \sqrt{2}\right) \tag{14}$$

are the frequencies which make the radicand that appears in Equation (12) equal to zero. This radicand becomes negative for frequencies in the intervals shown in (13), and therefore $k(\omega)$ becomes complex for these frequencies. In the next section we will investigate the consequences of the existence of these bands of forbidden frequencies.

The fact that the linear dispersion relation of Equation (4) is an elliptic curve suggests that this equation might be related to other subjects where these curves also appear. In particular, a relationship may exist between optical solitons and the theory of *modular forms*, which played a central role in the proof of Fermat's last theorem. For this reason in Appendix A we explain more precisely what an elliptic curve is, and in Appendix B we briefly review what was the role played by elliptic curves and modular forms in the proof of Fermat's last theorem.

To close this sub-section it is worth observing that if we introduced the mixed derivative u_{zt} in Equation (4), the resulting dispersion relation would still be an elliptic curve.

2.2. Dispersion Relation of Equation (5)

Now let us direct our attention to Equation (5). If we substitute the plane wave (9) into Equation (5) we arrive at the following dispersion relation:

$$c_0 k^2 + k + \left(c_2 \omega^2 - c_4 \omega^4\right) = 0 \tag{15}$$

Let us now study the form of this dispersion relation in more detail. To determine the range of wavenumbers permitted by this relation it is convenient to write it in the form:

$$\omega^2 = (2c_4)^{-1}\left\{c_2 \pm \sqrt{c_2^2 + 4c_4(c_0 k^2 + k)}\right\} \tag{16}$$

This equation implies that every real value of k will be permitted (i.e., every real value of k corresponds to a positive ω^2) if and only if:

$$c_0 c_2^2 > c_4 \tag{17}$$

In the opposite case, when:

$$c_0 c_2^2 < c_4 \tag{18}$$

the dispersion relation contains a band of forbidden wavenumbers. When the condition (18) is satisfied, the analysis of the function defined by Equation (16) shows that ω^2 turns out to be negative if k is in the interval:

$$k_1 < k < k_2 \tag{19}$$

where:

$$k_{1,2} = (2c_0)^{-1} \left\{ -1 \pm \sqrt{1 - c_0 c_2^2 / c_4} \right\} \tag{20}$$

Therefore, when the condition (18) holds, the inequalities (19) define a band of forbidden wavenumbers.

In Figure 3 we can see the shape of the dispersion relation $k(\omega)$ defined by Equation (15) when $c_0 = 1/30$, $c_2 = 1/2$ and $c_4 = 1/80$. As these values satisfy the condition (18), the gap of forbidden wavenumbers seen in the figure is explained. The dashed horizontal lines indicate the wavenumbers C_1 and C_2 of the soliton solutions of Equation (6) (with $\gamma_1 = 5$ and $\gamma_2 = 1$) that will be determined in Section 4.

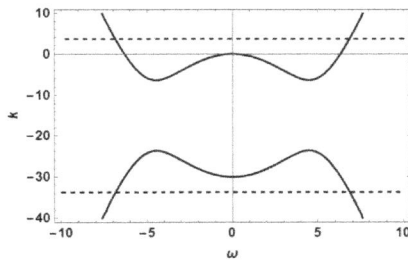

Figure 3. Dispersion relation of Equations (5) and (6), defined by Equation (15), with $c_0 = 1/30$, $c_2 = 1/2$ and $c_4 = 1/80$, which satisfy the condition (18). The dashed lines $k = C_{1,2}$ indicate the wavenumbers C_1 and C_2 of the solution of Equation (6) given by Equations (52)–(55) with $\gamma_1 = 5$ and $\gamma_2 = 1$.

Now let us study in more detail what happens when the condition (17) holds. In this case the linear dispersion relation presents *two bands of forbidden frequencies*, which is a novel situation in the study of optical solitons. To convince ourselves that such forbidden frequency gaps really exist, it is convenient to write the linear dispersion relation (15) in the form:

$$k = (2c_0)^{-1} \left\{ -1 \pm \sqrt{1 - 4c_0(c_2 \omega^2 - c_4 \omega^4)} \right\} \tag{21}$$

This equation implies that if $c_0 c_2^2 > c_4$ holds, there are frequencies for which k turns out to be complex. Analyzing the radicand which appears on the right-hand-side of Equation (21) we find that these forbidden frequencies are in the intervals:

$$-\omega_2 < \omega < -\omega_1 \text{ and } \omega_1 < \omega < \omega_2 \tag{22}$$

and the frequencies ω_1 and ω_2 are defined by:

$$\omega_{1,2}^2 = (2c_4)^{-1}\left\{c_2 \pm \sqrt{c_2^2 - c_4/c_0}\right\} \tag{23}$$

where the *minus* sign corresponds to ω_1, and the *plus* sign defines ω_2. In Figure 4 we can see the shape of the dispersion relation (21) in the particular case when $c_0 = 1/15$, $c_2 = 1/2$ and $c_4 = 1/256$. These coefficients satisfy the condition (17), and therefore there are two bands of forbidden frequencies, as we can see in the figure. The consequences of the presence of these forbidden frequency gaps will be investigated in the following section. The dashed horizontal lines shown in this figure indicate the values of the wavenumbers of the soliton solutions of Equation (6) (with $\gamma_1 = 5$ and $\gamma_2 = 1$) that will be found in Section 4.

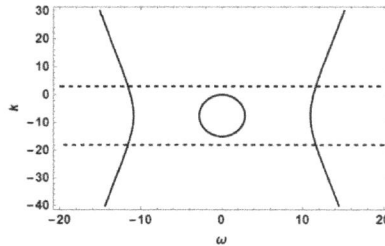

Figure 4. Dispersion relation of Equations (5) and (6), given by Equation (21), with $c_0 = 1/15$, $c_2 = 1/2$ and $c_4 = 1/256$, which satisfy the condition (17). The dashed lines $k = C_{1,2}$ indicate the wavenumbers C_1 and C_2 of the solution of Equation (6) given by Equations (52)–(55) with $\gamma_1 = 5$ and $\gamma_2 = 1$.

We now know that the dispersion relation (15) has interesting forms: when $c_0 c_2^2 < c_4$ there is a band of forbidden wavenumbers (as in Figure 3), and when $c_0 c_2^2 > c_4$ there are two bands of forbidden frequencies (as in Figure 4). The only case that we have not examined is when the coefficients c_0, c_2 and c_4 are on the surface defined by the equation:

$$c_0 c_2^2 = c_4 \tag{24}$$

Only in this case (intermediate between the inequalities (17) and (18)) does the dispersion relation (15) have no gaps. An example of this case is obtained when $c_0 = 1/20$, $c_2 = 1/2$ and $c_4 = 1/80$. With these coefficients the shape of the dispersion relation is shown in Figure 5.

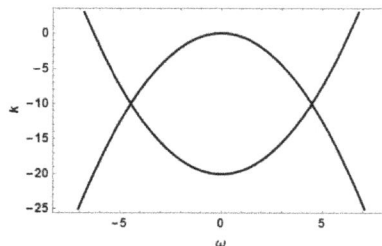

Figure 5. Dispersion relation of Equation (5), defined by Equation (15), with $c_0 = 1/20$, $c_2 = 1/2$ and $c_4 = 1/80$, which satisfy the condition (24).

We have thus seen that the dispersion relations of Equations (4) and (5) have interesting and unusual forms. Now we will investigate the behavior of solitary waves that obey these equations.

3. The Linear Equations (4) and (5)

3.1. The Issue of the Initial Conditions

In the study of optical solitons many variants of the nonlinear Shrödinger (NLS) equation have been considered in the past in order to describe the propagation of light pulses in optical fibers under different conditions. However, different as they are, most of these variants share a common denominator: they are *first-order equations with respect to the evolution variable z*. This characteristic implies that in all these cases the initial condition $u(z = 0, t)$ is sufficient to determine particular solutions of the corresponding equations. This fact seems to be consistent with physical reality: if we have an optical fiber (with well-defined characteristics: shape, composition, refraction index, type of cladding, etc.), and a laser beam (of known frequency) is sent into the fiber, the light intensity (as a function of time) at the beginning of the fiber (i.e., at $z = 0$) is all we need to determine completely the behavior of light along the fiber.

On the other hand, the situation with Equations (4)–(8) is completely different. These equations are of *second order in z*, and this characteristic has two important consequences:

CONSEQUENCE 1:

With these equations we require *two initial conditions*, $u(z = 0, t)$ and $u_z(z = 0, t)$, in order to determine a particular solution. This requirement is somewhat perplexing since physical intuition suggests that $u(z = 0, t)$ should be enough to determine the evolution of light along the fiber. Therefore, it seems as if we were in the presence of a contradiction: while mathematics indicates that Equations (4)–(8) require two initial conditions to define a particular solution, physics suggests that the knowledge of $u(0, t)$ should be enough to define completely the behavior of the system.

CONSEQUENCE 2:

Although formal proofs are still lacking, several arguments suggest that the Cauchy problems for Equations (4)–(8) might be *ill-posed* (in certain regions of the space of coefficients c_n). The following results permit us to understand why this ill-posedness is to be expected:

(i)   The presence of the non-paraxial term u_{zz} in the equation:

$$iu_z + \varepsilon u_{zz} + \frac{1}{2}\left(u_{xx} + u_{yy}\right) + |u|^2 u = 0 \tag{25}$$

makes the Cauchy problem for this equation ill-posed [21,41].

(ii) The presence of higher-order nonlinearities in equations of the form:

$$iu_z + u_{tt} + \lambda |u|^m u = 0 \tag{26}$$

makes their Cauchy problems ill-posed if $m \geq 4$ and $\lambda > 0$ [42].

(iii) The presence of higher-order derivative u_{4t} in the equation:

$$u_{zz} = u_{tt} + 2u_t^2 + 2uu_{tt} + u_{4t} \tag{27}$$

makes the Cauchy problem for this equation linearly ill-posed [43].

(iv) The solutions of the linear Equations (4) and (5) and the linear parts of Equations (6)–(8) contain decaying as well as growing modes of the form $\exp(\pm\sigma z - i\omega t)$, as we shall see in Sections 3.2 and 3.3. The existence of these modes makes the corresponding Cauchy problems linearly ill-posed [43].

(v)   Attemps of solving initial value problems for Equations (4)–(8) by finite differences in t, and Runge-Kutta algorithms in z, encounter difficulties that are typical of ill-posed problems.

At first sight these two consequences seem to be unrelated. However, as we explain in the following, there is a connection between them. Let us begin by paying attention to Consequence 2. If the Cauchy problems associated to Equations (4)–(8) are indeed ill-posed, then (as Hadamard first observed [44]) it would be necessary to impose certain compatibility conditions among the initial conditions, for these problems to have global solutions (i.e., solutions defined for all *t*). Consequently, these compatibility conditions will establish a link between the two initial conditions $u(z = 0, t)$ and $u_z(z = 0, t)$, thus implying that, essentially, the initial condition $u(z = 0, t)$ does indeed determine the future of the system, as physical intuition suggested. In this form, the compatibility conditions required by the ill-posedness of these problems (i.e., by Consequence 2), solve the apparent contradiction contained in Consequence 1 (i.e., the contrast between the mathematical requirement of two initial conditions, and the physical suggestion that only one condition should be necessary).

In the following two sub-sections we will examine different solutions of Equations (4) and (5), corresponding to different initial conditions $u(z = 0, t)$. We will see that the behavior of these solutions is particularly interesting if the initial condition is a narrow solitary wave.

3.2. Solutions of Equation (4)

Let us obtain the solution of Equation (4) corresponding to a given initial condition $u(0, t)$. We will see that we do not have to specify the value of $u_z(0, t)$.

Let us begin by calculating the Fourier transform (FT) of Equation (4) (with respect to *t*):

$$iU_z + c_0 U_{zz} - c_2 \omega^2 U + c_3 \omega^3 U = 0 \tag{28}$$

In this equation we have defined:

$$U(z, \omega) = \int_{-\infty}^{+\infty} u(z, t)\, e^{i\omega t}\, dt \tag{29}$$

The solution of Equation (28) can be obtained immediately:

$$U(z, \omega) = a(\omega)\, exp(\lambda_a z) + b(\omega)\, exp(\lambda_b z) \tag{30}$$

where λ_a and λ_b are the following functions of ω:

$$\lambda_{a,b} = \frac{-i \pm \sqrt{S(\omega)}}{2c_0} \tag{31}$$

where the *plus* sign in front of the radical corresponds to λ_a, the *minus* sign corresponds to λ_b, and the radicand $S(\omega)$ is given by:

$$S(\omega) = -1 - 4c_0\left(c_3\omega^3 - c_2\omega^2\right) \tag{32}$$

From Equation (30) it follows that:

$$U_z(z, \omega) = \lambda_a a(\omega)\, exp(\lambda_a z) + \lambda_b b(\omega)\, exp(\lambda_b z) \tag{33}$$

and from Equations (30) and (33) we can obtain the system:

$$U(0, \omega) = a(\omega) + b(\omega) \tag{34}$$

$$U_z(0, \omega) = \lambda_a a(\omega) + \lambda_b b(\omega) \tag{35}$$

and this system gives us the values of the coefficients $a(\omega)$ and $b(\omega)$ in terms of the functions $U(0, \omega)$ and $U_z(0, \omega)$:

$$a(\omega) = \frac{U_z(0, \omega) - \lambda_b U(0, \omega)}{\lambda_a - \lambda_b} \tag{36}$$

$$b(\omega) = U(0, \omega) - a(\omega) \tag{37}$$

Now we can observe something interesting. Let us consider, for example, the particular case when $c_0 = 1/40$, $c_2 = 1/2$ and $c_3 = 1/50$. In this case the radicand $S(\omega)$ (given in (32)) has the form shown in Figure 6.

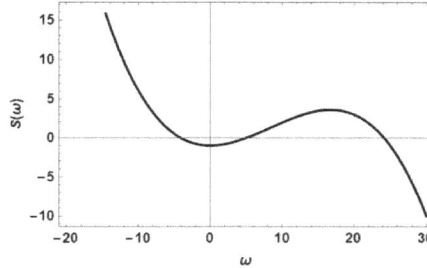

Figure 6. Function $S(\omega)$ given by Equation (32), with $c_0 = 1/40$, $c_2 = 1/2$ and $c_3 = 1/50$. $S(\omega) = 0$ at the values ω_n given in Equation (14).

We can see in this figure that $S(\omega) > 0$ if $\omega < \omega_1$ or $\omega_2 < \omega < \omega_3$, and $S(\omega) < 0$ if $\omega_1 < \omega < \omega_2$ or $\omega > \omega_3$, where the frequencies ω_n are the values shown in (14). Taking into account the signs of $S(\omega)$, we can write λ_a and λ_b in the following way:

$$\lambda_a = \begin{cases} \left[-i + \sqrt{S(\omega)} \right]/2c_0, & \omega < \omega_1 \\ \left[-i + i\sqrt{-S(\omega)} \right]/2c_0, & \omega_1 < \omega < \omega_2 \\ \left[-i + \sqrt{S(\omega)} \right]/2c_0, & \omega_2 < \omega < \omega_3 \\ \left[-i + i\sqrt{-S(\omega)} \right]/2c_0, & \omega > \omega_3 \end{cases} \tag{38}$$

$$\lambda_b = \begin{cases} \left[-i - \sqrt{S(\omega)} \right]/2c_0, & \omega < \omega_1 \\ \left[-i - i\sqrt{-S(\omega)} \right]/2c_0, & \omega_1 < \omega < \omega_2 \\ \left[-i - \sqrt{S(\omega)} \right]/2c_0, & \omega_2 < \omega < \omega_3 \\ \left[-i - i\sqrt{-S(\omega)} \right]/2c_0, & \omega > \omega_3 \end{cases} \tag{39}$$

We can see that λ_a has a positive real part when $\omega < \omega_1$ and $\omega_2 < \omega < \omega_3$, and therefore for frequencies within these two intervals the first term on the right-hand-side (rhs) of Equation (30) grows exponentially in z. As this growth is completely unphysical (as it would imply a fictitious energy growth), the coefficient $a(\omega)$ has to be equal to zero, and consequently from Equation (36) it follows that there is an unexpected relationship between $U_z(0, \omega)$ and $U(0, \omega)$:

$$U_z(0, \omega) = \lambda_b(\omega) U(0, \omega) \tag{40}$$

thus implying that:

$$u_z(0, t) = \frac{1}{2\pi} \iint\limits_{-\infty}^{+\infty} \lambda_b(\omega) \, u(0, \tau) \, e^{i\omega(\tau - t)} d\tau \, d\omega \tag{41}$$

Therefore, as far as Equation (4) is concerned, the perplexing situation associated to the Consequence 1 (mentioned in the previous sub-section) is completely clarified: physically realistic solutions of Equation (4) are indeed completely determined by the initial condition $u(0, t)$, as physics suggested. Once $u(0, t)$ is chosen, the function $u_z(0, t)$ is determined by Equation (41).

Now let us calculate the solution of Equation (4) corresponding to a soliton-like initial condition of the form:

$$u(0, t) = A \operatorname{sech}(Bt) \tag{42}$$

If we take the coefficients $c_0 = 1/40$, $c_2 = 1/2$ and $c_3 = 1/50$ in Equation (4), and the parameters $A = 1$ and $B = 1.5$ in Equation (42), the evolution of the pulse can be seen in Figure 7, which shows the form of $|u(z, t)|$ for $z = 0.5$, 1.0 and 1.5. We do not see anything special in this figure, which is somewhat disappointing, since we expected an unusual behavior due to the awkward form of the dispersion relation $k(\omega)$ shown in Figure 2.

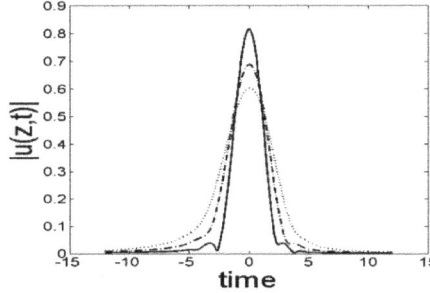

Figure 7. Modulus $|u(z, t)|$ of the solution of Equation (4) with $c_0 = 1/40$, $c_2 = 1/2$, $c_3 = 1/50$ and the initial condition (42) with $A = 1$ and $B = 1.5$. The shape of $|u(z, t)|$ is shown for $z = 0.5$ (continuous), 1.0 (dashed) and 1.5 (dotted).

If we now obtain the solution of Equation (4) (with the same coefficients indicated above) and using again an initial condition of the form (42), but with $A = 1/2$ and $B = 14$, the result is more interesting. In this case small ripples are generated on the leading edge of the pulse (i.e., the left hand side of the pulse), as shown in Figure 8.

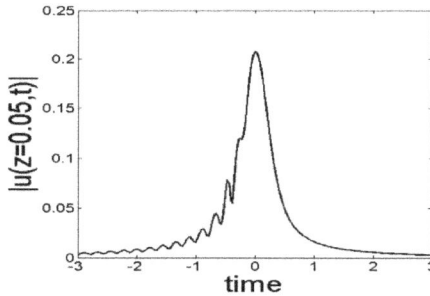

Figure 8. Modulus $|u(z = 0.05, t)|$ of the solution of Equation (4) with $c_0 = 1/40$, $c_2 = 1/2$, $c_3 = 1/50$ and the initial condition (42) with $A = 1/2$ and $B = 14$.

As the pulse advances along the fiber these ripples form a well-defined radiation wavetrain which propagates to the left of the pulse, as shown in Figures 9 and 10.

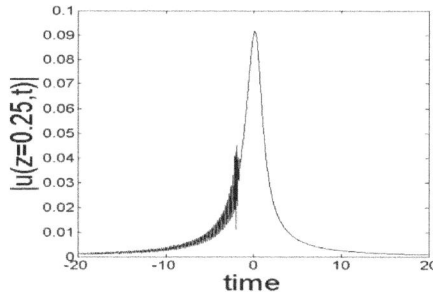

Figure 9. Same as Figure 8, but with $z = 0.25$.

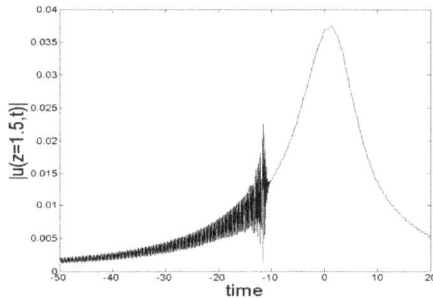

Figure 10. Same as Figure 8, but with $z = 1.5$.

At first sight, the radiation seen in Figure 9 might seem similar to the small-amplitude radiation waves emitted by solitary pulses which obey equations such as the following [1,2,11]:

$$iu_z + c_2 u_{tt} - ic_3 u_{ttt} + \gamma_1 |u|^2 u = 0 \tag{43}$$

$$iu_z + c_2 u_{tt} + c_4 u_{4t} + \gamma_1 |u|^2 u - \gamma_2 |u|^4 u = 0 \tag{44}$$

However, Figures 9 and 10 show that the radiation emitted by the solution of Equation (4) possesses a unique characteristic, that no other model of optical pulses has ever predicted: *the emission of radiation stops abruptly at a certain point along the retarded-time axis, and this point moves away from the pulse's peak at a constant velocity.* To understand clearly what this means we should remember that the independent variable "t" which appears in Equations (1), (4)–(8) and (26), and appears at the horizontal axes in Figures 8–10, is not the laboratory time, but the so-called *retarded time*. Therefore, the graphs shown in Figures 8–10 display the time-dependence of the square of the light intensity as measured by observers placed at different points (i.e., different values of z) along the optical fiber, but each of these observers has shifted his/her time origin in such a way that the time $t = 0$ corresponds to the instant when the pulse's maximum passes through the point z. Consequently, a graph such as the one shown in Figure 10 tells us that the radiation arrives at the observer placed at the position z *before the arrival of the pulse*. This implies that the radiation advances *faster* than the pulse. Moreover, it should be noticed that the point where we see an abrupt interruption of the radiation wave *corresponds to the endpoint of the radiation wavetrain*, since the first radiation waves that were emitted have already moved far away towards the left of the graph.

If we obtain numerical results similar to those shown in Figures 9 and 10, but for $z = 2$ and $z = 2.5$, we can see that if the observer advances a distance $\Delta z = 0.5$ along the optical fiber, the endpoint of the radiation wavetrain recedes to the left of the retarded time axis approximately $\Delta t \approx -5$, thus implying that this endpoint moves along the time axis with an "inverse velocity" $\Delta t / \Delta z \approx -10$. It might be worth emphasizing that in the study of light pulses along optical fibers *the evolution variable is not the time, but the distance along the fiber*, and therefore the "inverse velocity" $\Delta t / \Delta z$ is more adequate to describe the movement of radiation fronts (or pulses) such as those shown in Figures 9 and 10, when they travel along the retarded time axis.

To understand why the end point of the radiation wave moves to the left of the retarded time axis with an inverse velocity $\Delta t / \Delta z \approx -10$ we must remember that the radiation emitted by optical pulses is usually the result of a resonance between these pulses and the small-amplitude radiation waves capable of traveling along the fiber. This resonance occurs when an optical pulse has a wavenumber k_p that is contained in the range of wavenumbers permitted by the dispersion relation $k(\omega)$ of the system. In the case of Equation (4) we know that the dispersion relation is given by Equation (12), and the form of $k(\omega)$ is shown in Figure 2 for the coefficients $c_0 = 1/40$, $c_2 = 1/2$ and $c_3 = 1/50$. What we do not know is the wavenumber corresponding to the solution of Equation (4) whose profiles are shown in Figures 8–10. However, we can determine this wavenumber from the graphs of the real and imaginary parts of $u(z, t = 0)$ shown in Figures 11 and 12.

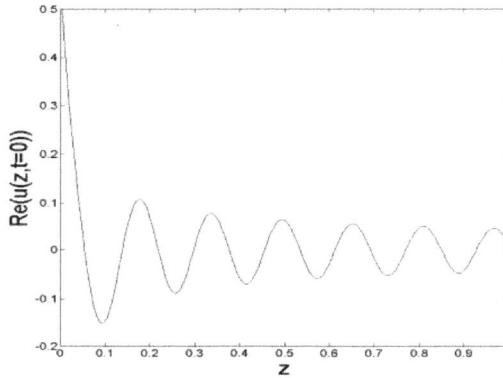

Figure 11. Real part of $u(z, t = 0)$, where $u(z, t)$ is the solution of Equation (4) with the same coefficients and the same initial condition used to obtain Figures 8–10.

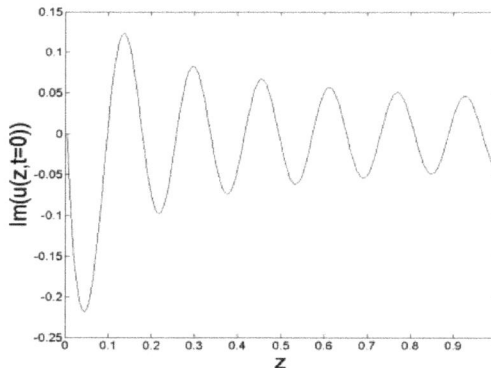

Figure 12. Imaginary part of $u(z, t = 0)$, where $u(z, t)$ is the solution of Equation (4) with the same coefficients and the same initial condition used to obtain Figures 8–10.

From these figures it follows that $k_p = -40$, and therefore a horizontal line placed at $k = -40$ intersects the dispersion relation $k(\omega)$ shown in Figure 2 at two points: $(k, \omega) = (-40, 0)$ and $(k, \omega) = (-40, 25)$. The first of these points does not correspond to a traveling wave (as its frequency is zero), but the second one corresponds to a linear wave with a resonant frequency $\omega = 25$, phase velocity $k/\omega = -1.6$ and group velocity $dk/d\omega = -12.5$ (i.e., in fact we are talking of *inverse velocities*, $\Delta t / \Delta z$, as mentioned in the previous paragraph). We can see, therefore, that the endpoint of the radiation wavetrain that we see in Figures 9 and 10 moves to the left of the retarded time axis at the group velocity corresponding to the resonant frequency $\omega = 25$.

It should be noticed that the value $k_p = -40$, which we obtained from Figures 11 and 12, could also be obtained from Equations (31) and (32). If we define $f(\omega) = c_3\omega^3 - c_2\omega^2$, we can approximate $\sqrt{S(\omega)} \cong i(1 + 2c_0 f(\omega))$, and therefore Equation (31) implies that $\lambda_b \cong -i/c_0 - if(\omega)$. Consequently, when $a(\omega) = 0$, Equation (30) takes the form $U(z, \omega) = b(\omega)exp(\lambda_b z)$, and from this equation it follows that $u(z, t)$ will contain the factor $exp(-iz/c_0)$, thus implying that $u(z, t)$ has a wavenumber $k_p = -1/c_0$. As we used $c_0 = 1/40$ to obtain Figures 11 and 12, the origin of the value $k_p = -40$ is now clear.

We now understand the origin of the velocity of the endpoint of the radiation wavetrain shown in Figures 9 and 10. However, we still do not know why this endpoint exists. In other words, we still do not know why the pulse stops emitting resonant radiation at a certain instant.

The clue to understand why the emission of radiation ends abruptly lies in the contrast between Figures 7 and 10. Figure 7 shows no radiation at all, while in Figure 10 the radiation is quite conspicuous. The reason for this difference must be necessarily in the initial conditions used to obtain the solutions of Equation (4) which led to these two figures. These initial conditions are shown in Figure 13, and the absolute values of their Fourier transforms are shown in Figures 14 and 15. We can see that the initial condition which lead to Figure 10 is much narrower than the initial condition corresponding to Figure 7, but the graphs which really suggest the answer to the presence of radiation in Figure 10, and its absence in Figure 7, are the Fourier transforms (FT) shown in Figures 14 and 15.

The FT shown in Figure 14 shows that the spectrum corresponding to the initial condition which lead to Figure 7 is essentially contained in the interval $-5 < \omega < 5$, and it is practically zero for $\omega > \omega_3 = 10\left(1 + \sqrt{2}\right) \approx 24.14$, which is the second interval where the dispersion relation $k(\omega)$ is well defined (remember, from Figure 2, that $k(\omega)$ is defined in two intervals: $\omega_1 < \omega < \omega_2$ and $\omega > \omega_3$) Therefore, it is impossible for the initial condition which lead to Figure 7 to resonate with the linear waves with frequencies $\omega > \omega_3$, since this initial condition has no frequency components in this interval. On the contrary, the spectrum shown in Figure 15 shows that the initial condition which leads to Figure 10 contains significant frequency components with $\omega > \omega_3$. Consequently, when this initial pulse starts traveling along the fiber, it resonates with the linear wave with wavenumber $k_p = -40$ (the pulse's wavenumber) and resonant frequency $\omega = 25$, and thus the emission of radiation begins. However, the pulse's width will begin to grow (and its spectrum will become narrower) as a result of the dispersive terms $c_2 u_{tt}$ and $ic_3 u_{ttt}$ which appear in Equation (4). As the dispersion of the pulse will continue indefinitely (since Equation (4) does not contain any nonlinear term which could cancel this dispersion), a moment will necessarily arrive when the pulse will be too wide, and its spectrum will be too narrow to have frequency components with $\omega > \omega_3$. Moreover, the components of the pulse with frequencies in the interval $\omega_2 < \omega < \omega_3$ will die quickly, since the dispersion relation (shown in Figure 2) does not permit the propagation of linear waves with these frequencies. Therefore, a moment will arrive when the pulse´s spectrum will be essentially contained in the interval $\omega_1 < \omega < \omega_2$ (where the dispersion relation allows the propagation of linear waves), and then the pulse will no longer be able to resonate with the wave with $\omega = 25$ (the resonant frequency), and the emission of radiation will stop. So, the interruption of the radiation is due to the fact that the pulse's spectrum is confined to live in the interval $\omega_1 < \omega < \omega_2$ (far from the resonant frequency $\omega = 25$), and the presence of the forbidden frequency gap $\omega_2 < \omega < \omega_3$ seen in Figure 2 strongly contributes to this confinement.

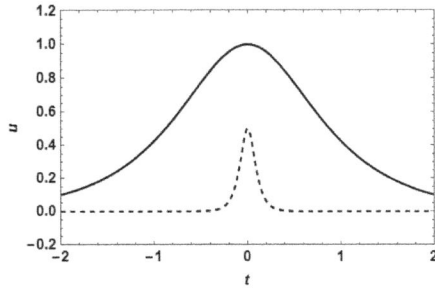

Figure 13. Initial conditions for Equation (4) used to obtain Figure 7 (continuous line) and Figures 8–10 (dashed line).

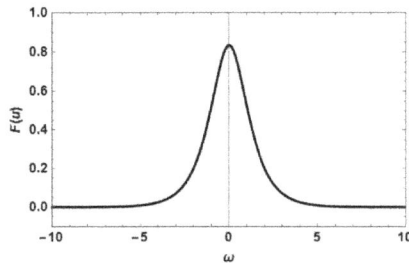

Figure 14. Fourier transform of the curve shown with the continuous line in Figure 13.

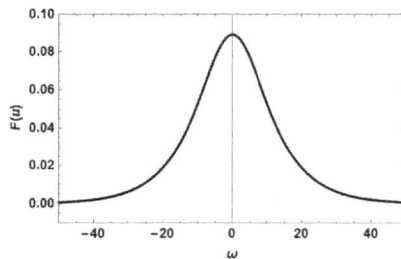

Figure 15. Fourier transform of the curve shown with the dashed line in Figure 13.

We have thus seen that a very original consequence of having a dispersion relation given by an elliptic curve such as that shown on Figure 2 is the abrupt interruption of the resonant radiation emitted by a narrow pulse which obeys the linear Equation (4). Therefore, this novel radiation process is the consequence of the *simultaneous* presence of the non-slowly-varying-envelope term u_{zz} and the third-order temporal derivative.

To close this sub-section it is worth observing that no qualitative changes are to be expected if we introduced the mixed derivative u_{zt} in Equation (4). The equation thus obtained could be solved by Fourier transforms, just as we solved Equation (4). The form of the Equation (30) would still be valid, although the form of the functions $\lambda_{a,b}(\omega)$ and $S(\omega)$ would be slightly different. In this case (when we take into account a term of the form cu_{zt} in Equation (4)), instead of Equations (31) and (32) we would have:

$$\lambda_{a,b} = \frac{-i(1 - c\omega) \pm \sqrt{S(\omega)}}{2c_0} \tag{45}$$

$$S(\omega) = -1 - 4c_0c_3\omega^3 + \left(4c_0c_2 - c^2\right)\omega^2 + 2c\omega \tag{46}$$

It should be emphasized that the key ingredient which produces the peculiar radiation process shown in Figures 9 and 10 is the presence of gaps of forbidden frequencies in the dispersion relation of Equation (4). It is interesting to investigate if these gaps survive if we introduce a term of the form cu_{zt} in Equation (4). Therefore, let us consider the equation:

$$iu_z + c_0u_{zz} + cu_{zt} + c_2u_{tt} - ic_3u_{ttt} = 0 \tag{47}$$

and let us take the same coefficients $c_0 = 1/40$, $c_2 = 1/2$ and $c_3 = 1/50$ used to obtain Figures 7–12. Concerning the coefficient c, one of the reviewers that revised this paper observed that in typical fibers $c < 0$ and the absolute value of the ratio c/c_0 is approximately equal to the dimensionless number $\delta = 2v_g^{-1}Z/T$, where Z and T are characteristic values of length and time, respectively, and v_g^{-1} is the reciprocal group velocity. In Figures 16–19 the dashed curves show the form of the dispersion relation of Equation (47) for three different values of c ($-c_0$, $-2.8\,c_0$ and $-3.2\,c_0$), and the continuous curve shows the dispersion relation when the term cu_{zt} is absent.

We can see that the finite gap of forbidden positive frequencies survives in Figures 16 and 17, but it disappears in Figure 18. Therefore, the presence of the term cu_{zt} in Equation (47) modifies the gaps of forbidden frequencies in a significant way. In fact, the gap of forbidden positive frequencies will disappear if the coefficients c, c_0, c_2 and c_3 are such that the equation $S(\omega) = 0$ has only one real root ($S(\omega)$ being the polynomial shown in Equation (46)). In particular, for pulses propagating in a typical optical fiber with a group velocity $v_g \cong 0.68c_L$ (c_L being the light speed in vacuum) the band of forbidden frequencies may not exist for processes whose characteristic dimensions are $Z \cong 1$ mm and $T \cong 1$ ps, as for these values $\delta \approx 10$, and Figures 16–18 show that the band of forbidden frequencies does not exist if $\delta \geq 3.2$ (and c_0, c_2 and c_3 have the values used to obtain Figures 16–18). However, the value of δ might be smaller in special fibers where v_g is closer to c_L (as in photonic fibers with air holes), and we consider processes where Z/T is also closer to c_L. In such conditions the band of forbidden frequencies might exist. In fact, the generation of a band of forbidden frequencies has already been observed by Fang et al. [38] during the propagation of very short optical pulses in photonic crystal fibers, a result that Fang et al. considered "the most distinctive feature" of this process.

It might be interesting to see how pulses evolve when the width of the band of forbidden frequencies shrinks, as in the dispersion relation shown by the dashed curve in Figure 17. In Figure 19 we can see the modulus of the solution (at $z = 1.5$) of Equation (47) corresponding to a narrow initial pulse of the form (42), with $A = 1/2$ and $B = 14$, and the coefficients used to obtain the dashed curve in Figure 17.

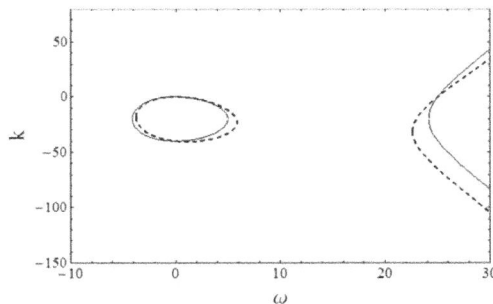

Figure 16. Dispersion relations of Equations (4) (continuous line) and (47) (dashed line) for $c_0 = 1/40$, $c_2 = 1/2$, $c_3 = 1/50$ and $c = -c_0$.

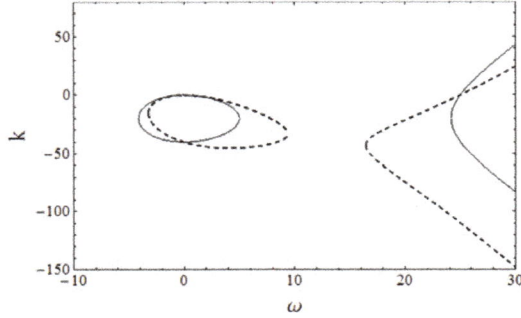

Figure 17. Dispersion relations of Equations (4) (continuous line) and (47) (dashed line) for $c_0 = 1/40$, $c_2 = 1/2$, $c_3 = 1/50$ and $c = -2.8c_0$.

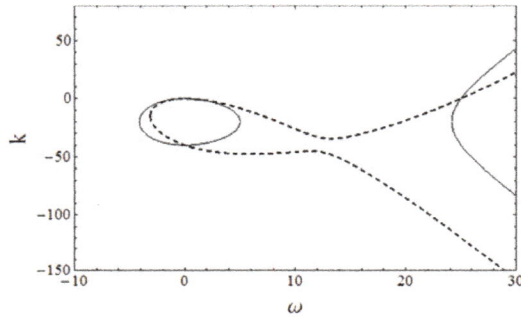

Figure 18. Dispersion relations of Equations (4) (continuous line) and (47) (dashed line) for $c_0 = 1/40$, $c_2 = 1/2$, $c_3 = 1/50$ and $c = -3.2c_0$.

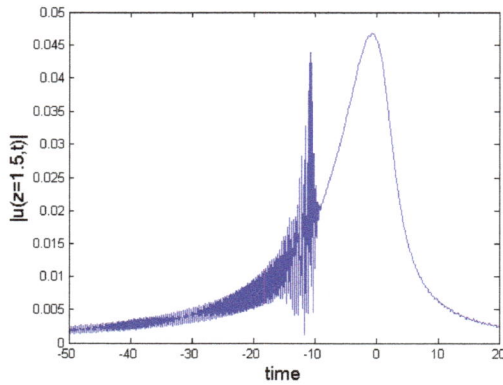

Figure 19. Modulus $|u(z = 1.5, \ t)|$ of the solution of Equation (47) with $c_0 = 1/40$, $c_2 = 1/2$, $c_3 = 1/50$, $c = -2.8c_0$ and the initial condition (42) with $A = 1/2$ and $B = 14$.

We can see that the graph shown in Figure 19 is similar to that shown in Figure 10, but the amplitude of the radiation is significantly higher. However, the radiation also stops at a definite time (as in Figure 10), due to the presence of the band of forbidden frequencies shown in Figure 17.

If we now calculate the solution of Equation (47) corresponding to the broader initial condition used to obtain Figure 7, and the same coefficients used in Figure 19, we obtain Figure 20. This figure shows that the pulse does not emit any radiation at all, as expected, since the initial condition is now a broad pulse, and its Fourier transform is too narrow to have frequency components which may resonate with the small-amplitude linear waves corresponding to the right branch of the dispersion relation seen in Figure 17. Figure 20 also shows that the position of the pulse has been shifted to the left, a result that is due to the presence of the term cu_{zt} in Equation (47), and the negative value of the coefficient c (Figure 7 shows that the pulse does not move to the left when the term cu_{zt} is absent).

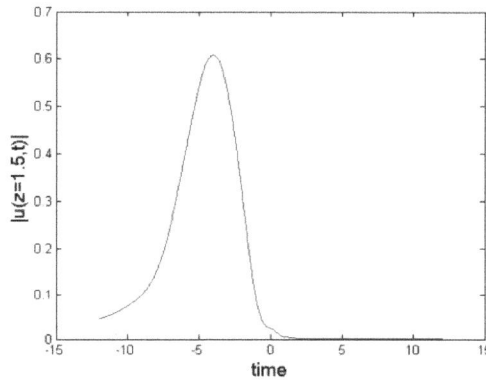

Figure 20. Modulus $|u(z = 1.5, \ t)|$ of the solution of Equation (47) with $c_0 = 1/40, c_2 = 1/2, c_3 = 1/50$ $c = -2.8c_0$ and the initial condition (42) with $A = 1$ and $B = 1.5$.

3.3. Solutions of Equation (5)

Now let us investigate how the solutions of Equation (5) behave. As we saw in Figure 4, when the coefficients c_0, c_2 and c_4 satisfy the condition (17), the dispersion relation presents two finite bands of forbidden frequencies (defined in (22)). Therefore, we expect that narrow solitons will exhibit a behavior similar to that of the solutions of the Equation (4). More precisely, we expect that if take an initial condition whose Fourier transform (FT) is wide enough to contain significant frequency components in the intervals $\omega < -\omega_2$ and $\omega > \omega_2$, the pulse will start radiating, but the radiation will stop at some time, when the pulse widens and its FT becomes too narrow. To verify this expectation we obtained the numerical solution of Equation (5) with coefficients $c_0 = 1/15, c_2 = 1/2$ and $c_4 = 1/256$, and using an initial condition $u(0, t)$ of the form (42) with $A = 1$ and $B = 7$. As in the case of Equation (4), we can obtain the solution of Equation (5) by calculating the Fourier transform (FT) of this equation, and the FT of the solution also has the form shown in Equation (30), with $\lambda_{a,b}(\omega)$ given by Equation (31), but with the function $S(\omega)$ which appears within the square root now defined as follows:

$$S(\omega) = -1 - 4c_0 \left(c_4 \omega^4 - c_2 \omega^2 \right) \tag{48}$$

As this function is positive on two regions of the ω axis, the function $U(z, \omega)$ (given in Equation (30)) would grow exponentially in z, unless we impose the restriction $a(\omega) = 0$. This restriction implies that Equation (41) is also valid in this case (but using Equation (48) in the expression (31) which defines the function $\lambda_b(\omega)$). Therefore, to solve Equation (5) we used $u(0, t)$ as defined in Equation (42), and we calculated $u_z(0, t)$ with Equation (41). In Figure 21 we can see the profile of $|u(z = 1.5, t)|$ corresponding to the initial condition (42) with $A = 1$ and $B = 7$. As expected, we see that the pulse starts emitting radiation, but afterwards the radiation stops. The only important difference in comparison with the behavior of the solutions of Equation (4) is that the solution of

Equation (5) emits radiation in both directions, due to the fact that its dispersion relation, shown in Figure 4, is now symmetrical with respect to the wavenumber axis.

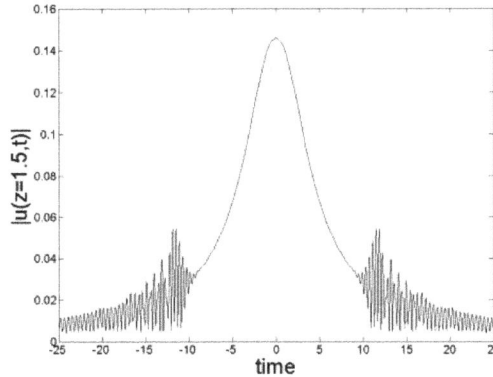

Figure 21. Modulus $|u(z = 1.5, t)|$ of the solution of Equation (5) with $c_0 = 1/15$, $c_2 = 1/2$, $c_4 = 1/256$ and the initial condition (42) with $A = 1$ and $B = 7$.

On the other hand, in Figure 22 we can see the profile of another solution of Equation (5) corresponding to an initial condition of the form (42), but now using $A = 1$ and $B = 2$ (and the same coefficients c_0, c_2 and c_4). In this case no radiation is observed because the initial pulse was very wide, and its FT was very narrow, and therefore the initial FT has no significant frequency components in the intervals $\omega < -\omega_2$ and $\omega > \omega_2$ (where the resonant frequencies are located).

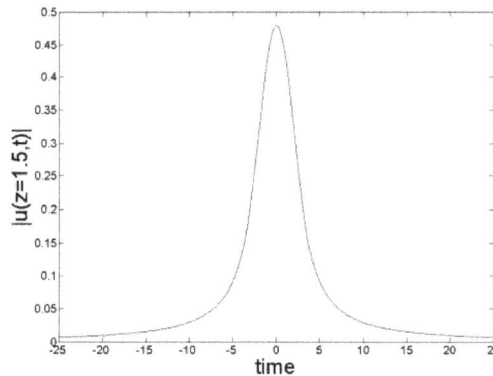

Figure 22. Same as Figure 21, but with $B = 2$.

We have thus seen that the behavior of the radiating pulses of Equations (4) and (5) is similar, except that the solutions of Equation (5) emit radiation in both directions, and those of Equation (4) only radiate to the left. However, Equations (4) and (5) differ in a significant aspect. In the case of Equation (4) the function $S(\omega)$ defined by Equation (32) is always positive in a portion of the real axis, and consequently is always necessary to impose the restriction $a(\omega) = 0$ which leads to the relationship between $u(0,t)$ and $u_z(0,t)$ given by Equation (41). On the other hand, in the case of Equation (5), depending on the values of the coefficients c_0, c_2 and c_4, the function $S(\omega)$ defined by Equation (48) *may* be positive over certain frequency intervals, and in these cases the restriction

$a(\omega) = 0$ is necessary, and the Equation (41) again defines a necessary relation between $u(0,t)$ and $u_z(0,t)$. However, for certain coefficients, the function $S(\omega)$ might be always negative, and then the values of the functions $\lambda_{a,b}(\omega)$ would be imaginary for all frequencies. In these cases there is no reason to impose the restriction $a(\omega) = 0$, and with the disappearance of this restriction also the relations (40) and (41) disappear. Therefore, in the case of Equation (5), when λ_a and λ_b are both imaginary, we face again the question: how do we specify the initial condition $u_z(0,t)$? In the following paragraph we address this question.

When $a(\omega) \neq 0$ the Equations (40) and (41) no longer hold, but we can obtain an approximate relation between $u_z(0,t)$ and $u(0,t)$ by considering what happens when we send a light pulse at the beginning of an optical fiber. As the pulse enters into the fiber, the wavelength of the carrier wave changes (since the light's velocity is different inside the fiber), and this change introduces a factor $\exp(i\Delta k\, z)$ in the mathematical form of the pulse. In other words, at the beginning of the fiber (inside the fiber, near the point $z = 0$) the form of the light pulse will be:

$$u(z,t) = f(t)\, e^{i\Delta k\, z} \tag{49}$$

where $f(t)$ defines the temporal profile of the pulse. Consequently, from Equation (49) it follows that:

$$u_z(0,t) = i\Delta k\, u(0,t) \tag{50}$$

and this equation, whose FT is similar to Equation (40), implies that the form of the function $u(0,t)$ determines the form of $u_z(0,t)$. It should be noticed, however, that we do not know the value of the parameter Δk. This parameter depends on the exact form of the refractive index of the fiber as a function of the light intensity. However, if we approximate the wavelength of the light inside the fiber in the form $\lambda \cong \lambda_0/n_0$, where λ_0 is the wavelength outside the fiber, and n_0 is the intensity-independent part of the refractive index, then we can approximate Δk in the form:

$$\Delta k \cong 2\pi(n_0 - 1)/\lambda_0 \tag{51}$$

Therefore, when there is no justification for using Equation (41) to calculate $u_z(0,t)$, this function might be calculated by means of the approximate Equations (50) and (51). In fact, in the following section we will see that the exact soliton solutions of Equation (6) are indeed consistent with a linear relationship between $u_z(0,t)$ and $u(0,t)$, such as that given by Equation (50).

To close this section, it is important to emphasize that the peculiar radiation behavior observed in Figures 9, 10, 19 and 21 is a direct consequence of the presence of *bands of forbidden frequencies* in the dispersion relations of Equations (4), (5) and (47), and these forbidden bands are the result of the interplay between the non-SVEA tem u_{zz} and higher-order dispersive terms such as $-iu_{ttt}$ or u_{4t}. Therefore, the principal lesson to be learnt from the study of Equations (4), (5) and (47) is the following: when a very short pulse is launched along an optical fiber, some of the frequencies which are contained in the initial pulse may not be permitted to travel along the fiber, since *bands of forbidden frequencies* may appear in the dispersion relation of the equation which controls the propagation of the pulse. Although these forbidden bands may be cancelled by the influence of the term cu_{zt} (which we did not include in Equations (4) and (5)), a band of this type has already been observed in a photonic crystal fiber, as mentioned before [38].

4. The Nonlinear Equations (6)–(8)

In this section we will show that the nonlinear Equations (6)–(8) have exact soliton solutions of different forms, some of them being *embedded solitons* (as we shall explain in the following).

4.1. The Solitons of Equation (6)

Equation (6) is an interesting generalization of Equation (5) because it accepts soliton solutions of two different types. The first of these solutions is the following:

$$u(z,t) = A \operatorname{sech}(Bt) \, e^{iCz} \tag{52}$$

where A, B and C are the following constants:

$$B^2 = \frac{\gamma_1}{20c_4} \left(\frac{24c_4}{\gamma_2} \right)^{1/2} - \frac{c_2}{10c_4} \tag{53}$$

$$A = \left(\frac{24c_4}{\gamma_2} \right)^{1/4} B \tag{54}$$

$$C = (2c_0)^{-1} \left[-1 \pm \sqrt{1 + 4c_0(c_4 B^4 + c_2 B^2)} \right] \tag{55}$$

Equation (55) defines two values for C, and consequently Equation (6) has two solutions of the form (52) with different wavenumbers C_1 and C_2 corresponding, respectively, to the signs "plus" and "minus" in front of the radical that appears in Equation (55).

It is worth observing that the solution defined by Equations (52)–(55) is the particular solution of Equation (6) corresponding to the initial conditions:

$$u(0,t) = A \operatorname{sech}(Bt) \tag{56}$$

$$u_z(0,t) = iC \, u(0,t) \tag{57}$$

with A, B and C given by Equations (53)–(55). We can see that Equation (57) has the same form as Equation (50), but with a particular value of the wavenumber. Therefore, the use of initial conditions of the form (50) in order to look for particular solutions of Equations (5) and (6) seems justified.

We must now remember that in Section 2 we saw that if the inequality (17) holds, the range of wavenumbers permitted by the linear dispersion relation of Equation (6) (given by Equation (15) or, alternatively, by Equation (21)), covers the entire real axis, and consequently, in this case (when $c_0 c_2^2 > c_4$) the soliton wavenumbers C_1 and C_2 are obviously contained within the range of wavenumbers permitted by the linear dispersion relation of the system. This characteristic implies that these solitons are not standard ones (since standard solitons have wavenumbers lying *outside* the linear spectrum of the system), but they are *embedded solitons*. It is worth remembering that prior to 1997 it was believed that soliton wavenumbers must necessarily lie *outside* the linear spectra of the systems. Otherwise (it was believed) the solitons would resonate with the small-amplitude linear waves capable of propagating in the system, and the solitons would emit resonant radiation. However, in 1997 it was discovered that *exact* radiationless soliton solutions with wavenumbers contained within the range of the linear dispersion relation of the system may indeed exist [11], and the term "embedded soliton" was coined two years later to distinguish these peculiar solitons from the standard ones [45,46].

In Figure 4 we showed the form of the dispersion relation, given by Equation (21), in the particular case when $c_0 = 1/15$, $c_2 = 1/2$ and $c_4 = 1/256$. This figure also shows the position of the soliton wavenumbers, C_1 and C_2, given by Equation (55) with $\gamma_1 = 5$ and $\gamma_2 = 1$. If we look at this figure an interesting question arises: how is the frequency spectrum of the exact soliton solution (52) in comparison to the dispersion relation? To answer this question, in Figure 23 we have superimposed the Fourier transform (FT) of the exact soliton (52) and the *scaled* dispersion relation $k(\omega)/8.6287$. The scaling factor $(8.6287)^{-1}$ was introduced for the amplitude of the FT and the scaled dispersion relation to have the same size at $\omega = 0$. This figure shows that the amplitude of the spectrum of the exact soliton solution is completely insignificant in the range of frequencies where the lateral branches of the dispersion relation are placed. Therefore, even though the two soliton solutions of

Equation (6) have wavenumbers C_1 and C_2 that intersect these branches (as the dashed lines show in Figure 4), and this fact favors a resonance between the solitons and the radiation modes, the soliton spectrum is practically zero at the frequencies where $k(\omega) = C_{1,2}$, thus implying that the solitons do not have frequency components capable of resonating with the radiation modes satisfying the resonance condition $k(\omega) = C_{1,2}$.

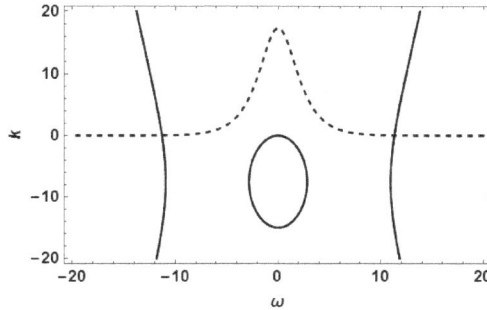

Figure 23. The continuous curve shows the scaled dispersion relation $k(\omega)/a$ of Equation (6) with $c_0 = 1/15$, $c_2 = 1/2$, $c_4 = 1/256$, $\gamma_1 = 5$, $\gamma_2 = 1$, $k(\omega)$ given by Equation (21) and $a = 8.6287$. The dashed line shows the Fourier transform of the initial form of the soliton solution (52), with A and B given by Equations (53) and (54).

This *qualitative* argument is helpful to understand why these *embedded solitons* do not resonate with the radiation modes, in spite of the fact of having wavenumbers contained in the range of the dispersion relation. However, if desired, it is also possible to use the procedure shown in Refs. [18,20] to construct a rigorous proof (not just a *qualitative* one) which shows that the absence of resonances between the solitons and the radiation modes is the consequence of a delicate balance between the linear and the nonlinear terms of Equation (6). This procedure consists in taking the FT of Equation (6), using a function of the form (52) to calculate the FT of the nonlinear terms. In this way, an expression for the FT of $u(z,t)$ is obtained, which contains a quotient of two polynomials in ω. The polynomial in the numerator depends on the nonlinear coefficients, and that in the denominator depends on the coefficients of the linear terms. Moreover, the roots of the polynomial that appears in the denominator determines the resonant frequencies. The next step is to prove that if the values of the parameters A, B and C which appear in (52) have the values given by the Equations (53)–(55), then the two polynomials cancel out, and the denominator disappears. With the disappearance of this denominator, the possibility of a resonance also disappears, thus explaining why the solution given by Equations (52)–(55) does not resonate with the radiation modes.

The above paragraph shows that the soliton solutions of Equation (6) (defined by Equations (52)–(55)) are *embedded solitons* if the inequality (17) holds. On the other hand, when the inequality (18) holds, the linear dispersion relation (21) presents a band of forbidden wavenumbers (defined by (19) and (20)), and in this case it is not evident if the soliton wavenumbers C_1 and C_2 are located within the forbidden band (in which case the solitons would be standard), or they lie outside this band (and they are contained within the linear spectrum of the system), in which case they would be *embedded solitons*. A careful comparison of the soliton wavenumbers C_1 and C_2 given by Equation (55), and the boundaries k_1 and k_2 of the band of forbidden frequencies (given by Equation (20)), show that:

$$C_2 > k_2 \text{ and } C_1 < k_1 \tag{58}$$

These inequalities imply that the soliton wavenumbers lie *outside* the forbidden band $k_1 < k < k_2$, and consequently there are contained within the range of wavenumbers permitted by dispersion relation, thus implying that also in this case (when the inequality (18) holds) the solitons defined by Equations (52)–(55) are *embedded solitons*. In Figure 3 we showed the dispersion relation $k(\omega)$ defined by Equation (15) when $c_0 = 1/30$, $c_2 = 1/2$ and $c_4 = 1/80$, and the dashed horizontal lines show the position of the soliton wavenumbers C_1 and C_2 (corresponding to $\gamma_1 = 5$ and $\gamma_2 = 1$), evidencing that these wavenumbers lie outside the forbidden band.

Now let us consider the second type of soliton solutions of Equation (6). Direct substitution shows that if the coefficients of the equation satisfy the following relation:

$$\gamma_1^2 c_4 = -\frac{18}{7}\gamma_2 c_2^2 \tag{59}$$

then Equation (6) has an exact solution of the following form:

$$u(z,t) = F\,\mathrm{sech}(Gt)e^{i[Pz+Q\,\ln\{\cosh(Gt)\}]} \tag{60}$$

where:

$$F^2 = -\frac{9}{4}\frac{c_2^2}{\gamma_1 c_4} \tag{61}$$

$$G^2 = \frac{c_2}{8c_4} \tag{62}$$

$$Q = \pm\sqrt{5} \tag{63}$$

and P is the solution of:

$$c_0 P^2 + P + \frac{9c_2^2}{16c_4} = 0 \tag{64}$$

which, if $c_0 \neq 0$, implies that:

$$P = (2c_0)^{-1}\left[-1 \pm \sqrt{1 - 9c_0 c_2^2/(4c_4)}\right] \tag{65}$$

The solution given by Equations (60)–(65) is an interesting one, as it contains an unusual nonlinear frequency shift. This type of frequency shift was also found in [17], and if $c_0 = 0$ our Equations (59)–(64) coincide with the results found in this reference.

It should be noticed that Equation (6) does not accepts simultaneously the solutions (52) and (60). The solution (52) exists for certain values of the coefficients c_0, c_2, c_4, γ_1 and γ_2, and the solution (60) exists for other values of these coefficients. For example, if c_2 and c_4 are both positive, the solution (52) only exists if γ_1 is also positive (as a consequence of Equation (53)), while the solution (60) will only exist if γ_1 is negative (as a consequence of Equation (61)).

The existence of two different types of solitons in Equation (6) suggests that this equation might have additional exact analytical solutions. It would be interesting to investigate if the new methods described in Refs. [47–49] and the references therein, might be successful in finding new solutions for Equation (6).

4.2. The Solitons of Equation (7)

Direct substitution shows that Equation (7) has *moving* soliton solutions of the following form:

$$u(z,t) = R\,\mathrm{sech}\left(\frac{t-Vz}{W}\right)e^{i(qz-rt)} \tag{66}$$

where r, q, V, R and W are constants whose values can be calculated as explained in the following. First we calculate r:

$$r = -\frac{c_3}{4c_4} \tag{67}$$

and then we define the following parameters:

$$Y = c_4 r^4 + c_3 r^3 - c_2 r^2 \tag{68}$$

$$M = \frac{3 c_0 \gamma_2}{20 c_4}\left(-8 c_4 r^3 - 2 c_2 r\right)^2 \left(\frac{24 c_4}{\gamma_2}\right)^{1/2} \tag{69}$$

$$L = \frac{3 \gamma_2}{40 c_4}\left(\frac{24 c_4}{\gamma_2}\right)^{1/2}\left[\gamma_1\left(\frac{24 c_4}{\gamma_2}\right)^{1/2} - \frac{3}{4}\frac{c_3^2}{c_4} - c_2\right] \tag{70}$$

Then we can obtain the possible values of q from the following equation:

$$(1 + 2 c_0 q)^4 \left[\gamma_1^2 - 6 \gamma_2\left(q + c_0 q^2 - Y\right)\right] = \left[L(1 + 2 c_0 q)^2 - M - \gamma_1(1 + 2 c_0 q)^2\right]^2 \tag{71}$$

and with each of the possible values of q we can calculate V, R and W with the following expressions:

$$V = \frac{-8 c_4 r^3 - 2 c_2 r}{1 + 2 c_0 q} \tag{72}$$

$$R^2 = \frac{1}{20 c_4}\left(\frac{24 c_4}{\gamma_2}\right)^{1/2}\left[\gamma_1\left(\frac{24 c_4}{\gamma_2}\right)^{1/2} - \frac{3}{4}\frac{c_3^2}{c_4} - 2 c_2 - 2 c_0 V^2\right] \tag{73}$$

$$W = \left(\frac{24 c_4}{\gamma_2}\right)^{1/4}\frac{1}{R} \tag{74}$$

As Equation (71) is a sixth-order equation for q, we may have up to six different values for q, and consequently Equation (7) might have six different soliton solutions for a given set of coefficients $\{c_0, c_2, c_3, c_4, \gamma_1, \gamma_2\}$. This result suggests that Equation (7) might have a continuous family of soliton solutions, in which case the solitons defined by Equations (66)–(74) would only be particular elements of this family that can be expressed in terms of hyperbolic secants.

Now let us investigate if the solitons defined by Equations (66)–(74) are standard or embedded. Apparently, in order to determine if the soliton (66) is embedded, we would only need to investigate if the soliton's wavenumber q (defined by Equation (71)) is contained in the range of wavenumbers permitted by the linear dispersion relation corresponding to Equation (7), which is given by:

$$c_0 k^2 + k = c_4 \omega^4 + c_3 \omega^3 - c_2 \omega^2 \tag{75}$$

However, this is not true. When we deal with a *moving soliton* (i.e., a soliton which moves along the retarded time axis, as the soliton in Equation (66)), the determination if the soliton is embedded is more involved. As explained in Ref. [18], in this case we must investigate if the *intrinsic* soliton's wavenumber defined as:

$$Q = q + Vr \tag{76}$$

is contained in the range of a *modified* dispersion relation, which is obtained by substituting a linear wave written in the form $u = \exp[kz - \omega(t - Vz)]$ into the linear part of the equation under study. In the case of Equation (7), this modified dispersion relation takes the form:

$$c_4 \omega^4 + c_3 \omega^3 - c_0(k + V\omega)^2 - (k + V\omega) = 0 \tag{77}$$

This equation permits us to write k as function of ω, but now something unexpected occurs. As Equation (77) involves V, and V depends on q (as shown in Equation (72)), the dispersion relation (77) will depend on the value of q. In other words: *we will have a different dispersion relation for each of the different real values of q defined by Equation (71).* Consider, for example, the following set of coefficients:

$$c_0 = 0.025, \; c_2 = 0.5, \; c_3 = 0.07, \; c_4 = 0.02, \; \gamma_1 = 5, \; \gamma_2 = 1 \tag{78}$$

Substituting these values in Equations (67)–(70) we obtain the values of Y, M and L. Then, substituting these values into Equation (71), and solving this equation, we find that there are two complex values for q, and the following four real values:

$$q_1 = -42.80, \; q_2 = -22.11, \; q_3 = -17.89, \; q_4 = 2.80 \tag{79}$$

Each of these values defines a dispersion relation given by Equation (77). The form of these dispersion relations can be seen in Figure 24. The dashed horizontal line which appears in each of the four figures indicates the position of the corresponding *intrinsic soliton's wavenumber* $Q_i = q_i + V(q_i)r$. The values of these Q_i are the following:

$$Q_1 = -42.05, \; Q_2 = -13.96, \; Q_3 = -26.04, \; Q_4 = 2.05 \tag{80}$$

In each of the four graphs shown in Figure 24 the dashed horizontal line intersects the dispersion relation $k(\omega)$, thus implying that the four solitons corresponding to the coefficients shown in (78) are embedded solitons.

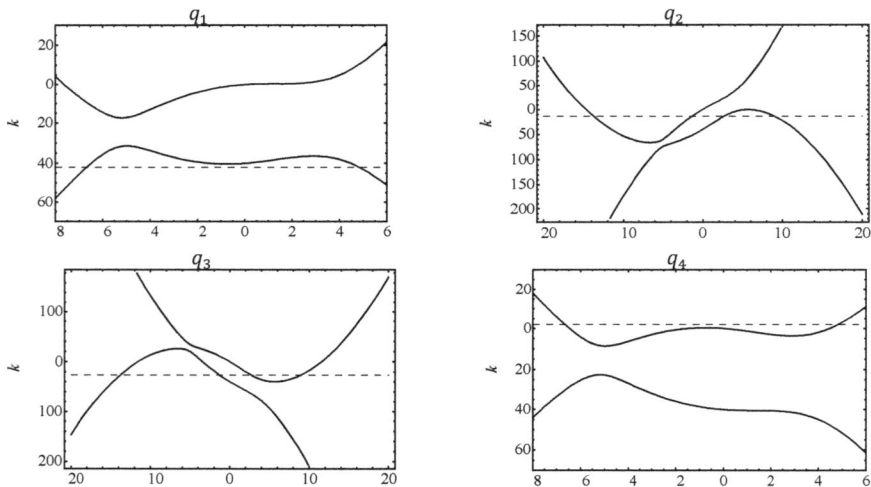

Figure 24. Dispersion relations of the form (77), corresponding to the four values of q_i shown in (79), which are the real roots of Equation (71) with the coefficients c_i and γ_i shown in (78). The dashed lines indicate the values of the corresponding intrinsic solitons' wavenumbers Q_i given in (80).

We may wonder if *all* the solitons of Equation (7) are embedded. It is difficult to answer such a question. We have examined the solitons corresponding to other coefficients, different from those shown in (78), and in all the cases that we have studied the solitons turned out to be embedded. However, it is difficult to prove that *in general*, for every set of coefficients which lead to real values of the parameters M, L, q, R and W (defined by Equations (69)–(74)), the corresponding solitons are all embedded. In fact, it might be impossible to generate such a proof, since we would need to compare the values of the *intrinsic* solitons' wavenumbers Q_i (which depend on the values of the solutions q_i of Equation (71)) with the boundaries of the forbidden bands that appear in the dispersion relations, and it is impossible to obtain analytical expressions for q_i in closed form, since Equation (71) is a sixth-order equation.

4.3. The Solitons of Equation (8)

Equation (8) is worth being studied because, in spite of being a particular case of Equation (6), it has soliton solutions which are *different* from those of Equation (6). Direct substitution shows that Equation (8) has exact soliton solutions of the following form:

$$u(z,t) = D \, sech^2(Et) \, exp(iH_{1,2}z) \tag{81}$$

where the parameters D, E and H are defined by following equations:

$$D^2 = -\frac{3}{10}\frac{c_2^2}{c_4\gamma_1} \tag{82}$$

$$E^2 = -\frac{c_2}{20c_4} \tag{83}$$

$$H_{1,2} = (2c_0)^{-1}\left(-1 \pm \sqrt{1 - \frac{16}{25}\frac{c_0 c_2^2}{c_4}}\right) \tag{84}$$

Optical solitons with profiles given by *squared* hyperbolic secants are not unknown in optics [50,51], but they are far less common than the typical *sech-type* solitons. Consequently, it is interesting that the solitons of Equation (8) have the form given by Equation (81). Moreover, it is also worth observing that even though Equation (6) reduces to Equation (8) when $\gamma_2 = 0$, neither of the two soliton solutions of Equation (6) (given either by Equations (52)–(55) or by Equations (60)–(65)) reduces to the soliton solution of Equation (8) given by Equations (81)–(84) in the limit when $\gamma_2 \to 0$. This is not an unusual situation. It occurs in many systems with embedded solitons. For example, the equation:

$$iu_z + \varepsilon_1 u_{tt} + \varepsilon_2 u_{4t} + \gamma_1|u|^2 u - \gamma_2|u|^4 u = 0 \tag{85}$$

reduces to the standard NLS equation when $\varepsilon_2 \to 0$ and $\gamma_2 \to 0$, but it is known that the solitons of Equation (85) do not reduce to NLS solitons in this limit [11]. In a similar way, Equation (8) reduces to the NLS equation when $c_0 = c_4 = 0$, but the solution defined by Equations (81)–(84) does not reduce to a NLS soliton in this case. In fact, as $c_4 \to 0$ the soliton's height (given by the parameter D defined by Equation (82)) tends to infinity, and the soliton's width (given by the parameter E^{-1} defined by Equation (83)) tends to zero.

Are the soliton solutions of Equation (8) *embedded solitons*? This is a question which deserves some attention. In the case when the inequality (17) holds, it is obvious that the solitons defined by Equations (81)–(84) are embedded, since in this case the wavenumbers permitted by the linear dispersion relation (15) cover the entire real axis (as explained in Section 2), and therefore the solitons' wavenumbers $H_{1,2}$ will necessarily be contained in the range of this dispersion relation. On the other hand, when the inequality (18) holds, the range of the dispersion relation will contain a band of forbidden wavenumbers, and the boundaries of this band are the values k_1 and k_2 given by Equation (20). In this case it is not evident if the solitons defined by Equations (81)–(84) are embedded or not. However, if we choose values of c_0 and c_4 which lead to real values of $H_{1,2}$ (when substituted in Equation (84)), an algebraic exercise shows that:

- if $c_4 > 0$ then: $H_1 < k_1$ and $H_2 > k_2 \Rightarrow$ the solitons are embedded
- if $c_4 < 0$ then: $k_1 < H_1 < H_2 < k_2 \Rightarrow$ the solitons are NOT embedded

Therefore, the soliton solutions of Equation (8) are interesting because, depending on the values of c_0, c_2 and c_4, they may be embedded solitons or standard ones.

5. Conclusions

In this paper we have studied five pulse propagation models (Equations (4)–(8)) which take into account the non-SVEA term u_{zz}, higher order dispersion, and a nonlinearity of the form $|u|^4 u$. Several terms which also participate in the propagation of short optical pulses along optical fibers have been discarded, not because they are negligible or unimportant, but to get a better understanding of the interplay between the terms u_{zz}, $-iu_{ttt}$, u_{4t} and $|u|^4 u$. The dismissal of significant terms implies that the models considered in this communication cannot aspire to provide quantitatively accurate descriptions of short light pulses propagating along an optical fiber. However, the analysis of the equations here considered (Equations (4)–(8)) reveals interesting aspects of the propagation of short pulses under conditions which are beyond the SVEA, and shows that the mathematical structure of these equations is interesting by itself, because it suggests that the study of optical solitons might be connected to other fields which seemed to be completely unrelated to optics.

The analysis of the nonlinear models (6)–(8) reveals that short solion-like pulses of different forms might be able to propagate under non-SVEA conditions. Equation (6) has soliton solutions of two different types: one of them with a phase linear in the propagation distance (as usual), and the second one with a nonlinear phase shift (see Equation (60)). Equation (7) has *moving* solitons of different heights and velocities, and the determination of the parameters that appear in these solitons involves the solution of a sixth-order algebraic equation (see Equations (66)–(74)). In the case of Equation (8), the profile of its solitons is given by a squared hyperbolic secant and, depending on the values of the coefficients of the equation, these solitons may be standard or *embedded* (i.e., with a wavenumber that is contained in the linear spectrum of the system).

On the other hand, the analysis of the linear models (4) and (5) shows that the interplay between u_{zz} and a higher-order dispersive term such as $-iu_{ttt}$ or u_{4t} generates bands of forbidden frequencies or forbidden wavenumbers. As a consequence of the existence of these forbidden bands, the behavior of short optical pulses which propagate along an optical fiber is particularly interesting. At the beginning of their journey along the fiber, the pulses start emitting radiation, due to a resonance between the pulses and the small-amplitude continuous waves (radiation modes) capable of propagating in the fiber. However, after some time, the radiation stops quite abruptly. This abrupt interruption of the radiation is a phenomenon that had never been predicted before. This unusual behavior is a consequence of the presence of bands of forbidden frequencies in the dispersion relations of Equations (4) and (5), as shown in Figures 2 and 4, respectively. The dispersion relation of Equation (4) (given by Equation (11)) is particularly interesting, because it is an *elliptic curve*, and these curves have had a profound influence in other fields, as in the proof of Fermat's last theorem, or in the development of new cryptographic protocols. As explained in Appendix B, in the process of proving Fermat's last theorem it was found that every elliptic curve is associated with a *modular form*. Consequently there must exist a certain relationship between Equation (4) (and its nonlinear extensions such as Equations (A12) and (A13)) and *modular forms*. The investigation of this hidden relationship may deserve further studies. On the other hand, as elliptic curves are useful in encrypting information, it is natural to wonder if a relation may be found between cryptography and equations such as Equations (4), (A12) or Equation (A13). Although the possibility of finding such a relation might seem remote, it is interesting to observe that it has already been pointed out that ideas related to ill-posed problems might be useful in cryptography [52]. Therefore, as Equations (4), (A12) and (A13) are related to elliptic curves, and are also related to ill-posed problems, it might be worth investigating if these equations may be used to design new cryptographic algorithms. It is worth mentioning that the new methods to obtain exact solutions for nonlinear and evolution equations described in Refs. [47–49], and references therein, might be helpful to investigate if Equations (A12) and (A13) can be indeed related to modular forms, or to Weierstrass $\wp$ functions, which are closely related to elliptic curves [39].

We would like to emphasize that the appearance of bands of forbidden frequencies in the dispersion relations of Equations (4)–(8), (47) (A12) and (A13) is the principal consequence of the interplay between u_{zz} and higher-order dispersive terms (such as $-iu_{ttt}$ or u_{4t}). We have seen (at the end of Section 3.2), that these forbidden bands may be cancelled by the effect of the term cu_{zt}, but there may exist systems where such bands appear. The experimental observation by Fang et al. [38] of a gap of forbidden frequencies in the spectral profile of very short optical pulses propagating in a photonic crystal fiber may be an example of a process of this type.

The main results of this communication are, therefore, the following:

- the discovery of a relationship between Equations (4), (47), (A12) and (A13) and *elliptic curves*,
- the discovery that the interplay between u_{zz} and higher-order dispersive terms generates bands of forbidden frequencies or forbidden wavenumbers in the dispersion relations of Equations (4)–(8) and (47),
- the discovery that the Cauchy problems associated to Equations (4)–(8) are probably *ill-posed*,
- the discovery that short pulses who evolve according to Equations (4), (5) and (47) radiate in a novel and peculiar way,
- the discovery of four different types of soliton solutions in Equations (6)–(8),
- the discovery that some of the solitons of Equation (6)–(8) are *embedded solitons*,
- the discovery that Equations (A12) and (A13) might be related to *modular forms* or Weierstrass $\wp$ functions,
- the discovery that we might find a relation between the equations studied in this paper and cryptography.

Several of these results suggest further topics of research, and consequently we hope that this article encourages the research along these lines.

Acknowledgments: We thank DGTIC-UNAM (Dirección General de Cómputo y Tecnologías de Información y Comunicación de la Universidad Nacional Autónoma de México) for granting us access to the computer *Miztli* through the Project SC16-1-S-6, in order to study the numerical solutions of Equations (4)–(8). We also thank Manuel Velasco Juan for detecting an error in one of the equations in a preliminary version of this article (the error is already corrected). Finally, we thank the reviewers who revised this paper, who generously dedicated a considerable amount of time to read the preliminary versions of this article, detected weak points of the paper, and suggested ways of improving it.

Author Contributions: J. Fujioka conceived the goal and the structure of the paper, and obtained the analytical solutions. A. Gómez-Rodríguez carried out the calculations needed to obtain all the figures. A. Espinosa-Cerón carried out numerical experiments to investigate if the Cauchy problems of Equations (4)–(8) are ill-posed.

Conflicts of Interest: The authors declare no conflicts of interest.

Appendix A. Elliptic Curves

Equation (11) is a particular case of equations of the form:

$$k^2 + a_1 k\omega + a_3 k = \omega^3 + a_2\omega^2 + a_4\omega + a_6 \tag{A1}$$

which are called *Weierstrass equations* [53]. The curves described by these equations may have interesting forms, and some examples are shown in Figure A1.

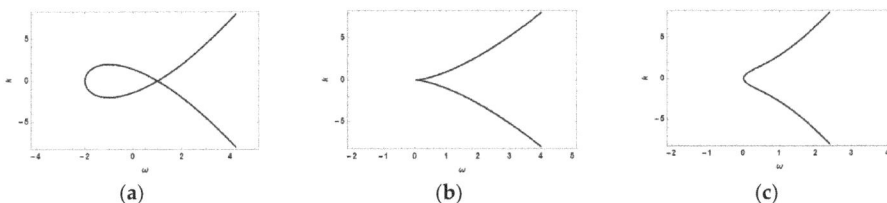

Figure A1. Different curves described by Equation (12): (a) $k^2 = \omega^3 - 3\omega + 2$; (b) $k^2 = \omega^3$; (c) $k^2 = 4\omega^3 + 4\omega$.

As we can see in this figure, these curves may have singular points where the curve intersects with itself (as in the curve $k^2 = \omega^3 - 3\omega + 2$ shown in Figure A1a), or where the curve presents cusps (as in the curve $k^2 = \omega^3$, shown in Figure A1b), but they may also describe regular curves without intersections or cusps (as in the curve $k^2 = 4\omega^3 + 4\omega$ shown in Figure A1c).

Weierstrass equations without singular points are termed *regular (nonsingular) Weierstrass equations*, and the curves described by these equations are so important in some contexts that they have received a special name: *elliptic curves*. Therefore, we have the following:

Definition A1. *An elliptic curve is a curve described by a regular Wierstrass equation.*

This definition implies that every elliptic curve is described by a Weierstrass equation. The converse, however, is not true: not every equation of the form (A1) defines an elliptic curve. Only the *regular* Weistrass equations describe elliptic curves.

Now, as the rescaled dispersion relation (11) is a Weierstrass equation, we would like to know if it describes an elliptic curve. In other words, we would like to know if the Weierstrass equation (11) is a *regular* one, or not. The answer to this query can be easily found, as there is a well-known criterion which tells us when a Weierstrass equation is *regular* [53]. This criterion can be easily modified to tell us when an equation of the form:

$$k^2 + a_1 k\omega + a_3 k = a_0 \omega^3 + a_2 \omega^2 + a_4 \omega + a_6 \tag{A2}$$

describes a *regular* curve without singular points (notice that (A2) only differs from (A1) by the presence of the coefficient a_0). The criterion is the following:

Criterion A1. *An equation of the form (A2) is regular if and only if the discriminant Δ defined below is not zero.*

To define the discriminant Δ it is useful to define the following quantities:

$$b_2 = \left(a_1^2 + 4a_2\right)/a_0 \tag{A3}$$

$$b_4 = (2a_4 + a_1 a_3)/a_0 \tag{A4}$$

$$b_6 = \left(a_3^2 + 4a_6\right)/a_0 \tag{A5}$$

$$b_8 = \left(a_1^2 a_6 + 4a_2 a_6 - a_1 a_3 a_4 + a_2 a_3^2 - a_4^2\right)/a_0^2 \tag{A6}$$

In terms of these quantities we can write the discriminant Δ in the following form:

$$\Delta = -b_2^2 b_8 - 8b_4^3 - 27b_6^2 + 9b_2 b_4 b_6 \tag{A7}$$

If we put $a_0 = 1$ in Equations (A3)–(A6), Criterion A1 tells us when a Weierstrass equation is regular. If we apply this criterion to the Weierstrass equations which describe the curves shown in Figure A1 we will see that only the curve (c) is an elliptic curve, since the discriminant Δ is zero for curves (a) and (b).

Now we would like to determine if our dispersion relation describes an elliptic curve. Equation (A2) reduces to Equation (10) if the coefficients a_n take the values:

$$a_1 = a_4 = a_6 = 0, \ a_0 = c_3/c_0, \ a_2 = -c_2/c_0, \ a_3 = 1/c_0, \tag{A8}$$

and in this case Equation (A7) takes the simpler form:

$$\Delta = 16c_2^2 / \left(c_0 c_3^4\right) \tag{A9}$$

Therefore, if c_0, c_2 and c_3 are different from zero, the condition $\Delta \neq 0$ will be satisfied, thus implying that the dispersion relation (10) is always a regular curve, and its rescaled form (11) is always an elliptic curve.

At this point it may be worth observing that if we apply in Equation (A2) the following changes of variables:

$$k = K - \frac{a_1}{2a_0^{1/3}}\Omega - \frac{a_3}{2} \ , \ \omega = \frac{\Omega}{a_0^{1/3}} \tag{A10}$$

Equation (A2) transforms into:

$$K^2 = \Omega^3 + \alpha_2\Omega^2 + \alpha_4\Omega + \alpha_6 \tag{A11}$$

where $\alpha_2 = \left(a_2 + a_1^2/4\right)/a_0^{2/3}$, $\alpha_4 = (a_4 - a_1a_3/2)/a_0^{1/3}$ and $\alpha_6 = a_3 + a_3^2/4$. For this reason we may find references where the curves defined by equations of the form $K^2 = f(\Omega)$, where $f(\Omega)$ is a cubic polynomial (with no repeated roots), are called *elliptic curves* [54]. However, this restricted definition is unable to recognize that Equation (11) describes an elliptic curve, since Equation (11) is not in the form (A11). For this reason it is more convenient to use Definition A1 (presented above) to identify elliptic curves.

Appendix B. Fermat's Last Theorem

The fact that the linear dispersion relation of Equation (4) is an elliptic curve suggests that this equation might be related to other subjects where these curves also play an important role. Among these subjects there is one which stands out among the others: Fermat's last theorem. The proof of this theorem rests, precisely, on the existence of a one-to-one correspondence between elliptic curves and a rare kind of complex functions: *modular forms*. Every elliptic curve is related to a modular form, and consequently, as Equation (4) is related to an elliptic curve, it should also bear some relationship with a modular form. Moreover, also the nonlinear extensions of Equation (4), such as the following:

$$iu_z + c_0u_{zz} + c_2u_{tt} - ic_3u_{ttt} + \gamma_1|u|^2u = 0 \tag{A12}$$

$$iu_z + c_0u_{zz} + c_2u_{tt} - ic_3u_{ttt} + \gamma_1|u|^2u - \gamma_2|u|^4u = 0 \tag{A13}$$

might be related to modular forms, since the three equations (Equations (4), (A12) and (A13)) share the same linear dispersion relation (the elliptic curve (10)). Therefore, as it might be important to advance in the study of the relationships between Equations (4), (A12) and (A13), elliptic curves and modular forms, in the following we describe the main steps which lead to the proof of Fermat's last theorem.

STEP 1: The Taniyama-Shimura conjecture.

In 1955 the young mathematician Yutaka Taniyama discovered that an unexpected (and surprising) relation seemed to exist between elliptic curves and *modular forms*. These "forms" are complex functions $F(\tau)$ which satisfy (among other requisites) the following condition [55]:

$$F\left(\frac{a\tau + b}{c\tau + d}\right) = (c\tau + d)^k F(\tau) \tag{A14}$$

where a, b, c, d and k are integers such that $ad - bc = 1$, and τ is a complex number with positive imaginary part. Taniyama, and a friend of him, Goro Shimura, studied various elliptic curves, and all of them turned out to be related to modular forms. Therefore, they conjectured that a one-to-one correspondence existed between elliptic curves and modular forms.

STEP 2: Hellegouarch's discovery.

In 1975 Yves Hellegouarch discovered that the elliptic curve [56]:

$$y^2 = x(x - a^p)(x - b^p) \tag{A15}$$

where a, b and $p > 2$ are integers, would have rather unusual properties if $a^p + b^p$ were also a p*th* power, i.e., if there existed an integer c such that:

$$a^p + b^p = c^p \tag{A16}$$

STEP 3: Frey's proposition.

In 1984 Gerhard Frey announced in a conference in Oberwolfach, Germany, that if there existed integers a, b, c and $p > 2$ satisfying (A16), then the elliptic curve (A15) would not have a modular form associated with it [57]. In other words: if Fermat's last theorem were false, then the Taniyama-Shimura would also be false, and this implied that:

If the Taniyama-Shimura is true, then Fermat's last theorem is also true!

Therefore, Frey had discovered that one way to prove Fermat's last theorem was to prove the Taniyama-Shimura conjecture. However, a problem remained: Frey's procedure to "prove" that the elliptic curve (A15) had not a corresponding modular form, contained an error. This problem was solved in 1986 by Ken Ribet, who presented a correct proof of Frey's proposition [57]. Therefore, the way to prove Fermat's last theorem was clearly indicated: it was necessary to prove the Taniyama-Shimura conjecture.

STEP 4: Wiles' proof.

In 1993 Andrew Wiles announced that he had proved a restricted form of the Taniyama-Shimura conjecture that was sufficient to prove Fermat's last theorem. It turned out that Wiles' proof had a mistake, but in 1994, in collaboration with Richard Taylor, the mistake was corrected, and Fermat's last theorem was finally proved. The full Taniyama-Shimura conjecture (now known as *modularity theorem*) was proved in 2001 by Christophe Breuil, Brian Conrad, Fred Diamond and Richard Taylor, thus completing the intellectual adventure started by Taniyama and Shimura 46 years before [58].

References

1. Kuehl, H.H.; Zhang, C.Y. Effects of higher-order dispersion on envelope solitons. *Phys. Fluids B* **1990**, *2*, 889. [CrossRef]
2. Wai, P.K.A.; Chen, H.H.; Lee, Y.C. Radiations by "solitons" at the zero group-dispersion wavelength of single-mode optical fibers. *Phys. Rev. A* **1990**, *41*, 426–439. [CrossRef] [PubMed]
3. Elgin, J.N.; Brabec, T.; Kelly, S.M.J. A perturbative theory of soliton propagation in the presence of third order dispersion. *Opt. Commun.* **1995**, *114*, 321–328. [CrossRef]
4. Wen, S.; Chi, S. Approximate solution of optical soliton in lossless fibres with third-order dispersion. *Opt. Quantum Electron.* **1989**, *21*, 335–341. [CrossRef]
5. Uzunov, I.M.; Gölles, M.; Lederer, F. Soliton interaction near the zero-dispersion wavelength. *Phys. Rev. E* **1995**, *52*, 1059–1071. [CrossRef]
6. Höök, A.; Karlsson, M. Ultrashort solitons at the minimum-dispersion wavelength: Effects of fourth-order dispersion. *Opt. Lett.* **1993**, *18*, 1388–1390. [CrossRef] [PubMed]
7. Karlsson, M.; Höök, A. Soliton-like pulses governed by fourth order dispersion in optical fibers. *Opt. Commun.* **1994**, *104*, 303–307. [CrossRef]
8. Karpman, V.I. Soliton-like pulses governed by fourth order dispersion in optical fibers. *Phys. Lett. A* **1994**, *193*, 355–358. [CrossRef]
9. Akhmediev, N.N.; Buryak, A.V.; Karlsson, M. Radiationless optical solitons with oscillating tails. *Opt. Commun.* **1994**, *110*, 540–544. [CrossRef]

10. Cavalcanti, S.B.; Cressoni, J.C.; da Cruz, H.R.; Gouveia-Neto, A.S. Modulation instability in the region of minimum group-velocity dispersion of single-mode optical fibers via an extended nonlinear Schrödinger equation. *Phys. Rev. A* **1991**, *43*, 6162–6165. [CrossRef] [PubMed]

11. Fujioka, J.; Espinosa, A. Soliton-Like Solution of an Extended NLS Equation Existing in Resonance with Linear Dispersive Waves. *J. Phys. Soc. Jpn.* **1997**, *66*, 2601–2607. [CrossRef]

12. Karpman, V.I. Evolution of solitons described by higher-order nonlinear Schrödinger equations. *Phys. Lett. A* **1998**, *244*, 397–400. [CrossRef]

13. Hayata, K.; Koshiba, M. Algebraic solitary-wave solutions of a nonlinear Schrödinger equation. *Phys. Rev. E* **1995**, *51*, 1499–1502. [CrossRef]

14. Fujioka, J.; Espinosa, A. Stability of the Bright-Type Algebraic Solitary-Wave Solutions of Two Extended Versions of the Nonlinear Schrödinger Equation. *J. Phys. Soc. Jpn.* **1996**, *65*, 2440–2446. [CrossRef]

15. Micallef, R.W.; Afanasjev, V.V.; Kivshar, Y.S.; Love, J.D. Optical solitons with power-law asymptotics. *Phys. Rev. E* **1996**, *54*, 2936–2942. [CrossRef]

16. Pelinovsky, D.E.; Afanasjev, V.V.; Kivshar, Y.S. Nonlinear theory of oscillating, decaying, and collapsing solitons in the generalized nonlinear Schrödinger equation. *Phys. Rev. E* **1996**, *53*, 1940–1953. [CrossRef]

17. Davydova, T.A.; Zaliznyak, Y.A. Schrödinger ordinary solitons and chirped solitons: Fourth-order dispersive effects and cubic-quintic nonlinearity. *Physics D* **2001**, *156*, 260–282. [CrossRef]

18. Espinosa-Cerón, A.; Fujioka, J.; Gómez-Rodríguez, A. Embedded Solitons: Four-Frequency Raiation, Front Propagation and Radiation Inhibition. *Phys. Scr.* **2003**, *67*, 314–324. [CrossRef]

19. Fujioka, J.; Espinosa, A.; Rodríguez, R.F. Fractional optical solitons. *Phys. Lett. A* **2010**, *374*, 1126–1134. [CrossRef]

20. Fujioka, J.; Espinosa, A. Radiationless Higher-Order Embedded Solitons. *J. Phys. Soc. Jpn.* **2013**, *82*, 034007. [CrossRef]

21. Kivshar, Y.S.; Agrawal, G.P. *Optical Solitons: From Fibers to Photonic Crystals*; Academic Press: San Diego, CA, USA, 2003; Sections 3.6.2 and 6.2.3.

22. Agrawal, G.P. *Nonlinear Fiber Optics*, 5th ed.; Academic Press: Oxford, UK, 2013; Section 4.4.1; pp. 116–119.

23. Fujioka, J.; Espinosa, A. Diversity of solitons in a generalized nonlinear Schrödinger equation with self-steepening and higher-order dispersive and nonlinear terms. *Chaos* **2015**, *25*, 113114. [CrossRef] [PubMed]

24. Porsezian, K.; Nakkeeran, K. Optical Solitons in Presence of Kerr Dispersion and Self-Frequency Shift. *Phys. Rev. Lett.* **1996**, *76*, 3955–3958. [CrossRef] [PubMed]

25. Karpman, V.I.; Rasmussen, J.J.; Shagalov, A.G. Dynamics of solitons and quasisolitons of the cubic third-order nonlinear Schrödinger equation. *Phys. Rev. E* **2001**, *64*, 026614. [CrossRef] [PubMed]

26. Pal, D.; Golam Ali, S.; Talukdar, B. Evolution of optical pulses in the presence of third-order dispersion. *Pramana J. Phys.* **2009**, *72*, 939–950. [CrossRef]

27. Wang, P.; Shang, T.; Feng, L.; Du, Y. Solitons for the cubic-quintic nonlinear Schrödinger equation with Raman effect in nonlinear optics. *Opt. Quantum Electron.* **2014**, *46*, 1117–1126. [CrossRef]

28. Christodoulides, D.N.; Joseph, R.I. Femtosecond solitary waves in optical fibers—Beyond the slowly varying envelope approximation. *Appl. Phys. Lett.* **1985**, *47*, 76–78. [CrossRef]

29. Leblond, H.; Mihalache, D. Optical solitons in the few-cycle regime: Recent theoretical results. *Romanian Rep. Phys.* **2012**, *63*, 1254–1266.

30. Leblond, H.; Mihalache, D. Models of few optical cycle solitons beyond the slowly varying envelope approximation. *Phys. Rep.* **2013**, *523*, 61–126. [CrossRef]

31. Leblond, H.; Mihalache, D. Ultrashort light bullets described by the two-dimensional sine-Gordon equation. *Phys. Rev. A* **2010**, *81*, 063815. [CrossRef]

32. Leblond, H.; Kremer, D.; Mihalache, D. Ultrashort spatiotemporal optical solitons in quadratic nonlinear media: Generation of line and lump solitons from few-cycle input pulses. *Phys. Rev. A* **2009**, *80*, 053812. [CrossRef]

33. Leblond, H.; Kremer, D.; Mihalache, D. Collapse of ultrashort spatiotemporal pulses described by the cubic generalized Kadomtsev-Petviashvili equation. *Phys. Rev. A* **2010**, *81*, 033824. [CrossRef]

34. Chamorro-Posada, P.; McDonald, G.S.; New, G.H.C. Non-paraxial beam propagation methods. *Opt. Commun.* **2001**, *192*, 1–12. [CrossRef]

35. Blair, S. Nonparaxial one-dimensional spatial solitons. *Chaos* **2000**, *10*, 570–583. [CrossRef] [PubMed]

36. Christian, J.M.; McDonald, G.S.; Chamorro-Posada, P. Helmholtz algebraic solitons. *J. Phys. A* **2010**, *43*, 085212. [CrossRef]
37. Tzoar, N.; Jain, M. Self-phase modulation in long-geometry optical waveguides. *Phys. Rev. A* **1981**, *23*, 1266–1270. [CrossRef]
38. Fang, X.; Karasawa, N.; Morita, R.; Windeler, R.S.; Yamashita, M. Nonlinear propagation of a Few-Optical-Cycle Pulses in a Photonic Crystal Fiber—Experimental and Theoretical Studies Beyond the Slowly Varying-Envelope Approximation. *IEEE Photonics Technol. Lett.* **2003**, *15*, 233–235. [CrossRef]
39. Koblitz, N. *Introduction to Elliptic Curves and Modular Forms*, 2nd ed.; Springer-Verlag: New York, NY, USA, 1993; p. 108.
40. Fúster Sabater, A.; Hernández Encinas, L.; Martín Muñoz, A.; Montoya Vitini, F.; Muñoz Masqué, J. *Criptografía, Protección de Datos y Aplicaciones*, 1st ed.; Alfaomega: México City, México, 2013; pp. 199–204.
41. Kivshar, Y.S.; Pelinovsky, D.E. Self-focusing and transverse instabilities of solitary waves. *Phys. Rep.* **2000**, *331*, 117–195. [CrossRef]
42. Birnir, B.; Kenig, C.E.; Ponce, G.; Svanstedt, N.; Vega, L. On the Ill-Posedness of the IVP for the Generalized Korteweg-De Vries and Nonlinear Schrödinger Equations. *J. Lond. Math. Soc.* **1996**, *53*, 551–559. [CrossRef]
43. Daripa, P. Some useful filtering techniques for illposed problems. *J. Comput. Appl. Math.* **1998**, *100*, 161–171. [CrossRef]
44. Hao, D.N.; Van, T.D.; Gorenflo, R. Towards the Cauchy problem for the Laplace equation. In *Partial Differential Equations*; Banach Center Publications, Institute of Mathematics, Polish Academy of Sciences: Warsaw, Poland, 1992; Volume 27, pp. 111–128.
45. Yang, J.; Malomed, B.A.; Kaup, D.J. Embedded Solitons in Second-Harmonic-Generating Systems. *Phys. Rev. Lett.* **1999**, *83*, 1958–1961. [CrossRef]
46. Fujioka, J.; Espinosa-Cerón, A.; Rodríguez, R.F. A survey of embedded solitons. *Rev. Mex. Fís.* **2006**, *52*, 6–14.
47. Ma, W.X.; Lee, J.-H. A transformed rational function method and exact solutions for the 3 + 1 dimensional Jimbo-Miwa equation. *Chaos Solitons Fractals* **2009**, *42*, 1356–1363. [CrossRef]
48. Ma, W.X.; Chen, M. Direct search for exact solutions to the nonlinear Schrödinger equation. *Appl. Math. Comput.* **2009**, *215*, 2835–2842. [CrossRef]
49. Ma, W.X. A refined invariant subspace method and applications to evolution equations. *Sci. China Math.* **2012**, *55*, 1769–1778. [CrossRef]
50. Piché, M.; Cormier, J.F.; Zhu, X. Bright optical soliton in the presence of fourth-order dispersion. *Opt. Lett.* **1996**, *21*, 845–847. [CrossRef] [PubMed]
51. Yang, J.; Malomed, B.A.; Kaup, D.J.; Champneys, A.R. Embedded solitons: A new type of solitary wave. *Math. Comput. Simul.* **2001**, *56*, 585–600. [CrossRef]
52. Hruby, J. On the Postquantum Cipher Scheme. Available online: https://eprint.iacr.org/2006/246.pdf (accessed on 28 October 2016).
53. Ivorra Castillo, C. Curvas Elípticas. Available online: https://www.uv.es/ivorra/Libros/Libros.htm (accessed on 30 July 2015).
54. Ellenberg, J.S. Arithmetic Geometry. In *The Princeton Companion to Mathematics*; Gowers, T., Barrow-Green, J., Leader, I., Eds.; Princeton University Press: Princeton, NJ, USA, 2008; Section IV.5; pp. 372–383.
55. Buzzard, K. Modular Forms. In *The Princeton Companion to Mathematics*; Gowers, T., Barrow-Green, J., Leader, I., Eds.; Princeton University Press: Princeton, NJ, USA, 2008; Section III.59; pp. 250–252.
56. Gowers, T.; Barrow-Green, J.; Leader, I. (Eds.) *The Princeton Companion to Mathematics*; Princeton University Press: Princeton, NJ, USA, 2008; Section V.10; pp. 691–693.
57. Singh, S. *Fermat's Last Theorem*; 4th Estate: London, UK, 1997.
58. Breuil, C.; Conrad, B.; Diamond, F.; Taylor, R. On the modularity of elliptic curves over Q: Wild 3-adic exercises. *J. Am. Math. Soc.* **2001**, *14*, 843–939. [CrossRef]

applied
sciences

MDPI

Review

Guided Self-Accelerating Airy Beams—A Mini-Review

Yiqi Zhang [1,2], Hua Zhong [1], Milivoj R. Belić [3] and Yanpeng Zhang [1,*]

[1] Key Laboratory for Physical Electronics and Devices of the Ministry of Education & Shaanxi Key Lab of Information Photonic Technique, Xi'an Jiaotong University, Xi'an 710049, China; zhangyiqi@mail.xjtu.edu.cn (Y.Z.); zhonghua@stu.xjtu.edu.cn (H.Z.)
[2] Department of Applied Physics, School of Science, Xi'an Jiaotong University, Xi'an 710049, China
[3] Science Program, Texas A&M University at Qatar, P.O. Box 23874 Doha, Qatar; Milivoj.belic@qatar.tamu.edu
[*] Correspondence: ypzhang@mail.xjtu.edu.cn; Tel.: +86-29-8266-8643 (ext. 2731)

Academic Editor: Boris Malomed
Received: 29 January 2017; Accepted: 27 March 2017; Published: 30 March 2017

Abstract: Owing to their nondiffracting, self-accelerating, and self-healing properties, Airy beams of different nature have become a subject of immense interest in the past decade. Their interesting properties have opened doors to many diverse applications. Consequently, the questions of how to properly design the spatial manipulation of Airy beams or how to implement them in different setups have become important and timely in the development of various optical devices. Here, based on our previous work, we present a short review on the spatial control of Airy beams, including the interactions of Airy beams in nonlinear media, beam propagation in harmonic potential, and the dynamics of abruptly autofocusing Airy beams in the presence of a dynamic linear potential. We demonstrate that, under the guidance of nonlinearity and an external potential, the trajectory, acceleration, structure, and even the basic properties of Airy beams can be adjusted to suit specific needs. We describe other fascinating phenomena observed with Airy beams, such as self-Fourier transformation, periodic inversion of Airy beams, and the appearance of spatial solitons in the presence of nonlinearity. These results have promoted the development of Airy beams, and have been utilized in various applications, including particle manipulation, self-trapping, and electronic matter waves.

Keywords: Airy beam; harmonic potential; dynamic linear potential; self-Fourier beam; phase transition; soliton

1. Introduction

Diffraction is a fundamental phenomenon in physical optics, due to which beams bend and spread, and the peak intensity of the beam decreases upon propagation. Sometimes it is beneficial, as in the diffraction gratings, sometimes a nuisance, as in the diffraction limit. In situations where diffraction is not desired and needs to be overcome, the nondiffracting beams come to the fore. These beams are a class of nondispersive solutions of the Helmholtz wave equation that display exotic characteristics: they are nondiffracting, self-accelerating, and self-healing, among other properties. The representative nondiffracting beams include the radially symmetric Bessel beams and the asymmetric Airy beams.

In comparison with the Bessel beam, the most remarkable feature of an Airy beam is the self-acceleration in free space [1–3]. The Airy beam concept originated from quantum mechanics. In 1979, Berry and Balazs demonstrated that the Airy function is an eigenmode of the linear Schrödinger equation [1] and that it is the only nontrivial solution that does not expand with time and it accelerates in space. The paraxial wave equation—the wave equation in the paraxial approximation—has the same form as the linear Schrödinger equation. Based on this mathematical similarity between optics and quantum mechanics, one can get not only Airy beams and Airy pulses, but also other types of waves which satisfy the paraxial wave equation, such as surface plasmon polaritons [4,5], acoustic waves [6] and water waves [7,8], among others. Notably, the nondiffracting feature of an Airy beam comes from its infinite transverse extension and power, since the ideal Airy function is not square integrable. This feature is similar to the simple plane wave. Hence, to possess finite energy and become a physical quantity, the Airy beam must be truncated. In optics, this is simply achieved by an exponential aperture, as first put forward in 2007, by Siviloglou et al. [2,9]. The truncated Airy beam can still propagate for a long distance, preserving major characteristics of an ideal Airy beam, but eventually it will diffract and lose its unique structure and properties. Nonetheless, this method of generation makes the Airy beam experimentally available and of wide interest for applications in optical beam manipulation. However, this method also limits the light energy utilization and the stability of beams in the nonlinear domain. Still, by exploiting these unique characteristics of Airy beams, various application possibilities have been explored or implemented, for example, for Airy plasma guiding [10], routing surface plasmon polaritons [11], image signal transmission [12,13], laser filamentation [14], optical micromanipulation [15,16], optical trapping [17–20], light bullet generation [21–24], electron acceleration [25], and other applications [25–29].

In addition to research on Airy beams in free space and linear media, work has also been extended to nonlinear (NL) media and regimes. In the most common optical NL media, i.e., the Kerr, saturable and quadratic media, the nonlinearity is spatially modulated, so that the beam diffraction can be effectively balanced by the nonlinearity through a soliton-like beam generation process. In 2009, Ellenbogen et al. produced an Airy beam in an asymmetrically modulated quadratic optical NL medium by the three-wave mixing process [30]. This novel nonlinear generation method not only produced an Airy beam at a new wavelength and a higher energy, but also provided new possibilities for manipulating the dynamics of Airy beams in NL media [31–35]. According to the nonlinear Schrödinger equation, the nonlinearity plays a nontrivial role in controlling the persistence as well as the breakdown of Airy beams [36–38]. It has been demonstrated that the main lobe of an Airy beam experiences self-phase modulation in NL media that results in the self-focusing or trapping of the beam [33]. And the field distribution of the Airy beam varies differently in different NL media [36,39].

As is well known, the beam focusing is an efficient way to improve laser power density. Thanks to the self-accelerating feature of Airy beams, in 2010, Efremidis and Christodoulides proposed a novel abruptly autofocusing (AAF) beam [40]. The beam possesses a ring-shaped initial transverse Airy amplitude pattern, and accelerates in either the inward or outward direction, determined by the Airy wave function tail [41]. During propagation, the AAF beam can keep a low intensity profile initially, and then abruptly converge to a focal point where the intensity grows by orders of magnitude [40,42–46]. This abruptly autofocusing property of an AAF beam avoids the possible interaction of the beam with the transmission medium before focusing. It can be used in biomedical treatments, bottle beams, light bullets, and in other nonlinear settings.

To date, immense research work has been devoted to Airy beams, from theoretical predications to experimental verifications, from fundamental research to potential applications. Without doubt, it has become one of the hottest developing fields in linear and nonlinear optics [47–50]. In this short review, based on our previously published work, we discuss some robust and flexible manipulation techniques applied to Airy beams. The organization of the paper is as follows. Section 2 provides results and discussion. In Section 2.1, we describe the interaction of two Airy beams as they propagate simultaneously in a NL medium; in Section 2.2, we summarize Airy beam propagation in a harmonic potential; in Section 2.3, we discuss the controllable spatial modulation of AAF beams, under an action of different dynamic linear potentials. In Section 3, we conclude the paper.

2. Results and Discussion

In quantum mechanics, the Schrödinger equation (SE) for a quantum particle moving in free space is written as:

$$i\hbar \frac{\partial \psi}{\partial t} + \frac{\hbar^2}{2m} \frac{\partial^2 \psi}{\partial x^2} = 0, \tag{1}$$

where $\hbar$ is the reduced Planck's constant and m is the particle mass. This equation describes the development of the wave function ψ of the particle, but in appropriate units it could also describe the development of an Airy wave packet.

In optics, the propagation of a scalar wave packet obeys the Helmholtz equation:

$$\left(\frac{\partial^2}{\partial x^2} + \frac{\partial^2}{\partial z^2} \right) \psi + k^2 \psi = 0, \tag{2}$$

where x and z are the transverse and longitudinal coordinates, and $k = 2\pi n / \lambda_0$ is the wavenumber (n is the index of refraction and λ_0 the wavelength in free space). Under the paraxial approximation $|\partial_z^2 \psi| \ll |2k\partial_z \psi|$, one obtains the paraxial wave equation,

$$i \frac{\partial \psi}{\partial z} + \frac{1}{2} \frac{\partial^2 \psi}{\partial x^2} = 0, \tag{3}$$

where now the variables x and z are the normalized transverse coordinate and the propagation distance, scaled by some characteristic transverse width x_0 and the corresponding Rayleigh range kx_0^2. Obviously, Equation (3) has the same form as Equation (1) in scaled units—it is just the SE without potential—so one of the accelerating solutions of Equation (3) is the well-known Airy function with the characteristic infinite oscillatory tail,

$$\psi(x,z) = \text{Ai} \left(x - \frac{z^2}{4} \right) \exp \left[\frac{i}{12} \left(6xz - z^3 \right) \right]. \tag{4}$$

From this solution, it is not hard to see that the trajectory is determined by the transverse accelerating term $x - z^2/4$, so the beam propagates along a parabolic curve. Note that the intensity of an ideal Airy wave packet remains invariant during propagation, as displayed in Figure 1a. However, the ideal Airy beam does not exist in reality, due to its infinite energy. To make it realizable in an experiment, an exponentially tapered Airy beam is introduced [2,9],

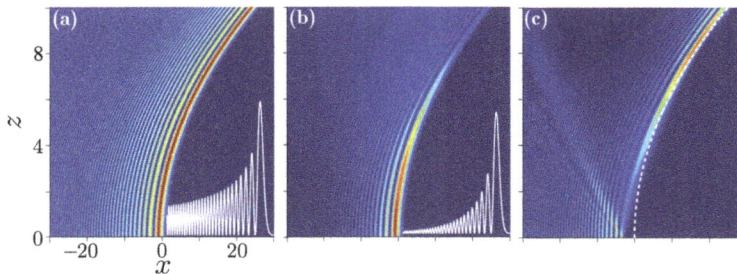

Figure 1. (**a**) Propagation of the ideal Airy beam according to Equation (3). The inset in the right corner represents the energy distribution of the beam at $z = 0$; (**b**) Same as (**a**), but for a truncated Airy beam, with $a = 0.1$; (**c**) Self-healing process of the truncated Airy beam. The white dashed line is the theoretical trajectory.

$$\psi(x) = \text{Ai}(x) \exp(ax), \tag{5}$$

where $a \geq 0$ is an arbitrary real decay parameter. In the momentum space, the corresponding Fourier transform is

$$\hat{\psi}(k) = \exp(-ak^2) \exp\left[\frac{a^3}{3} + \frac{i}{3}\left(k^3 - 3a^2k\right)\right], \tag{6}$$

which is of limited energy. So, according to the Parseval's theorem, the energy of the attenuated Airy beam is also limited. The propagation of the attenuated Airy beam is depicted by the solution

$$\psi(x,z) = \text{Ai}\left(x - \frac{z^2}{4} + iaz\right) \exp\left[\frac{i}{12}\left(6a^2z - 12iax + 6iaz^2 + 6xz - z^3\right)\right], \tag{7}$$

as shown in Figure 1b. Compared with the ideal Airy beam, the tail of the truncated beam quickly decays during propagation, which makes the nondiffracting and self-accelerating properties preserved only over a finite distance. In Figure 1c, the healing property of the Airy beam is displayed. The main lobe of the Airy beam is screened out initially, but it recovers quickly during propagation, due to the transfer of energy from the tail to the head of the beam [19,51,52].

Similarly, the AAF exponentially apodized radially symmetric Airy beam is written as:

$$\varphi_0(r) = \text{Ai}[\pm(r_0 - r)] \exp[\pm a(r_0 - r)], \tag{8}$$

where r_0 is the initial radius of the main lobe, and $\pm$ corresponds to the inward or outward going beams, respectively.

2.1. Nonlinear Guidance

Based on the above analysis, one finds that the properties of Airy beams are stable in free space. Naturally, one wonders whether these concepts can be extended to inhomogeneous or nonlinear media. Indeed, it has been confirmed that Airy beams can exist in photonic lattices and give rise to interesting phenomena, such as accelerating lattice solitons [53]. Furthermore, we have investigated the interactions of Airy beams in different NL media [34,35]. The governing nonlinear Schrödinger equation (NLSE) can now be written as

$$i\frac{\partial \psi}{\partial z} + \frac{1}{2}\frac{\partial^2 \psi}{\partial x^2} + \delta n \psi = 0, \tag{9}$$

where δn—a function of the intensity $|\psi(x)|^2$—is the refractive index change. It acts as a potential in the Schrödinger equation. The index change depends both on the light and the material, and it may vary widely. In Kerr media, for example, this change is proportional to $n_2 I$, where n_2 is the second-order nonlinear index and I is the intensity of the wave. For most materials, the value of n_2 is rather small, e.g., between 10^{-16} and 10^{-14} cm^2/W for glasses and transparent crystals, but in liquid crystals it can reach 10^{-4} cm^2/W. Thus, with a laser of 1 GW/cm^2, these nonlinear effects are easily observable over relatively short propagation distances (km in fibers and mm in liquid crystals).

For the sake of obtaining an accelerating solution of NLSE—a nonlinear accelerating beam—one introduces $x - z^2/4$ as a new variable, instead of x in Equation (9), to end up with

$$i\frac{\partial \psi}{\partial z} - i\frac{z}{2}\frac{\partial \psi}{\partial x} + \frac{1}{2}\frac{\partial^2 \psi}{\partial x^2} + \delta n \psi = 0. \tag{10}$$

Assuming the solution of Equation (10) of the form $\psi(x,z) = u(x)\exp[i(xz/2 + z^3/24)]$, allows Equation (10) to be recast into

$$\frac{\partial^2 u}{\partial x^2} + 2\delta n u - x u = 0, \tag{11}$$

which is simpler than Equation (9). We treat it as an initial value problem with a required asymptotic behavior $u(x) = \alpha \text{Ai}(x)$ and $u'(x) = \alpha \text{Ai}'(x)$ for large $x > 0$; here α represents the strength of the nonlinearity induced by the assumed solution. Similar to the ordinary Airy beams, the nonlinear accelerating beams are accelerating along parabolic trajectories.

To investigate the interaction of Airy beams, we take the initial beam as a superposition of two Airy components

$$\psi(x) = A_1 \text{Ai}[(x - B)] \exp[a(x - B)] + \exp(il\pi) A_2 \text{Ai}[-(x + B)] \exp[-a(x + B)], \qquad (12)$$

where B is the transverse position shift and l controls the phase shift. If $l = 0$, the two components are in-phase, while if $l = 1$, they are out-of-phase.

2.1.1. Kerr Medium

Initially, we consider the beam interaction in a Kerr NL medium, with $\delta n = |\psi(x)|^2$. Since the energy is mainly stored in the main lobe of the Airy beam, a large distance between components will lead to a weak interaction, so we just consider the interaction for relatively small distances. The results are shown in Figure 2.

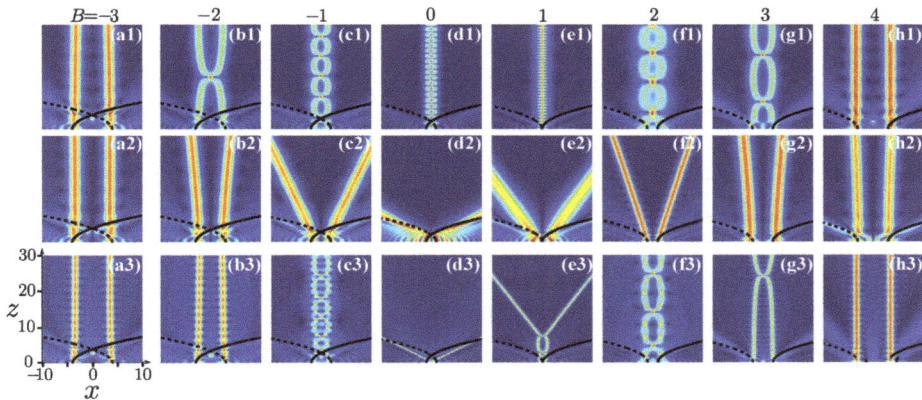

Figure 2. Soliton formation in the interaction of two in-phase (**a1–h1**) and out-of-phase (**a2–h2**) incident Airy beams with $A_1 = A_2 = 3$, in the Kerr medium. (**a3–h3**) The same as (**a1–h1**), but with $A_1 = A_2 = 4$. The distance between beams is chosen by varying B (on the top of the figure). Black solid and dashed curves represent the ideal accelerating trajectories of the main lobes. Reproduced with permission from [35], Copyright the Optical Society of America, 2014.

Obviously, the major difference between the first two rows is the attraction of beams when the beams are in-phase and the repulsion when they are out-of-phase. Also visible is the breathing or the filamentation of the beams when they strongly interact. In the in-phase case, for a large distance, the two Airy components form two parallel solitons, as depicted in Figure 2a1,h1. With decreasing distance between the beams, the attraction between components increases, and the bound breathing solitons form. In general, the smaller the distance, the stronger the attraction and the smaller the period of soliton breathing. Curiously, the intensity image shown in Figure 2e1 has a smaller period than the one in Figure 2d1, even though in that case $B = 0$. The reason is that the main lobe of the Airy beam with $B = 0$ is located at about -1, and there is still an interval between the two main lobes in the incidence, so the attraction is the biggest when $B = 1$ and the period of the formed soliton is then the smallest.

The results for the out-of-phase beams are shown in Figure 2a2–h2. One can see that the soliton pairs formed from the incidence actually repel each other; the smaller the distance, the stronger the repulsion, until the beams overlap. Considering that the two Airy components are out-of phase, the main lobes will balance each other out at $B = 1$, so the soliton pair shown in Figure 2e2 is generated from the secondary lobes, while the other two come from main lobes. This is why the repulsion of the soliton pair in Figure 2d2 is stronger than that in Figure 2e2. Notably, only two soliton pairs in Figure 2h2 are visible: the outer pair comes from the main lobes of the Airy components and the other from the secondary lobes. These results will be different when A is varied; for small A (less than 1), there will be no solitons generated; for large A ($\sim$10), multiple soliton pairs will be produced, but the propagation may become unstable because of the catastrophic self-focusing effect.

The results for the in-phase beams when $A_1 = A_2 = 4$ are shown in Figure 2a3–h3; from these, one finds repulsion between the two solitons, especially in the cases $B = 0$ and $B = 1$. As shown in Figure 2e3, when $B = 1$, the refractive index change will make the solitons attract each other, and the attraction is quite strong over a long distance, but eventually the repulsion overtakes the attraction. The intensity of the superposed main lobes is enhanced, while the width is suppressed, in comparison with the case $B = 0$. Thus, the two solitons generated in the splitting of the overlapping main lobes will experience a smaller repulsion force than in Figure 2d3. When B is further increased, the main lobe of one component will superpose with the high-order lobes of the other component, so the solitons will come from the overlapping main and high-order lobes, as shown in Figure 2h3. When the distance between two solitons is large, their interaction becomes weak, and they propagate in parallel, as in Figure 2a,h. In general, when the two interacting Airy beams are of different amplitudes, their energy distribution will be asymmetric, and the generated solitons will be of different intensities and mostly breathing.

2.1.2. Saturable Medium

In the saturable NL medium, the nonlinearity is of the form $\delta n = |\psi|^2/(1 + |\psi|^2)$. The behavior of interacting Airy beams is quite similar to the case of the Kerr medium, but the interactions also become "plastic". As a rule, the in-phase case can generate individual solitons that are positioned centrally. For small amplitudes A_1 and A_2, the individual solitons or soliton pairs cannot be formed in the interaction. Importantly, the repulsion between soliton pairs formed in the saturable NL medium is stronger than that in the Kerr medium. Different from the Kerr case, the propagation in saturable NL medium is stable for arbitrary A_1 and A_2.

2.1.3. Soliton and Kerr Case

Thus far, we have considered the interaction of Airy beams in different NL media; in this subsection, we investigate the interaction between a solitary beam and a Kerr nonlinear accelerating beam. As it is well known, in the Kerr medium, Equation (9) supports a stationary soliton solution of the form

$$\psi(x,z) = \operatorname{sech}(x)\exp(iz/2). \tag{13}$$

In principle, the emerging breathing soliton comes from the soliton component, modulated by the lobes of the Kerr accelerating beam. When the distance between two components is big, the soliton will collide with the relatively weak lobes of the nonlinear accelerating beam. In this case, the soliton will exhibit fluctuations and the main lobes will preserve the accelerating property of the beam, because of the insufficient interaction between beams to produce solitons from the main lobe. When the distance between the beams is small, the main lobes interact with the soliton, and the propagation properties depend on the profile of the superposed beam. Owing to the conservation laws and the stability of both beams, the properties of the soliton and nonlinear accelerating beam are quite immune to the collision, although the main lobe is affected both in amplitude and width, but it still conserves accelerating property, which is different from the cases mentioned above.

We would like to note that interactions of Airy beams have also been carried out in other types of nonlinear media, for example, nonlocal nonlinear media [54,55] and photorefractive nonlinear media [56,57]. In recent years, nonlinear modulation of Airy beams in the temporal domain [58–60] has attracted special attention.

2.2. Harmonic Potential Guidance

In optics and photonics, an external potential embedded in the medium's index of refraction is often used as an effective tool to modulate light beams. It comes in different forms, as exemplified by vastly different photonic crystal structures. In this subsection, we investigate the management of Airy beams by a harmonic potential added to the linear medium. Typically, such a potential is easily achieved in gradient-index (GRIN) media [61,62] and frequently utilized as a harmonic trap in Bose–Einstein condensates.

2.2.1. One-Dimensional Airy Beams

In the one-dimensional (1D) case, the paraxial propagation of a beam in a linear medium with an external harmonic potential is described by the following equation [63,64]:

$$i\frac{\partial\psi}{\partial z} + \frac{1}{2}\frac{\partial^2\psi}{\partial x^2} - \frac{1}{2}\alpha^2 x^2\psi = 0, \tag{14}$$

where α determines the width of the harmonic potential. The Fourier transformation (FT) of Equation (14) leads to the corresponding equation in the inverse space:

$$i\frac{\partial\hat{\psi}}{\partial z} + \frac{1}{2}\alpha^2\frac{\partial^2\hat{\psi}}{\partial k^2} - \frac{1}{2}k^2\hat{\psi} = 0. \tag{15}$$

Clearly, if $\alpha = 1$, Equations (14) and (15) have the same form, so that both equations have the same solutions but expressed in real (x) and inverse (k) spaces [63], respectively. As before, we are interested in the behavior of Airy beams. The propagation of a truncated Airy beam in the harmonic potential is shown in Figure 3.

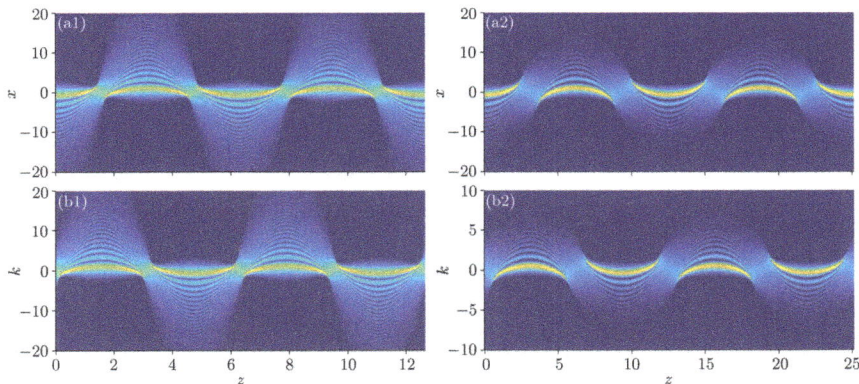

Figure 3. (Color online) Propagation of a finite energy Airy beam in a harmonic potential. (**a1,a2**) Real space; (**b1,b2**) Inverse space. Periodic inversion and an automatic Fourier transform of the beam are evident. The parameters are: $a = 0.1$, $\alpha = 1$ (**a1, b1**) and $\alpha = 0.5$ (**a2, b2**). Reproduced with permission from [63], Copyright Elsevier, 2015.

Generally, the solution of Equation (14) can be written as [63–69]:

$$\psi(x, z) = \int_{-\infty}^{+\infty} \psi(\xi, 0) \sqrt{\mathcal{H}(x, \xi, z)} d\xi, \tag{16}$$

where

$$\mathcal{H}(x, \xi, z) = -\frac{i}{2\pi} \alpha \csc(\alpha z) \exp\left\{ i\alpha \cot(\alpha z) \left[x^2 + \xi^2 - 2x\xi \sec(\alpha z) \right] \right\} \tag{17}$$

is associated with the corresponding kernel. Combining Equations (16) and (17), after some algebra one arrives at

$$\psi(x, z) = f(x, z) \int_{-\infty}^{+\infty} \left[\psi(\xi, 0) \exp\left(ib\xi^2 \right) \right] \exp(-iK\xi) d\xi, \tag{18}$$

where $b = \frac{\alpha}{2} \cot(\alpha z)$, $K = \alpha x \csc(\alpha z)$, and

$$f(x, z) = \sqrt{-\frac{i}{2\pi} \frac{K}{x}} \exp\left(ibx^2 \right).$$

One can see that the integral in Equation (18) is a Fourier transform of $\varphi(x, 0) \exp(ibx^2)$. In other words, the propagation of a beam in a harmonic potential is equivalent to an automatic FT, that is, to the periodic change from the beam to the FT of the beam with a parabolic chirp and back. It is worth mentioning that the same formula also represents a fractional Fourier transform of the initial beam [67,70,71], the "degree" of which is proportional to the propagation distance.

By choosing the input as $\psi(x, 0) = Ai(x) \exp(ax)$, the solution in Equation (18) can be found using the following steps:

(i)   Find the Fourier transforms of $\psi(x, 0) = Ai(x) \exp(ax)$ and $\exp(ibx^2)$, which can be written as [2,9,63,64]:

$$\hat{\psi}(k) = \exp\left(-ak^2 \right) \exp\left[\frac{a^3}{3} + \frac{i}{3} \left(k^3 - 3a^2 k \right) \right], \tag{19}$$

and

$$\sqrt{i\frac{\pi}{b}} \exp\left(-\frac{i}{4b} k^2 \right), \tag{20}$$

respectively.

(ii) Perform the convolution of the two Fourier transforms in Equations (19) and (20), and using the definition [72]

$$Ai(x) = \frac{1}{2\pi i} \int_{-i\infty}^{+i\infty} \exp\left(xt - \frac{t^3}{3} \right) dt,$$

find the inverse Fourier transform:

$$\psi(x, z) = - f(x, z) \sqrt{i\frac{\pi}{b}} \exp\left(\frac{a^3}{3} \right) Ai\left(\frac{K}{2b} - \frac{1}{16b^2} + i\frac{a}{2b} \right)$$
$$\times \exp\left[\left(a + \frac{i}{4b} \right) \left(\frac{K}{2b} - \frac{1}{16b^2} + i\frac{a}{2b} \right) \right] \exp\left[-i\frac{K^2}{4b} - \frac{1}{3} \left(a + \frac{i}{4b} \right)^3 \right]. \tag{21}$$

From this expression, one obtains the accelerating trajectory of the initial beam:

$$x = \frac{1}{4\alpha^2} \frac{\sin^2(\alpha z)}{\cos(\alpha z)}, \tag{22}$$

with the period $D = 2\pi/\alpha$. Here, $z \neq (2m+1)D/4$, where m is a positive integer. This trajectory is ideal, because it indicates that the Airy beam can accelerate all the way to infinity $x \to \pm\infty$, when $z \to (2m+1)D/4$. However, upon exponential apodization of the Airy beam, such an acceleration will stop when z is close to the points mentioned. At these points, the beam will turn around, accelerate in the opposite direction, and change the shape. We call these points the phase transition points, for the reasons explained below.

So, several issues concerning Equation (21) must be addressed:

(i) When $z = mD$, we have $\psi(x, z) = \psi(x, 0)$—an initial beam recurrence.
(ii) When $z = (2m+1)D/2$, we have $\psi(x, z) = \psi(-x, 0)$—an inversion of the initial beam.
(iii) When $z = (2m+1)D/4$, by directly solving Equation (18) for the FT of the initial beam, we have

$$\psi\left(x, z = \frac{2m+1}{4}D\right) = \sqrt{-i\frac{s\alpha}{2\pi}} \exp\left(-a\alpha^2 x^2\right) \exp\left[\frac{a^3}{3} + i\frac{s}{3}\left(\alpha^3 x^3 - 3a^2\alpha x\right)\right], \tag{23}$$

where $s = 1$ if m is even and $s = -1$ if m is odd. This field is unrelated to the initial Airy beam—that is, it represents a new "phase" of the propagating beam.

Equation (23) displays a Gaussian intensity profile, which is completely different from the intensity profiles elsewhere during propagation. It is similar to the propagating Gaussian pulse as it bounces off the harmonic potential wall—but the Gaussian beam remains Gaussian in propagation, whereas this pulse inverts and becomes an inverse Airy beam. On the other hand, it is different from a free Gaussian wave packet hitting an infinite potential wall—there, during the bounce, the packet becomes a rapidly oscillating multi-peaked structure, owing to the interference between the incoming and the reflected beam. Since the inversion introduces a discontinuity in the velocity and a singularity in the acceleration, for lack of a better word, we refer to the phenomenon as the phase transition of the finite energy Airy beam, due to the harmonic potential. Correspondingly, $z = (2m+1)D/4$ are the phase transition points.

When we introduce a transverse displacement x_0 of the beam, the initial beam is $\psi(x, 0) = Ai(x - x_0) \exp[a(x - x_0)]$, and the solution can be written as:

$$\psi(x, z) = -f(x, z)\sqrt{i\frac{\pi}{b}} \exp\left(\frac{a^3}{3}\right) Ai\left(\frac{K}{2b} - \frac{1}{16b^2} + i\frac{a}{2b} - x_0\right)$$

$$\times \exp\left[\left(a + \frac{i}{4b}\right)\left(\frac{K}{2b} - \frac{1}{16b^2} + i\frac{a}{2b} - x_0\right)\right] \exp\left[-i\frac{K^2}{4b} - \frac{1}{3}\left(a + \frac{i}{4b}\right)^3\right]. \tag{24}$$

The corresponding trajectory is:

$$x = \frac{1}{4\alpha^2}\frac{\sin^2(\alpha z)}{\cos(\alpha z)} + x_0 \cos(\alpha z), \tag{25}$$

At the transition points, we have:

$$\psi\left(x, z = \frac{2m+1}{4}D\right) = \sqrt{-i\frac{s\alpha}{2\pi}} \exp(-ix_0\alpha x) \exp\left(-a\alpha^2 x^2\right) \exp\left[\frac{a^3}{3} + i\frac{s}{3}\left(\alpha^3 x^3 - 3a^2\alpha x\right)\right]. \tag{26}$$

Comparing with the former case, the transverse displacement introduces a linear chirp at the transition points. There is no change of the period and phase transition points, so one can predict that the beam executes the same motion as before, but it is transversely stretched. One can also predict that with an increasing transverse displacement $|x_0|$, the beam will accelerate along an ever more elongated cosine curve.

To explore the phase transition region more clearly, we show the numerical simulations of trajectories, velocities, and accelerations of the beam during propagation in Figure 4, for different cases. One can observe that the accelerating trajectories are modulated by the transverse displacement; the beam acceleration with $x_0 < 0$ being opposite to the case with $x_0 > 0$. In addition, the beam inversion produces a discontinuity in the velocity and a singularity in the acceleration, which demonstrates nicely that the motion is not harmonic and that there exist two phase regions: the Airy phase and the single-peak phase. According to Equations (24) and (26), the single-peak structure only occurs at the phase transition points; before and after these points, the beam still exhibits multi-peak structure, having to reconnect the accelerating motion before the point with the decelerating motion after the point. The length of the single-peak phase is determined by the size of the decay parameter, as displayed in Figure 4b; the smaller a, the smaller the length, and the harder it is for the beam to make a sudden inversion, so the region in Figure 4c is narrower than that in Figure 4d. However, x_0 has no effect on the width of the single-peak phase region when a is fixed.

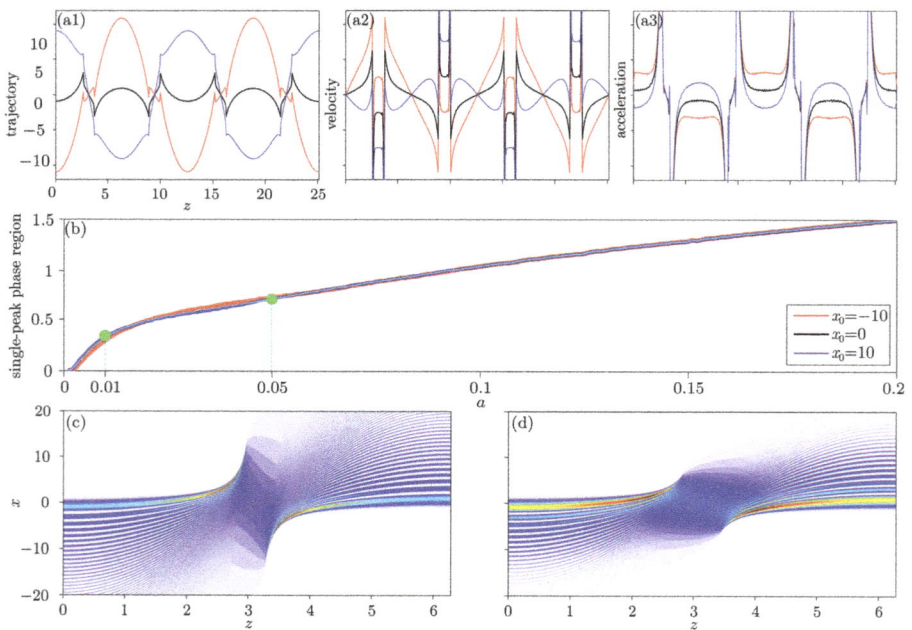

Figure 4. (**a**) Numerical trajectory (**a1**), velocity (**a2**) and acceleration (**a3**) of the Airy beam during propagation in a harmonic potential. Red, black, and blue curves correspond to the transversely displaced beams, with displacements $x_0 = -10, 0$, and 10, respectively, and with $a = 0.1$; (**b**) The width of the single-peak phase region versus the decay parameter a. (**c,d**) Corresponding to the green dots in (**b**). In the left panel, $a = 0.01$; in the right, $a = 0.05$. Other parameters: $\alpha = 0.5$. Reproduced with permission from [64], Copyright the Optical Society of America, 2015.

We next consider the initial beam with a linear chirp:

$$\psi(x, 0) = \text{Ai}(x - x_0) \exp[a(x - x_0)] \exp(i\beta x), \qquad (27)$$

with β being the constant wavenumber. Thus, the solution is written as:

$$\psi(x, z) = - f(x, z)\sqrt{i\frac{\pi}{b}} \exp\left(\frac{a^3}{3}\right) \text{Ai}\left(\frac{K'}{2b} - \frac{1}{16b^2} + i\frac{a}{2b} - x_0\right)$$
$$\times \exp\left[\left(a + \frac{i}{4b}\right)\left(\frac{K'}{2b} - \frac{1}{16b^2} + i\frac{a}{2b} - x_0\right)\right] \exp\left[-i\frac{K'^2}{4b} - \frac{1}{3}\left(a + \frac{i}{4b}\right)^3\right], \tag{28}$$

with $K' = K - \beta$. Clearly, the period $\mathcal{D}$ does not change and the phase transition point is still an odd integer multiple of the quarters of the period. Mathematically, the trajectory is given by

$$x = \frac{1}{4\alpha^2}\frac{\sin^2(\alpha z)}{\cos(\alpha z)} + x_0 \cos(\alpha z) + \frac{\beta}{\alpha}\sin(\alpha z),$$

and is modulated greatly by the linear chirp. At the phase transition points, we have:

$$\psi\left(x, z = \frac{2m+1}{4}\mathcal{D}\right) = \sqrt{-i\frac{s\alpha}{2\pi}} \exp[-ix_0(\alpha x - \beta)] \exp\left[-a(\alpha x - \beta)^2\right]$$
$$\times \exp\left\{\frac{a^3}{3} + i\frac{s}{3}\left[(\alpha x - \beta)^3 - 3a^2(\alpha x - \beta)\right]\right\}. \tag{29}$$

In this case, the beam is equivalent to an obliquely incident beam, but without the ballistic properties due to the harmonic potential [3,73].

If the initial finite energy Airy beam carries a quadratic chirp,

$$\psi(x, 0) = \text{Ai}(x - x_0) \exp\left[a(x - x_0)\right] \exp\left(i\beta x^2\right), \tag{30}$$

the analytical solution will be

$$\psi(x, z) = - f(x, z)\sqrt{i\frac{\pi}{b'}} \exp\left(\frac{a^3}{3}\right) \text{Ai}\left(\frac{K}{2b'} - \frac{1}{16b'^2} + i\frac{a}{2b'} - x_0\right)$$
$$\times \exp\left[\left(a + \frac{i}{4b'}\right)\left(\frac{K}{2b'} - \frac{1}{16b'^2} + i\frac{a}{2b'} - x_0\right)\right] \exp\left[-i\frac{K^2}{4b'} - \frac{1}{3}\left(a + \frac{i}{4b'}\right)^3\right]. \tag{31}$$

Here, $b' = b + \beta$, but the period in this case is still $\mathcal{D}$. However, the phase transition points are not the same as before; they are obtained as:

$$z = \frac{1}{\alpha}\arctan\left(-\frac{\alpha}{2\beta}\right) + \frac{m}{2}\mathcal{D}. \tag{32}$$

Concerning the trajectory, it is now:

$$x = \frac{\sin^2(\alpha z)}{4\alpha[\alpha \cos(\alpha z) + 2\beta \sin(\alpha z)]} + [\alpha \cos(\alpha z) + 2\beta \sin(\alpha z)]x_0, \tag{33}$$

and obviously, the influence from the quadratic chirp is not negligible.

Comparing these two chirped cases, we note that in both cases the trajectories are modulated greatly. In the linear chirp case, the phase transition points and the period do not change, but the beam at the phase transition point has a transverse displacement. While in the quadratic chirp case, the phase transition point is moved, but the beam profile is not affected.

By now, it is apparent that the propagation of beams according to the linear Schrödinger equation with parabolic potential is intimately connected with the self-Fourier (SF) transform. Generally, for an arbitrary $\psi(x)$ propagating to $\pi/4$ in a harmonic potential, an SF beam will be obtained at that point [74]. Here, when a truncated Airy beam $\varphi(x) = Ai(x)\exp(ax)$ propagates to $z = \pi/(4\alpha)$, we find the corresponding Fourier transform pair:

$$\mathcal{F}[\psi(x)](k) = \sqrt{\frac{2\pi}{\alpha}}\psi\left(-\frac{k}{\alpha}\right).$$ (34)

Therefore, the expression for the self-Fourier beam is:

$$\psi(x) = -\sqrt[4]{2}\,Ai\left(\sqrt{2}x - \frac{1}{4\alpha^2} + i\frac{a}{\alpha}\right)\exp\left[a\left(\sqrt{2}x - \frac{1}{2\alpha^2}\right)\right] \times$$
$$\exp\left[-\frac{i}{2}\left(2\alpha x^2 - \frac{\sqrt{2}}{\alpha}x + \frac{a^2}{\alpha} + \frac{1}{6\alpha^3}\right)\right].$$ (35)

In Figure 5, the intensity of the SF beam is shown by the black curve, and the corresponding intensity in Fourier space is shown by the red curve. One can see that the beam profiles are the same except for the inversion, which is in accordance with the theoretical result presented in Equation (34). This way of generating SF beams is universal, it does not depend on the form of the initial beam. Furthermore, we have recently demonstrated that the linear and nonlinear Talbot effects might be interpreted as a fractional SF or a regular SF transform phenomenon, respectively [75,76]. Such SF beams may find potential applications in optical information processing, routing, and switching.

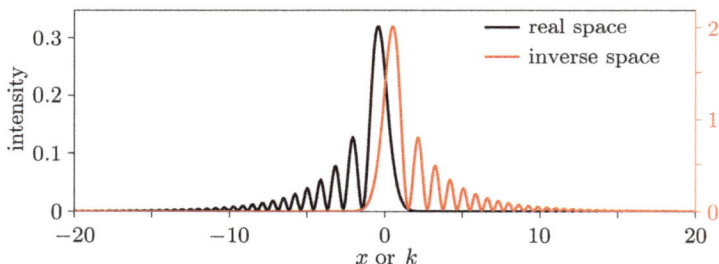

Figure 5. (Color online) Comparison of intensities of an Airy beam at $z = \pi/4$ in real space and inverse space, corresponding to Figure 3a. Intensities in real and frequency spaces refer to the left and right y scales, respectively. Reproduced with permission from [63], Copyright Elsevier, 2015.

2.2.2. Two-Dimensional Case

Naturally, the harmonic oscillator model in Equation (14) is easily extended to two, three or even four dimensions [64]. In 2D, it has the form:

$$i\frac{\partial\psi}{\partial z} + \frac{1}{2}\left(\frac{\partial^2\psi}{\partial x^2} + \frac{\partial^2\psi}{\partial y^2}\right) - \frac{1}{2}\alpha^2\left(x^2 + y^2\right)\psi = 0,$$ (36)

with the initial beam being:

$$\psi(x,y,z=0) = Ai(x)Ai(y)\exp[a(x+y)].$$ (37)

By the separation of variables, the 2D problem can be reduced to two 1D cases [64,77]. The result is displayed in Figure 6. Similar to the 1D case, the 2D Airy beam displays inversion and phase transition (the gaps represent the single-peak regions) during propagation. From Figure 6c,d, one can clearly see that the beam at $z = \pi/(4\alpha)$ is still an SF beam, just like in the 1D case. Thus, the wave function is a product of two finite-energy Airy beams: one along x and the other along the y direction. In a 2D parabolic potential, the wave exhibits all the properties of 1D Airy beams: periodic inversion, phase transition, and anharmonic oscillation.

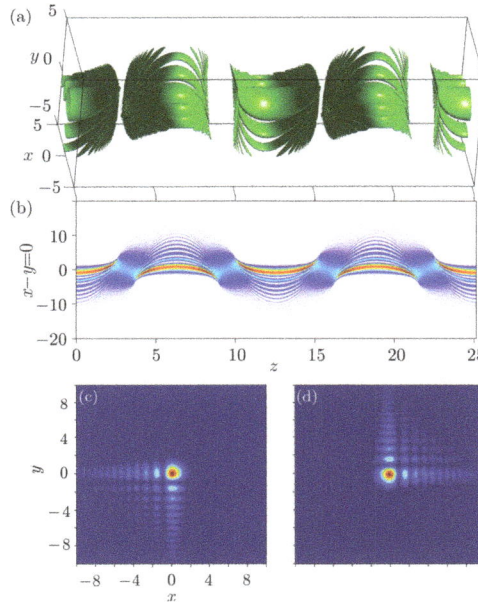

Figure 6. Propagation of a 2D finite energy Airy beam $\psi(x, y) = \exp(ax)\mathrm{Ai}(x)\exp(ay)\mathrm{Ai}(y)$ in a harmonic potential. (**a**) Iso-surface plot; (**b**) Intensity in the cross section $x - y = 0$; (**c**) Intensity of the beam at $\pi/(4\alpha)$ in the real spaces; (**d**) The corresponding intensity at $\pi/(4\alpha)$ in the inverse space. The parameters are $a = 0.1$ and $\alpha = 0.5$. Reproduced with permission from [64], Copyright the Optical Society of America, 2015.

For the cases that cannot be treated with the variable separation method, e.g., when the initial beam is a superposition of AAF beams carrying orbital angular momentum [18,41,42,45,78], in the radially symmetric case one can switch to the polar coordinates. Then, the input can be written as:

$$\psi(r,\theta) = \mathrm{Ai}[\pm(r_0 - r)]\exp[\pm a(r_0 - r)]\sum_{n=1}^{4}\exp(in\theta), \tag{38}$$

where $\pm$ represents the inward and outward AAF beams, r_0 determines the location of the main ring, and θ represents the spatial frequency in polar coordinates. An analytical propagating solution of Equation (38) is hard to obtain. However, using a fairly accurate approximation method developed in [18,79], the AAF beam propagation can be described as a superposition of Bessel beams:

$$\psi(r,\theta,z) \approx -A_0 f(r,\theta,z)\exp\left(ibr_0^2\right)\sum_{n=1}^{4} i^{1-n}\exp(in\theta)J_n(r_0 r), \tag{39}$$

where $A_0 \approx (1 - a^2/r_0)\exp(a^3/3)$. The results are depicted in Figure 7. The first row of panels presents the intensity of an outward AAF beam; the second row of panels presents the intensity of an inward AAF beam.

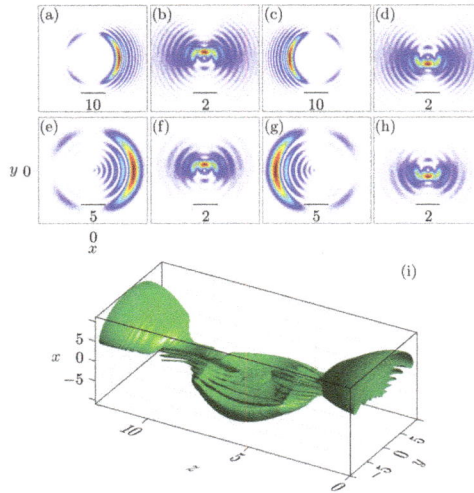

Figure 7. Propagation of circular Airy beams. From left to right: Intensity distributions at $z = 0$, $z = \mathcal{D}/4$, $z = \mathcal{D}/2$, and $z = 3\mathcal{D}/4$. Intervals inside give the relative measure of the beam size. (**a–d**) Outward abruptly autofocusing (AAF) beam; (**e–h**) Inward AAF beam; (**i**) Iso-surface plot of the propagation of the inward AAF beam. Parameters: $\alpha = 0.5$, $a = 0.1$ and $r_0 = 10$. Reproduced with permission from [79], Copyright the Optical Society of America, 2015.

In the 3D plot, one can see that the oscillation is more continuous, and there are no phase transition points.

2.3. Dynamic Linear Potential Guidance

As mentioned in the introduction, the AAF beams are radially symmetric beams that possess autofocusing property. It has been shown that the propagation trajectory as well as the positions of autofocusing points of the AAF beams can be controlled by potentials [41]. In our investigation [80], we theoretically analyzed the propagation and autofocusing effect of the AAF beams in a dynamic linear potential. We found that the linear potential may weaken (even eliminate) or strengthen the autofocusing effect of the AAF beams, depending on the form of the linear potential. In this case, the governing equation is written as [41,81,82]:

$$i\frac{d\psi}{dz} + \frac{1}{2}\left(\frac{\partial^2\psi}{\partial r^2} + \frac{1}{r}\frac{\partial\psi}{\partial r}\right) - \frac{d(z)}{2}r\psi = 0. \tag{40}$$

Here, the external potential is linear in r, with the scaling factor d that depends on the longitudinal coordinate; this is the so-called dynamical linear potential [81,82]. Again, it is hard to find an analytical solution of Equation (40), so we resort to an approximate analysis, by introducing an azimuthal modulation of the inward AAF beam:

$$\psi_{az}(x,y) = \mathrm{Ai}(r_0 - r)\exp[a(r_0 - r)]\exp\left(-\frac{(\theta - \theta_0)^2}{w_0^2}\right), \tag{41}$$

with w_0 being the width of the modulation, $\theta = \arctan(y/x)$ being the azimuthal angle, θ_0 representing the modulation direction. If the value of w_0 is small enough, the azimuthal modulation will result in a very narrow structure, so with $w_0 \to 0$ the result will be quite similar to the one-dimensional finite-energy Airy beam. In this way, the AAF beam can be transferred into the 1D finite-energy Airy beam [81,83], as:

$$\psi(x,y) = \sum_{\theta_0=-\pi}^{+\pi} \mathrm{Ai}(r_0 - x_p)\exp[a(r_0 - x_p)]. \tag{42}$$

Thus, by an analogy with the analytical result in [81], we describe the autofocusing effect and the propagation of the AAF beam as follows:

$$\psi(x,y,z) = C \sum_{\theta_0=-\pi}^{+\pi} \text{Ai}\left[iaz + \left(\frac{1}{2}f_1 - \frac{1}{4}z^2 + (r_0 - x_p)\right)\right] \exp\left[a\left(\frac{1}{2}f_1 - \frac{1}{2}z^2 + (r_0 - x_p)\right)\right] \times$$
$$\exp\left[i\left(\frac{1}{2}a^2z + \frac{1}{4}zf_1 - \frac{1}{8}f_2 - \frac{1}{12}z^3 - \frac{1}{2}(r_0 - x_p)g + \frac{1}{2}(r_0 - x_p)z\right)\right],$$

(43)

where, $g(z) = \int_0^z d(t)dt$, $f_1(z) = f_0 + \int_0^z g(t)dt$ and $f_2(z) = \int_0^z g^2(t)dt$. The trajectory of each component in a linear dynamic potential is

$$x_p = r_0 + \frac{1}{2}f_1 - \frac{1}{4}z^2.$$

(44)

Clearly, the AAF beam can be effectively manipulated by the linear potential.

If there is no autofocusing during propagation, Equation (43) can be reduced to the following form:

$$\psi(x,y,z) = \text{Ai}\left[iaz + \left(\frac{1}{2}f_1 - \frac{1}{4}z^2 + (r_0 - r)\right)\right] \exp\left[a\left(\frac{1}{2}f_1 - \frac{1}{2}z^2 + (r_0 - r)\right)\right] \times$$
$$\exp\left[i\left(\frac{1}{2}a^2z + \frac{1}{4}zf_1 - \frac{1}{8}f_2 - \frac{1}{12}z^3 - \frac{1}{2}(r_0 - r)g + \frac{1}{2}(r_0 - r)z\right)\right],$$

(45)

which becomes invalid when autofocusing happens. In this way, using Equations (43) and (45), the propagation of an AAF beam manipulated by a dynamic linear potential can now be reduced to a simple 1D case.

When we set $d(z) = 1$, the trajectory of the beam is $x_p = r_0$, that is, a straight line, so the autofocusing does not occur during propagation, and the propagation can be described by Equation (45). For this case, the linear potential exerts a "pulling" influence that can balance the virtual force which makes the beam focus. While, if the potential is a periodic function $d(z) = 1 + 4\pi^2 \cos(\pi z)$, as in the former case, the potential also exerts a pulling effect, and the trajectory is a cosine-like curve, which is also periodic. The corresponding results are displayed in Figure 8.

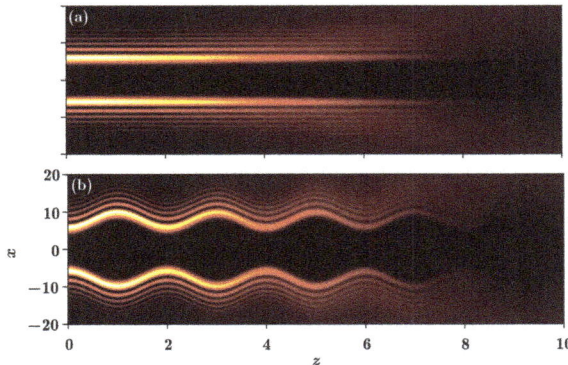

Figure 8. (a) Analytical intensity distribution of an AAF beam during propagation in the $x - z$ plane at $y = 0$, according to Equation (43), for $d = 1$; (b) Same as (a), but for $d(z) = 1 + 4\pi^2 \cos(\pi z)$. Reproduced with permission from [80], Copyright the Optical Society of America, 2016.

For the case with $d(z) = 13 - 12z$, from the corresponding trajectory $x_p = r_0 - z^3 + 3z^2$, we know that the beam will undergo autofocusing during propagation, so the process should be described by Equation (43). From the inset in Figure 9, the two components ($\theta_0 = 0$, $\theta_0 = \pi$) separate before the focusing point, and then converge. Since the slope of the components at the colliding point is much bigger than without a potential, as in [40], this can be viewed as the components acquiring a larger speed because of the "pushing" effect, hence the autofocusing is strengthened. Similar to the former study, the maximum of the beam intensity (MBI) is also a function of both r_0 and the propagation distance, as displayed in Figure 9b; the MBI first increases and then decreases with the increasing of r_0. Besides this effect, one can also see that the location of the MBI also changes with r_0, the reason being that the autofocusing effect requires a longer distance to establish itself when r_0 increases. As stated above, the transverse and longitudinal coordinates are normalized to some characteristic transverse width x_0 and the corresponding Rayleigh range kx_0^2. The unit of intensity of the beam is arbitrary.

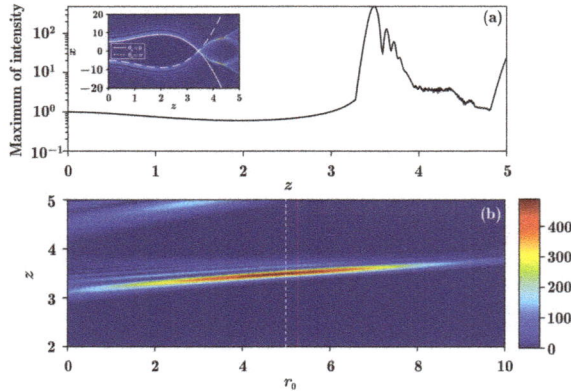

Figure 9. Manipulation of an AAF beam by a dynamical linear potential. (**a**) Maximum of the beam intensity during propagation for $d(z) = 13 - 12z$ and $r_0 = 5$. The maximum of the beam intensity at $z = 0$ is 1. Inset shows the propagation of the components corresponding to $\theta_0 = 0$ and $\theta_0 = \pi$, and the corresponding theoretical trajectories; (**b**) The maximum of the beam intensity as a function of r_0 and z. The white dashed line corresponds to the curve in (**a**). The decay parameter is $a = 0.05$. Reproduced with permission from [80], Copyright the Optical Society of America, 2016.

3. Conclusions

In conclusion, in this short review, we have briefly discussed the origin and the fundamental developments concerning Airy beams, and made a systematic review of the modulation of Airy beams under the guidance of nonlinearity and different potentials. This review is based on our recently published work. Our investigations will hopefully attract researchers who work on the related phenomena in other fields. Thus, the results presented here are not just limited to optics, but can lead to potential applications in biology, particle manipulation, microparticle trapping, Bose–Einstein condensates, signal processing and manipulating, and other disciplines. Actually, the Airy and other self-accelerating beams are among the hottest topics in optics. The investigations related to such beams are interesting and show impressive progress. Still, there are many unknown and important phenomena to be researched and new effects to be discovered concerning the accelerating beams.

Acknowledgments: The work was supported by China Postdoctoral Science Foundation (2016M600777), National Natural Science Foundation of China (11474228), and NPRP projects (6-021-1-005, 8-028-1-001) of the Qatar National Research Fund (a member of the Qatar Foundation). MRB acknowledges support by the Al Sraiya Holding Group.

Author Contributions: Y.Q.Z., H.Z. and M.R.B. wrote and organized the paper; Y.P.Z. supervised the project. All authors discussed the findings in paper.

Conflicts of Interest: The authors declare no conflict of interest.

References

1. Berry, M.V.; Balazs, N.L. Nonspreading wave packets. *Am. J. Phys.* **1979**, *47*, 264–267.
2. Siviloglou, G.A.; Christodoulides, D.N. Accelerating finite energy Airy beams. *Opt. Lett.* **2007**, *32*, 979–981.
3. Siviloglou, G.A.; Broky, J.; Dogariu, A.; Christodoulides, D.N. Ballistic dynamics of Airy beams. *Opt. Lett.* **2008**, *33*, 207–209.
4. Minovich, A.; Klein, A.E.; Janunts, N.; Pertsch, T.; Neshev, D.N.; Kivshar, Y.S. Generation and Near-Field Imaging of Airy Surface Plasmons. *Phys. Rev. Lett.* **2011**, *107*, 116802.
5. Li, L.; Li, T.; Wang, S.M.; Zhang, C.; Zhu, S.N. Plasmonic Airy Beam Generated by In-Plane Diffraction. *Phys. Rev. Lett.* **2011**, *107*, 126804.
6. Lin, Z.; Guo, X.; Tu, J.; Ma, Q.; Wu, J.; Zhang, D. Acoustic non-diffracting Airy beam. *J. Appl. Phys.* **2015**, *117*, 104503.
7. Fu, S.; Tsur, Y.; Zhou, J.; Shemer, L.; Arie, A. Propagation Dynamics of Airy Water-Wave Pulses. *Phys. Rev. Lett.* **2015**, *115*, 034501.
8. Bar-Ziv, U.; Postan, A.; Segev, M. Observation of shape-preserving accelerating underwater acoustic beams. *Phys. Rev. B* **2015**, *92*, 100301.
9. Siviloglou, G.; Broky, J.; Dogariu, A.; Christodoulides, D. Observation of Accelerating Airy Beams. *Phys. Rev. Lett.* **2007**, *99*, 213901.
10. Polynkin, P.; Kolesik, M.; Moloney, J.V.; Siviloglou, G.A.; Christodoulides, D.N. Curved plasma channel generation using ultraintense Airy beams. *Science* **2009**, *324*, 229–232.
11. Salandrino, A.; Christodoulides, D.N. Airy plasmon: A nondiffracting surface wave. *Opt. Lett.* **2010**, *35*, 2082–2084.
12. Liang, Y.; Hu, Y.; Song, D.; Lou, C.; Zhang, X.; Chen, Z.; Xu, J. Image signal transmission with Airy beams. *Opt. Lett.* **2015**, *40*, 5686–5689.
13. Jia, S.; Vaughan, J.C.; Zhuang, X. Isotropic three-dimensional super-resolution imaging with a self-bending point spread function. *Nat. Photon.* **2014**, *8*, 302–306.
14. Polynkin, P.; Kolesik, M.; Moloney, J. Filamentation of Femtosecond Laser Airy Beams in Water. *Phys. Rev. Lett.* **2009**, *103*, 123902.
15. Zheng, Z.; Zhang, B.F.; Chen, H.; Ding, J.; Wang, H.T. Optical trapping with focused Airy beams. *Appl. Opt.* **2011**, *50*, 43–49.
16. Cao, R.; Yang, Y.; Wang, J.; Bu, J.; Wang, M.; Yuan, X.C. Microfabricated continuous cubic phase plate induced Airy beams for optical manipulation with high power efficiency. *Appl. Phys. Lett.* **2011**, *99*, 261106.
17. Baumgartl, J.; Mazilu, M.; Dholakia, K. Optically mediated particle clearing using Airy wavepackets. *Nat. Photon.* **2008**, *2*, 675–678.
18. Zhang, P.; Prakash, J.; Zhang, Z.; Mills, M.S.; Efremidis, N.K.; Christodoulides, D.N.; Chen, Z. Trapping and guiding microparticles with morphing autofocusing Airy beams. *Opt. Lett.* **2011**, *36*, 2883–2885.
19. Schley, R.; Kaminer, I.; Greenfield, E.; Bekenstein, R.; Lumer, Y.; Segev, M. Loss-proof self-accelerating beams and their use in non-paraxial manipulation of particles' trajectories. *Nat. Commun.* **2014**, *5*, 5189.
20. Vettenburg, T.; Dalgarno, H.I.; Nylk, J.; Coll-Lladó, C.; Ferrier, D.E.; Čižmár, T.; Gunn-Moore, F.J.; Dholakia, K. Light-sheet microscopy using an Airy beam. *Nat. Meth.* **2014**, *11*, 541–544.
21. Abdollahpour, D.; Suntsov, S.; Papazoglou, D.G.; Tzortzakis, S. Spatiotemporal Airy Light Bullets in the Linear and Nonlinear Regimes. *Phys. Rev. Lett.* **2010**, *105*, 253901.
22. Chong, A.; Renninger, W.H.; Christodoulides, D.N.; Wise, F.W. Airy-Bessel wave packets as versatile linear light bullets. *Nat. Photon.* **2010**, *4*, 103–106.
23. Ament, C.; Polynkin, P.; Moloney, J.V. Supercontinuum Generation with Femtosecond Self-Healing Airy Pulses. *Phys. Rev. Lett.* **2011**, *107*, 243901.
24. Kim, K.Y.; Hwang, C.Y.; Lee, B. Slow non-dispersing wavepackets. *Opt. Express* **2011**, *19*, 2286–2293.
25. Voloch-Bloch, N.; Lereah, Y.; Lilach, Y.; Gover, A.; Arie, A. Generation of electron Airy beams. *Nature* **2013**, *494*, 331–335.
26. Gu, Y.; Gbur, G. Scintillation of Airy beam arrays in atmospheric turbulence. *Opt. Lett.* **2010**, *35*, 3456–3458.
27. Baumgartl, J.; Čižmár, T.; Mazilu, M.; Chan, V.C.; Carruthers, A.E.; Capron, B.A.; McNeely, W.; Wright, E.M.; Dholakia, K. Optical path clearing and enhanced transmission through colloidal suspensions. *Opt. Express* **2010**, *18*, 17130–17140.

28. Zhao, J.; Chremmos, I.D.; Song, D.; Christodoulides, D.N.; Efremidis, N.K.; Chen, Z. Curved singular beams for three-dimensional particle manipulation. *Sci. Rep.* **2015**, *5*, 12086.
29. Libster-Hershko, A.; Epstein, I.; Arie, A. Rapidly Accelerating Mathieu and Weber Surface Plasmon Beams. *Phys. Rev. Lett.* **2014**, *113*, 123902.
30. Ellenbogen, T.; Voloch-Bloch, N.; Ganany-Padowicz, A.; Arie, A. Nonlinear generation and manipulation of Airy beams. *Nat. Photon.* **2009**, *3*, 395–398.
31. Lotti, A.; Faccio, D.; Couairon, A.; Papazoglou, D.G.; Panagiotopoulos, P.; Abdollahpour, D.; Tzortzakis, S. Stationary nonlinear Airy beams. *Phys. Rev. A* **2011**, *84*, 021807.
32. Fattal, Y.; Rudnick, A.; Marom, D.M. Soliton shedding from Airy pulses in Kerr media. *Opt. Express* **2011**, *19*, 17298–17307.
33. Rudnick, A.; Marom, D.M. Airy-soliton interactions in Kerr media. *Opt. Express* **2011**, *19*, 25570–25582.
34. Zhang, Y.Q.; Belić, M.; Wu, Z.K.; Zheng, H.B.; Lu, K.Q.; Li, Y.Y.; Zhang, Y.P. Soliton pair generation in the interactions of Airy and nonlinear accelerating beams. *Opt. Lett.* **2013**, *38*, 4585–4588.
35. Zhang, Y.Q.; Belić, M.R.; Zheng, H.B.; Chen, H.X.; Li, C.B.; Li, Y.Y.; Zhang, Y.P. Interactions of Airy beams, nonlinear accelerating beams, and induced solitons in Kerr and saturable nonlinear media. *Opt. Express* **2014**, *22*, 7160–7171.
36. Kaminer, I.; Segev, M.; Christodoulides, D.N. Self-Accelerating Self-Trapped Optical Beams. *Phys. Rev. Lett.* **2011**, *106*, 213903.
37. Panagiotopoulos, P.; Abdollahpour, D.; Lotti, A.; Couairon, A.; Faccio, D.; Papazoglou, D.G.; Tzortzakis, S. Nonlinear propagation dynamics of finite-energy Airy beams. *Phys. Rev. A* **2012**, *86*, 013842.
38. Hu, Y.; Huang, S.; Zhang, P.; Lou, C.; Xu, J.; Chen, Z. Persistence and breakdown of Airy beams driven by an initial nonlinearity. *Opt. Lett.* **2010**, *35*, 3952–3954.
39. Chen, R.P.; Yin, C.F.; Chu, X.X.; Wang, H. Effect of Kerr nonlinearity on an Airy beam. *Phys. Rev. A* **2010**, *82*, 043832.
40. Efremidis, N.K.; Christodoulides, D.N. Abruptly autofocusing waves. *Opt. Lett.* **2010**, *35*, 4045–4047.
41. Hwang, C.Y.; Kim, K.Y.; Lee, B. Dynamic Control of Circular Airy Beams With Linear Optical Potentials. *IEEE Photon. J.* **2012**, *4*, 174–180.
42. Hwang, C.Y.; Kim, K.Y.; Lee, B. Bessel-like beam generation by superposing multiple Airy beams. *Opt. Express* **2011**, *19*, 7356–7364.
43. Chremmos, I.; Efremidis, N.K.; Christodoulides, D.N. Pre-engineered abruptly autofocusing beams. *Opt. Lett.* **2011**, *36*, 1890–1892.
44. Papazoglou, D.G.; Efremidis, N.K.; Christodoulides, D.N.; Tzortzakis, S. Observation of abruptly autofocusing waves. *Opt. Lett.* **2011**, *36*, 1842–1844.
45. Chremmos, I.; Zhang, P.; Prakash, J.; Efremidis, N.K.; Christodoulides, D.N.; Chen, Z. Fourier-space generation of abruptly autofocusing beams and optical bottle beams. *Opt. Lett.* **2011**, *36*, 3675–3677.
46. Chremmos, I.D.; Efremidis, N.K. Band-specific phase engineering for curving and focusing light in waveguide arrays. *Phys. Rev. A* **2012**, *85*, 063830.
47. Hu, Y.; Siviloglou, G.A.; Zhang, P.; Efremidis, N.K.; Christodoulides, D.N.; Chen, Z. Self-accelerating Airy Beams: Generation, Control, and Applications. In *Nonlinear Photonics and Novel Optical Phenomena*; Chen, Z., Morandotti, R., Eds.; Springer: New York, NY, USA, 2012; Volume 170, pp. 1–46.
48. Bandres, M.A.; Kaminer, I.; Mills, M.; Rodriguez-Lara, B.M.; Greenfield, E.; Segev, M.; Christodoulides, D.N. Accelerating Optical Beams. *Opt. Phot. News* **2013**, *24*, 30–37.
49. Chen, Z.; Xu, J.; Hu, Y.; Song, D.; Zhang, Z.; Zhao, J.; Liang, Y. Control and novel applications of self-accelerating beams. *Acta Opt. Sin.* **2016**, *36*, 1026009. (In Chinese)
50. Levy, U.; Derevyanko, S.; Silberberg, Y. Light Modes of Free Space. In *Progress in Optics*; Visser, T.D., Ed.; Elsevier: Amsterdam, The Netherlands, 2016; Volume 61, pp. 237–281.
51. Broky, J.; Siviloglou, G.A.; Dogariu, A.; Christodoulides, D.N. Self-healing properties of optical Airy beams. *Opt. Express* **2008**, *16*, 12880–12891.
52. Zhong, H.; Zhang, Y.; Zhang, Z.; Li, C.; Zhang, D.; Zhang, Y.; Belić, M.R. Nonparaxial self-accelerating beams in an atomic vapor with electromagnetically induced transparency. *Opt. Lett.* **2016**, *41*, 5644–5647.
53. Makris, K.G.; Kaminer, I.; El-Ganainy, R.; Efremidis, N.K.; Chen, Z.; Segev, M.; Christodoulides, D.N. Accelerating diffraction-free beams in photonic lattices. *Opt. Lett.* **2014**, *39*, 2129–2132.

54. Shen, M.; Gao, J.; Ge, L. Solitons shedding from Airy beams and bound states of breathing Airy solitons in nonlocal nonlinear media. *Sci. Rep.* **2015**, *5*, 9814.

55. Shen, M.; Li, W.; Lee, R.K. Control on the anomalous interactions of Airy beams in nematic liquid crystals. *Opt. Express* **2016**, *24*, 8501–8511.

56. Wiersma, N.; Marsal, N.; Sciamanna, M.; Wolfersberger, D. Spatiotemporal dynamics of counterpropagating Airy beams. *Sci. Rep.* **2015**, *5*, 13463.

57. Diebel, F.; Bokić, B.M.; Timotijević, D.V.; Savić, D.M.J.; Denz, C. Soliton formation by decelerating interacting Airy beams. *Opt. Express* **2015**, *23*, 24351–24361.

58. Driben, R.; Hu, Y.; Chen, Z.; Malomed, B.A.; Morandotti, R. Inversion and tight focusing of Airy pulses under the action of third-order dispersion. *Opt. Lett.* **2013**, *38*, 2499–2501.

59. Zhang, L.; Liu, K.; Zhong, H.; Zhang, J.; Li, Y.; Fan, D. Effect of initial frequency chirp on Airy pulse propagation in an optical fiber. *Opt. Express* **2015**, *23*, 2566–2576.

60. Hu, Y.; Tehranchi, A.; Wabnitz, S.; Kashyap, R.; Chen, Z.; Morandotti, R. Improved Intrapulse Raman Scattering Control via Asymmetric Airy Pulses. *Phys. Rev. Lett.* **2015**, *114*, 073901.

61. Lu, D.; Hu, W.; Zheng, Y.; Liang, Y.; Cao, L.; Lan, S.; Guo, Q. Self-induced fractional Fourier transform and revivable higher-order spatial solitons in strongly nonlocal nonlinear media. *Phys. Rev. A* **2008**, *78*, 043815.

62. Zhou, G.; Chen, R.; Ru, G. Propagation of an Airy beam in a strongly nonlocal nonlinear media. *Laser Phys. Lett.* **2014**, *11*, 105001.

63. Zhang, Y.Q.; Liu, X.; Belić, M.R.; Zhong, W.P.; Petrović, M.S.; Zhang, Y.P. Automatic Fourier transform and self-Fourier beams due to parabolic potential. *Ann. Phys.* **2015**, *363*, 305–315.

64. Zhang, Y.Q.; Belić, M.R.; Zhang, L.; Zhong, W.P.; Zhu, D.Y.; Wang, R.M.; Zhang, Y.P. Periodic inversion and phase transition of finite energy Airy beams in a medium with parabolic potential. *Opt. Express* **2015**, *23*, 10467–10480.

65. Agarwal, G.; Simon, R. A simple realization of fractional Fourier transform and relation to harmonic oscillator Green's function. *Opt. Commun.* **1994**, *110*, 23–26.

66. Bernardini, C.; Gori, F.; Santarsiero, M. Converting states of a particle under uniform or elastic forces into free particle states. *Eur. J. Phys* **1995**, *16*, 58–62.

67. Ozaktas, H.M.; Zalevsky, Z.; Kutay, M.A. *The Fractional Fourier Transform with Applications in Optics and Signal Processing*; Wiley: New York, NY, USA, 2001.

68. Bandres, M.A.; Gutiérrez-Vega, J.C. Airy-Gauss beams and their transformation by paraxial optical systems. *Opt. Express* **2007**, *15*, 16719–16728.

69. Kovalev, A.A.; Kotlyar, V.V.; Zaskanov, S.G. Diffraction integral and propagation of Hermite-Gaussian modes in a linear refractive index medium. *J. Opt. Soc. Am. A* **2014**, *31*, 914–919.

70. Mendlovic, D.; Ozaktas, H.M. Fractional Fourier transforms and their optical implementation: I. *J. Opt. Soc. Am. A* **1993**, *10*, 1875–1881.

71. Mendlovic, D.; Ozaktas, H.M.; Lohmann, A.W. Graded-index fibers, Wigner-distribution functions, and the fractional Fourier transform. *Appl. Opt.* **1994**, *33*, 6188–6193.

72. Vallée, O.; Soares, M. *Airy Functions and Applications to Physics*, 2nd ed.; Imperial College Press: Singapore, 2010.

73. Hu, Y.; Zhang, P.; Lou, C.; Huang, S.; Xu, J.; Chen, Z. Optimal control of the ballistic motion of Airy beams. *Opt. Lett.* **2010**, *35*, 2260–2262.

74. Caola, M.J. Self-Fourier functions. *J. Phys. A Math. Gen.* **1991**, *24*, L1143–L1144.

75. Zhang, Y.Q.; Belić, M.; Zheng, H.B.; Chen, H.; Li, C.B.; Song, J.P.; Zhang, Y.P. Nonlinear Talbot effect of rogue waves. *Phys. Rev. E* **2014**, *89*, 032902.

76. Zhang, Y.Q.; Belić, M.R.; Petrović, M.S.; Zheng, H.B.; Chen, H.X.; Li, C.B.; Lu, K.Q.; Zhang, Y.P. Two-dimensional linear and nonlinear Talbot effect from rogue waves. *Phys. Rev. E* **2015**, *91*, 032916.

77. Yang, B.; Zhong, W.P.; Belić, M.R. Self-Similar Hermite-Gaussian Spatial Solitons in Two-Dimensional Nonlocal Nonlinear Media. *Commun. Theor. Phys.* **2010**, *53*, 937–942.

78. Panagiotopoulos, P.; Papazoglou, D.G.; Couairon, A.; Tzortzakis, S. Sharply autofocused ring-Airy beams transforming into non-linear intense light bullets. *Nat. Commun.* **2013**, *4*, 2622.

79. Zhang, Y.Q.; Liu, X.; Belić, M.R.; Zhong, W.P.; Wen, F.; Zhang, Y.P. Anharmonic propagation of two-dimensional beams carrying orbital angular momentum in a harmonic potential. *Opt. Lett.* **2015**, *40*, 3786–3789.

80. Zhong, H.; Zhang, Y.; Belić, M.R.; Li, C.; Wen, F.; Zhang, Z.; Zhang, Y. Controllable circular Airy beams via dynamic linear potential. *Opt. Express* **2016**, *24*, 7495–7506.

81. Efremidis, N.K. Airy trajectory engineering in dynamic linear index potentials. *Opt. Lett.* **2011**, *36*, 3006–3008.

82. Chremmos, I.D.; Chen, Z.; Christodoulides, D.N.; Efremidis, N.K. Abruptly autofocusing and autodefocusing optical beams with arbitrary caustics. *Phys. Rev. A* **2012**, *85*, 023828.

83. Zhang, Z.; Liu, J.J.; Zhang, P.; Ni, P.G.; Prakash, J.; Hu, Y.; Jiang, D.S.; Christodoulides, D.N.; Chen, Z.G. Generation of autofocusing beams with multi-Airy beams. *Acta Phys. Sin.* **2013**, *62*, 034209.

applied sciences

MDPI

Article

Stability and Dynamics of Dark-Bright Soliton Bound States Away from the Integrable Limit

Garyfallia C. Katsimiga [1,*], Jan Stockhofe [1], Panagiotis G. Kevrekidis [2] and Peter Schmelcher [1,3]

[1] Zentrum für Optische Quantentechnologien, Universität Hamburg, Luruper Chaussee 149, 22761 Hamburg, Germany; Jan.Stockhofe@physnet.uni-hamburg.de (J.S.); pschmelc@physnet.uni-hamburg.de (P.S.)
[2] Department of Mathematics and Statistics, University of Massachusetts, Amherst, MA 01003-4515, USA; kevrekid5@gmail.com
[3] The Hamburg Centre for Ultrafast Imaging, Luruper Chaussee 149, 22761 Hamburg, Germany
* Correspondence: lkatsimi@physnet.uni-hamburg.de; Tel.: +49-40-8998-6511

Academic Editor: Boris Malomed
Received: 14 March 2017; Accepted: 7 April 2017; Published: 13 April 2017

Abstract: The existence, stability, and dynamics of bound pairs of symbiotic matter waves in the form of dark-bright soliton pairs in two-component mixtures of atomic Bose–Einstein condensates is investigated. Motivated by the tunability of the atomic interactions in recent experiments, we explore in detail the impact that changes in the interaction strengths have on these bound pairs by considering significant deviations from the integrable limit. It is found that dark-bright soliton pairs exist as stable configurations in a wide parametric window spanning both the miscible and the immiscible regime of interactions. Outside this parameter interval, two unstable regions are identified and are associated with a supercritical and a subcritical pitchfork bifurcation, respectively. Dynamical manifestation of these instabilities gives rise to a redistribution of the bright density between the dark solitons, and also to symmetry-broken stationary states that are mass imbalanced (asymmetric) with respect to their bright soliton counterpart. The long-time dynamics of both the stable and the unstable balanced and imbalanced dark-bright soliton pairs is analyzed.

Keywords: Bose–Einstein condensates; ultracold atoms; mixtures; Gross–Pitaevskii equation; nonlinear Schrödinger equation; symbiotic matter waves; dark-bright solitons

1. Introduction

After the experimental realization of Bose–Einstein condensates (BECs) in ultracold atoms, a plethora of studies has been devoted to examining and understanding the coherent structures that arise in them [1–4]. Among these structures, the formation, interactions and dynamics of matter wave dark [5–7] and bright solitons [2,8,9] have been a central focus of research both from the experimental and from the theoretical side. Such nonlinear waves were experimentally generated in single-component BECs over a decade ago [10–14]. The nature of nonlinear matter waves that can be created in a BEC background depends on the type of the interatomic interactions. Namely, dark solitons can be created in BECs with atom–atom repulsion resulting from a positive scattering length, while bright solitons exist in single-component settings with attractive interatomic interactions resulting from a negative scattering length.

In addition to the above single-component context, soliton states can arise also in multi-component settings. Such condensates have been created as mixtures of different spin states of ^{87}Rb [15,16] and ^{23}Na [17], and triggered numerous theoretical studies involving soliton complexes. A prototypical example of the latter is a coupled dark-bright (DB) soliton state in a highly elongated (quasi-one-dimensional) condensate cloud, consisting of a dark soliton in one component and a bright soliton in the second component of a binary BEC featuring intra- and inter-species repulsion. Since bright solitons are not self-sustained structures in repulsive (self-defocusing) media, DB solitons are often called symbiotic, that is, the dark soliton can be thought of as acting as an effective potential well trapping the bright soliton [18–22]. Such symbiotic entities were first observed and theoretically studied in the context of nonlinear optics [23–32]. However, their experimental realization in the atomic realm [33] opened a new and highly controllable direction towards a deeper understanding of the dynamics and interactions of these states both with each other as well as with external traps [34–38].

Additionally, current state-of-the art experiments offer the possibility of manipulating in a controllable fashion the nonlinear interactions via the well-established technique of Feshbach resonances [39–45]. This motivated recent theoretical activities where the static and dynamical properties of dark-bright symbiotic matter waves have been investigated on the level of mean field theory. Mathematically, tuning the interactions corresponds to deviating from the integrable limit [46–48] of the relevant nonlinear Schrödinger system, where nonintegrability is introduced when considering arbitrary nonlinearity (i.e., interaction) coefficients. The latter nonintegrable setting forms also the main focus of the present effort. In this context, despite the nonintegrability, analytical expressions of specific single-DB soliton states and lattices thereof have been obtained in [49]. Adding a parabolic trapping potential, it was revealed how the effective restoring force acting on the DB soliton depends on the inter-atomic interactions [49], verifying that the particle-like nature [34] of the symbiotic soliton is preserved. In the same spirit, the dependence of the binding energy of a DB soliton on the inter-species interactions was found analytically in [22], where, moreover, a proper bound on a phase imprinted in the bright soliton constituent was obtained (i.e., considering also moving single DB states), above which a breaking of the symbiotic entity was observed.

In our previous work [50], the interactions between DB matter waves and the consequent formation of bound states for out-of-phase (anti-symmetric) bright soliton components has been studied. Based on a two-soliton ansatz of the hyperbolic type [37,49], the full analytical expressions for the interaction energies between two DB solitons were obtained for arbitrary nonlinear coefficients, and in the absence of a confining potential. Furthermore, the key intuition that repulsion mediated by the dark solitons at short distances, and attraction mediated by anti-symmetric bright solitons at longer distances, would be counterbalanced, leading in turn to a bound state formation, has been enriched by taking into account the significant role of the cross-component interaction energy term. The crucial dependence of the latter on the inter-species interaction coefficient has been analyzed. It was shown that anti-symmetric stationary states exist and remain robust for a wide parametric window of inter-species repulsions. Importantly, an exponential instability of the anti-symmetric states was identified upon crossing a critical inter-species repulsion. The latter was found to be associated with a subcritical pitchfork bifurcation, giving rise to asymmetric stationary states with mass imbalanced bright soliton counterparts.

In the present work, we extend the aforementioned findings of [50] upon considering significant deviations from the integrable limit. By this, we mean that we always utilize a more realistic selection of the nonlinear inter- and intra-species repulsion coefficients motivated by ^{87}Rb (see details in Section 3), under which it is never possible that all coefficients are equal (which would represent the integrable limit). In our analysis, we vary the inter-species repulsion towards both the immiscible (i.e., dominated by inter-species repulsion and thus phase separated in the ground state) regime, but also towards the miscible (i.e., dominated by intra-species repulsion) regime. To this end, we investigate the stability and dynamics of the anti-symmetric states, the so-called "solitonic gluons" [24], as well as the above-mentioned asymmetric modes covering both the miscible and immiscible parameter regime. In particular, the full excitation spectrum of these symbiotic states is explored in detail by means of a Bogolyubov–de Gennes linearization analysis [1]. It is found that, for the asymmetric modes, the linearization spectrum is strongly affected when crossing the immiscibility-miscibility threshold, which is, in turn, directly connected with a rapid change of their density profiles as observed already in [50]. As a next step, the stability and dynamics of the anti-symmetric DB pairs are explored past the immiscibility-miscibility threshold. It is found that, in addition to the destabilization in the immiscible regime reported in [50], a second critical point occurs deep in the miscible regime, rendering the anti-symmetric state unstable once more. This instability scenario is found to be associated with a *supercritical* pitchfork bifurcation, giving rise to another family of mass imbalanced (asymmetric) symbiotic structures which are found to be stable. As the inter-species repulsion is decreased, the miscible character of this regime alters the bright soliton component, resulting into asymmetric pairs with the bright solitons "living" on top of a finite background. Comparing and contrasting the instability mechanisms in the different parameter regimes, the long-time evolution of the asymmetric and anti-symmetric states is performed numerically at different interaction ratios.

The presentation is structured as follows. In Section 2, a description of the theoretical model and prior results regarding the existence of the anti-symmetric DB soliton pairs are provided. Furthermore, we briefly comment on the methods to be used for the numerical analysis of our findings. Section 3 contains the results regarding both the stability and the dynamics of asymmetric and anti-symmetric DB soliton pairs beyond the integrable limit. Finally, in Section 4, we summarize our findings and discuss future perspectives.

2. Setup and Prior Background

2.1. Model and Theoretical Considerations

The system of interest is a two-component BEC strongly elongated along the x-direction, subject to a tight transverse harmonic trap of frequency $\omega_\perp$. Such a mixture can e.g., be composed of two different hyperfine states of the same alkali isotope, like ^{87}Rb. Within mean field theory and after integrating out the frozen transverse degrees of freedom, this mixture is described by the following two coupled (1+1)-dimensional Gross–Pitaevskii equations (GPEs) [51,52]:

$$i\hbar\partial_t\psi_j = \left(-\frac{\hbar^2}{2m}\partial_x^2 - \mu_j + \sum_{k=1}^{2} g_{jk}|\psi_k|^2\right)\psi_j. \qquad (j=1,2). \tag{1}$$

In the above equation, $\psi_j(x,t)$ denote the mean-field wave functions of the two components normalized to the numbers of atoms $N_j = \int_{-\infty}^{+\infty}|\psi_j|^2 dx$, while m and μ_j are the atomic mass (identical for both components) and chemical potentials, respectively. The effective one-dimensional coupling constants are given by $g_{jk} = 2\hbar\omega_\perp a_{jk}$, where a_{jk} denote the three s-wave scattering lengths that account for collisions between atoms belonging to the same (a_{jj}) or different ($a_{12} = a_{21}$) species. We restrict our considerations to purely repulsive interactions, i.e., all $g_{jk} > 0$, and consider an idealized homogeneous setting with no longitudinal trapping potential along the x-axis.

By measuring densities $|\psi_j|^2$, length, time and energy in units of $2a_{11}$, $a_\perp = \sqrt{\hbar/(m\omega_\perp)}$, $\omega_\perp^{-1}$, $\hbar\omega_\perp$, respectively, and in a second step rescaling space-time coordinates as $t \to \mu_1 t$, $x \to \sqrt{\mu_1}x$, and the densities $|\psi_{1,2}|^2 \to \mu_1^{-1}|\psi_{1,2}|^2$, the system of Equation (1) can be written in the following dimensionless form:

$$i\partial_t\psi_d + \frac{1}{2}\partial_x^2\psi_d - (|\psi_d|^2 + g_{12}|\psi_b|^2 - 1)\psi_d = 0, \tag{2}$$

$$i\partial_t\psi_b + \frac{1}{2}\partial_x^2\psi_b - (g_{12}|\psi_d|^2 + g_{22}|\psi_b|^2 - \mu)\psi_b = 0. \tag{3}$$

In the above system of equations, we slightly changed the notation, using $\psi_1 \equiv \psi_d$ and $\psi_2 \equiv \psi_b$, indicating this way that the component 1 (2) will be supporting dark (bright) solitons. Furthermore, $\mu \equiv \mu_2/\mu_1$ is the rescaled chemical potential, while the interaction coefficients are normalized to the scattering length a_{11}, i.e., $g_{12} \equiv g_{12}/g_{11}$, and $g_{22} \equiv g_{22}/g_{11}$. The system of Equations (2) and (3) conserves the total energy, E, the rescaled total number of atoms in each component (N_d and N_b, respectively) and the rescaled total number of atoms N, where:

$$E = \frac{1}{2} \int_{-\infty}^{+\infty} \left[|\partial_x \psi_d|^2 + |\partial_x \psi_b|^2 + (|\psi_d|^2 - 1)^2 + g_{22}|\psi_b|^4 - 2\mu|\psi_b|^2 + 2g_{12}|\psi_d|^2|\psi_b|^2 \right] dx, \quad (4)$$

$$N = N_d + N_b = \sum_{i=d,b} \int_{-\infty}^{\infty} dx |\psi_i|^2. \quad (5)$$

2.2. Interactions of Symbiotic Matter Waves beyond the Integrable Limit

In the special case where the nonlinear coefficients are all equal to each other, i.e., $g_{12} = g_{22} = 1$, the system of Equations (2) and (3) admits exact single-DB-soliton solutions of the form [34]:

$$\psi_d(x,t) = \cos\phi \tanh\left[D(x - x_0(t))\right] + i \sin\phi, \quad (6)$$

$$\psi_b(x,t) = \eta\mathrm{sech}\left[D(x - x_0(t))\right] \times \exp\left[ikx + i\theta(t) + i(\mu - 1)t\right], \quad (7)$$

subject to the boundary conditions $|\psi_d|^2 \to 1$, and $|\psi_b|^2 \to 0$ for $|x| \to \infty$. In the aforementioned solutions, ϕ is the so-called soliton's phase angle, which fixes the "grayness" of the dark soliton, while η denotes the amplitude of the bright component. Furthermore, $x_0(t)$ and D correspond to the soliton's center and inverse width, respectively, while $k = D\tan\phi$ is the wave-number of the bright soliton, associated with the speed $\dot{x}_0$ of the DB soliton, and $\theta(t)$ is its phase.

While such a general exact expression can only be obtained in the integrable limit, one can utilize an approximate ansatz based on the expressions given in Equations (6) and (7) and depart from the integrable limit. Such a method was employed in our very recent work of Ref. [50] in order to analyze effective interactions between symbiotic matter waves in the general case of different inter-atomic repulsion strengths within each and between the species. Our aim in what follows is to briefly comment on the key results obtained there, in order to be connected with the numerical findings that will be presented below. In particular, a pair of two equal-amplitude dark-bright solitons travelling in opposite directions was considered having the approximate form:

$$\psi_d(x,t) = (\cos\phi \tanh X_- + i \sin\phi) \times (\cos\phi \tanh X_+ - i \sin\phi), \quad (8)$$

$$\psi_b(x,t) = \eta \mathrm{sech} X_- \, e^{i[kx+\theta(t)+(\mu-1)t]} + \eta \mathrm{sech} X_+ \, e^{i[-kx+\theta(t)+(\mu-1)t]} \, e^{i\Delta\theta}. \quad (9)$$

In these expressions $X_\pm = D(x \pm x_0(t))$ where $2x_0(t)$ is the distance between the two DB solitons, while $\Delta\theta$ is the relative phase between the bright solitons. If $\Delta\theta = 0$, the bright solitons are in-phase (IP), while if $\Delta\theta = \pi$, the bright solitons are out-of-phase (OP) or anti-symmetric. Within a Hamiltonian variational approach [3,7,37,49], the approximate ansatz of Equations (8) and (9) is substituted into the total energy of the system given by Equation (4), and the relevant integrations are performed under the assumption that the soliton velocity is sufficiently small ($\phi \approx 0$, $k \approx 0$). Full analytical results for all the integrals contributing to E were obtained in [50]. At large distances between the solitons, $Dx_0 \gg 1$, simplified asymptotic expressions can be derived by expanding with respect to $\exp\left[-2Dx_0\right]$ (see also the earlier findings for the integrable case [37]). The total energy can be decomposed as $E = 2E_1 + E_{dd} + E_{bb} + E_{db}$, where E_1 corresponds to the energy of a single DB soliton, contributing twice to the total energy, while the remaining three terms, namely, E_{dd}, E_{bb} and E_{db}, account for the interaction between: the two dark solitons, the two bright solitons, and the cross-interaction of a dark soliton in the first component with the bright soliton of the second component, respectively.

Figure 1 illustrates our key findings regarding the interaction energy of two out-of-phase DB solitons as a function of their distance. Here and throughout this work, the chemical potential and the intra-species interaction coefficients are fixed to $\mu = 2/3$ and $g_{22} = 0.95$ [53,54], respectively, while the interatomic interaction coefficient g_{12} is left to widely vary. In panels (a)–(c) both the full and the approximate asymptotic, i.e., expanded to second order with respect to $\exp(-2Dx_0)$, forms of the variationally obtained total interaction energy, $E_{tot} = E_{dd} + E_{bb} + E_{db}$, are shown with solid green and dashed yellow lines, respectively. As it can be seen in all cases, the two results coincide at large distances as they should. Strikingly, in all three cases, a pronounced local energy maximum is identified. The identification of this extremum suggests, if we further take into account the effective negative mass of the DB soliton [7], the existence of an effective *stable* (with respect to variations in x_0) fixed point, thus a bound state for the two OP dark-bright solitons. Such an anti-symmetric two-DB soliton bound state can indeed be identified also in the full Gross–Pitaevskii system (for the numerical methods used, we refer the reader to the following section). The predictive strength of the variational approximation is directly checked by comparing with the full numerical computation of the respective equilibria x_0 on the level of Equations (2) and (3), and the outcome is illustrated in panel (d), showing good agreement.

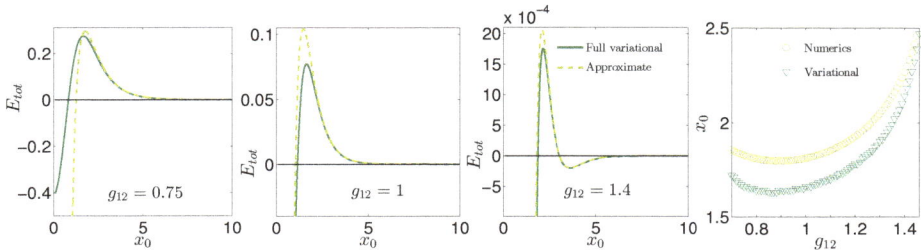

Figure 1. Variational estimate of the total interaction energy E_{tot} between two OP dark-bright solitons. From left to right in panels (**a**–**c**), the value of the inter-species interaction coefficient is increased from $g_{12} = 0.75$ to $g_{12} = 1.4$, while $g_{22} = 0.95$ and $\mu = 2/3$ throughout. In all cases, both the full and the asymptotic (valid for large x_0) expressions are shown with solid green and dashed yellow lines, respectively. The equilibrium separation x_0 of the stationary DB soliton pair as a function of g_{12} is shown in panel (**d**). In this latter panel, circles in yellow correspond to the equilibrium distance obtained by numerically solving the GPE system of Equations (2) and (3), while triangles in green denote the variational prediction of the equilibrium, i.e., the respective maximum of E_{tot}.

It is worth mentioning at this point that, in order to evaluate the energy functional, we need as input the width D and amplitude η occurring in Equations (8) and (9). These are obtained by numerically identifying the exact single-DB state at the given g_{12} and fitting to it the profile known from the integrable limit, i.e., Equations (6) and (7). There are thus two main sources of error of this scheme, namely: (i) the imperfect fit of the *tanh-sech* profile (strictly valid only in the integrable limit) to the single DB soliton mode and (ii) the limited accuracy of the two-DB soliton ansatz of Equations (8) and (9), especially at small separations x_0. For details regarding this and also a discussion of the case of in-phase bright solitons, we refer the interested reader to Ref. [50]. We note also that as g_{12} increases (while μ and g_{22} are kept fixed), the norm N_b of the bright species is increasingly suppressed and eventually vanishes [50], beyond which point there is no bound two-soliton state anymore, since the presence of the bright component is a necessary ingredient for holding the dark solitons together [24,37]. This suppression of the bright norm is compatible with the overall decrease of the total energies from panel (a) towards (b) and (c) in Figure 1.

Having variationally predicted and numerically confirmed the existence of anti-symmetric stationary dark-bright soliton bound states for a large interval of values of the inter-species interaction parameter g_{12} (typically, we study $0.75 \leq g_{12} \leq 1.5$ in the present work), our main goal in the following is to address their stability and (where applicable) decay dynamics.

2.3. Numerical Methods

In this section, we briefly comment on the numerical methods to be used so as to obtain stationary symbiotic states consisting of two dark-bright solitons and to determine their stability and time evolution. In the numerical computations that follow, we initially obtain stationary solutions of the system of Equations (2) and (3) in the form of $\psi_d(x,t) = u_d(x)$ and $\psi_b(x,t) = u_b(x)$ by means of a fixed-point numerical iteration scheme [55]. The linear stability of the latter is adressed by using the Bogolyubov–de Gennes (BdG) analysis [7,52]. Particularly to assess the stability of the obtained fixed points, we substitute the following ansatz into Equations (2) and (3):

$$\psi_d(x,t) = u_d(x) + \epsilon \left(a(x)e^{-i\omega t} + b^\star(x)e^{i\omega^\star t} \right), \tag{10}$$

$$\psi_b(x,t) = u_b(x) + \epsilon \left(c(x)e^{-i\omega t} + d^\star(x)e^{i\omega^\star t} \right). \tag{11}$$

In the above equations, the asterisk denotes complex conjugation while ϵ is the amplitude of infinitesimal perturbations. The resulting system of equations is then linearized, by keeping only terms of order $\mathcal{O}(\epsilon)$, and the eigenvalue problem for the eigenfrequencies ω and eigenmodes $(a(x), b(x), c(x), d(x))^T$ is solved numerically. Note that an instability occurs if modes with purely imaginary or complex eigenfrequencies are identified [1]. Since the linearization spectrum is invariant under $\omega \to -\omega$ and $\omega \to \omega^\star$, only results in the first quadrant of the complex plane will be shown below. For the simulation of the time evolution based on Equations (2) and (3), a fourth order Runge–Kutta algorithm is employed and a second-order finite differences method is used for the spatial derivatives. The grid spacing is fixed to $\Delta x = 0.08$, while the time step used is $\Delta t = 0.005$. In all cases, the numerical computations are performed on a spatial grid in the presence of an almost hard-wall super-Gaussian potential [50] that is chosen wide enough for boundary effects to be negligible on the small and intermediate time scales considered herein.

3. Numerical Results

Having verified that stationary anti-symmetric pairs of symbiotic matter waves can be found in a wide range of values of the inter-species repulsion coefficient g_{12}, a natural next step is to consider the fate of these solutions under small perturbations, providing information on their stability in the different regions of existence. The latter is explored by using the BdG linearization analysis discussed in Section 2.3. For the presentation of our results, we will distinguish the miscible ($g_{12} < g_{12_{th}}$) and immiscible ($g_{12} > g_{12_{th}}$) regimes, separated by the miscibility–immiscibility threshold [56] $g_{12_{th}} = \sqrt{g_{22}} = 0.975$ for our choice of $g_{22} = 0.95$.

In Figure 2, the BdG spectrum of the anti-symmetric stationary two dark-bright soliton states is shown as a function of g_{12}. In particular, both the real ($\Re(\omega)$) and the imaginary parts ($\Im(\omega)$) of the eigenfrequencies ω are depicted in the top (a) and middle (b) panels, respectively. Panels (c) and (d) depict profiles of the dark and bright wave functions at selected values of g_{12} in the miscible and immiscible regime, respectively. Two general comments can be made before examining in detail the excitation spectrum. The most significant one is that, within the background spectrum denoted with blue circles, there exist two distinguished modes. The trajectories of the latter are illustrated with red stars. These modes are the so-called anomalous modes since they possess a so-called negative Krein signature K [57], which, for the two-component system considered herein, is defined as:

$$K = \omega \int \left(|a|^2 - |b|^2 + |c|^2 - |d|^2 \right) dx. \tag{12}$$

The sign of this quantity is a topological property associated with the excited nature of this state, and the eigenvectors of such anomalous modes result in a variation of the solitary waves (as opposed to a variation of the system's background). Furthermore, we note in passing that each continuous symmetry of the system corresponds to a pair of zero modes, $\omega = 0$, in the BdG spectrum. Thus, we expect three pairs of such modes related, respectively, with the conservation of the particle number (or the $U(1)$ gauge invariance) in each of the two components, and with the translation invariance due to the absence of a confining potential. This is confirmed in the numerical data.

Just by inspecting the trajectories of the two anomalous modes that appear in the spectrum, their very different behavior becomes apparent. As it is observed in panel (a) of Figure 2, the higher frequency mode decreases almost monotonically upon increasing g_{12}. We were able not only to identify this mode but also to relate it with an out-of-phase vibration of the bound DB soliton pair around its equilibrium distance. The latter can be done by adding the corresponding BdG eigenvector, which turns out to be localized in the vicinity of the DB pair, to the relevant stationary anti-symmetric state (results not shown here). Next, and also in the same panel of Figure 2, let us closely follow the trajectory of the lower-lying anomalous mode. The trajectory of this mode is more complicated than the former one. In particular, starting from the aforementioned reference point, $g_{12_{th}} \approx 1$, and increasing the interspecies interactions towards the immiscible regime, i.e., for $g_{12} > g_{12_{th}}$, it can be seen that there exists a critical point $g_{12_{cr}}^{(1)} = 1.18$ where this mode destabilizes. This destabilization corresponds to an eigenvalue zero crossing and signals the instability of the anti-symmetric configuration deep in the immiscible regime. Notice the non-zero imaginary part that appears past this critical point shown in panel (b) of Figure 2. The existence and destabilization of this mode was found to be related with a symmetry breaking of the bright soliton component, being linked, in turn, to other stationary states that are mass imbalanced with respect to their bright soliton counterpart. The identification of these asymmetric states which exist below $g_{12_{cr}}^{(1)}$ and collide with the anti-symmetric branch at $g_{12_{cr}}^{(1)}$ in a subcritical pitchfork bifurcation was established in [50]. It is worth mentioning, at this point, that such asymmetric states were also analytically obtained in [31] in the integrable limit of the theory. However, to the best of our knowledge, the stability of such states has not been addressed. The following paragraphs will be devoted to a discussion of these asymmetric two-DB solutions, before we then return to an analysis of the BdG spectrum of the anti-symmetric pair, Figure 2, at small values of g_{12}.

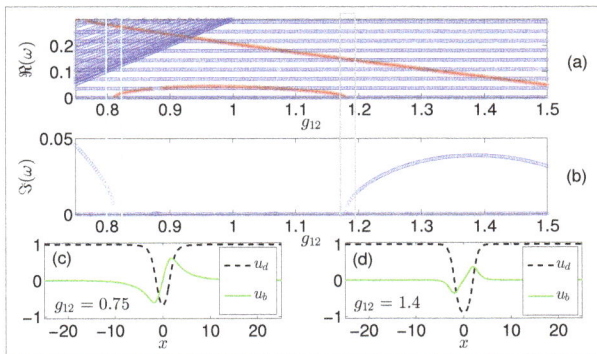

Figure 2. Full BdG spectrum of anti-symmetric two-DB soliton states, upon varying the inter-species interaction coefficient g_{12} within the interval $0.75 \leq g_{12} \leq 1.5$. (a) real part, $\Re(\omega)$, of the eigenfrequencies ω as a function of g_{12}. The corresponding imaginary part, $\Im(\omega)$, is shown in panel (b). Upon increasing g_{12}, there exist two critical values, $g_{12_{cr}}^{(2)} = 0.81$, and $g_{12_{cr}}^{(1)} = 1.18$, indicated with light blue and gray boxes, respectively, below and above which the anti-symmetric branch destabilizes. The trajectories of the two anomalous modes (see text) appearing in the spectrum are shown with red stars. (c,d) profiles of the anti-symmetric states for $g_{12} = 0.75$, and $g_{12} = 1.4$, i.e., deep in the miscible and immiscible regime, respectively.

As mentioned above, the asymmetric and the anti-symmetric two-DB states coincide at the bifurcation point $g_{12_{cr}}^{(1)}$. Below this critical g_{12}, the anti-symmetric state is stable and there are two symmetry-broken asymmetric solutions which are unstable (as is characteristic of a subcritical pitchfork bifurcation). Indeed, these asymmetric branches can themselves be continued towards much smaller values of g_{12} and exist both in the immiscible and the miscible regime (see the profiles shown in

Figure 3c,d). In both panels, dashed black lines denote the wavefunction of the dark soliton component, while solid green lines depict the corresponding bright soliton counterpart. To gain further insight regarding the nature of the instability of these asymmetric states, their full BdG spectrum is illustrated in Figure 3a,b. Once more, both the real $\Re(\omega)$ and the imaginary $\Im(\omega)$ parts of the eigenfrequencies ω are shown as a function of the interspecies interaction. As expected from the above discussion, only one of the anomalous modes appears in the excitation spectrum of these mass imbalanced states and is depicted with stars in red. Replacing the second anomalous mode, there is now throughout a purely imaginary frequency signaling the instability of the asymmetric branch. Remarkably, as the immiscibility-miscibility threshold is crossed, the growth rate of the instability is drastically decreased, rendering these states only weakly unstable for $g_{12} < g_{12_{th}}$. This latter observation is in close contact with the change in the spatial character of these asymmetric states, and also in line with our previous findings [50]. Namely, as the interspecies interactions decrease towards the miscible region, the asymmetric states change gradually from only weakly asymmetric (i.e., weakly mass imbalanced with respect to their bright amplitudes), to maximally asymmetric (i.e., almost purely dark/dark-bright bound states). The difference is clearly seen in the profiles provided in panels (c) and (d) of Figure 3. The existence of such maximally asymmetric bound states can be intuitively understood as follows. We have found in [50] that the effective interaction between dark solitons (E_{dd}) is repulsive while the interaction between OP bright solitons (E_{bb}) with fixed $g_{22} = 0.95$ and $\mu = 2/3$ is attractive. Furthermore, the cross term effective interaction (E_{db}) is also attractive within the g_{12} interval considered herein (again, we refer the reader to the relevant analytical expressions obtained in [50]). Thus, in the limiting case of maximal asymmetry, while the bright-bright interaction is absent, the dark-bright one is still present. It is this latter cross term attraction that counterbalances the dark-dark repulsion, leading in turn to the observed bound state formation.

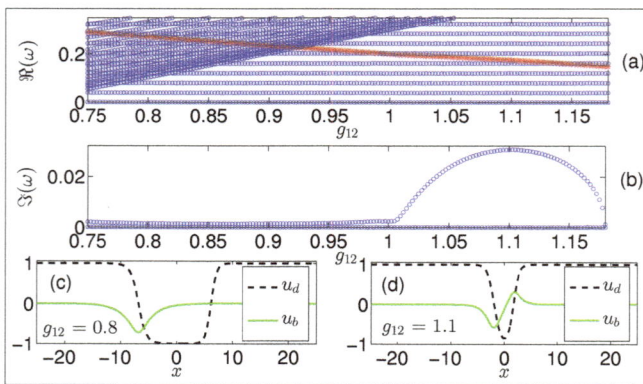

Figure 3. BdG spectrum for the asymmetric two-DB soliton states. (**a**) real part, $\Re(\omega)$, of the eigenfrequencies ω as a function of g_{12}. The corresponding imaginary part, $\Im(\omega)$, is shown in panel (**b**). Notice in this case the significant modification of the stability properties (i.e., a nearly vanishing imaginary part of the eigenfrequency in panel (**b**)), as the miscibility-immiscibility threshold $g_{12_{th}} \approx 1$ is crossed. (**c,d**) profiles of the asymmetric states for $g_{12} = 0.8$, and $g_{12} = 1.1$, i.e., in the miscible and immiscible regime, respectively.

Having identified the unstable linearization eigenmodes, we now turn to the associated decay dynamics. In particular, Figure 4 shows the long-time dynamics of the above-obtained stationary asymmetric states for different values of the nonlinear coefficient g_{12}. Panels (a)–(d) correspond to the dark soliton component while the respective bright wave functions are depicted in panels (e)–(h). For $g_{12} < 1$ shown in panels (a)–(c) and (e)–(g), respectively, it is observed that, as a result of the instability, the solitons move towards each other forming breathing-like structures. This is accompanied by a redistribution of the bright "filling" component between the two dark solitons, in the form of almost complete tunneling back and forth. For parameter values slightly above $g_{12_{th}}$, as e.g., the one depicted in panels (c) and (g), a strong beating phenomenon is observed, clearly evident in the dark soliton component, with the solitons oscillating around a fixed distance from each other forming an almost stationary breather that persists until the end of the propagation. Below the above-mentioned threshold and towards the miscible region, the picture becomes progressively more dramatic. The beating gets much more pronounced with the solitons experiencing more frequent collisions as shown in panels (b) and (f). Finally, when entering even deeper into the miscible side illustrated in panels (a) and (e), eventually the bound pair fully splits into an essentially empty dark and a dark-bright soliton that are released (i.e., are no longer bound by each other) and propagate towards the outer parts of the simulation domain (where they are ultimately reflected by the boundaries). However, for $g_{12} > 1$, i.e., upon increasing g_{12} towards the immiscible side depicted in panels (d) and (h) of Figure 4, a rather different picture is painted by the symbiotic entities when compared to the asymmetric states presented above. In particular, slightly after the beginning of the propagation, where the asymmetric entity looked quite robust, a dramatic redistribution of the bright soliton's mass occurs. The latter results in a strong repulsion between the single DB soliton formed and the almost empty dark one, leading in turn to their subsequent separation. Note that a similar decaying mechanism was also observed in [50] but for the unstable anti-symmetric states (see also later on in the text).

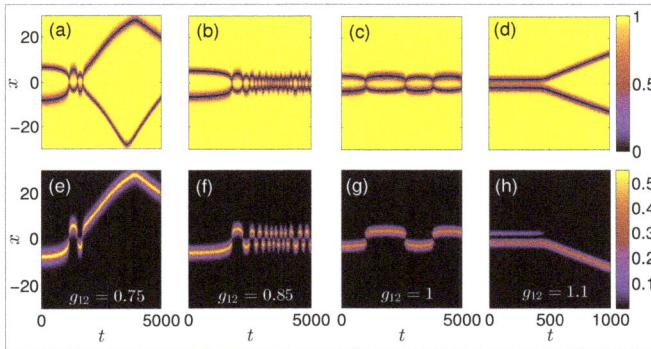

Figure 4. Space-time evolution of the unstable asymmetric two-DB soliton state for different values of the inter-species interaction coefficient g_{12}. From left to right, the interaction is increased from $g_{12} = 0.75$ to $g_{12} = 1.1$, while panels (**a–d**) ((**e–h**)) correspond to the densities $|\psi_d|^2$ ($|\psi_b|^2$) of the dark (bright) component.

We have now characterized the branch of asymmetric two-DB soliton modes, which were seen to bifurcate from the anti-symmetric two-DB soliton branch in a subcritical pitchfork bifurcation in the immiscible regime, at $g_{12_{cr}}^{(1)} = 1.18$. Let us now return to the anti-symmetric branch itself and study its fate in the miscible regime. This regime is also covered in the full excitation spectrum depicted in Figure 2, with a typical profile of the state being shown in panel (c). Departing from $g_{12} \approx 1$ towards the miscible regime, and in particular by following once again the lower-lying anomalous BdG mode, we observe that as g_{12} decreases a second critical point that exists deep within the miscible side. The eigenvalue zero crossing occurs at $g_{12_{cr}}^{(2)} = 0.81$, and is indicated by the solid light blue box. We note here that this latter critical point was not discussed in our previous work [50], where only larger values of g_{12} were studied. The destabilization of the aforementioned mode suggests the existence of a second pitchfork bifurcation. This is, once again, related with a symmetry breaking of the bright soliton component, resulting in mass imbalanced (i.e., asymmetric) two DB states. Indeed, we were able to identify these *new* pairs. In contrast to the previous instability scenario, these mass imbalanced states exist past the critical point but are *stable*, i.e., the pitchfork deep in the miscible domain is found to be *supercritical*. The corresponding bifurcation diagram is shown in panel (a) of Figure 5. In order to obtain this diagram, we measure the relative bright imbalance defined as $\Delta N_b \equiv \left(\int_{-\infty}^{0} |u_b|^2 dx - \int_{0}^{\infty} |u_b|^2 dx \right) / \int_{-\infty}^{\infty} |u_b|^2 dx$ upon varying g_{12}. Notice that four branches are identified, i.e., three stable ones consisting of two asymmetric and an anti-symmetric branch all denoted by solid blue lines, and one unstable anti-symmetric branch illustrated with a dashed green line. To further demonstrate the stability of the new asymmetric symbiotic pairs in panels (b)–(d), the BdG spectra are shown for different values of the inter-species interactions below the associated critical point $g_{12_{cr}}^{(2)} = 0.81$. Two anomalous modes appear in the linearization spectra of these asymmetric states illustrated with red stars. The respective stationary wave profiles are depicted in panels (e)–(g), where the dark (bright) soliton wavefunction is shown with a dashed black (solid green) line. We observe, that upon decreasing g_{12}, the asymmetry between the bright solitons increases, as was also the case for the respective asymmetric but unstable states found in the immiscible regime. Furthermore, for $g_{12} < 0.7$, a background gradually builds up for the bright solitons as is evident in the stationary state shown in panel (e) of Figure 5, revealing the miscible character of the regime supporting these states. Our detailed BdG analysis indicates that, for all values within the miscible regime that we have checked, the asymmetric states exist as stable configurations and should remain dynamically robust. This result is verified and highlighted in panels (h)–(m) of Figure 5, where we use as initial condition the stationary asymmetric states depicted in panels (e)–(g). We note that, in all three cases, panels (h)–(j) ((k)–(m)) show the evolution of the dark (bright) soliton component. Having studied both the static and the dynamical properties of the new asymmetric structures that bifurcate from the OP dark-bright states, we now turn our attention to the dynamics of the anti-symmetric DB waveforms.

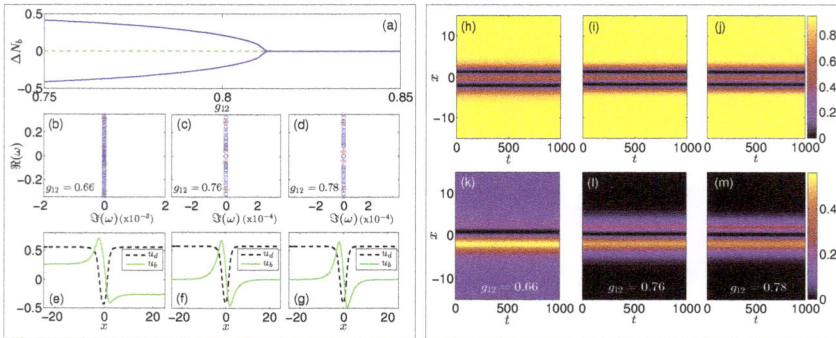

Figure 5. (**a**) bifurcation diagram obtained by measuring the relative bright imbalance ΔN_b as a function of the inter-species interaction coefficient g_{12}. The stable anti-symmetric and asymmetric branches are denoted with solid blue lines, while the unstable anti-symmetric one is shown with a dashed green line. Notice that the asymmetric branches exist before the critical point, $g_{12_{cr}}^{(2)} = 0.81$, but are stable verifying the supercritical nature of the bifurcation. (**b–d**) BdG spectra at three different values of g_{12} showcasing the stability of the asymmetric states. In all cases, the anomalous modes are illustrated with red stars. The associated stationary DB profiles are depicted in panels (**e–g**). (**h–m**) spatio-temporal evolution of the stationary asymmetric states shown in panels (**e–g**) showing the dynamical stability of these symbiotic structures. Panels (**h–j**) ((**k–m**)) correspond to the dark (bright) soliton component.

In particular, we explore the long time evolution of the OP DB states so as to reveal the decay mechanisms that such a pair suffers from. The dynamics at different values of g_{12} are summarized in Figure 6, all initiated at equilibrium. As before, the upper row of panels (a)–(d) shows the spatio-temporal evolution of the dark soliton counterpart, while panels (e)–(h) depict the corresponding bright component. From the stability analysis presented above, it is expected that for all values of the interspecies interaction coefficient $g_{12_{cr}}^{(2)} < g_{12} < g_{12_{cr}}^{(1)}$, the anti-symmetric two dark-bright soliton states exist as stable configurations, and as such should be robust throughout the propagation. This latter result is confirmed in panels (c) and (g) of Figure 6, for $g_{12} = 0.82$, which is slightly above the lower critical point. However, and as anticipated, a very different picture is found below the critical point $g_{12_{cr}}^{(2)}$ depicted in panels (b) and (f), (a) and (e), respectively. Starting with the former, we observe that slightly below $g_{12_{cr}}^{(2)}$ the initial stationary state quickly decays and in (b) and (f) we observe periodic tunneling of the bright component between the two dark solitons, while the dark solitons are only relatively weakly affected here. It is worth mentioning that similar tunneling dynamics have been identified and interpreted in terms of a bosonic Josephson junction model in [58], but with the crucial difference that the soliton pair was further supported by the restoring force of a harmonic trap in that work, while, in our present setup, there is no external potential that would keep the dark solitons in place. Remarkably, in this regime, despite the mass exchange, the bound soliton pair does not disintegrate. Further decreasing g_{12} in panels (a) and (e), the effects of the instability are more drastic. The time scale of the bright component oscillations decreases and the dark solitons vibrate more strongly. After some time of almost periodic oscillations, a more irregular type of motion sets in, with both dark solitons (one filled by most of the bright component, the other almost empty) eventually moving towards positive x while separating and recolliding in the process, suggesting still a kind of effective attractive interaction between them. This decay mechanism deep in the miscible regime is to be contrasted to the unstable dynamics in the immiscible regime above $g_{12_{cr}}^{(1)}$. In panels (d) and (h), we show the time evolution of the anti-symmetric two-DB soliton state at $g_{12} = 1.4$. While initially almost no dynamics are visible in the densities, especially no oscillations within the bright component, on intermediate time scales, a strong asymmetry in the bright filling builds up (see again [50] for a more detailed discussion) and subsequently the filled and the empty dark soliton split, showing no sign of effective attraction. Notice that the above described decaying mechanism is rather similar to the one observed for the unstable asymmetric states for $g_{12} > 1$ (see panels (d) and (h) of Figure 4).

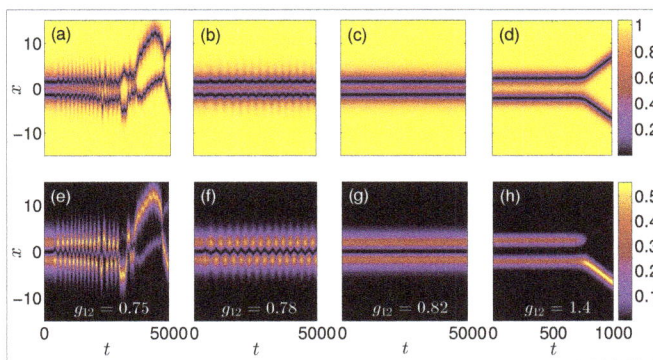

Figure 6. Space-time evolution of the anti-symmetric stationary two-DB state for different values of the inter-species interaction coefficient g_{12}. From left to right, the interaction is increased from $g_{12} = 0.75$ to $g_{12} = 0.82 > g_{12_{cr}}^{(2)}$ and then to $g_{12} = 1.4 > g_{12_{cr}}^{(1)}$. Panels (**a–d**) ((**e–h**)) correspond to the densities $|\psi_d|^2$ ($|\psi_b|^2$) of the dark (bright) component.

4. Discussion

In the present contribution, we investigated in detail the stability and dynamics of bound pairs of dark-bright symbiotic solitons, which arise as nonlinear matter wave excitations in mixtures of Bose–Einstein condensates featuring inter-atomic repulsion. In particular, we explored the scenario of differently weighted inter- and intra-species interaction coefficients, breaking the integrability of the relevant nonlinear Schrödinger model. It was argued by means of a recently proposed variational approach [50] and shown numerically that, upon departing from the integrable limit, bound states of such symbiotic entities exist for anti-symmetric bright soliton counterparts, the so-called solitonic gluons. These anti-symmetric states were found to be robust within a bounded interval of the inter-species repulsion coefficient g_{12}, limited by critical points both in the miscible and in the immiscible regime of the model, associated with a supercritical and a subcritical pitchfork bifurcation, respectively. Below and above these boundaries, i.e., deep in the miscible and the immiscible regime, respectively, the anti-symmetric pair becomes unstable. Long-time propagation revealed differences, but also common characteristics of the decay mechanisms in the two domains of instability. Specifically, a striking common feature is the relevance of bright mass transfer between the two dark-bright solitons. In particular, in the miscible domain, we identified new stationary asymmetric states that bifurcate from the anti-symmetric ones in a supercritical pitchfork. The stability of the new asymmetric states was also dynamically confirmed. Moreover, it was shown that a further decrease of the interspecies repulsion results in asymmetric states with the bright solitons living on top of a finite BEC background, highlighting in this way the miscible character of such bound pairs. In contrast to the above picture, upon entering the immiscible regime, we had found in our recent work [50] that the destabilization of the anti-symmetric dark-bright soliton pair is caused by a subcritical pitchfork bifurcation involving an unstable stationary, mass-imbalanced dark-bright soliton pair mode. In the present work, we further explored the range of existence and stability of this asymmetric branch, demonstrating its overall instability and its substantial deformation upon entering the miscible regime.

There are many directions that are worth considering in the future along the lines of this work. In particular, by fixing the intra-species interactions $g_{22} = 0.95$ in this work, we have not addressed the fate of the dark-bright soliton pair states when actually approaching the integrable limit, which would require $g_{12} = g_{22} = 1$. In this respect, it would be particularly interesting to see if the asymmetric state fully stabilizes upon restoring integrability or maintains its weakly unstable nature. Towards this direction, and since in the integrable limit exact solutions are available [31,59], one could link the anti-symmetric and asymmetric states obtained here with the exact families of dark-bright soliton solutions known in the integrable case. Establishing a possible connection of this type would also open a new direction of exploration and understanding of the symbiotic soliton pairs, since, in such a case, one could depart once more from the integrable limit, but having at hand exact analytical expressions for the two-soliton problem rather than the approximate ones constructed only from the single soliton solution, as used in our present work. Furthermore, and also in this direction, in the integrable model, exact closed form expressions exist not only for static symbiotic states like the ones considered here, but also for moving and scattering ones. Based on these, one could hope to get insights into the collisional dynamics of symbiotic entities at least in the vicinity of the integrable limit, paving the way for a detailed understanding of features like the breathing state formation observed here. Studies along these lines are left for future work.

5. Conclusions

In the present work, the existence, stability, and dynamics of symbiotic matter wave solitons that exist as nonlinear excitations in $(1 + 1)$-dimensional repulsively interacting mixtures of BECs has been investigated. Significant deviations from the integrable limit of the theory were considered, by utilizing a more realistic selection of the inter- and intra- species repulsion coefficients. In particular, by fixing the latter while varying the former thus examining the aforementioned system mono-parametrically, bound pairs consisting of two DB solitons have been identified for OP bright soliton counterparts. Our detailed stability analysis revealed that such anti-symmetric states exist as stable configurations in a wide window of inter-species repulsions that spans both the miscible and the immiscible regime of interactions. Below and above this domain, the anti-symmetric pair destabilizes giving rise to new symmetry-broken states that are asymmetric (mass imbalanced) with respect to their bright soliton counterpart. In particular, the destabilization of the OP pair deep in the miscible side was found to be connected with a supercritical pitchfork bifurcation resulting in stable asymmetric bound pairs. Deep in the immiscible regime of interactions, asymmetric states have been also identified, but were found to bifurcate from the anti-symmetric ones in a subcritical pitchfork bifurcation, and as such are unstable. In all cases our stability analysis results were verified by considering the long-time dynamics of both the stable and unstable anti-symmetric and asymmetric soliton pairs.

Acknowledgments: P. S. gratefully acknowledges financial support by the Deutsche Forschungsgemeinschaft (DFG) in the framework of the grant SCHM 885/26-1. P.G.K. gratefully acknowledges the support of NSF-PHY-1602994, the Alexander von Humboldt Foundation, the Stavros Niarchos Foundation via the Greek Diaspora Fellowship Program and the ERC under FP7, Marie Curie Actions, People, International Research Staff Exchange Scheme (IRSES-605096).

Author Contributions: G.C.K. performed the analytical and numerical computations; G.C.K., J.S. and P.G.K. conducted the interpretation of the results; P.S. conceived the idea of the present work; and all authors contributed to the writing of the paper.

Conflicts of Interest: The authors declare no conflict of interest.

References

1. Kevrekidis, P.G.; Frantzeskakis, D.J. Solitons in coupled nonlinear Schrödinger models: A survey of recent developments. *Rev. Phys.* **2016**, *1*, 140–153.
2. Kevrekidis, P.G.; Frantzeskakis, D.J.; Carretero-González, R. *Emergent Nonlinear Phenomena in Bose–Einstein Condensates: Theory and Experiment*; Springer: Heidelberg, Germany, 2008.
3. Carretero-González, R.; Kevrekidis, P.G.; Frantzeskakis, D.J. Nonlinear waves in Bose–Einstein condensates: Physical relevance and mathematical techniques. *Nonlinearity* **2008**, *21*, R139–R202.
4. Abdullaev, F.K.; Gammal, A.; Kamtchatnov, A.M.; Tomio, L. Dynamics of matter wave bright solitons in a Bose–Einstein condensate. *Int. J. Mod. Phys. B* **2005**, *19*, 3415–3473.
5. Weller, A.; Ronzheimer, J.P.; Gross, C.; Esteve, J.; Oberthaler, M.K.; Frantzeskakis, D.J.; Theocharis, G.; Kevrekidis, P.G. Experimental Observation of Oscillating and Interacting Matter Wave Dark Solitons. *Phys. Rev. Lett.* **2008**, *101*, 130401-1–130401-4.
6. Theocharis, G.; Weller, A.; Ronzheimer, J.P.; Gross, C.; Oberthaler, M.K.; Kevrekidis, P.G.; Frantzeskakis, D.J. Multiple atomic dark solitons in cigar-shaped Bose–Einstein condensates. *Phys. Rev. A* **2010**, *81*, 063604-1–063604-15.
7. Frantzeskakis, D.J. Dark solitons in atomic Bose–Einstein condensates: From theory to experiments. *J. Phys. A* **2010**, *43*, 213001-1–213001-68.
8. Sakaguchi, H.; Malomed, B.A. Matter-wave solitons in nonlinear optical lattices. *Phys. Rev. E* **2005**, *72*, 046610.
9. Borovkova, O.V.; Kartashov, Y.V.; Torner, L.; Malomed, B.A. Bright solitons from defocusing nonlinearities. *Phys. Rev. E* **2011**, *84*, 035602(R)-1–035602(R)-5.
10. Burger, S.; Bongs, K.; Dettmer, S.; Ertmer, W.; Sengstock, K.; Sanpera, A.; Shlyapnikov, G.V.; Lewenstein, M. Dark Solitons in Bose–Einstein Condensates. *Phys. Rev. Lett.* **1999**, *83*, 5198–5201.
11. Denschlag, J.; Simsarian, J.E.; Feder, D.L.; Clark, C.W.; Collins, L.A.; Cubizolles, J.; Deng, L.; Hagley, E.W.; Helmerson, K.; Reinhardt, W.P.; et al. Generating Solitons by Phase Engineering of a Bose–Einstein Condensate. *Science* **2000**, *287*, 97–101.
12. Anderson, B.P.; Haljan, P.C.; Regal, C.A.; Feder, D.L.; Collins, L.A.; Clark, C.W.; Cornell E.A. Watching Dark Solitons Decay into Vortex Rings in a Bose–Einstein Condensate. *Phys. Rev. Lett.* **2001**, *86*, 2926–2929.
13. Khaykovich, L.; Schreck, F.; Ferrari, G.; Bourdel, T.; Cubizolles, J.; Carr, L.D.; Castin, Y.; Salomon, C. Formation of a Matter-Wave Bright Soliton. *Science* **2002**, *296*, 1290–1293.
14. Strecker, K.E.; Partridge, G.B.; Truscott, A.G.; Hulet, R.G. Formation and propagation of matter-wave soliton trains. *Nature* **2002**, *417*, 150–153.
15. Myatt, J.C.; Burt, E.A.; Ghrist, R.W.; Cornell, E.A.; Wieman, C.E. Production of Two Overlapping Bose–Einstein Condensates by Sympathetic Cooling. *Phys. Rev. Lett.* **1997**, *78*, 586–589.
16. Hall, D.S.; Matthews, M.R.; Ensher, J.R.; Wieman, C.E.; Cornell, E.A. Dynamics of Component Separation in a Binary Mixture of Bose–Einstein Condensates. *Phys. Rev. Lett.* **1998**, *81*, 1539–1542.
17. Stamper-Kurn, D.M.; Andrews, M.R.; Chikkatur, A.P.; Inouye, S.; Miesner, H.-J.; Stenger, J.; Ketterle, W. Optical Confinement of a Bose–Einstein Condensate. *Phys. Rev. Lett.* **1998**, *80*, 2027–2030.

18. Nistazakis, H.E.; Frantzeskakis, D.J.; Kevrekidis, P.G.; Malomed, B.A.; Carretero-González, R. Bright-dark soliton complexes in spinor Bose–Einstein condensates. *Phys. Rev. A* **2008**, *77*, 033612-1–033612-12.

19. Yan, D.; Chang, J.J.; Hamner, C.; Hoefer, M.; Kevrekidis, P.G.; Engels, P.; Achilleos, V.; Frantzeskakis, D.J.; Cuevas, J. Beating dark-dark solitons in Bose–Einstein condensates. *J. Phys. B At. Mol. Opt. Phys.* **2012**, *45*, 115301-1–115301-11.

20. Achilleos, V.; Yan, D.; Kevrekidis, P.G.; Frantzeskakis, D.J. Dark-bright solitons in Bose–Einstein condensates at finite temperatures. *New J. Phys.* **2012**, *14*, 055006-1–055006-23.

21. Pérez-García, V.M.; Beitia, J.B. Symbiotic Solitons in Heteronuclear Multicomponent Bose–Einstein condensates. *Phys. Rev. A* **2005**, *72*, 033620-1–033620-5.

22. Majed, O.; Alotaibi, D; Carr, L.D. Dynamics of Vector Solitons in Bose–Einstein Condensates. *arXiv* **2016**, arXiv:1607.00108v2.

23. Chen, Z.; Segev, M.; Coskun, T.H.; Christodoulides, D.N.; Kivshar, Y.S. Coupled photorefractive spatial-soliton pairs. *J. Opt. Soc. Am. B* **1997**, *14*, 3066–3077.

24. Ostrovskaya, E.A.; Kivshar, Y.S.; Chen, Z.; Segev, M. Interaction between vector solitons and solitonic gluons. *Opt. Lett.* **1999**, *24*, 327–329.

25. Kivshar, Y.S.; Agrawal, G.P. *Optical Solitons: From Fibers to Photonic Crystals*; Academic Press: San Diego, CA, USA, 2003.

26. Christodoulides, D.N. Black and white vector solitons in weakly birefringent optical fibers. *Phys. Lett. A* **1988**, *132*, 451–452.

27. Afanasjev, V.V.; Kivshar, Y.S.; Konotop, V.V.; Serkin, V.N. Dynamics of coupled dark and bright optical solitons. *Opt. Lett.* **1989**, *14*, 805–807.

28. Kivshar, Y.S.; Turitsyn, S.K. Vector dark solitons. *Opt. Lett.* **1993**, *18*, 337–339.

29. Radhakrishnan, R.; Lakshmanan, M. Bright and dark soliton solutions to coupled nonlinear Schrödinger equations. *J. Phys. A Math. Gen.* **1995**, *28*, 2683–2692.

30. Buryak, A.V.; Kivshar, Y.S.; Parker, D.F. Coupling between dark and bright solitons. *Phys. Lett. A* **1996**, *215*, 57–62.

31. Sheppard, A.P.; Kivshar, Y.S. Polarized dark solitons in isotropic Kerr media. *Phys. Rev. E* **1997**, *55*, 4773–4782.

32. Park, Q.H.; Shin H.J. Systematic construction of multicomponent optical solitons. *Phys. Rev. E* **2000**, *61*, 3093–3106.

33. Becker, C.; Stellmer, S.; Soltan-Panahi, P.; Dörscher, S.; Baumert, M.; Richter, E.-M.; Kronjäger, J.; Bongs, K.; Sengstock, K. Oscillations and interactions of dark and dark-bright solitons in Bose–Einstein condensates. *Nat. Phys.* **2008**, *4*, 496–501.

34. Busch, T.; Anglin, J.R. Dark-Bright Solitons in Inhomogeneous Bose–Einstein Condensates. *Phys. Rev. Lett.* **2001**, *87*, 010401-1–010401-4.

35. Hamner, C.J.; Chang, J.; Engels, P.; Hoefer, M.A. Generation of Dark-Bright Soliton Trains in Superfluid-Superfluid Counterflow. *Phys. Rev. Lett.* **2011**, *106*, 065302-1–065302-4.

36. Middelkamp, S.; Chang, J.J.; Hamner, C.; Carretero-González, R.; Kevrekidis, P.G.; Achilleos, V.; Frantzeskakis, D.J.; Schmelcher, P.; Engels, P. Dynamics of dark-bright solitons in cigar-shaped Bose–Einstein condensates. *Phys. Lett. A* **2011**, *375*, 642–646.

37. Yan, D.; Chang, J.J.; Hamner, C.; Kevrekidis, P.G.; Engels, P.; Achilleos, V.; Frantzeskakis, D.J.; Carretero-González, R.; Schmelcher, P. Multiple dark-bright solitons in atomic Bose–Einstein condensates. *Phys. Rev. A* **2011**, *84*, 053630-1–053630-10.

38. Álvarez, A.; Cuevas, J.; Romero, F.R.; Hamner, C.; Chang, J.J.; Engels, P.; Kevrekidis, P.G.; Frantzeskakis, D.J. Scattering of atomic dark-bright solitons from narrow impurities. *J. Phys. B* **2013**, *46*, 065302-1–065302-11.

39. Inouye, S.; Andrews, M.R.; Stenger, J.; Miesner, H.-J.; Stamper-Kurn, D.M.; Ketterle, W. Observation of Feshbach resonances in a Bose–Einstein condensate. *Nature* **1998**, *392*, 151–154.

40. Roberts, J.L.; Claussen, N.R.; Burke, J.P., Jr.; Greene, C.H.; Cornell, E. A.; Wieman, C.E. Resonant Magnetic Field Control of Elastic Scattering in Cold ^{85}Rb. *Phys. Rev. Lett.* **1998**, *81*, 5109–5112.

41. Donley, E.A.; Claussen, N.R.; Cornish, S.L.; Roberts, J.L.; Cornell, E.A.; Wieman, C.E. Dynamics of collapsing and exploding Bose–Einstein condensates. *Nature* **2001**, *412*, 295–299.

42. Thalhammer, G.; Barontini, G.; de Sarlo, L.; Catani, J.; Minardi, F.; Inguscio, M. Double Species Bose–Einstein condensate with Tunable Interspecies Interactions. *Phys. Rev. Lett.* **2008**, *100*, 210402-1–210402-4.

43. Papp, S.B.; Pino, J.M.; Wieman, C.E. Tunable Miscibility in a Dual-Species Bose–Einstein condensate. *Phys. Rev. Lett.* **2008**, *101*, 040402-1–040402-4.
44. Chin, C.; Grimm, R.; Julienne, P.; Tiesinga, E. Feshbach resonances in ultracold gases. *Rev. Mod. Phys.* **2010**, *82*, 1225–1286.
45. Nguyen, J.H.V.; Dyke, P.; Luo, D.; Malomed, B.A.; Hulet, R.G. Collisions of matter-wave solitons. *Nat. Phys.* **2014**, *10*, 918–922.
46. Manakov, S.V. On the theory of two-dimensional stationary self-focusing of electromagnetic waves. *Sov. Phys. JETP* **1974**, *38*, 248–253.
47. Gerdzhikov, V.S.; Kulish, P.P. Multicomponent nonlinear Schrödinger equation in the case of nonzero boundary conditions. *J. Sov. Math.* **1983**, *85*, 2261–2269.
48. Prinari, B.; Ablowitz, M.J.; Biondini, G. Inverse scattering transform for the vector nonlinear Schrödinger equation with nonvanishing boundary conditions. *J. Math. Phys.* **2006**, *47*, 063508-1–063508-33.
49. Yan, D.; Tsitoura, F.; Kevrekidis, P.G.; Frantzeskakis, D.J. Dark-bright solitons and their lattices in atomic Bose–Einstein condensates. *Phys. Rev. A* **2015**, *91*, 023619-1–023619-12.
50. Katsimiga, G.C; Stockhofe, J.; Kevrekidis, P.G.; Schmelcher, P. Dark-bright soliton interactions beyond the integrable limit. *Phys. Rev. A* **2017**, *95*, 013621-1–013621-11.
51. Pitaevskii, L.P.; Stringari, S. *Bose–Einstein Condensation*; Oxford University Press: Oxford, UK, 2003.
52. Kevrekidis, P.G.; Frantzeskakis, D.J.; Carretero-González, R. *The Defocusing Nonlinear Schrödinger Equation*; SIAM: Philadelphia, PA, USA, 2015.
53. Mertes, K.M.; Merrill, J.W.; Carretero-González, R.; Frantzeskakis, D.J.; Kevrekidis, P.G.; Hall, D.S. Non-Equilibrium Dynamics and Superfluid Ring Excitations in Binary Bose–Einstein Condensates. *Phys. Rev. Lett.* **2007**, *99*, 190402-1–190402-4.
54. Egorov, M.; Opanchuk, B.; Drummond, P.; Hall, B.V.; Hannaford, P.; Sidorov, A.I. Measurement of s-wave scattering lengths in a two-component Bose–Einstein condensate. *Phys. Rev. A* **2013**, *87*, 053614-1–053614-10.
55. Kelley, C.T. *Solving Nonlinear Equations with Newton's Method*; Society for Industrial and Applied Mathematics: Philadelphia, PA, USA, 1995.
56. Ao, P.; Chui, S.T. Binary Bose–Einstein condensate mixtures in weakly and strongly segregated phases. *Phys. Rev. A* **1998**, *58*, 4836–4840.
57. Skryabin, D.V. Instabilities of vortices in a binary mixture of trapped Bose–Einstein condensates: Role of collective excitations with positive and negative energies. *Phys. Rev. A* **2000**, *63*, 013602-1–013602-10.
58. Karamatskos, E.T.; Stockhofe, J.; Kevrekidis, P.G.; Schmelcher, P. Stability and tunneling dynamics of a dark-bright soliton pair in a harmonic trap. *Phys. Rev. A* **2015**, *91*, 043637-1–043637-10.
59. Garrett, D.; Klotz, T.; Prinari, B.; Vitale, F. Dark-dark and dark-bright soliton interactions in the two-component defocusing nonlinear Schrödinger equation. *Appl. Anal.* **2013**, *92*, 379–397.

![applied sciences logo] *applied sciences*

MDPI

Article

Polarization Properties of Laser Solitons

Pedro Rodriguez [1,*,†], Jesus Jimenez [2,†], Thierry Guillet [2,3,†] and Thorsten Ackemann [2,†]

1   Physics Department, University of Cordoba, Cordoba, 14071 Andalusia, Spain
2   SUPA and Department of Physics, University of Strathclyde, Glasgow G4 0NG, Scotland, UK;
    jesusj78@gmail.com (J.J.); thierry.guillet@strath.ac.uk (T.G.); thorsten.ackemann@strath.ac.uk (T.A.)
3   Laboratoire Charles Coulomb, UMR 5221 CNRS, Universite de Montpellier, 163 Rue Auguste Broussonnet,
    34090 Montpellier, France
*   Correspondence: pm1rogap@uco.es; Tel.: +34-957-212-551
†   The authors contributed equally to this work.

Academic Editor: Boris Malomed
Received: 13 March 2017; Accepted: 22 April 2017; Published: 27 April 2017

Abstract: The objective of this paper is to summarize the results obtained for the state of polarization in the emission of a vertical-cavity surface-emitting laser with frequency-selective feedback added. We start our research with the single soliton; this situation presents two perpendicular main orientations, connected by a hysteresis loop. In addition, we also find the formation of a ring-shaped intensity distribution, the vortex state, that shows two homogeneous states of polarization with very close values to those found in the soliton. For both cases above, the study shows the spatially resolved value of the orientation angle. It is important to also remark the appearance of a non-negligible amount of circular light that gives vectorial character to all the different emissions investigated.

Keywords: cavity solitons; vortex beams; vectorial light; spatially resolved polarization

1. Introduction

The appearance of stable structures, solitons, in laser emission is an important topic in the field of nonlinear optics. Depending on their nature, they can be classified as temporal [1] and spatial solitons [2].

The spatial solitons are self-localized states of light capable of beating diffraction through nonlinearities; this attribute makes them useful for potential application in all optical processing and switching operations [3,4]. Among them, cavity solitons (CS) are bistable spatially self-localized waves that exist in the transverse aperture of broad-area nonlinear optical resonators. Examples of spatial CS have been found in many different experiments, e.g., laser with a saturable absorber [5], Kerr media [6], liquid crystal light valve [7] or vertical-cavity surface-emitting laser (VCSEL) [8].

The VCSEL is one of the most used and investigated kinds of solid state laser [9]. Since its development, it has been widely utilized in different applications [10]. In conditions of normal emission, above the threshold value for the injected current, a well characterized linear polarization is observed [11]. Depending on the state of polarization, the CS with only one component is called scalar; otherwise these structures are called vector CSs.

In this contribution, we create the solitonic structure in a very simple system—a VCSEL with a frequency-selective feedback setup as is shown in different publications of some of the authors [12,13]. In the aforementioned conditions, a rich collection of spatially self-localized intensity structures can be found, from the single spot CS to much more complex formations, i.e., the appearance of a ring-shaped structure, usually made by a circular intensity pattern with peaks—the vortex beam. In this particular experiment, a novel feature associated with the state of polarization of the single soliton is discovered, i.e., the existence of a hysteresis loop that connects two orthogonal orientations. Also remarkable is the discovery of a significant amount of circular light in all the cases studied, for the first time in this system, in accordance to other authors' results [14] for the VCSEL emission, making it a full Poincare beam [15].

Our objective is to resolve the spatial distribution of the polarization in all these situations; starting from the case of homogeneous polarization found in the CS and applying also our analysis to the vortex.

The organization of the paper is as follows. In Section 2, the experimental setup and the method are included; in Section 3, the results are presented; in Section 4, we make an interpretation of the results and the future directions of our work.

2. Materials and Method

The experimental setup and the instrumentation used in this experiment are very similar to those used in previous works on laser solitons in our group [8,12]. A detailed representation of this arrangement is shown in Figure 1. The VCSEL on this experiment is a large aperture device with a diameter of 200 mm. The emission takes place in the 975–980 nm region through the n-doped Bragg reflector and the transparent substrate, the so-called bottom emitter [16]. A Peltier element with a feedback circuit is utilized to stabilize the VCSEL temperature at 20 °C.

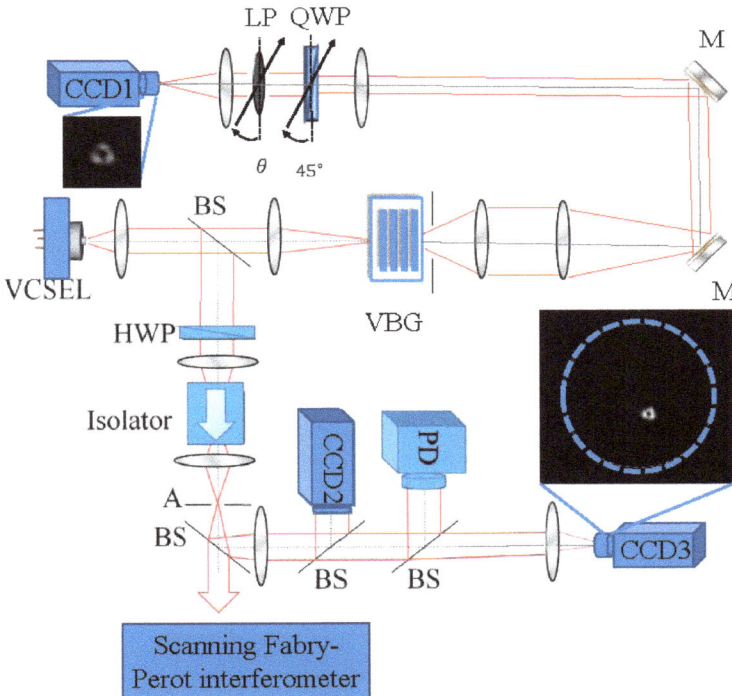

Figure 1. Experimental setup: A volume Bragg grating (VBG) provides frequency-selective feedback to a vertical-cavity surface-emitting laser (VCSEL). BS: beam sampler, CCD: charge-coupled device camera, PD: photo-detector, LP: linear polarizer, QWP: quarter-wave plate, HWP: half-wave plate. The upper arm is used to measure the spatially resolved Stokes parameter at high magnification (CCD1), the lower monitors power (PD) and near (CCD3) and far field (CCD2) distributions of potentially the whole laser.

The output of the VCSEL is collimated by an aspheric lens of $f_1 = 8$ mm focal length. A second lens with $f_2 = 50$ mm is used to focus the light onto the frequency-selective element, a volume Bragg grating (VBG). These two lenses are adjusted to form an a focal telescope, i.e., the external cavity is self-imaging after a round trip. The VBG has a narrow-band reflection peak of 95% at $\lambda_g = 978.1$ nm, with a reflection bandwidth of 0.1 nm full-width half-maximum (FWHM).

For monitoring the output, a wedged glass plate with an uncoated facet at the front and an anti-reflection coated facet at the back serves as an outcoupler or beam sampler (BS). The reflection is relying on Fresnel reflection and therefore is polarization dependent. The reflectivity is on the order of 10% for s-polarized light and 1% for p-polarized light. Note that the polarization asymmetry is much smaller (1:1.1) in transmission. Via a half-wave plate and an optical isolator, polarization resolved light-current (LI) characteristics as well as near and far field intensity distributions can be obtained with a photo-diode and CCD cameras, respectively.

The use of the intra-cavity BS also allows measurements without feedback. However, the main results for the polarization distributions are obtained by observation after the VBG, as the intra-cavity polarization state can be accessed directly from there. The light that goes through the VBG is re-imaged onto another CCD-camera (CCD1) by two telescopic systems, providing enough magnification to accurately resolve the different polarization zones. Within the collimated range between the two lenses of the second telescope, a linear polarizer (LP) and a quarter-wave plate (QWP) are used with different combinations in its orientation.

To analyze the state of polarization of the VCSEL's emission, all our work relies on the spatially resolved measurement of the well-known Stokes parameters [17]. The data needed to accomplish this task are the different intensity patterns obtained for the different orientations of the polarizer's axis. The polarization orientation is defined with respect to the propagation direction of the beam, i.e., in the plane orthogonal (x, y) to its wave vector.

The measurements necessary to determine the Stokes parameters are:

- I_x: Horizontally polarized component of the intensity.
- I_y: Vertically polarized component of the intensity
- I_{45}: Intensity component diagonally polarized.

Usually, this study is restricted to the linear components of the polarization but it is important to comment that in our case the results obtained confirm the existence of a non-negligible quantity of circular light. To measure this factor, a fourth value of the intensity is taken:

- I_{circ}: Circular component of the emission. In this case, a QWP is used in addition to the linear polarizer. including the S_3 factor associated with this component, the Stokes parameters are calculated from the following set of equations:

$$S_0 = I_x + I_y \tag{1}$$

$$S_1 = \frac{(I_x - I_y)}{S_0} \tag{2}$$

$$S_2 = \frac{2 \cdot I_{45}}{S_0} - 1 \tag{3}$$

$$S_3 = \frac{2 \cdot I_{circ}}{S_0} - 1 \tag{4}$$

where S_0 represents the total intensity; S_1 the degree of horizontally (positive values of S_1) or vertically (negative values of S_1) polarized light; in the same way, S_2 accounts for the polarization degree across the diagonals (positive for 45°, negative for −45°); and S_3 represents the degree of circular polarization (the sign denotes the direction of rotation). Furthermore, two additional calculations have been done—the fractional polarization (FP):

$$FP = \sqrt{S_1^2 + S_2^2 + S_3^2} \tag{5}$$

in order to ensure the validity of our results, they are checked against the ideal value of 1, corresponding to the radius of the Poincare sphere.

The last parameter calculated is the value of the direction of polarization ψ, calculated with:

$$\psi = 0.5 \cdot atan2 \frac{S_2}{S_1} \tag{6}$$

Once the different intensity patterns for the VCSEL emission have been measured and recorded, the analysis is carried out using specific software developed in our group that allows us to have access to every individual pixel of the camera (CCD1 in Figure 1). This tool permits us to know not the mean value or average of the aforementioned parameters but the complete, spatially resolved description of these. In particular, we can obtain $\psi(x, y)$ as the polarization state which is the main goal of our research.

3. Results

We start our experiment with a low value for the injection current, well below the lasing threshold, that places the device in the spontaneous emission regime. When the current applied to the VCSEL is increased, it switches up abruptly; then, we can observe the appearance of different bright single spots (CS) or the formation of more complicated structures.

From here on, the experiment is completed by decreasing the current again. In this part, two main aspects are observed in the L–I diagram: a significant grade of hysteresis in the emission and the existence of abrupt transitions; this last effect happens when the whole structure simplifies as shown in Figure 2.

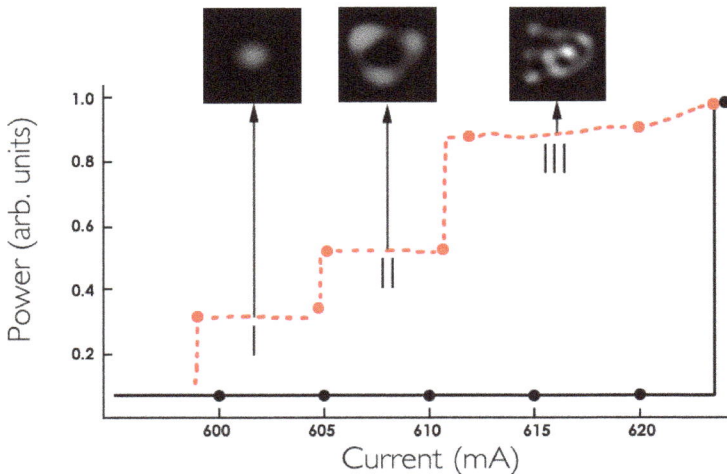

Figure 2. Typical L–I characteristic curve obtained by monitoring the output power of the VCSEL. The black dots account for regular spaced samples taken when the current is rising (continuous line); the constant value of the power measured is due to the absence of any output apart from the spontaneous emission. Once the system switches on, a complicated structure is formed (zone III); as we lower the injection current (dotted line), this structure changes and simplifies as can be seen in the second upper image (zone II) corresponding to the vortex beam. The last picture shows the single soliton (Zone I).

In order to achieve a better understanding of the results, we split them in two parts. The first group deals with the treatment of the results obtained for the single soliton, its different polarization states and the transitions between them; the second part explains the polarization distribution obtained in the vortex pattern observed in the VCSEL's emission. In both cases, a clear homogeneous polarization across all the intensity structure is found. The intensity characteristics for the soliton and vortex profile appear in Figure 3 .

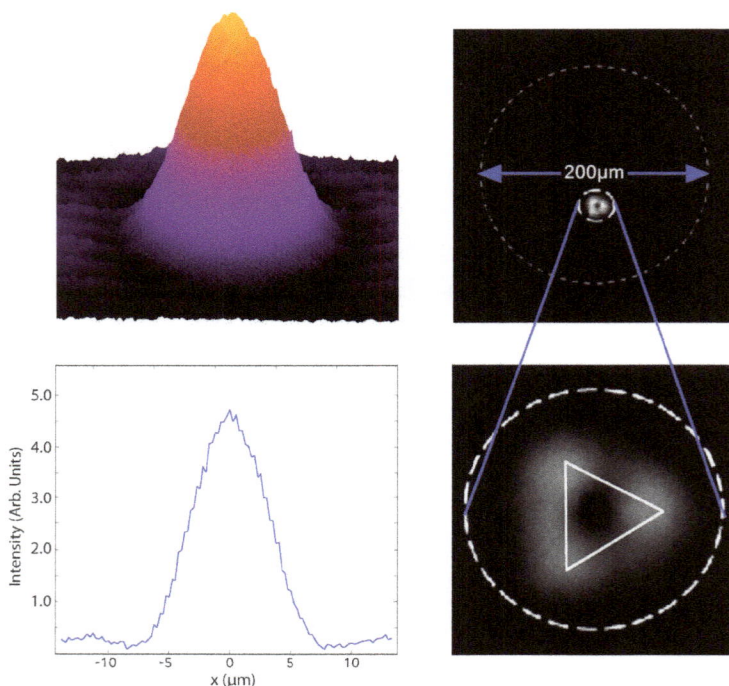

Figure 3. The left column depicts the intensity profile and dimensions of the single soliton. The right column accounts for the shape of the ring distribution; in this case, the distance between peaks is about ten microns. The intensity pattern resembles the shape of an optical vortex, including the phase singularity at the center.

3.1. Single Soliton Case. The Cavity Soliton

To start with, in the analysis, we are going to focus on the simplest result obtained—the single soliton. The existence of CS in semiconductor microcavities has been predicted theoretically [18] and experimentally achieved in VCSEL's below lasing threshold and proposed as pixels in all-optical systems [19].

Our case of study corresponds to those solitons formed in a cavity (CS) with a frequency-selective feedback term obtained from a VBG as is described in the second section. It is important to note that although it is possible for localized states to exist at any position of a homogeneous device, a strong pinning at preferred locations—traps—is found [20]. This effect is depicted in Figure 2; the complex cluster that appears for the highest value of the current evolves to the single soliton when the current decreases, but this behavior is restricted to a specific area on the VCSEL's surface.

The intensity distribution of the single-peaked spots is reasonably Gaussian, with sizes ranging from 4.8 μm to 5.8 μm radius at $1/e^2$ point of intensity, as can be seen in the left column of Figure 3.

In this situation, two possible directions for the polarization are found, mutually orthogonal. Both orientations are homogeneous across all the soliton area, therefore we can assign just a single value to represent it, as is shown in Figure 4.

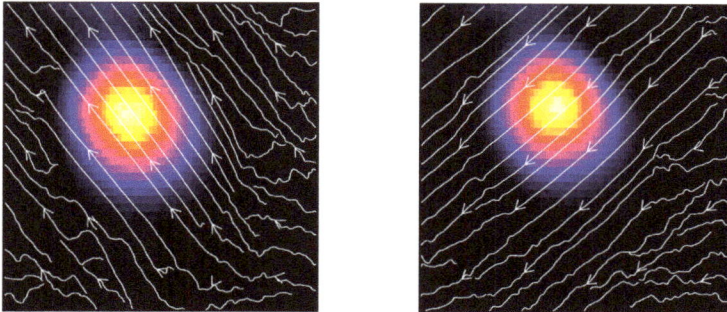

Figure 4. Results obtained for the single soliton showing the existence of two orthogonal directions clearly represented by streamlines with constant orientation.

The aforementioned states are connected by a switch when the injection current makes a cycle for a particular range of values. The change between orthogonal polarization orientations in a VCSEL is a well-known topic [21]. In our work, this effect appears for the first time between CS through a hysteresis region, as can be seen in the right part of Figure 5, where this effect is analyzed for three different linear polarizer orientations. Another important element in this part of our experiment is the measurement of the frequency for both orientations. These data are taken out of the cavity by the BS and clearly exhibit the existence of two different modes, each one corresponding to the two polarizations observed, as is shown in the left part of Figure 5.

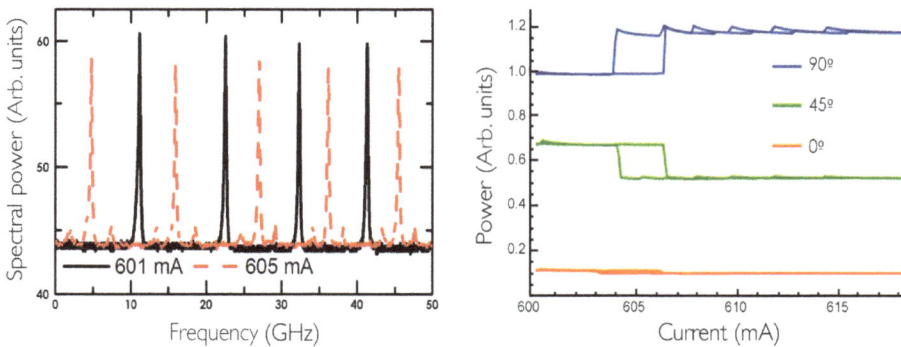

Figure 5. Results obtained for the frequency spectrum (**left**), and the I-L diagram for a cycle centered in the zone where the change in the soliton polarization orientation happens (**right**) using a linear polarizer with different orientations as the analyzer.

The amount of circular light in the emission is also studied by the analysis of the S3 parameter; in all the cases measured, this quantity represents about one-tenth of the total emitted intensity. The origin of this component is explained in the framework of the spin–flip model [22] in recent experiments for VCSELs [14]. For the polarization case investigated here, a theoretical treatment is currently under development in our group. Also remarkable is the association between the linear polarization and the circular component, i.e., the circular component changes its sense of rotation when the linear component flips, giving a complete vectorial character to both polarization states. This fact could be utilized in the aforementioned all-optical systems [19] to improve their performance by the use of the complete set of Stokes values as parameters in the information processing.

In order to verify the accuracy of our results, the FP factor is measured for both polarization orthogonal states. The results obtained are very close to the ideal value of one—radius of the Poincare sphere—ranging in the interval [0.96–0.98] for both orientations.

3.2. The Ring-Shaped Structure. The Optical Vortex Beam

A more complicated structure than the single soliton appears at higher values for the injection current, as can be seen in the picture corresponding to zone II in Figure 2. This emission pattern is normally made by a three-peak ring with a central hole in the middle and we can call it an optical vortex [23,24]. Its dimensions are of about ten microns for the distance between peaks; a plot of this structure is shown in Figure 3.

These were observed recently in coupled VCSELs—one operated as a gain device and one as a saturable absorber [25]. Vortices in self-focusing Ginzburg–Landau models were predicted [26]. References [27,28] investigate a nonlinear cavity with a saturable or cubic, i.e., Ginzburg–Landau-like nonlinearity, coupled to an additional linear filter which provides a minimal model for a VCSEL with frequency-selective feedback, similar to our experiment.

The homogeneous polarization, in the same way as in the previous section, describes a constant orientation for the whole area of the emission, obtaining one value or its orthogonal depending on the injection current value, with a very similar intensity pattern (S_0) in both cases. This structure is essentially identical to the vortex soliton observed in a prior work of our group [12], including a phase singularity in the dark center.

Corresponding generalized vortex solitons were predicted [29] and indications observed [30] in single-pass conservative systems and termed azimuthons. Theoretical predictions also exist for dissipative systems [31].

Figure 6 shows the states of polarization for the vortex. The angle takes again a very similar value to that already shown for the single soliton. In addition, the linear and circular components have the same behavior already seen in the soliton case, i.e., a particular sense of rotation is associated to one of the linear orientations.

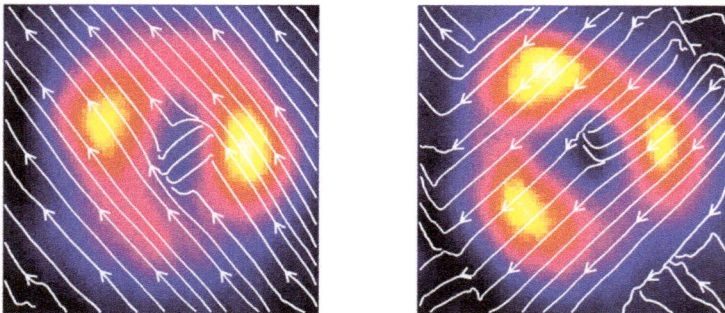

Figure 6. Total intensity S_0 for the vortex case showing two-peak structure in the left and three-peak structure in the right part of the figure. The polarization streamline representation for the two orthogonal polarization orientations that appear in the vortex beam reaches values very close to those of the single soliton.

4. Discussion

The observations presented establish the existence of solitons and vortex solitons in a cavity with a self-focusing medium. This phenomenon happens through a direct transition from the off-state, with only spontaneous emission, to the vortex appearance when we rise the injection current; this structure simplifies to the single soliton once we lower the current.

The setup described here is much simpler than the typical schemes to create vectorial vortex beams; this fact opens the possibility for the monolithic implementation of the external cavity in order to build a complete useful device to generate them.

In this work, the spatially resolved analysis of the polarization of single solitons and vortex structure is shown; it reveals two main directions for the polarization state in both cases. In addition, a relevant quantity of circular light is also found, which gives a full vectorial character to this light emission.

Another set of experiments, intended for the investigation of the polarization state for the soliton at different temperatures, is currently under way. Finally, the possible existence of more complex polarization structures for the vortex is another line of research in our group.

Acknowledgments: We are grateful to Gian-Luca Oppo and William J. Firth for useful discussions and to Roland Jäger (Ulm Photonics) for supplying the devices. P.R. acknowledges Universidad de Cordoba for its support and Jacobo Muniz-Lopez for his help in the configuration of some figures. J.J. gratefully acknowledges support from CONACYT.

Author Contributions: T.A. and J.J. conceived and designed the experiments; J.J. and T.G. performed the experiments; P.R. and T.G. analyzed the data; P.R. wrote the manuscript.

Conflicts of Interest: The authors declare no conflict of interest.

References

1. Leo, F.; Coen, S.; Kockaert, P.; Gorza, S.-P.; Emplit, P.; Haelterman, M. Temporal cavity solitons in one-dimensional Kerr media as bits in an all-optical buffer. *Nat. Photonics* **2010**, *4*, 471–476. [CrossRef]
2. Schapers, B.; Feldmann, M.; Ackemann, T.; Lange, W. Interaction of localized structures in an optical pattern-forming system. *Phys. Rev. Lett.* **2000**, *85*, 748–751. [CrossRef] [PubMed]
3. Firth, W.J.; Weiss, C.O. Cavity and feedback solitons. *Opt. Photonics News* **2002**, *13*, 54–58. [CrossRef]
4. Akhmediev, N.; Ankhiewicz, A. (Eds.) *Dissipative Solitons*; Lecture Notes in Physics; Springer: Berlin, Germany, 2005; Volume 336.
5. Bazhenov, V.Y.; Taranenko, V.B.; Vasnetsov, M.V. Transverse optical effects in bistable active cavity with nonlinear absorber on bacteriorhodopsin. *Proc. SPIE* **1992**, *1840*, 183–193.
6. Odent, V.; Tlidi, M.; Clerc, M.G.; Glorieux, P.; Louvergneaux, E. Experimental observation of front propagation in a negatively diffractive inhomogeneous kerr cavity. *Phys. Rev. A* **2014**, *90*, 011806. [CrossRef]
7. Schreiber, A.; Thüring, B.; Kreuzer, M.; Tschudi, T. Experimental investigation of solitary structures in a nonlinear optical feedback system. *Opt. Commun.* **1997**, *136*, 415–418. [CrossRef]
8. Radwell, N.; Ackemann, T. Characteristics of laser cavity solitons in a vertical-cavity surface-emitting laser with feedback from a volume Bragg grating. *IEEE J. Quantum Electron. Phys. Lett.* **2009**, *45*, 1388–1395. [CrossRef]
9. Wilmsen, C.; Temkin, H.; Coldren, L. (Eds.) *Vertical-Cavity Surface-Emitting Lasers. Design, Fabrication, Characterization and Applications*; Cambridge Studies in Modern Optics; Cambridge University Press: Cambridge, UK, 2001.
10. Michalzik, R. (Ed.) *VCSEL's: Fundamentals, Technology and Applications of Vertical-Cavity Surface-Emitting Lasers*; Springer Series in Optical Sciences; Springer: Berlin, Germany, 2013.
11. Webb, C.; Jones, J. (Eds.) *Handbook of Laser Technology and Applications. Volume II: Laser Design and Laser Applications*; Institute of Physics Publishing: Bristol, UK, 2004.
12. Jimenez, J.; Noblet, Y.; Paulau, P.V.; Gomila, D.; Ackemann, T. Observation of laser vortex solitons in a self-focusing semiconductor laser. *J. Opt.* **2013**, *15*, 044011. [CrossRef]
13. Jimenez, J.; Oppo, G.-L.; Ackemann, T. Temperature dependence of spontaneous switch-on and switch-off of laser cavity solitons in vertical-cavity surface-emitting lasers with frequency-selective feedback. *J. Phys. D* **2016**, *49*, 095110. [CrossRef]
14. Averlant, E.; Tlidi, M.; Thienpont, H.; Ackemann, T.; Panajotov, K. Vector cavity solitons in broad area Vertical-Cavity Surface-Emitting Lasers. *Sci. Rep.* **2016**, *6*, 20428. [CrossRef] [PubMed]

15. Beckley, A.M.; Brown, T.G.; Alonso, M.A. Full Poincare Beams. *Opt. Exp.* **2010**, *18*, 10777–10785. [CrossRef] [PubMed]

16. Grabher, M.; Jager, R.; Miller, M.; Thalmaier, C.; Herlein, J.; Ebeling, K.J. Bottom emitting VCSELs for High-CW optical output power. *IEEE Photonics Technol. Lett.* **1998**, *10*, 1061–1603. [CrossRef]

17. Hecht, E. *Optics*; Pearson Education Ltd.: Harlow, UK, 2012.

18. Brambilla, M.; Lugiato, L.A.; Prati, F.; Spinelli, L.; Firth, W. Spatial soliton pixels in semiconductor devices. *Phys. Rev. Lett.* **1997**, *79*, 2042–2045. [CrossRef]

19. Barland, S.; Tredicce, J.R.; Brambilla, M.; Lugiato, L.A.; Balle, S.; Guidici, M.; Maggipinto, T.; Spinelli, L.; Tissoni, G.; Knodel, T.; et al. Cavity solitons as pixels in semiconductors. *Nature* **2002**, *419*, 699–702. [CrossRef] [PubMed]

20. Firth, W.J.; Scroggie, A.J. Optical Bullet Holes: Robust Controllable Localized States of a Nonlinear Cavity. *Phys. Rev. Lett.* **1996**, *76*, 1623–1626. [CrossRef] [PubMed]

21. Panajotov, K.; Prati, F. *Polarization Dynamics in VCSELs; Chapter 6 in "VCSEL's: Fundamentals, Technology and Applications of Vertical-Cavity Surface-Emitting Lasers"*; Springer Series in Optical Sciences; Springer: Berlin, Germany, 2013.

22. San Miguel, M.; Feng, Q.; Moloney, J.V. Light polarization dynamics in surface-emitting semiconductor lasers. *Phys. Rev. A* **1995**, *52*, 1728–1739. [CrossRef] [PubMed]

23. Nye, J.F.; Berry, M.V. Dislocations in wave trains. *Proc. R. Soc. A* **1974**, *336*. [CrossRef]

24. Kivshar, Y.S.; Ostrovskaya, E.A. Optical vortices: Folding and twisting waves of light. *Opt. Photonics News* **2001**, *12*, 24–28. [CrossRef]

25. Genevet, P.; Barland, S.; Giudici, M.; Tredicce, J.R. Bistable and Addressable Localized Vortices in Semiconductor Lasers. *Phys. Rev. Lett.* **2010**, *104*, 223902. [CrossRef] [PubMed]

26. Mihalache, D.; Mazilu, D.; Lederer, F.; Leblond, H.; Malomed, B.A. Collisions between coaxial vortex solitons in the three-dimensional cubic-quintic complex Ginzburg-Landau equation. *Phys. Rev. A* **2008**, *77*, 033817. [CrossRef]

27. Paulau, P.V.; Gomila, D.; Colet, P.; Loiko, N.A.; Rosanov, N.N.; Ackemann, T.; Firth, W.J. Vortex solitons in lasers with feedback. *Opt. Express* **2010**, *18*, 8859–8866. [CrossRef] [PubMed]

28. Paulau, P.V.; Gomila, D.; Colet, P.; Malomed, B.A.; Firth, W.J. From one- to two-dimensional solitons in the Ginzburg-Landau model of lasers with frequency-selective feedback. *Phys. Rev. E* **2011**, *84*, 036213. [CrossRef] [PubMed]

29. Desyatnikov, A.S.; Sukhorukov, A.A.; Kivshar, Y.S. Azimuthons: Spatially Modulated Vortex Solitons. *Phys. Rev. Lett.* **2005**, *95*, 203904. [CrossRef] [PubMed]

30. Minovich, A.; Neshev, D.N.; Desyatnikov, A.S.; Krolikowski, W.; Kivshar, Y.S. Observation of optical azimuthons. *Opt. Express* **2009**, *17*, 23610–23616. [CrossRef] [PubMed]

31. Fedorov, S.V.; Rosanov, N.N.; Shatsev, A.N.; Veretenov, N.A.; Vladimorov, A.G. Topologically multicharged and multihumped rotating solitons in wide-aperture lasers with a saturable absorber. *IEEE J. Quantum Electron.* **2003**, *39*, 197–205. [CrossRef]

Article

Perfect Light Absorbers Made of Tungsten-Ceramic Membranes

Masanobu Iwanaga

National Institute for Materials Science (NIMS), 1-1 Namiki, Tsukuba 305-0044, Japan;
iwanaga.masanobu@nims.go.jp

Academic Editor: Boris Malomed
Received: 8 March 2017; Accepted: 26 April 2017; Published: 29 April 2017

Abstract: Plasmonic materials are expanding their concept; in addition to noble metals that are good conductors even at optical frequencies and support surface plasmon polaritons at the interface, other metals and refractory materials are now being used as plasmonic materials. In terms of complex permittivity at optical frequencies, these new plasmonic materials are, though not ideal, quite good to support surface plasmons. Numerical investigations of the optical properties have been revealing new capabilities of the plasmonic materials. On the basis of the precise computations for electromagnetic waves in artificially designed nanostructures, in this article, we address membrane structures made of tungsten and silicon nitride that are a typical metal and ceramic, respectively, with high-temperature melting points. The membranes are applicable to low-power-consuming thermal emitters operating at and near the visible range. We numerically substantiate that the membranes serve as perfect light absorbers, in spite of the subwavelength thickness, that is, 200–250 nm thickness. Furthermore, we clarify that the underlying physical mechanism for the unconventional perfect absorption is ascribed to robust impedance matching at the interface between air and the membranes.

Keywords: perfect light absorber; membrane; tungsten; ceramics; impedance matching; guided mode; plasmonic resonances

---

1. Introduction

Membrane structures can be used to manipulate light in outstanding ways. One example is photonic crystals made of high-refractive index materials such as semiconductors [1]. However, such membrane structures are generally fragile and difficult to handle since they are free-standing in air and have thickness less than the wavelength of light in vacuum, that is, subwavelength. Today, ceramic membranes can be realized, thanks to reliable thin-film fabrication techniques such as plasma-enhanced chemical vapor deposition. Here, we focus on membrane structures based on free-standing thin films made of ceramics. To make the application clear, we design perfect light absorbers (PLAs) suitable for thermal emitters in the short-wavelength infrared range of 1–2 µm. Thermal emitters made of nanostructured materials are currently optimized for the mid-infrared range of 2–5 µm, using metal-insulator-metal structures that typically comprise noble metals [2–5], aluminum [6], etc., as constituent materials; however, these metals cannot withstand high temperatures.

A basic requirement exists for the thermal emitters, that is, high melting points exceeding 2000 K. From this point of view, we chose silicon nitride (SiN) as the ceramic and tungsten (W) as the metal in this study. Both of them are stable at high temperatures. Considering practical application, the thickness of SiN membranes and W are assumed to be 100–200 nm and 50–100 nm, respectively, which are usual values in commercial products [7] and laboratory use. A 3D photonic crystal made of only W was once produced with a similar goal as mentioned above and worked as a thermal emitter at about 2 µm [8]. However, production of thermal emitters based on such structures still has a high demand and has not been standardized to date; thus, much simpler structures are highly preferred from the viewpoint of practical application.

Light absorbers in infrared ranges have a long history. Originally, the idea to obtain PLAs was simple; that is, to use materials as analogous to black bodies as possible. The absorbed infrared light has been converted to electric signals since the 1940s [9]; the devices are well known as conventional bolometers. However, it has been difficult to attain PLAs over the whole wavelength range. Accordingly, PLAs based on the artificially designed structures were conceived for the particular wavelength ranges [2–6]; in fact, the PLAs do not look like black bodies. Thus, there are distinct differences between black-body-based and artificial-structure-based PLAs.

Figure 1 illustrates schematics of W-SiN membrane structures studied in this work. Figure 1a–c shows simply stacked W and SiN layers. The two-layer membrane structure is formed using a SiN membrane on a Si substrate. The section view is shown in Figure 1b. The image of the original SiN membrane is shown in Figure 1c; the central square domain of 2×2 mm^2 is the SiN membrane. Once deposition of W is performed, the W-SiN membrane structure in Figure 1a,b is obtained. The white dashed line indicates a section corresponding to Figure 1b. Figure 1a also indicates the coordinate axes and optical configuration of p polarization, which is defined such that the incident polarization vector is parallel to the plane of incidence. Without loss of generality, we set the plane of incidence to be parallel to the xz plane.

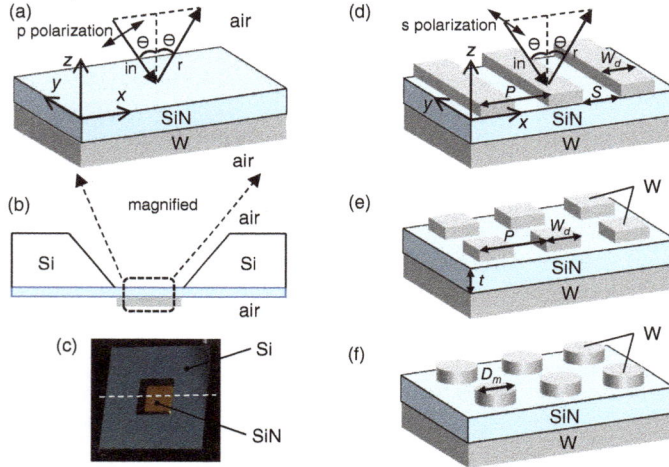

Figure 1. Schematics of metal-ceramic membrane structures. The metals are W (gray) and the ceramics are SiN (pale blue) in this study. (**a**) simply stacked W-SiN membranes, which are assumed to be formed on SiN membrane grown on a Si substrate. Both-end arrows represent p polarization where the incident electric-field vector is parallel to the plane of incidence; (**b**) a section view of the whole SiN-membrane substrate; (**c**) a photograph of SiN membrane supported by a Si substrate. White dashed line shows a section corresponding to (**b**). The SiN membrane is located at the center and has a dimension of 2×2 mm^2; (**d**) W-SiN-W membranes with 1D periodic structures on the top. The plane of incidence is set to be parallel to the xz plane. Both-end arrows show s polarization where the incident electric-field vector is perpendicular to the xz plane; (**e,f**) W-SiN-W membranes of 2D periodic structures with on-top rectangular and circular W patches, respectively.

Figure 1d–f shows schematics of W-SiN-W membrane structures, whose top layers are set to be periodic. Figure 1d also shows the coordinate axes and s polarization. When the plane of incidence is parallel to the xz plane, s polarization means that the incident polarization is perpendicular to the xz plane and parallel to the y-axis. Figure 1d shows a 1D periodic structure along the x axis. Figure 1e,f shows 2D periodic structures of square arrays of W patches; we set the coordinate axes to be similar to Figure 1d. In Figure 1d–f, we define notations P, S, W_d, and D_m for the periodic lengths, slit length, width of the W bars or square patches, and diameter of the circular W patches, respectively.

In the optical linear response regime, light absorptance A is evaluated using the following equation:

$$A = 1 - \sum_{m,n} (R_{mn} + T_{mn}),$$ (1)

where R_{mn} and T_{mn} denote the mn-th reflective and transmissive diffraction, respectively. R_{00} and T_{00} are ordinary reflectance (R) and transmittance (T), respectively. As we describe in detail in Section 3.2, the linear optical responses of R_{mn} and T_{mn} are directly computed in a numerically precise way, based on Maxwell equations. Note that Equation (1) is represented in the $(0, 1)$ range where 0 and 1 denote 0% and 100%, respectively.

Perfect light absorption is related to perfect emittance by reciprocity [10]; therefore, when thermal emitters were produced, researchers tried to attain perfect-absorption structures [2–6]. Recently, nearly perfect absorption was observed in stacked complementary structures [11], which show significant enhancement of electric-dipole emission loaded on the outmost surfaces [12–14]. In addition, it was lately reported that high-emittance plasmonic cavities substantially enhance both electric- and magnetic-dipole emissions of Er ions [15]. Thus, it has been quite common to relate efficient light absorption with high emittance.

In this study, we aim at designing PLAs made of a set of representative refractory metal and ceramic, based on precise numerical calculations. To propose realistic designs, we conduct the numerical calculations using reliable material parameters. As for the metal W, the complex relative permittivity is taken from the literature [16], and, as for the ceramic SiN, we adopt a representative value, that is, permittivity of 4.3. Further details are provided in Section 3.1. The representative value is sufficient to extract fundamental properties of the W–ceramic systems. We show that the W-SiN membranes are able to serve as PLAs even when they have subwavelength thickness, suggesting that they can function as low-power-consumption thermal emitters near the visible range. We also clarify that the working mechanism of the PLAs is unconventional, robust, and different from that of noble-metal–insulator systems studied so far [2–6].

2. Results and Discussion

In this section, we show numerical results for PLAs made of W and SiN. Periodic structures are introduced to absorb incident light efficiently in the very thin membrane structures of subwavelength thickness. Below, we confirm the basic optical properties of the constituent materials.

Figure 2 shows the basic optical properties of W and SiN. Figure 2a presents R and T spectra of W membrane with red solid and dashed lines, respectively. The W membrane is assumed to be free standing in air. The spectra are shown in the $(0, 1)$ scale. Note that the T spectrum is close to 0 in the whole wavelength range shown. Figure 2a also shows R spectrum of bulk W of 1 mm thickness with blue curve, which is almost identical to the R spectrum of the W membrane. This means that the W membrane of 100 nm thickness is optically thick enough, being almost equivalent to bulk W in terms of the optical properties.

As for the spectral shapes of the W membrane in Figure 2a, R decreases as the wavelength becomes shorter. For example, at 500 nm, R is 0.48 and, at 1000 nm, R is 0.58. In comparison with a noble metal Ag, the R values are quite small. This means that W is associated with substantially large optical loss even at the near-infrared range.

Figure 2b shows R and T spectra of SiN membranes of thickness $t = 100, 200$, and 300 nm with red, green, and blue curves, respectively. The R and T spectra are shown with solid and dashed curves, respectively. The SiN membranes were assumed to be free-standing in air. The spectral shapes manifest the Fabry–Perot-type interference that is determined by the refractive index and thickness.

Since we are assuming loss-less refractive index, light absorption does not take place at all in the SiN membranes.

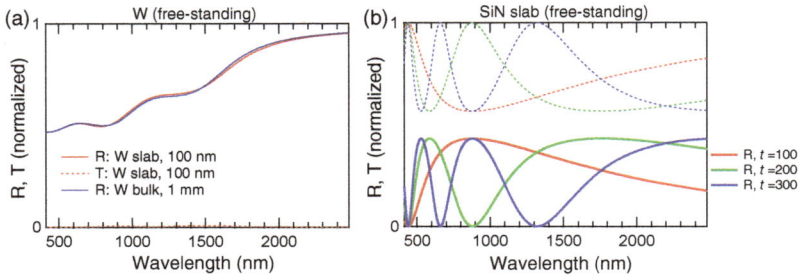

Figure 2. Basic optical properties of constituent materials. (**a**) R and T spectra of W slab (red solid and dotted curves, respectively) and bulk W (blue solid curve). The slab thickness was set to 100 nm and the bulk thickness was 1 mm; (**b**) R and T spectra of SiN membranes. Solid curves denote R spectra and dotted curves T spectra. The thickness *t* was set such as *t* = 100 (red), 200 (green), and 300 nm (blue). All the spectra in this figure were computed at the normal incidence.

2.1. Nearly PLA of Simply Stacked W-SiN Membrane

Figure 3 shows typical results on light absorption and electromagnetic (EM) fields in a simply stacked W-SiN membrane. The schematic is shown in Figure 1a. We fix the thickness of the W and SiN layers to be 100 nm because we intend to elucidate representative properties of the two-layer membrane.

Figure 3a,b shows 2D color and contour plots of light absorptance under p and s polarizations, respectively. The horizontal and vertical axes represent incident angles and photon energy in eV, respectively. Note that the normal incidence of 0° is in common with the two polarizations. The absorptance was evaluated using Equation (1). High absorptance of more than 0.8 (or 80%) appears both at p and s polarizations in the 1.0–1.5 eV (that is, 1239.5–826.3 nm in wavelength) range including the normal incidence. Perfect absorption also appears both at p and s polarizations; under p polarization, it is located around 78° and 2.8 eV, and, under s polarization, it is around 60° and 1.2 eV.

Figure 3a,b explicitly shows perfect absorption by the two-layer W-SiN membrane. Multilayers of noble metals and insulators have been studied since the 1960s [17,18]. Although several papers reported experimental observations of the optical responses [19–23], perfect light absorption, to our best knowledge, has not been observed in the noble-metal–insulator multilayer membranes. Thus, the optical response of the two-layer W-SiN membrane turns out to be unconventional.

Figure 3c,d shows snapshots of electric-field distributions under conditions of large light absorption; each condition is indicated by open circle (o) and cross (×), respectively, in Figure 3b. The corresponding incident angles are 0° and 60°. Under both conditions, the incident polarization was s polarization; accordingly, the E_y component is plotted with blue-to-red colors in Figure 3c,d. We set the incident $|E_y|$ equal to 1.0. The incident light propagates from the top. The interfaces of air/SiN, SiN/W, and W/air are shown with green lines.

Figure 3e,f corresponds to Figure 3c, and shows the absolute value of electric field $|E|$ and the profile along the z-axis, respectively. The green lines indicates the interfaces, similarly to Figure 3c. We set the interface of air/SiN at $z = 0$. Note that Figure 3f is plotted in decreasing order on the horizontal axis. Obviously, the electric field rapidly decays inside the two-layer W-SiN membrane; in other words, it is efficiently absorbed in the membrane. From Equation (1), perfect absorption ($A = 1$) requires R = 0, which is often called impedance matching. Under the open-circle condition in Figure 3b, although the absorptance is not exactly equal to 1, the electric-field distribution is qualitatively understood to be due to impedance matching and does not arise from a plasmonic resonance.

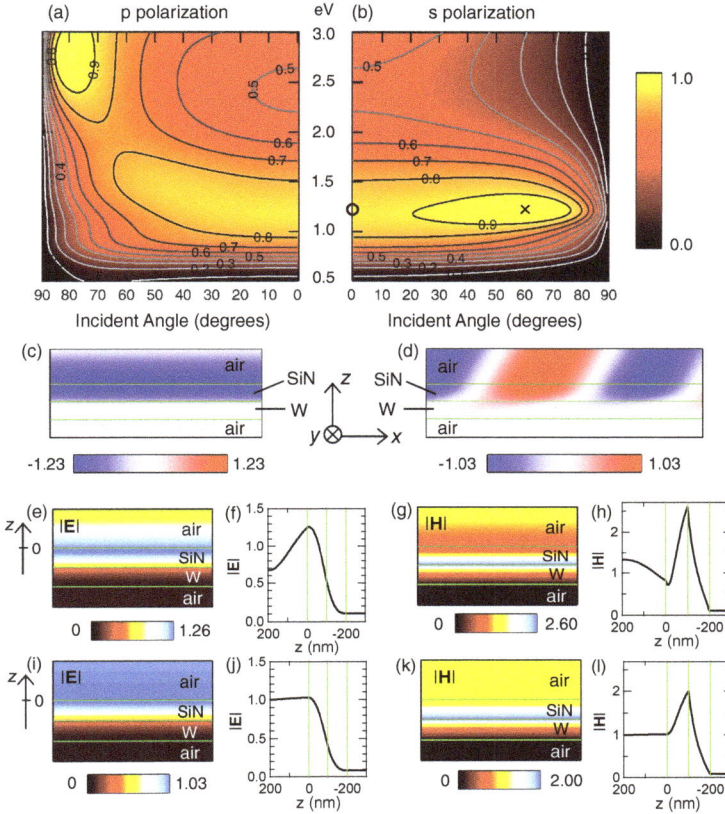

Figure 3. Light absorption properties and electromagnetic (EM)-field distribution at nearly perfect-absorption conditions. (**a**,**b**) 2D color and contour plots of light absorptance under p and s polarizations, respectively. In (**a**,**b**), the vertical axis is in common, representing photon energy in eV, and the horizontal axes represent incident angles. In addition, the common color bar is presented at the right-hand side. (**c**,**d**) snapshots of E_y components near the W-SiN membrane at the normal incidence and incident angle of 60°, respectively. The photon energy was set to 1.208 eV. These conditions are indicated by open circle (∘) and cross (×) in (**b**). The incident electric-field vector $\mathbf{E}_{in}$ was set to be parallel to the y-axis and $|\mathbf{E}_{in}| = 1$; (**e**) $|\mathbf{E}|$ distribution at the condition ∘ in (**b**); (**f**) profile of the $|\mathbf{E}|$ distribution in (**e**) along the z-axis; (**g**) $|\mathbf{H}|$ distribution at the condition ∘ in (**b**); (**h**) profile of the $|\mathbf{H}|$ distribution in (**g**) along the z-axis; (**i**–**l**) a set of EM-field distribution and the z profiles at the condition indicated by × in (**b**), shown in a similar way to (**e**–**h**), respectively. The green lines in (**c**–**l**) denote the interfaces of the hetero-materials.

Figure 3g,h corresponds to Figure 3e,f, and shows the absolute value of magnetic field $|\mathbf{H}|$ and the profile along the z axis, respectively. The $|\mathbf{H}|$ profile takes the maximum value at the W/SiN interface. We mention that the magnetic-field distribution is uniquely determined by the boundary-connection conditions in the membrane structure and that light absorption mostly takes place in the W layer through one of Maxwell equation,

$$\nabla \times \mathbf{H} = \varepsilon_0 \varepsilon_W \frac{\partial \mathbf{E}}{\partial t}, \tag{2}$$

where ε_0 is permittivity in vacuum because spatial distribution of **H** is directly affected by the complex relative permittivity in the W layer (ε_W); the permittivity ε_W has a large imaginary part [16] that is associated with optical loss.

Figure 3i,j corresponds to Figure 3d and shows the |**E**| distributions in a similar way to Figure 3e,f. Also under the oblique incidence ($\times$ in Figure 3b), the |**E**| profile indicates efficient light absorption.

Figure 3k,l corresponds to Figure 3i,j, and shows the |**H**| distribution and the z profile, respectively. Qualitatively, the descriptions for Figure 3g,h are true for Figure 3k,l as well.

In short, the large light absorption at 1.0–1.5 eV and wide incident angles including 0° is ascribed to impedance matching of the two-layer membrane structure. The efficient absorption primarily originates from the property of W permittivity. We point out that the single layer of W does not realize efficient light absorption as shown in Figure 2a. In this sense, the two-layer membrane structure is the minimal requirement for large light absorption in W–ceramic systems.

2.2. 1D Periodic PLAs Made of W-SiN-W Membranes

Figure 4 shows absorptance spectra of on-top 1D periodic W-SiN-W membranes. The schematic is shown in Figure 1d. We set the periodicity P to be 1000 nm and the incidence was set to be normal; we found signatures of diffraction at 1000 nm such as small steps in Figure 4a and kinks in Figure 4b. Various absorptance spectra represent light absorption in the 1D periodic W-SiN-W membranes with different widths of air slit S, which was varied from 150 to 850 nm. Accordingly, the width of metal, W_d, was varied from 850 to 150 nm because $P = S + W_d$. The linear optical responses (R_{mn} and T_{mn}) were numerically computed and the absorptance was evaluated using Equation (1).

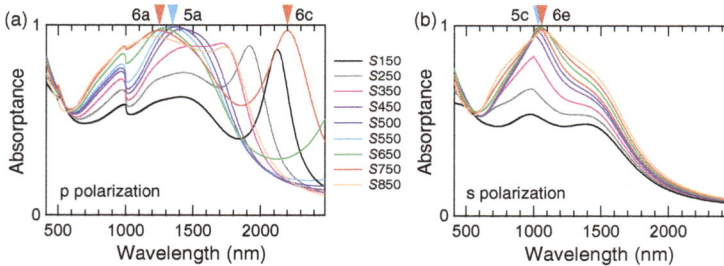

Figure 4. (**a,b**) Absorptance spectra of 1D periodic W-SiN-W membranes under p and s polarizations, respectively. The schematic is given in Figure 1d. The incidence was set to be normal. Width of air slit, S, was varied from 150 to 850 nm, which is symbolized as $S150$ to $S850$, respectively, while the periodicity was fixed at 1000 nm. The color usage is in common with (**a,b**).

As S increases under p polarization in Figure 4a, the absorptance spectra for $S = 450$ (purple curve) to 750 nm (red curve) have perfect-absorption peaks around 1300 nm. For $S = 750$ nm, another nearly perfect-absorption peak appears at 2200 nm, whose signature is seen for $S = 650$ nm (green curve) at the longest wavelength edge.

As S increases under s polarization in Figure 4b, the absorptance gradually increases and reaches almost 1 for $S = 500$ nm (blue curve), and keeps the perfect absorption at approximately 1050 nm for larger S.

Color cones in Figure 4 indicate some of the perfect-absorption peaks for which we next examine the EM fields. Symbols such as 5a and 6a associated with the color cones denote the corresponding EM distributions shown in Figures 5a and 6a, respectively.

Figure 5 presents EM-field distributions at perfect-light-absorption conditions of 1D W-SiN-W membranes with air slit $S = 550$ nm and width of the metal $W_d = 450$ nm. The incidence conditions of polarization and wavelength are indicated in Figure 4 with the sky-blue cones. The absolute values |**E**| and |**H**| are shown in linear scale in the minimum-to-maximum-value manners.

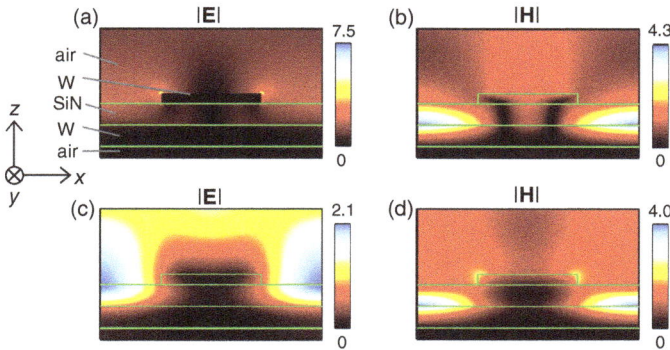

Figure 5. EM-field distributions of the 1D W-SiN-W membrane of *S*550 in Figure 4. (**a**,**b**) absolute values of electric- and magnetic-field distributions, |**E**| and |**H**|, respectively; the incident polarization and wavelength are indicated by the sky-blue cone in Figure 4a; (**c**,**d**) |**E**| and |**H**| distributions, respectively; the incidence condition is indicated by the sky-blue cone in Figure 4b. The green lines commonly represent the boundaries of the 1D W-SiN-W structure. The coordinate axes are shown, in common, at the left-hand side. All the panels are shown in linear scale in the minimum-to-maximum manners.

Figure 5a,b shows an *xz*-section view of |**E**| and |**H**| distributions under the incidence condition indicated by the sky-blue cone in Figure 4a (p polarization and 1350.2 nm). The incidence was normal, being normalized such that the incident $|\mathbf{E}| = |E_x| = 1.0$ and $|\mathbf{H}| = |H_y| = 1.0$. The boundaries of the involved materials are shown with green lines, except for the air/W boundary in Figure 5a because of the high contrast and to avoid concealing small hot spots by the lines. The electric field is locally amplified at the W corners, exhibiting the small hot spots, whereas the magnetic field is localized at the flat SiN/W interface. The |**H**| profile is quite similar to the two-layer system in Figure 3. We point out that the so-called gap mode in the W-SiN-W structure is not responsible for the perfect absorption.

Figure 5c,d shows the |**E**| and |**H**| distributions in the *xz* section under another perfect-absorption condition, indicated by the sky-blue cone in Figure 4b (s polarization and 1032.1 nm). Then, the incident polarization vector is parallel to the *y*-axis and satisfies $|\mathbf{E}| = |E_y| = 1.0$; accordingly, the incident $|\mathbf{H}| = |H_x| = 1.0$. Figure 5c shows that there is no hot spot and the electric field reaches the bottom W layer unless the top W bar interrupts the propagation. Figure 5d shows that the magnetic field is mainly enhanced at the flat SiN/W interface, similarly to Figure 5b. Thus, in this 1D periodic system of $S = 550$ nm, the perfect absorption takes place with a mechanism similar to that in the two-layer membrane structure. It is to be stressed that, even though the structure is certainly anisotropic in the 1D periodic W-SiN-W membrane, the perfect-absorption mechanism is primarily independent of the incident polarizations and the structural anisotropy.

Figure 6 displays EM-field distributions at perfect-light-absorption conditions of 1D W-SiN-W membranes with air slit $S = 750$ nm and width of the metal $W_d = 250$ nm. The corresponding absorptance spectra are shown with the red curves in Figure 4 and the three nearly perfect-absorption peaks are indicated by the red cones. Green lines indicate the boundaries of the 1D periodic membrane, similarly to Figure 5.

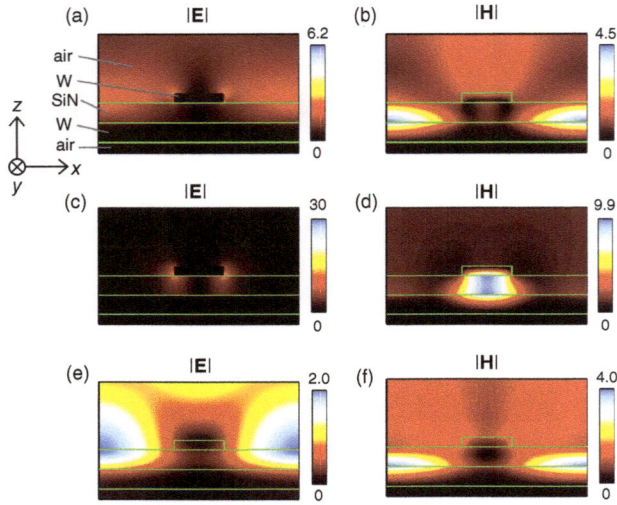

Figure 6. EM-field distributions of the 1D W-SiN-W membrane of S750 in Figure 4. (**a,b**) |**E**| and |**H**| distributions, respectively; the p-polarized incidence condition is indicated by the red cone with the mark 6a in Figure 4a; (**c,d**) |**E**| and |**H**| distributions, respectively; the p-polarized incidence corresponds to the longer wavelength peak in Figure 4a, indicated by the red cone with the mark 6c. (**e,f**) |**E**| and |**H**| distributions, respectively; the s-polarized incidence condition is indicated in Figure 4b with the red cone with the mark 6e. The green lines commonly represent the boundaries of the 1D W-SiN-W structure. The coordinate axes are shown in common. All the panels are shown in linear scale in the minimum-to-maximum manners.

Figure 6a,b shows the |**E**| and |**H**| distributions, respectively, under p polarization and 1252.0 nm. The EM-field distributions are qualitatively similar to those in Figure 5a,b, irrespective of the substantial difference in air slit S.

Figure 6c,d shows the |**E**| and |**H**| distributions, respectively, under the condition of p polarization and 2201.6 nm, which correspond to nearly perfect absorption sensitive to S. We note that only the scale in Figure 6c is shown in a peak-cut way; that is, although the maximum value of |**E**| in Figure 6c is 64, the scale is set to have the maximum at 30 and the locations exceeding 30 are represented in sky blue. Qualitatively, the EM-field distributions are distinct from those in Figure 6a. The electric- and magnetic fields show strong localization around the W-SiN-W gap. It is obvious that the |**H**| distribution takes the maximum value in the gap. These EM-field distributions are peculiar to the gap mode or plasmonic waveguide mode [12]. Thus, it is found that the plasmonic gap mode appears at a particular condition, being sensitive to the structural parameters.

Figure 6e,f shows the |**E**| and |**H**| distributions, respectively, under s polarization and 1060.2 nm. The EM-field distributions are quite similar to those in Figure 5c,d, strongly suggesting that the perfect absorption owing to the impedance matching is quite robust to the structural modifications.

2.3. 2D Periodic PLAs Made of W-SiN-W Membranes

Figure 7a,b shows absorptance spectra of 2D periodic W-SiN-W membranes of on-top rectangular and circular patches, respectively. The schematics are shown in Figure 1e,f. Each panel in Figure 7 presents two spectra for the 2D periodic membranes with periodicity $P = 1000$ nm and the middle SiN-layer thickness $t = 100$ and 200 nm, which are shown with black and green curves, respectively. One side of the W square was set to $W_d = 450$ nm and the diameter of the circular W was $D_m = 504$ nm, where D_m was chosen to keep the volume of the circular patch equivalent to that of the square one.

The thickness of the top and bottom layers of W was set to 50 and 100 nm, respectively. Quantitatively, they exhibit similar absorptance spectra, irrespective of the difference of the patch shapes.

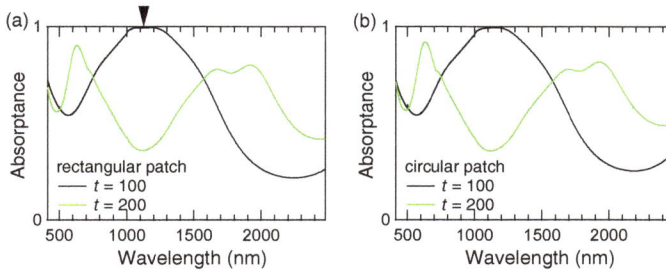

Figure 7. Absorptance spectra of 2D periodic W-SiN-W membranes at the normal incidence. (**a,b**) on-top rectangular (side $W_d = 450$ nm) and circular patches (diameter $D_m = 504$ nm), whose schematics are illustrated in Figure 1e,f, respectively. The black and green curves denote the 2D membranes of SiN thickness (*t*) 100 and 200 nm, respectively. The periodicity *P* was set to $P = 1000$ nm in these cases.

One common feature in Figure 7a,b is that perfect-absorption bands appear at 1050–1200 nm. We note that, in the 1D periodic membranes in Figure 4, the perfect absorption appears as peaks (not band). This feature is due to the 2D periodic structures and probably originate from the dimensionality. The cone at 1126.8 nm in Figure 7a indicates the wavelength corresponding to EM-field distributions in the next Figure.

Figure 8 shows a set of EM-field distributions of the 2D periodic W-SiN-W membrane at a perfect-absorption condition indicated by the cone (1126.8 nm) in Figure 7a. The incidence was set to be normal and to have $|\mathbf{E}_{in}| = 1.0$ and $|\mathbf{H}_{in}| = 1.0$. Note that the scales for $|\mathbf{E}|$ and $|\mathbf{H}|$ are in common to provide systematic views. Green lines indicate the boundaries in the 2D membrane.

Figure 8a,b shows the $|\mathbf{E}|$ and $|\mathbf{H}|$ distributions, respectively, in a *xz* section that cut through the center of the on-top square W patch. The incident polarization vector $\mathbf{E}_{in}$ is parallel to the *x*-axis and induces hot spots at the corner of the W patch. The $|\mathbf{H}|$ distribution takes larger values than the incidence at the interfaces of air/W and SiN/W.

Figure 8c,d shows the corresponding *yz*-section view through the center of the on-top square W patch. The $|\mathbf{E}|$ distribution does not include any hot spot whereas the $|\mathbf{H}|$ distribution is strongly localized at the SiN/W interface, being scattered at the corner of the on-top W patch.

Figure 8e displays an *xy*-view $|\mathbf{E}|$ distribution at a representative *z* position, which was set to be 1 nm above the air/W interface and indicated by the cones in Figure 8a,c. The position of the square W patch is shown with green lines for better visibility. The $|\mathbf{E}|$ distribution takes the maximum value at the corners of the W patch and mainly localizes around the W patch.

Figure 8f shows an *xy*-view $|\mathbf{H}|$ distribution at a representative *z* position, which was set to be 1 nm above the interface of the SiN/W and indicated by the cones in Figure 8b,d. The $|\mathbf{H}|$ distribution is mostly enhanced at the positions between the on-top W patches and localizes at the SiN/W interface. Thus, it is found that the perfect-absorption band is realized by the combination of local electric-field scattering by the on-top W patch and magnetic-field localization at the SiN-W interfaces. The local scattering can be interpreted as local plasmon at the rectangular W nanostructures and includes substantial longitudinal components that are usually excited as continuum [12] and are considered to realize the perfect-absorption continuum. In terms of the structures, the 1D structures in Figure 4 do not have the rectangular corners in the *xy* plane and therefore do not tend to form perfect-absorption continuum.

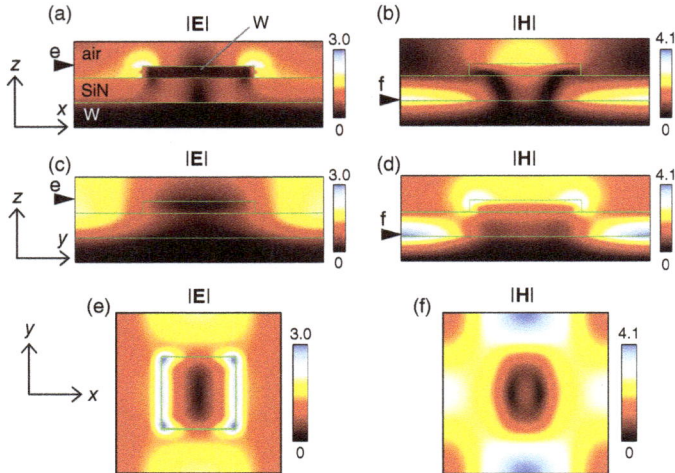

Figure 8. EM-field distributions of the 2D W-SiN-W membrane at the perfect-absorption condition indicated by the black cone in Figure 7a. (**a,b**) the *xz*-section views of |**E**| and |**H**| distributions, respectively. The *xz* section cuts through the center of the rectangular W patch; (**c,d**) the *yz*-section views, corresponding to (**a,b**), respectively; (**e,f**) the *xy*-section views of |**E**| and |**H**| distributions, respectively. The *z* positions are indicated by the cones in (**a–d**), respectively. The green lines represent the boundaries of the 2D periodic W-SiN-W structure. The relevant coordinate axes are shown on the left-hand side. All of the panels are shown in linear scale in the minimum-to-maximum-value manners.

3. Materials and Methods

3.1. Material Parameters

As for the metal W, the complex relative permittivity is taken from the literature [16], in which the measured data were well fitted by the Lorentz–Drude model over a wide wavelength range including the range of interest in this study. Therefore, we use the complex permittivity approximated by the model.

As for the ceramic SiN, although literature is available [24], the measured data are quite sparse in the near-infrared range of interest. Therefore, we adopt a representative value, that is, a permittivity of 4.3. Although there is slight wavelength dispersion in the range of present interest, the representative value is sufficient to extract fundamental properties of the W–ceramic systems. In addition, the representative value is a good approximation for other ceramics such as Ta_2O_5 and allows us to consider other alternatives for ceramics.

3.2. Numerical Method

Here, we employed the rigorous coupled-wave analysis (RCWA) method to compute the optical responses and EM field distributions of the different PLA structures. To evaluate the optical responses of the periodic structures in the frequency domain, the RCWA method is suitable and one of the most reliable methods because it directly solves Maxwell equations for the periodic structures in a numerical manner without any simplification and modelling. The RCWA method was originally formulated in the frequency domain, using Fourier transformation from the spatial coordinate to the wavenumber space and was substantially improved in the truncation order of the Fourier expansion [25].

To handle stacked structures, it is important to incorporate the scattering-matrix algorithm in order to avoid numerical divergence that makes the numerical calculations impossible [26]. The optical spectra and the resultant absorptance spectra were obtained based on the RCWA method

Appl. Sci. **2017**, *7*, 458

incorporating the scattering-matrix algorithm, which was implemented in a multi-parallel manner on supercomputers.

4. Conclusions

We have numerically examined several W-SiN membrane structures and demonstrated that they are able to serve as PLAs in the short-wavelength infrared range. The unconventional mechanism of perfect light absorption has been revealed and was ascribed to the robust impedance matching of the membranes to air, irrespective of the dimensionality. We also showed that the minimal requirement for the nearly perfect absorption is the two-layer W-SiN membrane. In addition, the plasmonic waveguide mode was observed in the 1D periodic W-SiN-W membrane with a particular structural parameter. The 1D and 2D periodic structures were found to be useful to obtain perfect absorption though their fabrication is more complicated, compared to the simple two-layer membrane. Overall, the W-SiN membranes are found to be practical solutions for thermal emitters working near the visible range. On the basis of the present designs, we would fabricate low-power-consuming thermal devices emitting visible and near-infrared light.

Acknowledgments: The author is grateful to Hideki T. Miyazaki (NIMS) for discussions and suggestions, and thanks financial support from KAKENHI Grant (No. 26706020) and from the 4th mid-term research project in NIMS. The author also thanks the High-Performance Computing Infrastructure (HPCI) system research project (ID: hp160035) for the support with the numerical implementations at Cyberscience Center, Tohoku University.

Conflicts of Interest: The author declares no conflict of interest.

Abbreviations

The following abbreviations are used in this manuscript:

1D	One-Dimensional
2D	Two-Dimensional
3D	Three-Dimensional
EM	Electromagnetic
PLA	Perfect Light Absorber
R	Reflectance
RCWA	Rigorous Coupled-Wave Analysis
T	Transmittance

References

1. Lourtioz, J.M.; Benisty, H.; Berger, V.; Gérard, J.M.; Maystre, D.; Tchelnokov, A. *Photonic Crystals Towards Nanoscale Photonic Devices*; Springer: Berlin, Germany, 2005.
2. Hendrickson, J.; Guo, J.; Zhang, B.; Buchwald, W.; Soref, R. Wideband perfect light absorber at midwave infrared using multiplexed metal structures. *Opt. Lett.* **2012**, *37*, 371–373.
3. Bouchon, P.; Koechlin, C.; Pardo, F.; Haïdar, R.; Pelouard, J.L. Wideband omnidirectional infrared absorber with a patchwork of plasmonic nanoantennas. *Opt. Lett.* **2012**, *37*, 1038–1040.
4. Miyazaki, H.T.; Kasaya, T.; Iwanaga, M.; Choi, B.; Sugimoto, Y.; Sakoda, K. Dual-band infrared metasurface thermal emitter for CO_2 sensing. *Appl. Phys. Lett.* **2014**, *105*, 121107.
5. Miyazaki, H.T.; Kasaya, T.; Oosato, H.; Sugimoto, Y.; Choi, B.; Iwanaga, M.; Sakoda, K. Ultraviolet-nanoimprinted packaged metasurface thermal emitters for infrared CO_2 sensing. *Sci. Technol. Adv. Mater.* **2015**, *16*, 035005.
6. Dao, T.D.; Chen, K.; Ishii, S.; Ohi, A.; Nabatame, T.; Kitajima, M.; Nagao, T. Infrared perfect absorbers fabricated by colloidal mask etching of Al-Al_2O_3-Al trilayers. *ACS Photonics* **2015**, *2*, 964–970.
7. SiN Membrane. Available online: http://www.ntt-at.com/product/membrane/ (accessed on 28 April 2017).
8. Lin, S.Y.; Moreno, J.; Fleming, J.G. Three-dimensional photonic-crystal emitter for thermal photovoltaic power generation. *Appl. Phys. Lett.* **2003**, *83*, 380–382.
9. Golay, M.J.E. Theoretical consideration in heat and infra-reded detection, with particular reference to the pneumatic detector. *Rev. Sci. Instrum.* **1947**, *18*, 347–356.

10. Greffet, J.J.; Nieto-Vesperinas, M. Field theory for generalized bidirectional reflectivity: Derivation of Helmholtz's reciprocity principle and Kirchhoff's law. *J. Opt. Soc. Am. A* **1998**, *15*, 2735–2744.
11. Iwanaga, M.; Choi, B. Heteroplasmon hybridization in stacked complementary plasmo-photonic crystals. *Nano Lett.* **2015**, *15*, 1904–1910.
12. Iwanaga, M. *Plasmonic Resonators: Fundamentals, Advances, and Applications*; Pan Stanford Publishing: Singapore, 2016.
13. Choi, B.; Iwanaga, M.; Miyazaki, H.T.; Sugimoto, Y.; Ohtake, A.; Sakoda, K. Overcoming metal-induced fluorescence quenching on plasmo-photonic metasurfaces coated by a self-assembled monolayer. *Chem. Commun.* **2015**, *51*, 11470–11473.
14. Iwanaga, M.; Choi, B.; Miyazaki, H.T.; Sugimoto, Y. The artificial control of enhanced optical processes in fluorescent molecules on high-emittance metasurfaces. *Nanoscale* **2016**, *8*, 11099–11107.
15. Choi, B.; Iwanaga, M.; Sugimoto, Y.; Sakoda, K.; Miyazaki, H.T. Selective plasmonic enhancement of electric- and magnetic-dipole radiations of Er ions. *Nano Lett.* **2016**, *16*, 5191–5196.
16. Rakić, A.D.; Djurušić, A.B.; Elazar, J.M.; Majewski, M.L. Optical properties of metallic films for vertical-cavity optoelectronic devices. *Appl. Opt.* **1998**, *37*, 5271–5283.
17. Swihart, J.C. Field solution for a thin-film superconducting strip transmission line. *J. Appl. Phys.* **1961**, *32*, 461–469.
18. Economou, E.N. Surface plasmons in thin films. *Phys. Rev.* **1969**, *182*, 539–554.
19. Scalora, M.; Bloemer, M.J.; Pethel, A.S.; Dowling, J.P.; Bowden, C.M.; Manka, A.S. Transparent, metallo-dielectric, one-dimensional, photonic band-gap structures. *J. Appl. Phys.* **1998**, *83*, 2377–2383.
20. Bennink, R.S.; Yoon, Y.K.; Boyd, R.W.; Sipe, J.E. Accessing the optical nonlinearity of metals with metal–dielectric photonic bandgap structures. *Opt. Lett.* **1999**, *24*, 1416–1418.
21. Scalora, M.; Vincenti, M.A.; de Ceglia, D.; Roppo, V.; Centini, M.; Akozbek, N.; Bloemer, M.J. Second- and third-harmonic generation in metal-based structures. *Phys. Rev. A* **2010**, *82*, 043828.
22. Iwanaga, M. Hyperlens-array-implemented optical microscopy. *Appl. Phys. Lett.* **2014**, *105*, 053112.
23. Iwanaga, M. Toward super-resolution imaging at green wavelengths employing stratified metal-insulator metamaterials. *Photonics* **2015**, *2*, 468–482.
24. Philipp, H.R. Silicon nitride (Si_3N_4) (noncrystalline). In *Handbook of Optical Constants of Solids*; Palik, E.D., Ed.; Academic Press: San Diego, CA, USA, 1985; Volume 2, pp. 771–774.
25. Li, L. New formulation of the Fourier modal method for crossed surface-relief gratings. *J. Opt. Soc. Am. A* **1997**, *14*, 2758–2767.
26. Li, L. Formulation and comparison of two recursive matrix algorithm for modeling layered diffraction gratings. *J. Opt. Soc. Am. A* **1996**, *13*, 1024–1035.

![applied sciences logo]
applied
sciences

MDPI

Article

Near-Field Coupling and Mode Competition in Multiple Anapole Systems

Valerio Mazzone, Juan Sebastian Totero Gongora and Andrea Fratalocchi *

PRIMALIGHT, Faculty of Electrical Engineering, Applied Mathematics and Computational Science,
King Abdullah University of Science and Technology, Thuwal 23955-6900, Saudi Arabia;
valerio.mazzone@kaust.edu.sa (V.M.); js.totero@kaust.edu.sa (J.S.T.G.)
* Correspondence: andrea.fratalocchi@kaust.edu.sa; Tel.: +966-(2)-808-0348

Academic Editor: Boris Malomed
Received: 17 April 2017; Accepted: 11 May 2017; Published: 24 May 2017

Abstract: All-dielectric metamaterials are a promising platform for the development of integrated photonics applications. In this work, we investigate the mutual coupling and interaction of an ensemble of anapole states in silicon nanoparticles. Anapoles are intriguing non-radiating states originated by the superposition of internal multipole components which cancel each other in the far-field. While the properties of anapole states in single nanoparticles have been extensively studied, the mutual interaction and coupling of several anapole states have not been characterized. By combining first-principles simulations and analytical results, we demonstrate the transferring of anapole states across an ensemble of nanoparticles, opening to the development of advanced integrated devices and robust waveguides relying on non-radiating modes.

Keywords: anapole; silicon nanoparticles; integrated photonics; FDTD; near-field coupling

1. Introduction

Dielectric nanostructures at optical frequencies are characterized by an extremely complex landscape of interacting resonant states. By finely tuning the material and geometrical properties of the nanostructures, it is possible to engineer advanced functionalities and applications such as, e.g., anti-reflection surfaces [1] and integrated waveguides based on chains of nanoparticles [2]. One of the most fascinating manifestations of multi-mode interaction in dielectric nanoparticles is the formation of radiationless states known as anapoles. These states have recently been demonstrated in silicon nanoparticles [3]. Anapoles are characterized by a strong reduction of the scattering from the nanoparticle at the anapole wavelength, which acquires the character of a fully-cloaked structure [4,5]. The mechanism underlying the formation of anapole states is the superposition of internal multi-mode components of the nanoparticle, which cancel each other in the far-field and which produce a radiation pattern confined to the near-field. As a result, anapole states are not the result of any resonant process (as in the case of dark resonances), but their origin lies entirely in modal competition and superposition [6]. While the theoretical description of anapole states in single dielectric nanoparticles is well established, the mutual coupling and interaction along a chain of anapole nanoparticles have yet to be investigated. Arrays of nanoparticles have been the subject of intense study, starting from the remarkable guiding properties demonstrated in plasmonic nanochains [7,8]. To mitigate the effect of metal losses, researchers have focused on all-dielectric solutions based on high refractive index nanoparticles [9]. As discussed in [10], the guiding properties of nanoparticle arrays can be strongly enhanced by minimizing the electromagnetic scattering from each nanoparticle in the array. As a result, anapole states could represent—in principle—a perfect candidate for efficient integrated waveguides based on silicon nanoparticles. However, due to their intrinsic non-resonant nature, anapoles should not manifest mutual coupling (as would be expected in the case of standard resonant states). Therefore, it should not be possible to transfer radiationless states among closely-packed dielectric structures. As recently demonstrated in two-dimensional cylinders, however, anapole states can be re-interpreted as the result of Fano interference between two or more, overlapping resonant states in the proximity of the anapole wavelength [11]. If such analysis could be extended to the three-dimensional case, it would be possible to describe and tailor the mutual coupling of an ensemble of invisible nanostructures. In this work, we address this problem by combining first-principles simulations and analytical theory, showing how the anapole state can be effectively transferred among distinct nanoparticles. Our results play a key role in the development of integrated optical circuitry based on non-radiating states (e.g., anapole-based wave-guides). Due to the near-field confinement produced by the anapole excitation and the suppression of the scattered field, wave-guides based on non-radiating states are extremely robust against physical bending and splitting, opening to the realization of high-density optical circuitry entirely based on silicon.

2. Results

2.1. Ab-Initio Analysis of Multiple Anapole Systems

We performed finite-differences time-domain (FDTD) analysis by considering an ensemble of three-dimensional anapole nanoparticles, each composed of a silicon nanodisk of radius $R = 150$ nm and height $h = 50$ nm. Each nanodisk is aligned with the z-axis, and it is illuminated by an E_x polarized plane wave propagating along the cylinder axis. In order to characterize the mutual coupling of closely-packed anapole nanoparticles, we consider a system of two slightly displaced nanodisks (see Figure 1a). The relative displacement of the identical nanoparticles is described by a centre-to-centre distance d and by a rotation angle α measured from the x-axis (Figure 1a-inset).

The anapole wavelength λ_{an} is identified by analysing the scattering cross-section C_{sca} of the isolated structures, as reported in Figure 1b (blue line). In the proposed configuration, the anapole state corresponds to $\lambda_{an} = 568$ nm (green dashed line). Despite the strong reduction of the scattering cross-section, the anapole state is associated with a strong enhancement of the electric field inside the nanodisk. The field enhancement is measured by integrating the electric intensity inside the resonator (Figure 2b, dashed orange line), which exhibits a strong peak at the anapole wavelength λ_{an}. The strong field enhancement associated to the anapole state is a counter-intuitive feature of the non-radiating state, and it has recently been exploited to amplify light–matter processes in semiconductor nanostructures [12,13]. To characterize the mutual coupling between anapole states, we performed a set of simulations for different rotation angles α and distances d, whose results are reported in Figure 1c,d. In our numerical experiments, we selectively excited one of the two nanoparticles at the anapole wavelength λ_{an}, and we measured the steady-state electric field intensity $|\mathbf{E}|$ in the second resonator. In terms of angular displacement α, the anapole coupling is characterized by a symmetric dipolar profile (Figure 1c). The near-field coupling is maximum in the direction of the three-lobe profile of the anapole state ($\alpha = 0, 2\pi$), while it is negligible in the orthogonal direction ($\alpha = \pi/2, 3\pi/2$). Conversely, in terms of the mutual distance d, the anapole coupling exhibits a more complex profile, with a sharp 20% reduction in less than 150 nm total displacement (Figure 1d). Interestingly, the spatial decay of the anapole coupling does not follow a power law decay, as would be expected from near-field dipolar states. To verify this, we compared the anapole coupling distribution from Figure 1c,d to the scattered field of an isolated nanoparticle (see Supplementary Figure S1). Even in the case of an isolated nanoparticle, the scattered field is mostly dipolar, as multipole-components are negligible at the anapole wavelength [3]. The dipolar angular distribution—associated with a complex decay profile—can be considered as a first signature of a complex modal interaction between closely-packed anapole nanoparticles.

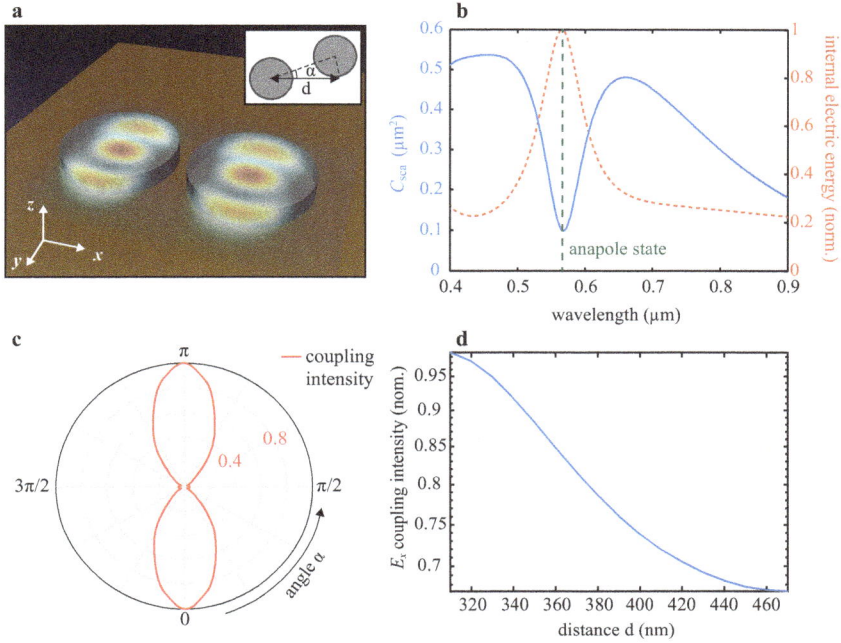

Figure 1. Mutual coupling of non-radiating anapole states. (a) Near-field coupling between two silicon nanodisks ($n = 3.5$) excited at the anapole wavelength $\lambda_{an} = 568$ nm. (inset) The dielectric resonators are mutually displaced by a centre-to-centre distance d and by an angle α; (b) Scattering cross-section C_{sca} (blue line) and internal electric energy (orange dotted line) as a function of the incident wavelength λ. The anapole state (green-dashed line) is characterized by the simultaneous suppression of the scattering cross-section, and by a strong enhancement of the internal field intensity; (c) Coupled electric energy as a function of the rotation angle α ($d = 450$ nm). The mutual coupling is maximum at $\alpha = 0, \pi$ and negligible at $\alpha = \pi/2, 3\pi/2$; (d) Coupled electric energy as a function of the mutual distance d. The results correspond to the angular condition of maximum scattering $\alpha = 2\pi$.

2.2. Fano–Feshbach Analysis of the Internal Modes

The counter-intuitive coupling properties of anapole–anapole systems can be explained by analysing the internal resonances of the system. The resonant properties of the single nanostructures can be extracted from the integrated density of states (DOS) $\rho(\lambda)$, which can be directly computed from FDTD simulations [14]. The integrated DOS is defined as:

$$\rho(\lambda) = \int d\mathbf{x}\, \rho(\mathbf{x}, \lambda) \tag{1}$$

where $\rho(\mathbf{x}, \lambda)$ is the local DOS, which is a function of the spatial position inside the resonator. In the FDTD framework, the local DOS corresponds to the spectral response to a single pulse excitation, and it can be defined separately for each component of the electromagnetic fields. For a generic component E_j of the electric field, as an example, the integrated DOS reads:

$$\rho_{E_j}(\omega) = \int d\mathbf{x}\, \left| \mathcal{F}\left\{ E_j(\mathbf{x}, t) \right\} \right|^2, \tag{2}$$

where $\mathcal{F}[\cdots]$ stands for Fourier transform and $E_j(\mathbf{x},t)$ is the electric field along the j-th direction ($j = x,y,z$). An analogue definition holds for the magnetic field components $H_j(\mathbf{x},t)$.

As can be easily verified by solving Maxwell's equations in cylindrical coordinates (r,ϕ), the response from the anapole resonator can be fully represented in terms of the E_z and H_z field components [15]. In Figure 2, we report the integrated DOS for the (a) H_z and (b) E_z field components. The integrated DOS are computed by considering a single pulse excitation centred at $\lambda = 700$ nm, and the spectral responses are normalized to the source spectrum. Interestingly, the transverse electric (TE) spectrum exhibits a strong peak exactly at the anapole wavelength (red vertical line), while the transverse magnetic (TM) spectrum shows only a slight enhancement at the anapole frequency (Figure 2b).

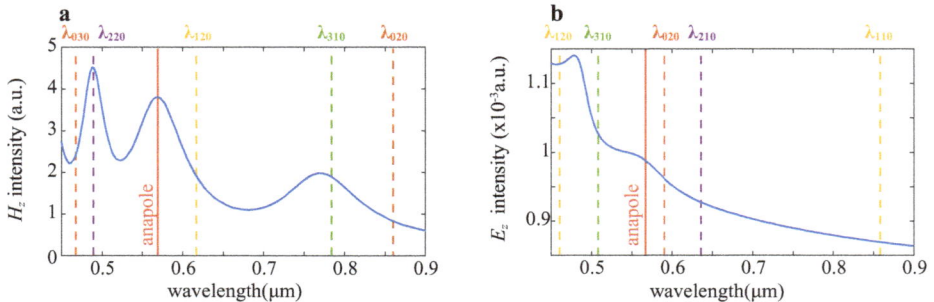

Figure 2. **Local density of states and interacting resonant modes**. (a,b) Local density of state for the (a) H_z and (b) E_z field components, corresponding to the transverse electric (TE) and transverse magnetic (TM) modes of the silicon nanostructure, respectively. The resonant wavelengths (vertical dashed lines) in both configurations are computed by means of Equation (4).

Further insights on the resonant properties of the anapole state can be obtained by decomposing the electromagnetic fields into a set of orthogonal eigenmodes. By definition, however, the silicon resonator represents an open cavity, and the definition of a set of orthogonal resonator modes is a challenging task [16,17]. As recently shown in [11], a possible way to circumvent such difficulties is the introduction of a Fano–Feshbach partitioning of the system [18]. In a nutshell, the Fano–Feshbach partitioning consists of the mathematical splitting of the total system into two orthogonal eigenspaces, corresponding to the resonator and environment regions. Maxwell's equations are rewritten in the partitioned subspaces, providing a rigorous description of light–matter interaction in open resonators. For a cylinder aligned along the z-axis, the Fano–Feshbach internal resonances for the fields H_z and E_z correspond to the TE and TM modes of a perfect electric conductor (PEC) cavity [19]. They are expressed as:

$$\begin{cases} \mathrm{TE}_{mpq} = N_{mpq} J_m(\frac{\chi'_{mp}}{R}r)\sin(\frac{q\pi}{hn})\exp(im\phi) \\ \mathrm{TM}_{mpq} = M_{mpq} J_m(\frac{\chi_{mp}}{R}r)\cos(\frac{q\pi}{hn})\exp(im\phi) \end{cases} \tag{3}$$

where J_m is a Bessel function of the first-kind of order m; N_{mpq}, M_{mpq} are normalization constants; and where χ_{mp},χ'_{mp} denote the p-th zero of the Bessel function J_m and its derivative $J'_m = \partial J_m/\partial\rho$, respectively. The resonance frequencies of the internal modes are defined as:

$$\begin{cases} \omega_{mpq}^2 = \dfrac{c^2}{n^2}\left[\left(\dfrac{q\pi}{hn}\right)^2 + \left(\dfrac{\chi'_{mp}}{R}\right)\right] & \text{TE modes,} \\[3mm] \omega_{mpq}^2 = \dfrac{c^2}{n^2}\left[\left(\dfrac{q\pi}{hn}\right)^2 + \left(\dfrac{\chi_{mp}}{R}\right)\right] & \text{TM modes.} \end{cases} \tag{4}$$

The choice of PEC boundary conditions for the internal modes is not fixed, as it is only dictated by the mathematical partitioning scheme adopted. As thoroughly discussed in [11,18,20], they can be easily exchanged with perfect magnetic conductor (PMC) boundary conditions, which are usually employed to analyse high refractive index nanoparticles [21,22]. A preliminary analysis of the system in terms of Fano–Feshbach internal resonances allows us to unveil some fundamental properties of the anapole excitation. By solving Equation (4), we computed the Fano–Feshbach resonant frequencies ω_{mpq} of the system, which are reported as sets of dashed vertical lines on Figure 2. Interestingly, the anapole state is generated in the proximity of a few orthogonal resonances. Among all the available candidates, the characteristic three-lobes mode profile of the anapole state can be obtained by superimposing two TM modes (Figure 3): a cylindrically symmetric TM_{020} and a quadrupolar TM_{210}. Remarkably, these modes constitute the three-dimensional version of the eigenmodes composing the anapole state in two-dimensional cylinders (cfr. Figure 3 of [11]).

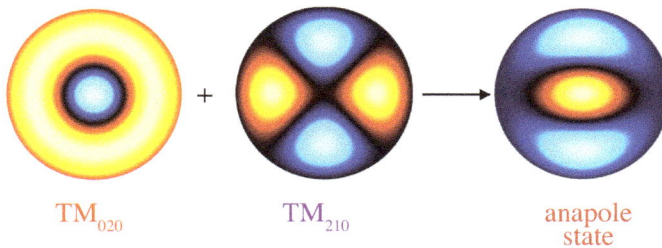

TM_{020} $\qquad$ TM_{210} $\qquad$ anapole state

Figure 3. **Fano–Feshbach partitioning of the anapole state.** The characteristic mode-profile of the anapole state is originated by the superposition of a cylindrically symmetric TM_{020} and a quadrupolar TM_{210}.

3. Discussion and Conclusions

By combining FDTD simulations and analytical theory, we have shown how anapole states can be coupled and transferred among closely-packed nanoparticles. This result—which recalls the coupling properties of standard resonant modes—is at first counter-intuitive due to the non-resonant character of the anapole state. However, as can be demonstrated by performing a Fano–Feshbach partitioning of the anapole resonator, the three-dimensional state is characterized by the superposition of several distinct resonances of the system, which collectively produce a scattering-suppression state at the anapole wavelength and which mutually couple among the ensemble of nanoparticles. The Fano–Feshbach analysis of three-dimensional anapole states—including a detailed analysis of the scattered fields—goes beyond the scope of this work, and it will be the subject of a future specialized work on the topic.

The ability to control and transfer non-radiating states along optical circuitry, however, opens to intriguing possibilities in the development of advanced integrated photonics platforms. As shown in Figure 4, anapole nanoparticles can in fact be arranged in a compact nanochain with remarkable guiding properties. Due to the efficient near-field coupling between adjacent nanoparticles, the nanochain can support guided modes which propagate without radiative losses at distances of several µm. In these simulations, we selectively excited the first anapole state in the nanochain (not shown in the figure) and we characterized the propagation of energy along the chain. Such excitation can be achieved experimentally, as single anapole nanoparticles can be excited by means of a near-field scanning optical microscopy (NSOM) setup [3]. As an appealing alternative, anapole-based waveguides could be combined with integrated nanolasers emitting at the anapole frequency [13].

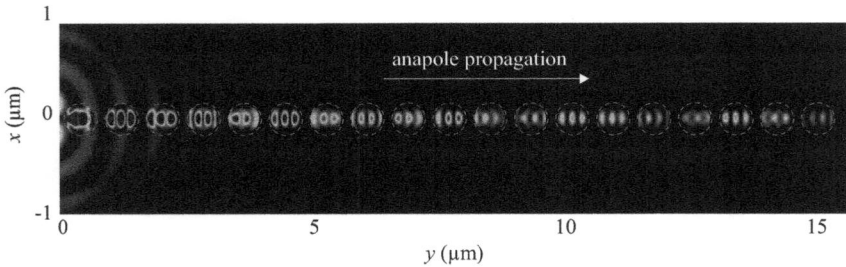

Figure 4. Anapole nanochain. Steady-state electromagnetic energy distribution along a chain of anapole nanoparticles. The centre-to-centre distance is $d = 400$ nm. The external excitation is restricted to the first anapole of the chain (not included in the panel).

As the anapole states are tightly confined in the near-field, optical nano-circuitry based on non-radiating modes is extremely robust to bending and splitting, as shown in Figure 5. In standard photonics applications, wave-guide deformation produces significant radiation losses, in particular when considering 90-degree bends and turns [23]. Conversely, in the case of an anapole nanochain, the near-field properties of the non-radiating state allow for efficient transmission of the guided mode across deformations and bends, such as in the case of wave-guide splitting (Figure 5a) or 90-degree bending and re-routing (Figure 5b). These illustrative examples strongly suggest the possibility of integrating anapole nanoparticles with state-of-the-art integrated photonics applications.

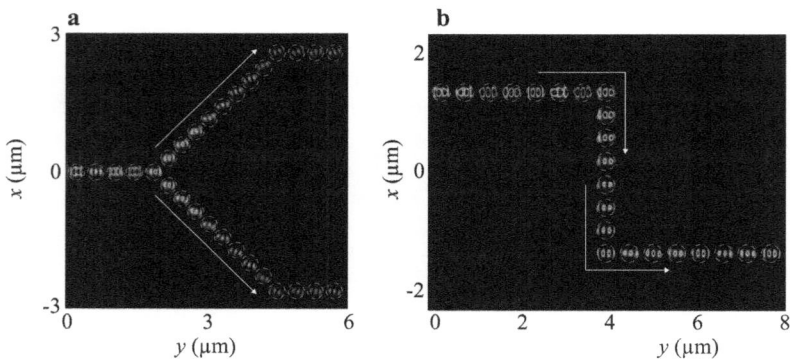

Figure 5. Robust sub-wavelength guiding via near-field transfer of anapole states. Due to the near-field confinement produced by the anapole state, the anapole nanochain is robust against bending and splitting of the integrated wave-guide. This opens to the realization of (**a**) integrated splitters and (**b**) 90-degree bends without introducing radiation losses.

4. Materials and Methods

4.1. FDTD Simulations

We performed fully-dispersive three-dimensional FDTD simulations using our home-made simulator NANOCPP [24–29]. In our simulations, the computational domain was organized as follows:

the z-aligned nanodisks were placed at the centre of a 2 μm × 2 μm × 1 μm box, with uniaxial perfectly matched layer (UPML) boundary conditions emulating an open system [30]. The numerical resolution was set as $\Delta x = 2$ nm, corresponding to 81 points per internal wavelength at the anapole frequency λ_{an}. The system was illuminated by plane wave, implemented according to the total-field scattered-field (TFSF) formalism [31]. The scattering cross-section (Figure 1b) was computed by integrating the Poynting vector along a three-dimensional surface surrounding the objects and entirely placed in the scattered-field region of the TFSF. In order to characterize the anapole coupling among distinct nanoparticles, we included one of the resonators in the total-field region of the TFSF, while the other resonators were placed in the scattered-field region.

Supplementary Materials: The following are available online at http://www.mdpi.com/2076-3417/7/6/542/s1, Figure S1: Scattered field from an isolated anapole state.

Acknowledgments: This research is supported by KAUST (Award No. OSR-2016-CRG5-2995). For the computer time, we have used the resources of the KAUST Supercomputing Laboratory and the Redragon cluster of the PRIMALIGHT group.

Author Contributions: V.M. and J.S.T.G. performed first principle parallel simulations and analytical modelling. A.F. supervised the research. All authors contributed to the analysis of data. All authors contributed equally to manuscript preparation.

Conflicts of Interest: The authors declare no conflict of interest.

Abbreviations

The following abbreviations are used in this manuscript:

FDTD	Finite-Differences Time-Domain
DOS	Density Of States
UPML	Uniaxial Perfectly Matched Layer
TFSF	Total-Field Scattered-field
TM	Transverse Magnetic
TE	Transverse Electric
PEC	Perfect Electric Conductor
PMC	Perfect Magnetic Conductor
NSOM	Near-field scanning optical microscopy

References

1. Wang, K.X.; Yu, Z.; Sandhu, S.; Liu, V.; Fan, S. Condition for perfect antireflection by optical resonance at material interface. *Optica* **2014**, *1*, 388–395.
2. Liu, W.; Miroshnichenko, A.E.; Neshev, D.N.; Kivshar, Y.S. Broadband Unidirectional Scattering by Magneto-Electric Core-Shell Nanoparticles. *ACS Nano* **2012**, *6*, 5489–5497.
3. Miroshnichenko, A.E.; Evlyukhin, A.B.; Yu, Y.F.; Bakker, R.M.; Chipouline, A.; Kuznetsov, A.I.; Luk'yanchuk, B.; Chichkov, B.N.; Kivshar, Y.S. Nonradiating anapole modes in dielectric nanoparticles. *Nat. Commun.* **2015**, *6*, 8069.
4. Liu, W.; Zhang, J.; Lei, B.; Hu, H.; Miroshnichenko, A.E. Invisible nanowires with interfering electric and toroidal dipoles. *Optics Lett.* **2015**, *40*, 2293–2296.
5. Luk'yanchuk, B.; Paniagua-Domínguez, R.; Kuznetsov, A.I.; Miroshnichenko, A.E.; Kivshar, Y.S. Suppression of scattering for small dielectric particles: anapole mode and invisibility. *Philos. Trans. R. Soc. Lond. A Math. Phys. Eng. Sci.* **2017**, *375*, 20160069.
6. Wei, L.; Xi, Z.; Bhattacharya, N.; Urbach, H.P. Excitation of the radiationless anapole mode. *Optica* **2016**, *3*, 799–802.
7. Brongersma, M.L.; Hartman, J.W.; Atwater, H.A. Electromagnetic energy transfer and switching in nanoparticle chain arrays below the diffraction limit. *Phys. Rev. B* **2000**, *62*, R16356–R16359.
8. Totero Gongora, J.S.; Fratalocchi, A. Harnessing Disorder at the Nanoscale. In *Encyclopedia of Nanotechnology*; Bhushan, B., Ed.; Springer Netherlands: Heidelberg, Germany, 2015; pp. 1–13.

9. Bakker, R.M.; Yu, Y.F.; Paniagua-Domínguez, R.; Luk'yanchuk, B.; Kuznetsov, A. Silicon Nanoparticles for Waveguiding. *Frontiers in Optics*; Optical Society of America: Washington, DC, USA, 2015; p. FM1B.2.

10. Evlyukhin, A.B.; Reinhardt, C.; Seidel, A.; Luk'yanchuk, B.S.; Chichkov, B.N. Optical response features of Si-nanoparticle arrays. *Phys. Rev. B* **2010**, *82*, 045404.

11. Gongora, J.S.T.; Favraud, G.; Fratalocchi, A. Fundamental and high-order anapoles in all-dielectric metamaterials via Fano–Feshbach modes competition. *Nanotechnology* **2017**, *28*, 104001.

12. Grinblat, G.; Li, Y.; Nielsen, M.P.; Oulton, R.F.; Maier, S.A. Efficient Third Harmonic Generation and Nonlinear Subwavelength Imaging at a Higher-Order Anapole Mode in a Single Germanium Nanodisk. *ACS Nano* **2017**, *11*, 953–960.

13. Totero Gongora, J.S.; Miroshnichenko, A.E.; Kivshar, Y.S.; Fratalocchi, A. Anapole nanolasers for mode-locking and ultrafast pulse generation. *Nat. Commun.* **2017**, in press.

14. Taflove, A.; Oskooi, A.; Johnson, S.G. *Advances in FDTD Computational Electrodynamics: Photonics and Nanotechnology*; Artech House Antennas and Propagation Series; Artech House: Norwood, MA, USA, 2013.

15. Yariv, A.; Yeh, P. *Photonics: Optical Electronics in Modern Communications (The Oxford Series in Electrical and Computer Engineering)*; Oxford University Press, Inc.: New York, NY, USA, 2006.

16. Türeci, H.E.; Ge, L.; Rotter, S.; Stone, A.D. Strong Interactions in Multimode Random Lasers. *Science* **2008**, *320*, 643–646.

17. Kristensen, P.T.; Hughes, S. Modes and Mode Volumes of Leaky Optical Cavities and Plasmonic Nanoresonators. *ACS Photonics* **2014**, *1*, 2–10.

18. Viviescas, C.; Hackenbroich, G. Field quantization for open optical cavities. *Phys. Rev. A* **2003**, *67*, 013805.

19. Jin, J.M. *Theory and Computation of Electromagnetic Fields*; John Wiley & Sons: Hoboken, NJ, USA, 2011; Google-Books-ID: D6SqmxJVV5wC.

20. Viviescas, C.; Hackenbroich, G. Quantum theory of multimode fields: Applications to optical resonators. *J. Opt. B Quantum Semiclass. Opt.* **2004**, *6*, 211.

21. Van Bladel, J. Radiation in Free Space. In *Electromagnetic Fields*; Wiley-IEEE Press: Hoboken, NJ, USA, 2007; pp. 277–356.

22. Tribelsky, M.I.; Miroshnichenko, A.E. Giant in-particle field concentration and Fano resonances at light scattering by high-refractive-index particles. *Phys. Rev. A* **2016**, *93*, 053837.

23. Cherchi, M.; Ylinen, S.; Harjanne, M.; Kapulainen, M.; Aalto, T. Dramatic size reduction of waveguide bends on a micron-scale silicon photonic platform. *Opt. Express* **2013**, *21*, 17814.

24. Gentilini, S.; Fratalocchi, A.; Angelani, L.; Ruocco, G.; Conti, C. Ultrashort pulse propagation and the Anderson localization. *Opt. Lett.* **2009**, *34*, 130–132.

25. Crosta, M.; Fratalocchi, A.; Trillo, S. Bistability and instability of dark-antidark solitons in the cubic-quintic nonlinear Schrödinger equation. *Phys. Rev. A* **2011**, *84*, 063809.

26. Huang, J.; Liu, C.; Zhu, Y.; Masala, S.; Alarousu, E.; Han, Y.; Fratalocchi, A. Harnessing structural darkness in the visible and infrared wavelengths for a new source of light. *Nat. Nanotechnol.* **2016**, *11*, 60–66.

27. Totero Gongora, J.S.; Miroshnichenko, A.E.; Kivshar, Y.S.; Fratalocchi, A. Energy equipartition and unidirectional emission in a spaser nanolaser. *Laser Photonics Rev.* **2016**, *10*, 432–440.

28. Labelle, A.J.; Bonifazi, M.; Tian, Y.; Wong, C.; Hoogland, S.; Favraud, G.; Walters, G.; Sutherland, B.; Liu, M.; Li, J.; et al. Broadband Epsilon-near-Zero Reflectors Enhance the Quantum Efficiency of Thin Solar Cells at Visible and Infrared Wavelengths. *ACS Appl. Mater. Interfaces* **2017**, *9*, 5556–5565.

29. Galinski, H.; Favraud, G.; Dong, H.; Totero Gongora, J.S.; Favaro, G.; Döbeli, M.; Spolenak, R.; Fratalocchi, A.; Capasso, F. Scalable, ultra-resistant structural colors based on network metamaterials. *Light Sci. Appl.* **2017**, *6*, e16233.

30. Berenger, J.P. A Perfectly Matched Layer for the Absorption of Electromagnetic-Waves. *J. Comput. Phys.* **1994**, *114*, 185–200.

31. Taflove, A.; Hagness, S.C. *Computational Electrodynamics: The Finite-Difference Time-Domain Method*, 3rd ed.; Artech House Antennas and Propagation Library, Artech House: Boston, MA, USA, 2005.

![applied sciences logo] *applied sciences*

MDPI

Review

Ultrafast Optical Signal Processing with Bragg Structures

Yikun Liu [1], Shenhe Fu [2], Boris A. Malomed [3,4], Iam Choon Khoo [5] and Jianying Zhou [1,*]

[1] State Key Laboratory of Optoelectronic Materials and Technologies, Sun Yat-sen University, Guangzhou 510275, China; liuyk6@mail.sysu.edu.cn

[2] Department of Optoelectronic Engineering, Jinan University, Guangzhou 510632, China; fushenhe@jnu.edu.cn

[3] Department of Physical Electronics, School of Electrical Engineering, Faculty of Engineering, Tel Aviv University, Tel Aviv 69978, Israel; malomed@post.tau.ac.il

[4] Laboratory of Nonlinear-Optical Informatics, ITMO University, St. Petersburg 197101, Russia

[5] Electrical Engineering Department, Pennsylvania State University, University Park, PA 16802, USA; ick1@psu.edu

* Correspondence: stszjy@mail.sysu.edu.cn; Tel.: +86-20-8411-0277

Academic Editor: Takayoshi Kobayashi
Received: 15 March 2017; Accepted: 2 May 2017; Published: 27 May 2017

Abstract: The phase, amplitude, speed, and polarization, in addition to many other properties of light, can be modulated by photonic Bragg structures. In conjunction with nonlinearity and quantum effects, a variety of ensuing micro- or nano-photonic applications can be realized. This paper reviews various optical phenomena in several exemplary 1D Bragg gratings. Important examples are resonantly absorbing photonic structures, chirped Bragg grating, and cholesteric liquid crystals; their unique operation capabilities and key issues are considered in detail. These Bragg structures are expected to be used in wide-spread applications involving light field modulations, especially in the rapidly advancing field of ultrafast optical signal processing.

Keywords: Bragg structure; ultrafast; optical signal process

1. Introduction

In the next generation of high-speed information networks, the direct processing of optical signals is required. On the other hand, the basic signal-processing capabilities, such as switching, logic operations, and buffering, are still lacking in practically usable forms. Useful for achieving these objectives should be the deceleration of optical signals and the creation of standing ones. These effects have been demonstrated with the help of various techniques, such as electromagnetically-induced transparency [1,2], but those interference-based techniques are often not suitable for broadband signal processing, when the carrier waves are represented by picosecond or even femtosecond pulses [2]. The light can also be retarded in optical fibers by a stimulated Brillouin scattering effect [3], and in photonic-crystal waveguides by manipulating the dispersion [4,5]. Light-matter interactions can be enhanced by the retardation of light. Since the light-matter interaction time t is inversely proportional to the group velocity v_g, the use of slow light with a small v_g implies longer interaction times, and consequently, a more efficient energy conversion [6]. Slow light also offers the possibility to compress optical signals in space, thus reducing the device size [7,8].

In this article, we review both theoretical and experimental results concerning the processing of ultrafast optical signals in one-dimensional (1D) Bragg gratings, which exhibit a 1-D photonic bandgap that makes it possible to significantly reduce the speed of light launched at a carrier frequency close to the bandgap. We discuss both artificially engineered Bragg structures, made of optoelectronic materials, and those produced by natural self-assembly, such as cholesteric liquid crystals. We provide a critical review of the performance of the Bragg gratings and the limitations in their use. In particular, we describe how some of the inevitable deleterious effects that accompany a strong dispersion experienced by light at the band-edge, can be balanced by the material nonlinearity, which provides the laser-induced self-phase modulation of the optical field. Theoretical modeling of these processes in some Bragg structures have shown that ultrafast laser pulses can be decelerated, stopped, and buffered; as a result, stationary nonlinear optical frequency conversion can be very efficient, even with a very thin resonant absorption Bragg reflector sample [9–13]. In a tailored optical structure, such as one into which defects and spatial chirp are integrated, various optical logic operations can be efficiently realized [14–16]. Recently, BG structures consisting of cholesteric liquid crystals (CLCs), which possess extraordinarily large ultrafast optical nonlinearities due to photonic crystal band-edge enhancement, have also been shown to be highly effective for direct-action compression, or for the stretch and recompression of pico- and femto-second pulses, opening up new possibilities for efficient broadband optical-signal processing [17,18].

2. The Bragg-Grating (BG) Structure

Bragg structures are basically 1D photonic crystals (PCs) [19,20] comprising materials with periodic refract index modulation along one direction, taken as:

$$n(z) = n_0[1 + a_1 \cos(2k_c z)], \tag{1}$$

with constant k_c, a_1, and n_0 values. The fabrication, characterization, and optical properties of PCs have been thoroughly investigated since the original works of John and Yablonovitch [19,20]. It is well known that the BG structure gives rise to a bandgap, with its central wavelength located at $\lambda_0 = \frac{2\pi n_0}{k_c}$ and a bandwidth of $\Delta\lambda = \frac{2\pi n_0 a_1}{k_c}$. Strong dispersion and velocity reduction occur at the photonic band edges [21]; in some materials, losses can also be significant. Both a low group velocity and low dispersion can be obtained by designing the structures of BG. In Reference [22], a low velocity of $0.02c$ (c is light speed in vacuum), in combination with a 10 nm-bandwidth and low dispersion, was demonstrated by changing the structure of a PC's waveguide [22].

A frequently studied 1-D Bragg structure is the fiber Bragg grating, which can be fabricated by interference lithography in fibers [23]. Fiber Bragg gratings are widely used in sensors [24], optical telecommunications [25], and for dispersion compensation [26]. 1D Bragg structures can also be fabricated in other solid materials, such as silicon [27] and AlGaAs [28], for optical switching and limiting operations [29]. Besides such artificial Bragg structures, self-assembled Bragg structures can also be found in liquid-crystal materials, such as cholesteric liquid crystals, whose optical properties can be modulated by the light field [30]. Cholesteric liquid crystals can also be used as temperature sensors [31], due to their temperature-dependent pitch.

By introducing nonlinear optical effects, many interesting discoveries have been made in one-dimension (1D) fiber Bragg grating. These include multistability, a zero velocity, and the creation of wobbling or oscillating solitons [32–36]. Using Kerr nonlinearity to balance the strong dispersion near the bandgap's edge, BG solitons with speeds of $0.5\,c$ [35,37] and as low as $c/7$ [33] were observed in experiments. Furthermore, it was predicted that standing light can be created using BG fibers with defects [38–40], Bragg reflectors combined with resonant nonlinearity [41], and the collision of BG solitons [42,43]. However, high input power densities ($>10\,\mathrm{GW/cm^2}$) and long propagation lengths are required to achieve strong nonlinear effects, which may pose serious problems in experiments and applications.

Several Bragg structures have been evoked to address the power and interaction length issues, as detailed in the following sections.

3. Optical Signal Processing in Resonantly Absorbing Bragg Reflector (RABR)

3.1. Theoretical Considerations

RABR is produced by adding narrow stripes, doped by two-level atoms resonantly interacting with the transmitted electromagnetic fields, to a BG structure, as schematically shown in Figure 1 [9–11].

Figure 1. A scheme of RABR with black stripes representing thin layers of two-level atoms. White and gray bands represent a periodically structured nonabsorbing medium (After Reference [11]).

Light-matter interaction in RABR, built of infinitely thin atomic layers, is modeled by the Maxwell-Bloch equations for electric-field components E^+ and E^- of the forward- and backward-propagating waves, and population W [12]:

$$\frac{\partial \Sigma^\pm}{\partial \tau} \pm \frac{\partial \Sigma^\pm}{\partial \zeta} = i\eta\Sigma^\pm + P, \tag{2}$$

$$\frac{\partial P}{\partial \tau} = -i\delta P + (\Sigma^+ + \Sigma^-)W \tag{3}$$

$$\frac{\partial W}{\partial \tau} = -\text{Re}\left[(\Sigma^+ + \Sigma^-)P^*\right] \tag{4}$$

where $\Sigma^\pm \equiv (2\tau_c\mu/\hbar)E^\pm$, $\tau_c \equiv (2\hbar n_0/\mu_0 c^2\omega_c\mu\rho)^{1/2}$ is the cooperative time (ranging from pico- to femto-seconds)which is determined by the presence of the two-level atoms; n_0 is the average refraction index; ω_c is the resonant frequency of the two-level atom; μ is the magnitude of the dipole matrix element ρ is the density of the dopant atoms; $\eta \equiv (n_1\omega\tau_c)/4$ is the dimensionless coupling constant; P and W are the material polarization and population inversion density, respectively; $\delta \equiv (\omega - \omega_c)/\tau_c$ is the dimensionless detuning; $\tau \equiv t/\tau_c$ and $\zeta \equiv (n_0/c\tau_c)z$ are the dimensionless time and spatial coordinates, respectively; and ω is the frequency of the incident light. Here, τ_c is equal to 300 fs [13].

Solutions of the Maxwell-Bloch equations, some being available in an analytical form [10], produce a vast family of stable gap solitons, of both a standing and moving nature. Compared to the self-induced-transparency (SIT) in uniform media, the solitons generated in RABR may have an arbitrary pulse area, while in uniform media solitons, are only created by pulses with an area exactly equal to 2π [44,45]. Stable dark solitons can also be excited in RABR [11]. Thus, the theoretical analysis demonstrates that light signals with any velocity can indeed exist in the RABR.

An optical pulse with a hyperbolic-secant shape, generated by the input with a small area, undergoes complete Bragg reflection in the RABR. With an increase in the input intensity, the SIT solitons can be excited, making it possible for light pulses to propagate without loss. If the intensity is still higher, the splitting of the pulse occurs. Most remarkably, with a suitable incident pulse, an oscillating gap soliton trapped in the RABR as a standing wave can be created. Figure 2a shows the evolution of the optical energy, which is represent by the population distribution W [12,13], in this case. Furthermore, we have found that multiple gap solitons can be simultaneously created as standing modes, thus predicting a possibility of efficient storage of the optical energy in the RABR structure (Figure 2b).

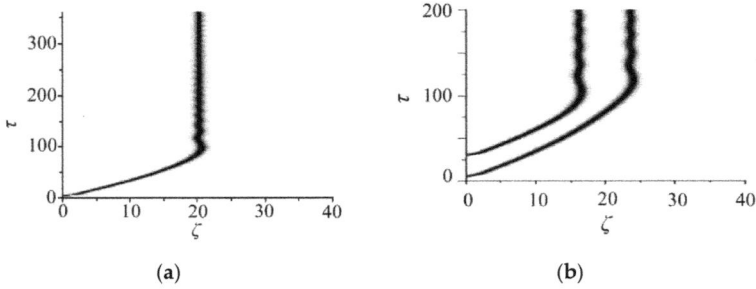

Figure 2. (a) The evolution of the population inversion, *W*, illustrating the creation of an oscillatory standing soliton in the RABR from a sech-shaped input. The pulse width is $\tau_0 = 0.5\tau_c$, and the injected amplitude is $\Sigma_0^+ = 4.3$; (b) The generation of two standing oscillatory solitons, with the pulse widths $\tau_0 = 0.5\tau_c$ and injected amplitude $\Sigma_0^+ = 3.6$ [12].

For an input pulse with a suitable area, the numerical simulations have shown that the pulse can evolve into a standing gap soliton that can be stored as a stable self-localized state. The minimum length of the sample, necessary for the realization of the storage of the standing light pulse, is 1200 BG periods [12,13]. The self-trapping dynamics of the laser pulse may be considered as the motion of a quasi-particle in an effective potential representing the nonlinear interaction between the input pulse and the two-level atomic medium [46]. As a result, the pulse decelerates and eventually comes to a halt inside the structure, under the action of the trapping potential.

By colliding with another input pulse, the stored pulse can be released; both pulses can be released from the RABR with an efficiency of up to 96% (Figure 3) [13].

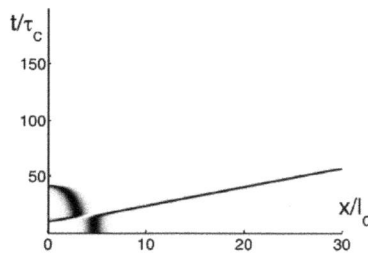

Figure 3. A contour plot illustrating the release of a laser pulse stored in the RABR by dint of its collision with an additional pulse injected into the structure [13]. The pulse width is $\tau_0 = 0.5\tau_c$, and the injected amplitude is $\Sigma_0^+ = 3.5$, for the stopped pulse, and $\Sigma_0^+ = 3.84$ for the incident one. Both input pulses have a standard sech profile.

The nonlinear frequency conversion of zero-velocity (standing) short light pulses, based on the stimulated Raman scattering (SRS), has also been studied [7]. To reduce the walk-off effects, both the pump and Stokes waves must be phase-matched, traveling with equal speeds. This can be realized in a 1D doubly resonant Bragg reflector (DRBR), where "doubly" implies supporting the resonant interactions at two different wavelengths simultaneously, as specified below [7].

The DRBR structure consists of 1D periodically arranged layers of two-level atoms and a passive BG. The period of the array of the two-level-atom layers is equals to the half-wavelength of the pump pulse. The bandgap center of the passive BG is equal to the wavelength of the Stokes pulse. In the DRBR, the pump pulse can be decelerated and stopped by thin atomic layers [41]. On the other hand, the Stokes pulse can be generated as a standing or slowly oscillating soliton by the passive Bragg reflector [14]. Due to the interaction of these two kinds of stopped light pulses, the energy of the pump pulse can be efficiently transferred to the Stokes pulse. Furthermore, the energy of the Stokes pulse will eventually leak out from both edges of the finite-length of DRBR in the form of Raman solitons.

The propagation of light in the DRBR is modeled by the Maxwell-Bloch equations, which take into account the Raman and Kerr nonlinearities [7]:

$$\pm\frac{\partial \Sigma_p^\pm}{\partial \zeta} + \frac{\partial \Sigma_p^\pm}{\partial \tau} = -\frac{G_p}{2}\Sigma_p^\pm(|\Sigma_s^+|^2 + |\Sigma_s^-|^2) + P + i\Gamma_p\Sigma_p^\pm\left[\left|\Sigma_p^\pm\right|^2 + \left|\Sigma_p^\mp\right|^2 + (2-f_R)(|\Sigma_s^+|^2 + |\Sigma_s^-|^2)\right] \quad (5)$$

$$\pm\frac{\partial \Sigma_s^\pm}{\partial \zeta} + \frac{\partial \Sigma_s^\pm}{\partial \tau} = -\frac{G_s}{2}\Sigma_s^\pm\left(\left|\Sigma_p^+\right|^2 + \left|\Sigma_p^-\right|^2\right) + iK\Sigma_s^\mp + i\Delta\Sigma_s^\pm + i\Gamma_s\Sigma_s^\mp\left[\left|\Sigma_s^\pm\right|^2 + |\Sigma_s^\mp|^2 + (2-f_R)\left(\left|\Sigma_p^+\right|^2 + \left|\Sigma_p^-\right|^2\right)\right] \quad (6)$$

$$\frac{\partial P}{\partial \tau} = (\Sigma_p^+ + \Sigma_p^+)W, \quad (7)$$

$$\frac{\partial W}{\partial \tau} = -\text{Re}\left[(\Sigma_p^+ + \Sigma_p^+)P^*\right], \quad (8)$$

where $\Sigma_{p,s}^\pm \equiv (2\tau_c\mu/\hbar)E_{p,s}^\pm$ represents the forward- and backward-propagating pump and Stokes waves; $K \equiv \kappa c\tau_c/n_0$ and $\Delta \equiv \delta c\tau_c/n_0$ are the dimensionless coupling constants and detuning, respectively; f_R is the fraction of the nonlinearity arising from molecular vibrations with a typical value of 0.18 [45]; and $G_{p,s} \equiv (\varepsilon_0 c^2\hbar^2/8\mu^2\tau_c)g_{p,s}$ and $\Gamma_{p,s} \equiv \omega_{p,s}n_2\varepsilon_0 c\hbar^2/8\mu^2\tau_c$ are the dimensionless nonlinearity and gain coefficients for the pump and Stokes waves, respectively.

When the time of the interaction of the standing Raman active medium and soliton is large enough, the intensity of the zero-velocity Stokes pulse starts to increase at the expense of the pump field. Following the efficient power exchange between the pump and Stokes pulses, the energy of the Stokes pulse can leak out from both edges of the DRBR. By comparing the output energy of the Stokes pulses with the energy of the standing pump pulse, the efficiency of the Raman shift can be estimated (Figure 4). Remarkably, it may exceed 85%, i.e., higher than the efficiency in the bulk medium [47]. In practice, such an efficient conversion can be achieved by using a periodically arranged multi-quantum-well structure, [48]. The required length and power are only a few millimeters and 100 µJ, respectively.

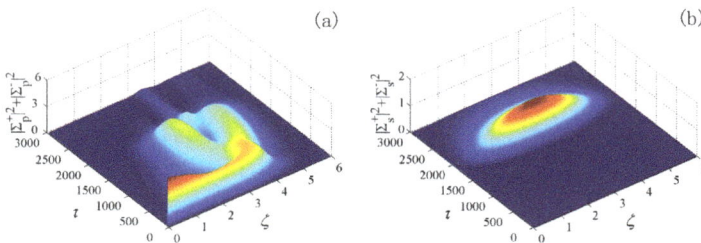

Figure 4. The evolution of the energy density of (**a**) pump and (**b**) stokes pulses in the DRBR under the action of the SRS [7].

RABR can also be used to reshape and compress ultrashort laser pulses. From the Maxwell-Bloch Equation (1), one can show that the RABR may compress the sech-shaped input, with a pulse width ranging from $2\tau_c$ to $5\tau_c$ (τ_c is the cooperative time), into a 2π SIT soliton [49]. Furthermore, input pulses with a multiple-peak shape can be re-shaped to produce a single-peak output. Figure 5 shows the transformation of the three-peak pulse into a single-peak pulse in the RABR structure. In fact, SIT in bulk media can also be applied to compressed optical pulses. However, in that case, the input pulse with single peak is split into multiple ones, rather than morphed into a single-peak SIT soliton, as in RABR. The difference is explained by the fact that reshaping in the RABR originates from multiple reflections in the Bragg structure, while no reflections occur in the the bulk SIT medium. Other interesting phenomena, such as the negatively refracted light in the RABR, have also been discovered [50].

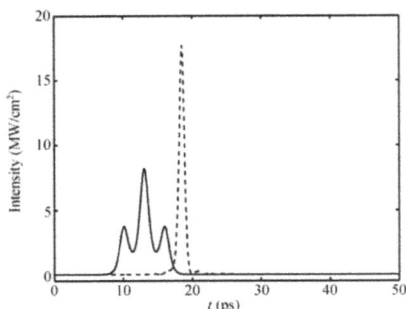

Figure 5. The input (solid line) and output (dash line) from the RABR, when a three-peak pulse is used as the input [40].

The validity of the numerical predictions was also checked by the comparison with a more realistic finite-atomic-width approach [51]. The result shows that stable moving solitons can be generated, but the zero-velocity soliton no longer exists if the thickness of the two-level atom layer exceeds a critical value (1.2 nm).

3.2. Experimental Work on RABR

Many methods, such as photo lithography, e-beam, and self-assembling, can be used to fabricate 1D photonic structures. Due to the extremely thin layers filled by dopant atoms, the molecular-beam epitaxy technique is particularly suitable for fabricating RABRs. In Reference [51], InGaAs/GaAs multi-quantum-well structures, which may be seen as typical RABRs, have been made by using this technique. In a 200 layer InGaAs/GaAs multi-quantum-well pattern, ultrafast switching based on the ac Stark has been observed [52].

For the purpose of buffering the optical pulse, the number of layers should be increased to about 1000 [43]. In this case, the sample will be very easy to peel off from the substrate with the help of demoulding. Normally, for measuring the nonlinear effect of the InGaAs/GaAs multi-quantum-well, the experiment should be performed at a low temperature (4 K–10 K) [48,52]. These strict experimental conditions limit the application of InGaAs/GaAs multi-quantum-wells in optical signal processing.

4. Optical Signal Processing in Chirped Bragg Structures

Chirped fiber BGs with a gradually varying local BG period, have been widely used to provide a strong positive/negative dispersion in a fiber system to stretch and compress the optical pulse [26]. Recently, the nonlinear propagation of the optical pulse in a silicon chirp Bragg structure was investigated [14,15]. The simulation result shows that the optical pulse can be buffered and released in the silicon chirp Bragg structure.

Generally, for the purpose of creating very slow BG solitons, two conditions should be met. First, the grating-induced dispersion must be balanced by the nonlinearity [53]. Second, since the optical field in the 1D Bragg structure is represented by forward-traveling and backward-traveling waves, an initial configuration should have nearly-equal powers in the two waves [54]. The former condition can be achieved by using a high-power input, and therefore, sufficient nonlinearity [12]. The latter condition is much harder to meet with the single incident pulse [55], as it does not initially contain any backward component.

In other studies [14,15], concatenated BGs were used to generate slow ultrashort pulses, or even standing ones. Such concatenated BGs are built by linking a linearly chirped grating with a uniform one. The main advantage of this setting is that the initial conditions for pulses at the input edge of the uniform BG segment may be manipulated by means of the preceding chirped BG, which provides a possibility of preparing the right mix of forward and backward fields.

As in Bragg superstructures, light propagation in chirped gratings can be described by the standard coupled-mode theory [53,56–59]. For slowly varying envelopes of forward and backward waves, E_f and E_b, the coupled-mode equations are written as [14,56]:

$$\pm i\frac{\partial E_{f,b}}{\partial z} + \frac{i}{v_g}\frac{\partial E_{f,b}}{\partial t} + \delta(z)E_{f,b} + \kappa(z)E_{b,f} + \gamma\left(\left|E_{f,b}\right|^2 + 2\left|E_{b,f}\right|^2\right)E_{f,b} = 0 \tag{9}$$

where t is the time; $v_g = c/n_0$ is the group velocity in the material of which the BG is fabricated; and $\gamma = n_2\omega/c$ is the nonlinearity strength, where ω is the frequency of the carrier wave and n_2 is the Kerr coefficient. Further, z is the propagation distance, L is the total length of the grating, n_0 is the average refractive index with modulation depth Δn, Λ_0 is the BG period at the input edge, and C is the chirp. The wavenumber-detuning parameter δ (Figure 6), and the coupling between the forward and backward field, $\kappa = \pi\Delta n/\Lambda_0$, are functions of the propagation distance.

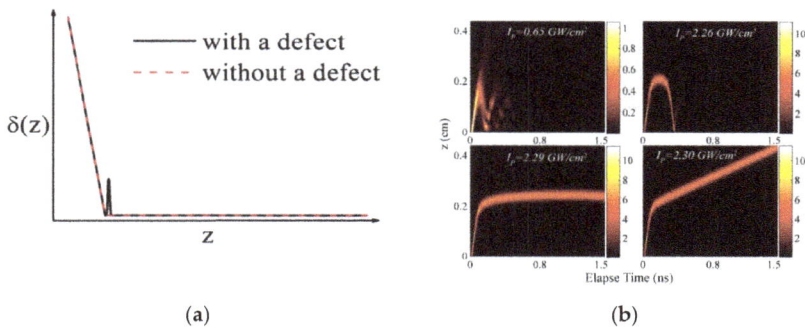

(a)   (b)

Figure 6. (**a**) The relation between wavenumber detuning and propagation distance in the concatenated BG, which does not include the local defect; (**b**) The evolution of the pulse with peak intensity equal to (d_1) 0.65 GW/cm^2, (d_2) 2.26 GW/cm^2, (d_3) 2.29 GW/cm^2, and (d_4) 2.30 GW/cm^2 [14].

The pulse-propagation properties are first investigated with a different input-pulse peak intensity (Figure 6). The analysis leads to the following conclusions.

(i)   At low intensities, e.g., 0.65 GW/cm^2, the pulse is almost totally reflected by the Bragg structure due to the presence of the photonic bandgap (Figure 6bd$_1$). Since the pulse's dispersion is not compensated by the weak nonlinearity, pulse stretching is observed.

(ii)   At higher intensities, an unstable standing light pulse trapped at the interface is generated. In particular, at $I_P = 2.26$ GW/cm^2, the pulse will be reflected after stopping for a short time, as seen in panel (d_2) of Figure 6b. Slightly increasing I_P to 2.29 GW/cm^2, in the range of 2.26 to 2.29 GW/cm^2, gives rise to a stopping time of ~1.3 ns for the pulse that is eventually reflected, as shown in panel (d_3).

(iii) A slow moving Bragg soliton can be observed, as shown in panel (d_4), for $I_P = 2.30$ GW/cm^2. In this case, the velocity of the pulse is equal to 0.005 c. Further simulations demonstrate that the velocity of the moving soliton increases with a further increase of I_P.

Since the standing pulse is unstable, a defect is introduced between the chirped and uniform BG segments, as shown in Figure 6a. By introducing such a defect, the stability of pulse trapping can be much improved. Additionally, the intensity range of achieving a standing pulse can be made several times larger than without the defect (Figure 7).

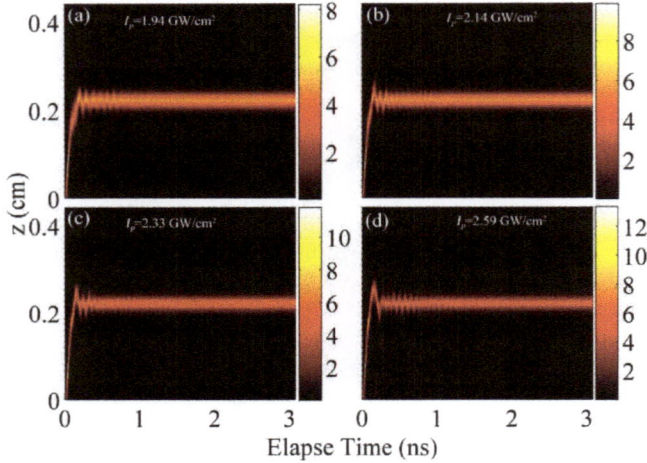

Figure 7. Simulations of the pulse propagation in the concatenated BG, with a defect located at the conjunction of the chirped and uniform segments, for different injected peak intensities: (a) 1.94 GW/cm^2; (b) 2.14 GW/cm^2; (c) 2.33 GW/cm^2; and (d) 2.59 GW/cm^2 [14].

To control the trapping position, a periodic set of defects is introduced into the uniform part of the BG structure, as shown in Figure 8, which shows $\delta_d(z)$ as a function of z. Here, ε is the strength of each defect and d_w is its width [15].

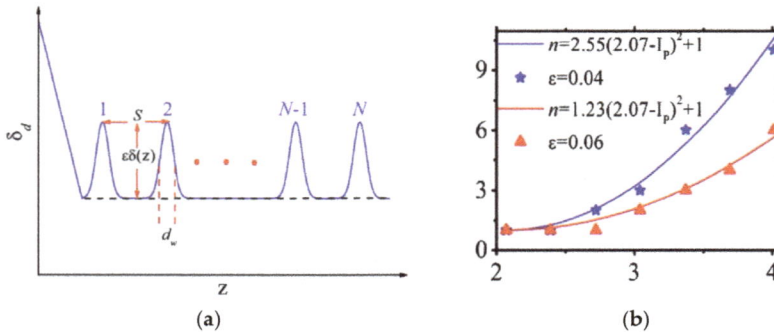

Figure 8. (a) A schematic of the system, built of the linearly chirped BG segment (on the left-hand side) followed by the uniform grating with an inserted periodic array of local defects. The system can be described by parameters (S; d_w; ε), and the definition of S, d_w, and ε are shown in (a); (b) Relations between the trapping position and the initial pulse's intensity I_P for (S; d_w; ε) = (0.132 cm; 50 μm; 0.04) and (0.132 cm; 50 μm; 0.06) (blue and red curves, respectively) [15].

The results of the simulations show that, at different input-pulse intensities, the pulse can be trapped at different defects. Such a defect array can also trap several pulses at different defects (Figure 9). In addition to that, the trapped pulses can interact with each other.

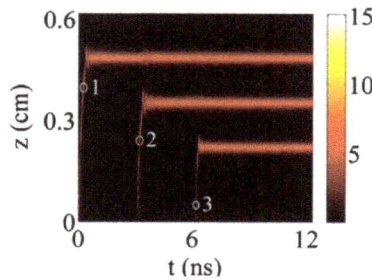

Figure 9. The simulation result of the trapping of three pulses in the BG structure. Here $(S; d_w; \varepsilon)$ = (0.132 cm; 50 μm; 0.04). The peak intensities of the three pulses are 3.04, 2.78, and 2.07 GW/cm². Such pulses are launched into the gratings at $t = 0$, $t = 3$, and $t = 6$ ns, respectively [15].

An all-optical femtosecond soliton diode can also be designed, using a tailored BG heterojunction. Highly nonreciprocal transmission, with an extinction ratio of up to 120, was produced by simulations of this setting (see Figure 10).

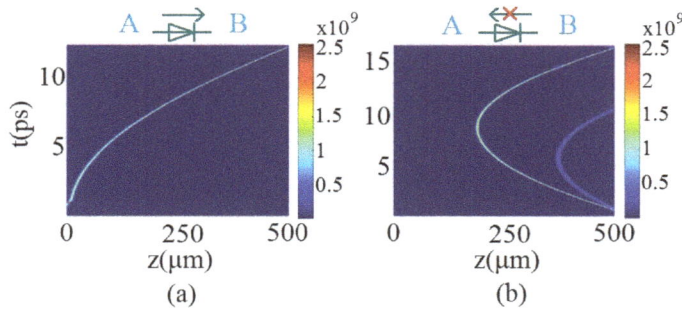

Figure 10. A typical example of the predicted femtosecond diode effect. The pulses with a width of 100 fs are injected from the left (**a**) and right (**b**) hand sides, respectively. The incident peak intensity of the pulse is $I = 0.35$ GW/cm². (**a**) The femtosecond soliton can propagate from A to B; (**b**) the soliton bounces back when it is injected from B [16].

5. Ultrafast Pulse Modulations Based on Cholesteric Liquid Crystal (CLCs) Bragg Gratings

For modulating the optical signal in a compact photonic device, the following properties of the device materials are required: (i) Fast response to process a high speed optical signal; and (ii) Strong dispersion and nonlinearity to achieve a shorter dispersion and nonlinear length to decrease the required propagation length. Many works on the pulse modulation have been completed in fiber Bragg grating, and for which, the typical nonlinear coefficient is on the order of 10^{-16} cm²/W; in conjunction with a peak power of kilowatts [26], the resulting nonlinear effective length is on the millimeter scale [60]. In this section, we show that naturally occurring Bragg grating in a highly nonlinear CLCS can enable the same ultrafast (femtoseconds-picoseconds) pulse modulation operations in sub-mm interaction lengths.

After decades of studies, liquid crystals have emerged as highly versatile and nonlinear optical materials, due to their organic molecular constituents and unique liquid crystalline properties [61].

In liquid crystals, the underlying mechanisms generally fall into two classes, as depicted in Figure 11 for the nematic phase (which include the chiral nematic phase often called cholesteric liquid crystals (CLCs)): (i) Macroscopic crystalline responses including the director axis, reorientation by light fields or a light/DC field induced photorefractive effect, index/birefringence change by laser induced order parameter and/or temperature changes, and flow reorientation; these bulk effects generally respond in the millisecond to a sub-microsecond regime [62]; and (ii) Individual molecule's nonlinear polarization associated with single- and multiple photonic transitions within the molecular energy levels; these respond on the sub-picoseconds to a femtoseconds scale [63–67].

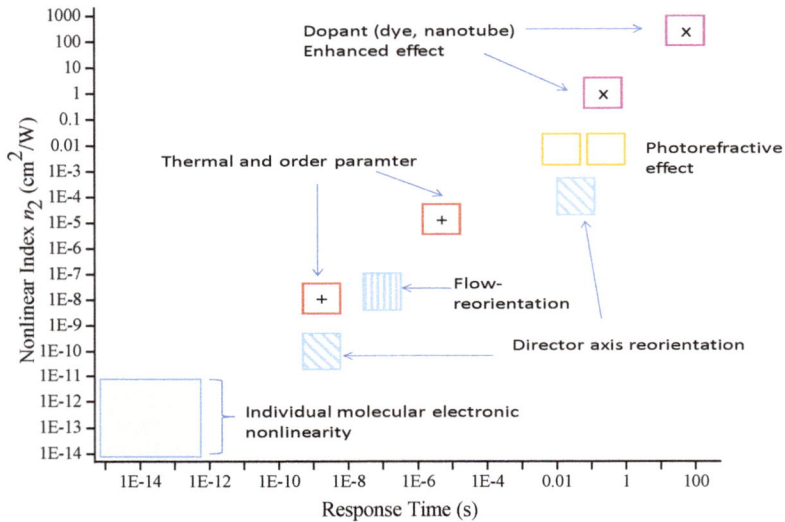

Figure 11. Observed optical nonlinearity in terms of the nonlinear index coefficients of nematic liquid crystals (including chiral nematics) for several mechanisms and the characteristic relaxation time constants.

CLCs are formed by introducing a chiral agent in standard nematic liquid crystals used in ubiquitous liquid crystal display devices. In CLC, the birefringent molecules self-assemble with the director (crystalline) axis arranged in a spiral, with a pitch on the order of the optical wavelength stretching from the UV to the infrared regime, as seen in Figure 12. As a result, CLC's possess not only the advantageous features of liquid crystals such as fabrication ease and low cost, and a very wide spectral dynamic range of operation [68,69], but also the 1-D photonic crystals' unique ability to enhance the nonlinear ultrafast all-optical responses of the CLC constituent molecules [mostly due to the nematic constituent], in addition to the dispersion effect at the photonic band-edges. A typical magnitude of the non-resonant nonlinear index coefficient n_2 is in the order of 10^{-14}–10^{-13} cm^2/W, but owing to the photonic crystal band-edge enhancement, the magnitude of the effective n_2 can be as large as 10^{-12}–10^{-11} cm^2/W. In comparison to other materials [70] used for ultrafast pulsed laser modulation applications, these n_2 values are orders of magnitude larger, and thus, one can envision a tremendous miniaturization possibility.

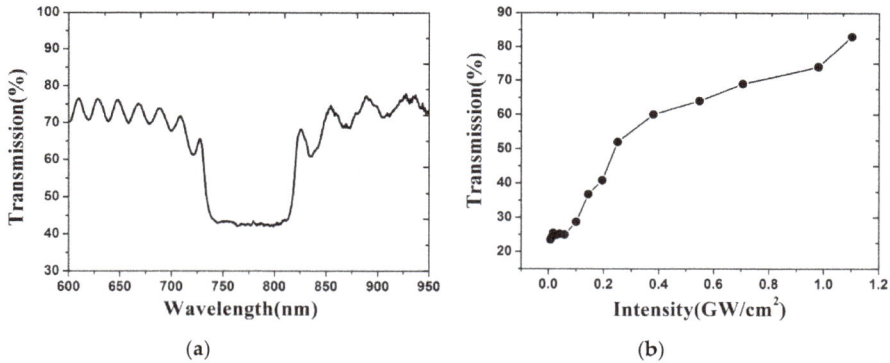

Figure 12. (a) Observed transmission spectrum of linearly polarized light through the CLC cell used in the experiment; (b) Observed dependence of the transmission of a left-handed circularly polarized laser pulse (λ = 815 nm) on the peak intensity [17].

Depending on the electronic resonances of the nematic constituent used in synthesizing a CLC and the wavelength of the laser under study, the sign of n_2 can be positive [18] or negative [17]—a common feature found in the electronic nonlinearities of most materials [70]. As shown in Figure 12b for the nematic compound used in [18], the refractive index change induced by the laser causes the bandgap to shift towards the shorter wavelength region (i.e., blue-shit), and consequently gives rise to an increasing intensity dependence since the laser wavelength is located at the long wavelength edge of the CLC (Figure 12b). In Reference [17], the CLC sample with self- defocusing nonlinear properties is used to compress the femtosecond pulse (Figure 13). The nonlinear coefficient of the CLC used in that study was measured to be about -10^{-11} cm^2/W using a similar intensity dependent transmission measurement. This is four orders of magnitude higher than silica [12]; as a result, the required the thickness of the CLC sample is merely 6 microns in order to compress a 100 fs laser pulse to about 50 fs [17]. If the constituent nematic molecules possess a lower optical nonlinearity, thicker CLC cells of several 100's microns are required [18], which are nevertheless thin/short compared to other materials used for ultrafast laser pulse modulations.

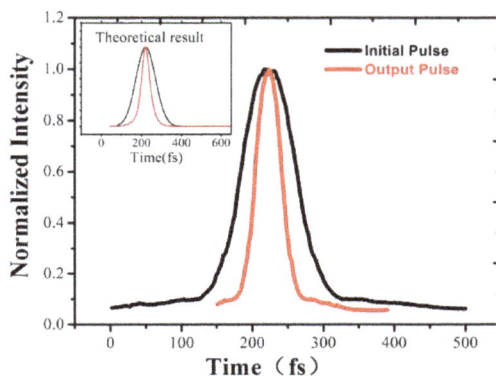

Figure 13. Observed direct pulse compression effect on an initial transform-limited 100 fs pulse (black line) to a 48 fs output pulse (red line); the inset figure corresponds to the simulation results using the measured experimental parameters [17].

The pulse propagation can be simulated by the nonlinear coupld-mode theory presented in Part 4. In [17,18], it is shown that the simulations for the pulse width and the spectra are in good agreement with the experimental results, as seen in Figures 13 and 14. The compression ability for the pulse with different widths is determined by the propagation length, since the differences of dispersion for different widths lead to a difference in the dispersion length. Another work reported in Reference [18] shows that by using a CLC sample with a 500 μm cell gap, the sub-picosecond pulse with 800 fs can be compressed to 286 fs. The corresponding spectra broadening can also be observed (Figure 15). This result clearly indicates that by cascading the CLC sample with a different cell gap, the pulse can be compressed in a wide temporal range.

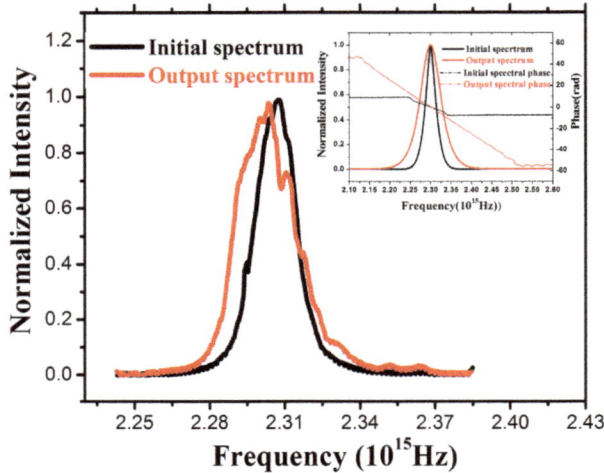

Figure 14. Observed spectral broadening effect due to the pulse compression for an input pulse peak intensity of 1.04 GW/cm^2; the inset depict simulation results for the spectrum (solid lines) and spectral phase (dashed lines) [17].

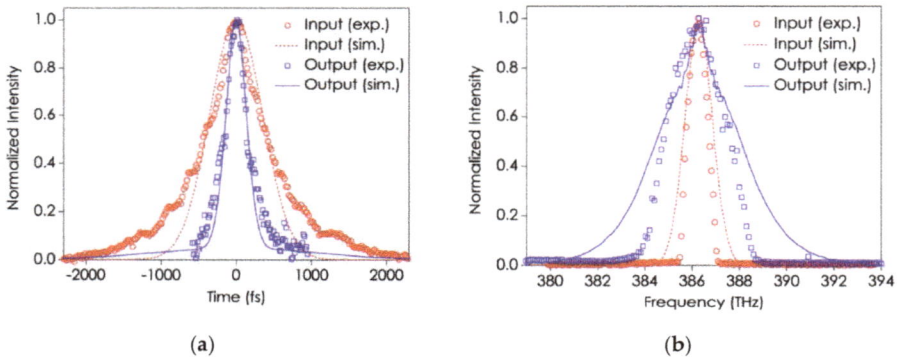

(a) (b)

Figure 15. (a) Observed direct pulse compression effect on an initial transform-limited 847 fs pulse (open circles) to a 286 fs output pulse (open squares); the inset figure corresponds to the simulation results using the measured experimental parameters; (b) Observed and simulated spectra for the input and output (compressed) pulses [18].

Another possible pulse-modulation device is based on two tandem CLC cells that are tailor-made so that their respective blue/red photonic band-gap edges match the operating laser wavelength, and to utilize the opposite linear dispersions (i.e., without involving the optical nonlinearity) from these band edges to impart an opposite chirping effect on these pulses. As a result, an incident 100 femtoseconds laser pulse can be stretched to nearly 2 picoseconds by the first CLC cell, and then recompressed to the original pulse length by the second CLC cell, as seen in Figure 16. Such pulse stretching and recompressing operations are commonly employed in chirped pulse amplification (CPA) systems [71] that require an intermediate pulse stretching process to prevent amplification saturation. In this case, the entire stretch-recompress operation can be done all-optically with CLC cells measuring <1 mm in interaction length, contrasting greatly with the bulky optics employed in conventional CPA systems.

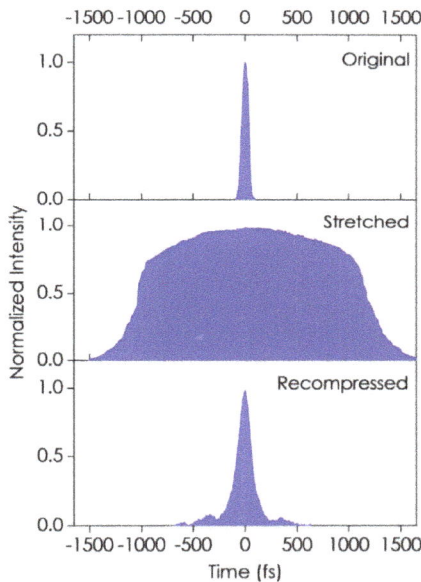

Figure 16. Results of pulse modulations by two tandem 550 µm thick CLC cells. Observed pulse shape for the input (upper trace), output after traversing the first cell (middle), and output after traversing the second cell (bottom trace). Initial pulse width: 100 fs laser pulse; wavelength: 780 nm [18].

6. Conclusions

We have discussed several methods to realize an ultrafast optical signal process in 1D Bragg structures, including resonantly absorbing Bragg reflectors (RABR), chirped BGs (Bragg gratings), and transparent but highly nonlinear CLCs (cholesteric liquid crystals) that exhibit properties of 1D photonic crystals. The result shows that an ultrafast pulse can be buffered and subsequently released in the RABR Additionally, based on the reduction of the light speed, a nonlinear frequency conversion can be achieved over a short propagation distance. Using a linearly chirped BG concatenated with a uniform one, we have demonstrated the possibility of achieving the efficient slow-down of light with the right mix of forward and backward propagating fields. Such chirped gratings can also function as all-optical photodiodes. Finally, we have discussed a unique class of self-assembled BGs based on CLCs. They feature extraordinarily large and ultrafast-responding optical nonlinearities, which enable direct compressions, as well as the stretching and recompressions of femtoseconds laser pulses over very short (sub-mm) propagation lengths.

Acknowledgments: This work is supported by the Chinese National Natural Science Foundation (61505265, 11374067, 11534017). Yikun Liu is also supported by the Fundamental Research Funds for the Central Universities. Iam Choon Khoo's travel and local support is provided by the William E. Leonhard Professorship of PSU.

Author Contributions: J.Z. coordinated the work on the paper. Y.L. collected the data and wrote the draft. S.F., B.A.M., and I.C.K. took part in drafting the paper and contributed to the critical analysis of the data.

Conflicts of Interest: The authors declare no conflict of interest.

References

1. Hau, L.V.; Harris, S.E.; Dutton, Z.; Behroozi, C.H. Light speed reduction to 17 metres pe rsecond in an ultra cold atomic gas. *Nature* **1999**, *397*, 594–598. [CrossRef]
2. Liu, C.; Dutton, Z.; Behroozi, C.H.; Hau, L.V. Observation of coherent optical information storage in an atomic medium using halted light pulses. *Nature* **2001**, *409*, 490–493. [CrossRef] [PubMed]
3. Song, K.Y.; Herraez, M.G.; Thevenaz, L. Observation of pulse delaying and advancement in optical fibers using stimulated Brillouin scattering. *Opt. Express* **2005**, *13*, 82–88. [CrossRef] [PubMed]
4. BaBa, T. Slow light in photonic crystals. *Nat. Photonic* **2008**, *2*, 465–473. [CrossRef]
5. Li, J.T.; White, T.P.; O'Faolain, L.; Gomez-Iglesias, A.; Krauss, T.F. Systematic design of flat band slow light in photonic crystal waveguides. *Opt. Express* **2008**, *16*, 6227–6232. [CrossRef] [PubMed]
6. Li, J.; O'Faolain, L.; Schulz, S.A.; Krauss, T.F. Low loss propagation in slow light photonic crystal waveguides at group indices up to 60. *Hydrometallurgy* **2012**, *10*, 589–593. [CrossRef]
7. Li, J.T.; Zhou, J.Y. Nonlinear optical frequency conversion with stopped short light pulses. *Opt. Express* **2006**, *14*, 2811–2816. [CrossRef] [PubMed]
8. Li, J.; O'Faolain, L.; Krauss, T.F. Four-wave mixing in slow light photonic crystal waveguides with very high group index. *Opt. Express* **2012**, *20*, 17474–17479. [CrossRef] [PubMed]
9. Kozhekin, A.; Kurizki, G. Self-induced transparency in Bragg Reflectors: Gap Solitons near absorption resonances. *Phys. Rev. Lett.* **1995**, *74*, 5020–5023. [CrossRef] [PubMed]
10. Kozhekin, A.; Kurizki, G.; Malomed, B. Standing and moving gap solitons in resonantly absorbing gratings. *Phys. Rev. Lett.* **1998**, *81*, 3647–3650. [CrossRef]
11. Opatrny, T.; Kurizki, G.; Malomed, B. Dark and bright solitons in resonantly absorbing gratings. *Phys. Rev. E* **1999**, *60*, 6137–6149. [CrossRef]
12. Xiao, W.N.; Zhou, J.Y.; Prineas, J.P. Storage of ultrashort optical pulses in a resonantly absorbing Bragg reflector. *Opt. Express* **2003**, *11*, 3277–3283. [CrossRef] [PubMed]
13. Zhou, J.Y.; Shao, H.G.; Zhao, J.; Yu, X.; Wong, K.S. Storage and release of femtosecond laser pulses in a resonant photonic crystal. *Opt. Lett.* **2005**, *30*, 1560–1562. [CrossRef] [PubMed]
14. Fu, S.; Liu, Y.; Li, Y.; Song, L.; Li, J.; Malomed, B.A.; Zhou, J. Buffering and trapping ultrashort optical pulses in concatenated Bragg gratings. *Opt. Lett.* **2013**, *38*, 5047–5050. [CrossRef] [PubMed]
15. Fu, S.; Li, Y.; Liu, Y.; Zhou, J.; Malomed, B.A. Tunable storage of optical pulses in a tailored Bragg-grating structure. *J. Opt. Soc. Am.* **2015**, *32*, 534–539. [CrossRef]
16. Deng, Z.; Lin, H.; Li, H.; Fu, S.; Liu, Y.; Xiang, Y.; Li, Y. Femtosecond soliton diode on heterojunction Bragg-grating structure. *Appl. Phys. Lett.* **2016**, *109*, 121101. [CrossRef]
17. Song, L.; Fu, S.; Liu, Y.; Zhou, J.; Chigrinov, V.G.; Khoo, I.C. Direct femtosecond pulse compression with miniature-sized Bragg cholesteric liquid crystals. *Opt. Lett.* **2013**, *38*, 5040–5042. [CrossRef] [PubMed]
18. Liu, Y.; Wu, Y.; Chen, C.W.; Zhou, J.; Lin, T.H.; Khoo, I.C. Ultrafast pulse compression, stretching-and-recompression using cholesteric liquid crystals. *Opt. Express* **2016**, *24*, 10458. [CrossRef] [PubMed]
19. John, S. Strong localization of photons in certain disordered dielectric superlattices. *Phys. Rev. Lett.* **1987**, *58*, 2486–2489. [CrossRef] [PubMed]
20. Yablonovitch, E. Inhibited Spontaneous emission in solid-state physics and electronics. *Phys. Rev. Lett.* **1987**, *58*, 2059–2062. [CrossRef] [PubMed]
21. Vlasov, Y.A.; O'Boyle, M.; Hamann, H.F.; McNab, S.J. Active control of slow light on a chip with photonic crystal waveguides. *Nature* **2005**, *438*, 65–69. [CrossRef] [PubMed]
22. Frandsen, L.H.; Lavrinenko, A.V.; Fage-Pedersen, J.; Borel, P.I. Photonic crystal waveguides with semi-slow light and tailored dispersion properties. *Opt. Express* **2006**, *14*, 9444–9450. [CrossRef] [PubMed]

23. Hill, K.O.; Malo, B.; Bilodeau, R.; Johnson, D.C.; Albert, J. Bragg gratings fabricated in monomode photosensitive optical fiber by UV exposure through a phase mask. *Appl. Phys. Lett.* **1993**, *62*, 1035. [CrossRef]

24. Patrick, H.J.; Williams, G.M.; Kersey, A.D.; Pedrazzani, J.R.; Vengsarkar, A.M. Hybrid fiber Bragg grating/long period fiber grating sensor for strain/temperature discrimination. *IEEE Photonics Technol. Lett.* **1996**, *8*, 1223–1225. [CrossRef]

25. Hill, K.O.; Meltz, G. Fiber Bragg grating technology fundamentals and overview. *J. Light. Technol.* **1997**, *15*, 1263–1276. [CrossRef]

26. Agrawal, G.P. *Nonlinear Fiber Optics*; Academic Press: Waltham, MA, USA, 2007.

27. Sankey, N.D.; Prelewitz, D.F.; Brown, T.G. All-optical switching in a nonlinear periodic-waveguide structure. *Appl. Phys. Lett.* **1992**, *60*, 1427–1429. [CrossRef]

28. Millar, P.; De La Rue, R.M.; Krauss, T.F.; Aitchison, J.S.; Broderick, N.G.R.; Richardson, D.J. Nonlinear propagation effects in an AlGaAs Bragg grating filter. *Opt. Lett.* **1999**, *24*, 685–687. [CrossRef] [PubMed]

29. Scalora, M.; Dowling, J.P.; Bowden, C.M.; Bloemer, M.J. Optical limiting and switching of ultrashort pulses in nonlinear photonic band gap materials. *Phys. Rev. Lett.* **1994**, *73*, 1368–1371. [CrossRef] [PubMed]

30. Chilaya, G.S. Light-controlled change in the helical pitch and broadband tunable cholesteric liquid-crystals lase. *Crystallogr. Rep.* **2006**, *51*, S108–S118. [CrossRef]

31. Moreira, M.F.; Carvalho, I.C.; Cao, W.; Bailey, C.; Taheri, B. Cholesteric liquid-crystal laser as an optic fiber-based temperature sensor. *Appl. Phys. Lett.* **2004**, *85*, 2691–2693. [CrossRef]

32. Mantsyzov, B.I.; Kuz'min, R.N. Coherent interaction of light with a discrete periodic resonant medium. *Sov. Phys. JETP* **1986**, *64*, 37–44.

33. Conti, C.; Assanto, G.; Trillo, S. Self-sustained trapping mechanism of zero- velocity parametric gap solitons. *Phys. Rev. E* **1999**, *59*, 2467–2470. [CrossRef]

34. Rossi, A.D.; Conti, C.; Trillo, S. Stability, multistability, and wobbling of optical gap solitons. *Phys. Rev. Lett.* **1998**, *81*, 85–88. [CrossRef]

35. Mantsyzov, B.I. Gap 2π pulse with an inhomogeneously broadened line and an oscillating solitary wave. *Phys. Rev. E* **1995**, *51*, 4939–4943. [CrossRef]

36. Mak, W.C.K.; Malomed, B.A.; Chu, P.L. Slowdown and splitting of gap solitons in apodized Bragg gratings. *J. Mod. Opt.* **2004**, *51*, 2141–2158. [CrossRef]

37. Eggleton, B.J.; Slusher, R.E.; Sterke, C.M.; Krug, P.A.; Sipe, J.E. Bragg grating solitons. *Phys. Rev. Lett.* **1996**, *76*, 1627–1630. [CrossRef] [PubMed]

38. Mok, J.T.; de Sterke, C.M.; Littler, I.C.M.; Eggleton, B.J. Dispersionless slow light using gap solitons. *Nat. Phys.* **2006**, *2*, 775–780. [CrossRef]

39. Goodman, R.H.; Slusher, R.E.; Weinstein, M.I. Stopping light on a defect. *J. Opt. Soc. Am. B* **2002**, *19*, 1635–1652. [CrossRef]

40. Mak, W.C.K.; Malomed, B.A.; Chu, P.L. Interaction of a soliton with a local defect in a fiber Bragg grating. *J. Opt. Soc. Am. B* **2003**, *20*, 725–735. [CrossRef]

41. Kurizki, G.; Kozhekin, A.E.; Opatrny, T.; Malomed, B.A. Optical solitons in periodic media with resonant and off-resonant nonlinearities. *Prog. Opt.* **2001**, *42*, 93–146.

42. Mak, W.C.K.; Malomed, B.A.; Chu, P.L. Formation of a standing-light pulse through collision of gap solitons. *Phys. Rev. E* **2003**, *68*, 026609. [CrossRef] [PubMed]

43. Zhou, J.; Zeng, J.; Li, J. Quantum coherent control of ultrashort laser pulses. *Chin. Sci. Bull.* **2008**, *53*, 652–658. [CrossRef]

44. McCall, S.L.; Hahn, E.L. Self-induced transparency. *Phys. Rev.* **1969**, *183*, 457–485. [CrossRef]

45. McCall, S.L.; Hahn, E.L. Pulse-area-pulse-energy description of a traveling-wave laser amplifier. *Phys. Rev. A* **1970**, *2*, 861–870. [CrossRef]

46. Mantsyzov, B.I.; Silnikov, R.A. Unstable excited and stable oscillating gap 2π pulses. *J. Opt. Soc. Am. B* **2002**, *19*, 2203–2207. [CrossRef]

47. Shen, Y.R. *The Principles of Nonlinear Optics*; Wiley-Interscience: New York, NY, USA, 1984.

48. Prineas, J.P.; Ell, C.; Lee, E.S.; Khitrova, G.; Gibbs, H.M.; Koch, S.W. Exciton-polariton eigenmodes in light-coupled In0.04Ga0.96As/GaAs semiconductor multiple-quantum-well periodic structures. *Phys. Rev. B* **2000**, *61*, 13863–13872. [CrossRef]

49. Zhao, J.; Li, J.T.; Shao, H.G.; Wu, J.; Zhou, J.; Wong, K.S. Reshaping ultrashort light pulses in resonant photonic crystals. *J. Opt. Soc. Am. B* **2006**, *23*, 1981–1987. [CrossRef]

50. Zhao, J.; Lan, Q.; Zhang, J.; LI, J.; Zeng, J.; Cheng, J.; Friedler, I.; Kurizki, G. Nonlinear dynamics of negatively refracted light in a resonantly absorbing Bragg reflectors. *Opt. Lett.* **2007**, *32*, 1117–1119. [CrossRef]
51. Shao, H.G.; Zhao, J.; Wu, J.W.; Zhou, J. Moving and zero-velocity gap soliton in resonantly absorbing Bragg reflector of finite atomic widths. *Acta Phys. Sin.* **2005**, *54*, 1420–1425.
52. Prineas, J.P.; Zhou, J.Y.; Kuhl, J.; Gibbs, H.M.; Khitrova, G.; Koch, S.W.; Knorr, A. Ultrafast ac Stark effect switching of the active photonic band gap from Bragg-periodic semiconductor quantum wells. *Appl. Phys. Lett.* **2002**, *81*, 4332–4334. [CrossRef]
53. De Sterke, C.M.; Sipe, J.E. Gap solitons. *Prog. Opt.* **1994**, *33*, 203–260.
54. Eggleton, B.J.; de Sterke, C.M.; Slusher, R.E. Nonlinear pulse propagation in Bragg gratings. *J. Opt. Soc. Am. B* **1997**, *14*, 2980–2993. [CrossRef]
55. Shnaiderman, R.; Tasgal, R.S.; Band, Y.B. Creating very slow optical gap solitons with a grating-assisted coupler. *Opt. Lett.* **2011**, *36*, 2438–2440. [CrossRef] [PubMed]
56. Poladian, L. Graphical and WKB analysis of non-uniform Bragg gratings. *Phys. Rev. E* **1993**, *48*, 4758. [CrossRef]
57. Tsoy, E.N.; de Sterke, C.M. Propagation of nonlinear pulses in chirped fiber gratings. *Phys. Rev. E* **2000**, *62*, 2882. [CrossRef]
58. Janner, D.; Galzerano, G.; Valle, G.D.; Laporta, P.; Longhi, S. Slow light in periodic superstructure Bragg gratings. *Phys. Rev. E* **2005**, *72*, 056605. [CrossRef] [PubMed]
59. Yu, M.; Wang, L.; Nemati, H.; Yang, H.; Bunning, T.; Yang, D.K. Effects of polymer network on electrically induced reflection band broadening of cholesteric liquid crystals. *J. Polym. Sci. Part B Polym. Phys.* **2017**. [CrossRef]
60. Broderick, N.G.R.; Taverner, D.; Richardson, D.J.; Ibsen, M.; Laming, R.I. Optical pulse compression in fiber Bragg gratings. *Phys. Rev. Lett.* **1997**, *79*, 4566. [CrossRef]
61. Khoo, I.C. Nonlinear optics, active plasmonics and metamaterials with liquid crystals. *Prog. Quant. Electron.* **2014**, *38*, 77–117. [CrossRef]
62. Khoo, I.C. Nonlinear optics of liquid crystalline materials. *Phys. Rep.* **2009**, *471*, 221–267. [CrossRef]
63. Khoo, I.C.; Webster, S.; Kubo, S.; Youngblood, W.J.; Liou, J.D.; Mallouk, T.E.; Lin, P.; Haganb, D.J.; Strylandb, E.W.V. Synthesis and characterization of the multi-photon absorption and excited-state properties of a neat liquid 4-propyl 4'-butyl diphenyl acetylene. *J. Mat. Chem.* **2009**, *19*, 7525–7531. [CrossRef]
64. Khoo, I.C. Nonlinear Organic Liquid Cored Fiber Array for All- Optical Switching and Sensor Protection against Short Pulsed Lasers. *IEEE J. Sel. Top. Quantum Electron.* **2008**, *14*, 946–951. [CrossRef]
65. Khoo, I.C.; Diaz, A. Multiple-time-scale dynamic studies of nonlinear transmission of pulsed lasers in a multiphoton-absorbing organic material. *J. Opt. Soc. Am. B* **2011**, *28*, 1702–1710. [CrossRef]
66. Hwang, J.; Ha, N.Y.; Chang, H. J.; Park, B.; Wu, J.W. Enhanced optical nonlinearity near the photonic bandgap edges of a cholesteric liquid crystal. *Opt. Lett.* **2004**, *29*, 2644–2646. [CrossRef] [PubMed]
67. Hwang, J.; Wu, J.W. Determination of optical Kerr nonlinearity of a photonic bandgap structure by Z-scan measurement. *Opt. Lett.* **2005**, *30*, 875–877. [CrossRef] [PubMed]
68. Ha, N.Y.; Ohtsuka, Y.; Jeong, S.M.; Nishimura, S.; Suzaki, G.; Takanishi, Y.; Ishikawa, K.; Takezoe, H. Fabrication of a simultaneous red–green–blue reflector using single-pitched cholesteric liquid crystals. *Nat. Mater.* **2008**, *7*, 43–47. [CrossRef] [PubMed]
69. White, T.J.; Bricker, R.L.; Natarajan, L.V.; Tabiryan, N.V.; Green, L.; Li, Q.; Bunning, T.J. Phototunable azobenzene cholesteric liquid crystals with 2000 nm range. *Adv. Funct. Mater.* **2009**, *19*, 3484–3488. [CrossRef]
70. Christodoulides, D.N.; Khoo, I.C.; Salamo, G.J.; Stegeman, G.I.; Stryland, E.W.V. Nonlinear refraction and absorption: Mechanisms and magnitudes. *Adv. Opt. Photonics* **2010**, *2*, 60–200. [CrossRef]
71. Maine, P.; Strickland, D.; Bado, P.; Pessot, M.; Mourou, G. Generation of ultrahigh peak power pulses by chirped pulse amplification. *IEEE J. Quantum Electron.* **1988**, *24*, 398–403. [CrossRef]

![applied sciences logo] *applied sciences*

MDPI

Article

Rogue Wave Modes for the Coupled Nonlinear Schrödinger System with Three Components: A Computational Study

Hiu Ning Chan * and Kwok Wing Chow

Department of Mechanical Engineering, University of Hong Kong, Pokfulam, Hong Kong, China; kwchow@hku.hk
* Correspondence: hnchan06@connect.hku.hk; Tel.: +852-3917-2641; Fax: +852-2858-5415

Academic Editor: Boris Malomed
Received: 13 April 2017; Accepted: 16 May 2017; Published: 29 May 2017

Abstract: The system of "integrable" coupled nonlinear Schrödinger equations (Manakov system) with three components in the defocusing regime is considered. Rogue wave solutions exist for a restricted range of group velocity mismatch, and the existence condition correlates precisely with the onset of baseband modulation instability. This assertion is further elucidated numerically by evidence based on the generation of rogue waves by a single mode disturbance with a small frequency. This same computational approach can be adopted to study coupled nonlinear Schrödinger equations for the "non-integrable" regime, where the coefficients of self-phase modulation and cross-phase modulation are different from each other. Starting with a wavy disturbance of a finite frequency corresponding to the large modulation instability growth rate, a breather can be generated. The breather can be symmetric or asymmetric depending on the magnitude of the growth rate. Under the presence of a third mode, rogue wave can exist under a larger group velocity mismatch between the components as compared to the two-component system. Furthermore, the nonlinear coupling can enhance the maximum amplitude of the rogue wave modes and bright four-petal configuration can be observed.

Keywords: rogue waves; manakov system; modulation instability; numerical simulation

---

1. Introduction

Rogue waves or freak waves are extreme events in the ocean which are characterized by the emergence of large waves from an otherwise tranquil background [1–3]. The unexpectedly large displacements of the sea surface pose threats to maritime activities and offshore structures. Intensive research efforts are conducted to understand the physics of rogue wave and to develop measures to predict or detect such waves [4,5]. Although rogue waves originate from the context of water waves [1–3,6–9], these studies have been extended to other physical contexts like optical fibers [10–12] and Bose-Einstein condensates [13]. Moreover, it has been demonstrated that optical rogue waves are related to supercontinuum generation [10,11]. The high repetition rate of optical experiments is an advantage in the study of these rare events.

For hydrodynamic surface waves, the nonlinear Schrödinger (NLS) equation governs the slow evolution of a weakly nonlinear wave packet [2]. The NLS equation can also describe the dynamics of temporal pulses in an optical fiber [10]. The Peregrine breather of the NLS equation is localized in both space and time, and is a widely utilized model for rogue waves [14,15]. This solution is only nonsingular in the focusing regime unless higher order terms are considered [16].

When there are two or more wave trains present, the governing model is then the system of coupled nonlinear Schrödinger equations [17,18]. The special case where all the coefficients of cubic nonlinearities are identical is known as the Manakov system. In optics, the Manakov equations provide an analytical account on the propagation of two optical beams in a photorefractive medium [19]. Moreover, when polarization effects are taken into account, the evolution of slowly varying, mutually perpendicular electric fields along an optical fiber with birefringence can be described by a system of coupled NLS equations too [20]. Another relevant example in optical physics is the issue of data communication using multi-mode fibers. Spatial multiplexing can relieve the capacity constraints of single-mode fibers. Random coupling of degenerate or quasi-degenerate modes for multi-mode fibers in the nonlinear propagation regime can then lead to generalized Manakov equations with many components [21]. Subsequently, transmissions in multi-mode fibers exhibiting rapidly varying birefringence will also give rise to Manakov equations upon averaging [22], and intermodal modulation instabilities for coupled Schrödinger systems can be analyzed [23]. Finally, the Manakov system is also a useful model in the dynamics of cold atomic species studied in Bose-Einstein condensates [13,24].

Theoretically, the Manakov equations are obtained by considering the evolution of slowly varying electric fields in temporal or spatial waveguides. For temporal waveguides, group velocity dispersion is balanced by Kerr (cubic) nonlinearity. Rogue waves of the two-component Manakov system in the focusing regime have been studied intensively [25–27]. In contrast to the single component case, rogue wave modes had been discovered for the coupled NLS equations in the defocusing regime [28–31] with group velocity mismatch between the two components. This scenario is closely related to new ranges of modulation instability in the defocusing regime. Theoretically, baseband instability is associated with the unstable behavior of disturbances in the low frequency regime. Remarkably these theoretical studies are amply supported by experimental efforts. Indeed, wavelength-division-multiplexed systems have been extended in the laboratory setting beyond the soliton formation regime, or more precisely to baseband and passband regimes for polarization modulation instabilities and the existence of rogue wave modes of the Manakov equations [32]. In another multi-component investigation, optical dark rogue waves are demonstrated by a suitable injection of two colliding and modulated pump beams with orthogonal states of polarization [33]. Furthermore, rogue waves for the three-component Manakov system in the focusing regime had been derived. Four-petal patterns are possible [34,35]. Such wave profiles are otherwise inadmissible for the one-component and two-component counterparts. Interactions of rogue waves with solitons and breathers were investigated by utilizing the Darboux transformation [36].

In this work, we extend the study to the three-component Manakov system in the defocusing regime. For mathematical convenience, the system is taken in the non-dimensional and normalized form. We incorporate symmetrically placed group velocity differences between the complex valued components u, v, and w:

$$
\begin{aligned}
iu_z + i\delta u_t + u_{tt} - \sigma\left(|u|^2 + |v|^2 + |w|^2\right)u &= 0, \\
iv_z - i\delta v_t + v_{tt} - \sigma\left(|u|^2 + |v|^2 + |w|^2\right)v &= 0, \\
iw_z + w_{tt} - \sigma\left(|u|^2 + |v|^2 + |w|^2\right)w &= 0,
\end{aligned}
\tag{1}
$$

where δ describes the group velocity mismatch and σ measures the coefficient of cubic nonlinearity.

The objective of this paper is to investigate the dynamics of rogue waves and breathers in a multi-component system through a combination of theoretical perspective and computational approach. Theoretically, the formation of rogue wave and breathers can be explained in terms of modulation instability (MI). A detailed numerical investigation on the evolution of a plane wave perturbed by a single wavy disturbance would supplement the theoretical framework. In particular, a disturbance from the baseband of the MI spectrum would generate a rogue wave. For a mode with a finite frequency, both symmetric and asymmetric breathers are observed from the computational study.

The outline of the paper is as follows. In Section 2, the rogue wave mode is derived by the Hirota bilinear method. The coupling effect is discussed in Section 3. The generation of rogue wave mode from baseband modulation instability is confirmed directly by computer simulation in Section 4. Similarly, the generation of symmetric and asymmetric breathers is demonstrated in Section 5. In Section 6, a numerical method for finding rogue wave modes in non-integrable systems is proposed. The conclusion is drawn in Section 7.

2. Formulation of the Rogue Wave Modes

Under the transformations, $u = \rho_1 \exp(-i\omega_1 z)\frac{g_1}{f}$, $v = \rho_2 \exp(-i\omega_2 z)\frac{g_2}{f}$, $w = \rho_3 \exp(-i\omega_3 z)\frac{g_3}{f}$, Equation (1) can be rewritten in terms of the Hirota bilinear operator [37] as

$$\left[iD_z + i\delta D_t + D_t^2 \right] g_1 \cdot f = 0,$$

$$\left[iD_z - i\delta D_t + D_t^2 \right] g_2 \cdot f = 0,$$

$$\left(iD_z + D_t^2 \right) g_3 \cdot f = 0,$$

$$\left(D_t^2 - C \right) f \cdot f = -\sigma \left(\rho_1^2 |g_1|^2 + \rho_2^2 |g_2|^2 + \rho_3^2 |g_3|^2 \right),$$

where $\omega_1 = \omega_2 = \omega_3 = C$, and $C = \sigma(\rho_1^2 + \rho_2^2 + \rho_3^2)$.

The methodology in deriving the rogue wave solution is similar to that used in our earlier works [31] and thus the details are omitted here. The amplitude of the plane wave background, ϱ, is taken to be identical for all three waveguides. Basically a breather is first derived using a two-soliton expression with complex conjugate wavenumbers. By taking the small frequency limit of the breather, the rogue wave mode is given by

$$u = \rho \exp\left(-3i\sigma\rho^2 z\right) \left\{ 1 - 4 \frac{1 + i(b - \delta)t + i\left(a^2 - b^2 + b\delta\right)z}{\left[a^2 + (b - \delta)^2\right]\left[(t - bz)^2 + a^2 z^2 + \frac{1}{a^2}\right]} \right\},$$

$$v = \rho \exp\left(-3i\sigma\rho^2 z\right) \left\{ 1 - 4 \frac{1 + i(b + \delta)t + i\left(a^2 - b^2 - b\delta\right)z}{\left[a^2 + (b + \delta)^2\right]\left[(t - bz)^2 + a^2 z^2 + \frac{1}{a^2}\right]} \right\}, \tag{2}$$

$$w = \rho \exp\left(-3i\sigma\rho^2 z\right) \left\{ 1 - 4 \frac{1 + ibt + i\left(a^2 - b^2\right)z}{\left(a^2 + b^2\right)\left[(t - bz)^2 + a^2 z^2 + \frac{1}{a^2}\right]} \right\},$$

where a and b are the real part and imaginary part of Ω_0: $\Omega_0 = a + ib$. The parameter Ω_0 is the leading order term in the asymptotic expansion of the wavenumber, which satisfies the dispersion relation,

$$\Omega_0^6 + 2\left(\delta^2 + 3\sigma\rho^2\right)\Omega_0^4 + \delta^4 \Omega_0^2 + 2\delta^4 \sigma\rho^2 = 0. \tag{3}$$

This cubic polynomial in Ω_0^2 will dictate the dynamics and profiles of the rogue waves, to be highlighted in the following section.

3. The Effect of Coupling

3.1. Extension of Existence Regime

The rogue wave solution in Equation (2) is nonsingular if and only if a never vanishes. Since Ω_0 cannot be real, this is equivalent to having a non-real root for the cubic polynomial $p_3(x) = x^3 + 2\left(\delta^2 + 3\sigma\rho^2\right)x^2 + \delta^4 x + 2\delta^4\sigma\rho^2$. By considering the discriminant of $p_3(x)$, complex roots (and hence rogue waves) will exist for $0 < \delta^2 < 16.9\,\sigma\rho^2$. One highlight of the present work is that this constraint is much less restrictive than the corresponding existence condition for the Manakov system with two components, namely, $0 < \delta^2 < 4\sigma\rho^2$ [31]. We conjecture that the addition of more components to coupled systems may in general induce further modulation instabilities and enhance the existence of rogue waves. This hypothesis obviously must be tested for other dynamical systems in the future.

Moreover, the incorporation of the third component increases the complexity of the geometry of the wave profiles. Analytically, the dispersion relation is expressed as a higher order algebraic polynomial and will allow multiple rogue wave solutions under the same input physical parameters. From Equation (3), if $a + ib$ is a root of the dispersion relation, then $a - ib$ will also be admissible and provides another rogue wave solution. This phenomenon was also observed in other multi-component system [38–41]. Such multi-rogue-wave scenarios are not allowed in the two-component Manakov system.

3.2. Enhancement of Amplitude

For the two-component Manakov system in the defocusing regime [31], the range of amplitude and configurations of the rogue waves in the two components are identical. Either both wave profiles are eye-shaped dark rogue waves (EDRW) with the main displacement below the background, or both patterns are four-petal-shaped rogue waves (FPRW) with two local maxima and two local minima. Moreover, the maximum and minimum values attained are identical in both components.

For the three components case, the scenarios are drastically different as the various components in Equation (1) can exhibit distinct forms of rogue waves and the maximum displacements can be different (Figure 1). Interestingly, there must be at least one component in the form of an EDRW where the minimum amplitude is bounded below by about 0.4 (See Proof in Appendix A). However, the rogue wave solution cannot take the form of an eye-shaped bright rogue wave with the main displacement above the background (See Proof in Appendix A). As compared to the two-component system, the nonlinear coupling of the third component can enhance the rogue wave in two ways: increasing the maximum amplitude and 'squeezing' a bright rogue wave.

The maximum amplitude can be greater than $\sqrt{2}\rho$ (Figure 1a), which is the upper bound of amplitude for the two-component Manakov system [31]. Similar increment in amplitude due to coupling was also found in other coupled systems such as the long wave-short wave resonance model with two short wave components [41] and a system of coupled derivative nonlinear Schrödinger equations [38].

Although the formation of eye-shaped bright rogue wave is also prohibited [31], a tendency towards the formation of bright rogue wave can be observed in the three-component system. Such bright type rogue wave has a four-petal configuration where the saddle point is above the background and is closer to the maximum than the minimum. The u-component in Figure 1a exhibits a bright four-petal configuration where the amplitude at the saddle point is about 1.3 and the rogue wave ranges from 0 to about 1.5. Such geometry closely resembles the widely studied eye-shaped bright rogue wave.

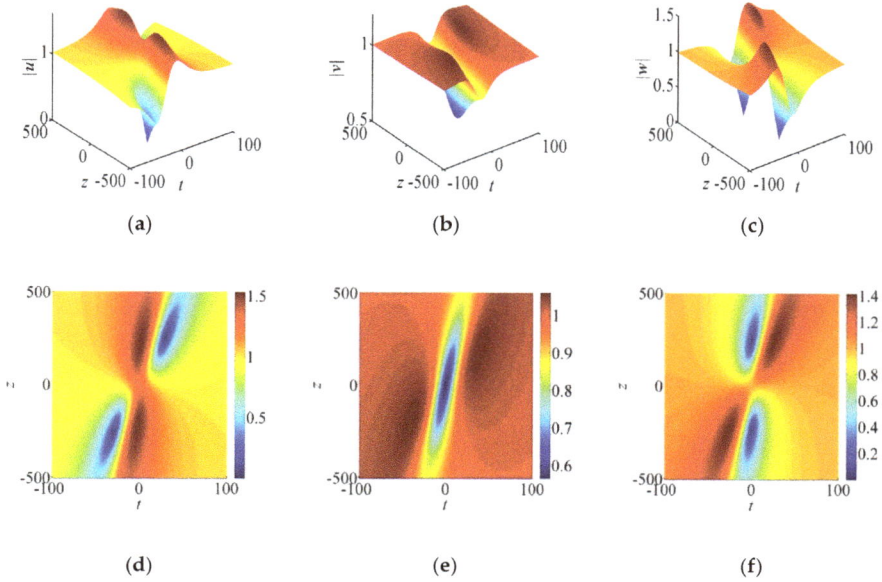

Figure 1. The rogue wave solution given by Equations (2) and (3) with $\varrho = 1$, $\sigma = 1$, $\delta = 0.1$ and $\Omega_0 = 0.0537 + 0.0537i$: (**a**) u-component (bright FPRW); (**b**) v-component (EDRW); (**c**) w-component (FPRW); (**d**) the top view of (**a**); (**e**) the top view of (**b**); (**f**) the top view of (**c**).

4. Baseband Modulation Instability

Baseband modulation instability, the instability due to low frequency disturbances, has been shown to be intimately related to the existence condition of rogue waves [29]. The connection was established theoretically in several dynamical systems [38,39,41,42]. This section focuses on the role of baseband modulation instability in the formation of rogue wave.

4.1. Analytical Approach

To study the correlation between rogue waves and modulation instability, plane waves with identical amplitude are considered,

$$
\begin{aligned}
u_0 &= \rho \exp\left[i\left(-3\sigma\rho^2 z\right)\right], \\
v_0 &= \rho \exp\left[i\left(-3\sigma\rho^2 z\right)\right], \\
w_0 &= \rho \exp\left[i\left(-3\sigma\rho^2 z\right)\right].
\end{aligned}
\tag{4}
$$

Small perturbations of the form $\exp[i(Kt - Wz)]$ would be governed by

$$
\begin{aligned}
&W^6 - \left[3K^4 + 2\left(\delta^2 + 3\sigma\rho^2\right)K^2\right]W^4 + \left(3K^8 + 12\sigma\rho^2 K^6 + \delta^4 K^4\right)W^2 \\
&- \left[K^{12} + 2\left(-\delta^2 + 3\sigma\rho^2\right)K^{10} + \delta^2\left(\delta^2 - 8\sigma\rho^2\right)K^8 + 2\sigma\rho^2\delta^4 K^6\right] = 0.
\end{aligned}
\tag{5}
$$

Focusing on the instability of low frequency disturbances ($K \to 0$), $c = W/K = O(1)$ is determined from $c^6 - 2\left(\delta^2 + 3\sigma\rho^2\right)c^4 + \delta^4 c^2 - 2\sigma\rho^2\delta^4 = 0$.

This is identical to the dispersion relation Equation (3) with a slight change in variable: $(i\Omega_0)^6 - 2\left(\delta^2 + 3\sigma\rho^2\right)(i\Omega_0)^4 + \delta^4(i\Omega_0)^2 - 2\delta^4\sigma\rho^2 = 0$, confirming again the relation between baseband modulation instability and rogue waves.

4.2. Computational Approach

The generation of localized modes resembling rogue waves starting from a chaotic field initial condition was studied in the literature [14]. Such modes can only be generated in parameter regimes with baseband modulation instability [42]. Here we demonstrate the emergence of rogue waves from a plane wave perturbed by one single mode instead of a random noise. More precisely, we consider the initial condition

$$
\begin{aligned}
u(t,0) &= [1 + 0.05 \exp(iKt)]u_0(t,0), \\
v(t,0) &= [1 + 0.05 \exp(iKt)]v_0(t,0), \\
w(t,0) &= [1 + 0.05 \exp(iKt)]w_0(t,0),
\end{aligned}
\tag{6}
$$

where u_0, v_0 and w_0 are the plane waves given in Equation (4). Equation (1) is numerically solved with a combination of pseudospectral method and a fourth-order Runge-Kutta scheme [43].

The result is illustrated with the typical case of $\sigma = 10$, $\varrho = 1$ and $\delta = 5$. The modulation instability gain spectra exhibit multiple bands due to the existence of multiple complex roots of Equation (5) (Figure 2). For small frequency K, patterns resembling rogue waves are observed (Figure 3, only patterns for u are shown, as profiles for the other two waveguides are similar). If there exists a pair of rogue wave modes for the same input parameters, the rogue wave with a higher baseband growth rate would dominate the other mode [39]. However, both modes here share the same baseband growth rate because they correspond to a pair of complex conjugate roots of Equation (3). Similar co-existence of rogue waves in a chaotic wave field was also reported earlier in the literature [40].

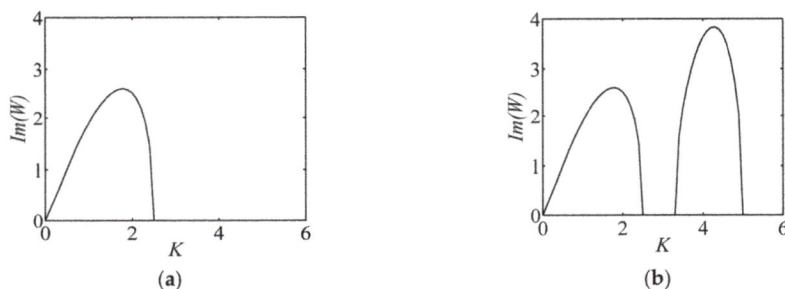

Figure 2. Multiple bands of the modulation instability gain spectra with $\sigma = 10$, $\varrho = 1$ and $\delta = 5$: (**a**) the first band; (**b**) the second band.

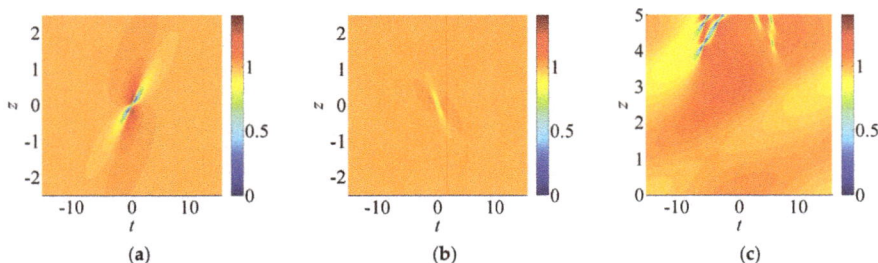

Figure 3. With $\sigma = 10$, $\varrho = 1$ and $\delta = 5$, (**a**) the *u*-component of the first rogue wave mode given by Equations (2) and (3); (**b**) the *u*-component of the second RW mode; (**c**) the simulated results with the initial condition given by Equation (6) and $K = 0.2$.

5. Asymmetric Breathers

Breathers can be generated through these simulations by starting with a disturbance of higher frequency in the unstable band. For typical values of $\sigma = 10$, $\varrho = 1$ and $\delta = 5$, there exist two pairs of complex conjugate roots for Equation (5) at the baseband and the instability growth rates are identical. Two breathers can be generated concurrently and superposition leads to an asymmetric breather (Figure 4). For most rogue waves studied in the literature, the local extrema are usually symmetric with respect to the main displacement and attain the same value. For an asymmetric rogue wave or breather [27,44], symmetry is broken, e.g., the four-petal arrangement is destroyed and one of the minimum points splits into two minima (Figure 4a).

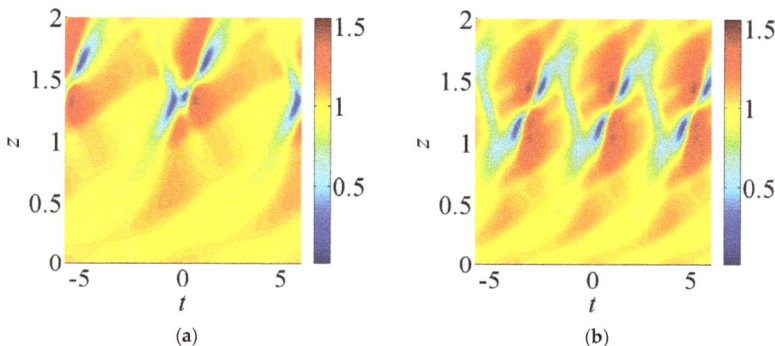

Figure 4. The *u*-component of the asymmetric breathers generated by simulation with the initial condition given by Equation (6) at (**a**) $K = 1$; (**b**) $K = 1.5$. In both cases, $\sigma = 10$, $\varrho = 1$ and $\delta = 5$.

6. Rogue Wave in Non-integrable Systems

In many physical applications, the coefficients for the self-phase modulation (SPM) and cross-phase modulation (XPM) are distinct from each other, analytical schemes will usually fail for such coupled nonlinear Schrödinger equations [17,18]. It will be instructive to apply the mechanism developed here for such 'non-integrable' equations. Based on the numerical solution, wave profile and amplification ratio of the rogue wave can be calculated.

We demonstrate the idea with a special case where the ratio of SPM to XPM is a constant:

$$
\begin{aligned}
iq_z + i\delta' q_t + q_{tt} - \sigma' \left(|q|^2 + \gamma|r|^2 + \gamma|s|^2 \right) q &= 0, \\
ir_z - i\delta' r_t + r_{tt} - \sigma' \left(\gamma|q|^2 + |r|^2 + \gamma|s|^2 \right) r &= 0, \\
is_z + s_{tt} - \sigma' \left(\gamma|q|^2 + \gamma|r|^2 + |s|^2 \right) s &= 0,
\end{aligned}
\tag{7}
$$

where γ is the ratio of XPM to SPM. Under the initial condition

$$
\begin{aligned}
q(t,0) &= [1 + 0.05\exp(iKt)]\rho, \\
r(t,0) &= [1 + 0.05\exp(iKt)]\rho, \\
s(t,0) &= [1 + 0.05\exp(iKt)]\rho,
\end{aligned}
\tag{8}
$$

where K is a low frequency of the wavy disturbance, rogue wave modes are generated (Figures 5 and 6). Four-petal RW-like waveforms can be observed with $\gamma = 2$ (Figure 5). As compared to Figure 3, the minima get closer and the saddle point attains a smaller intensity. Dark RW-like patterns are observed for the case where XPM is half of SPM (Figure 6). Apparently, only one type of RW waveform can be observed.

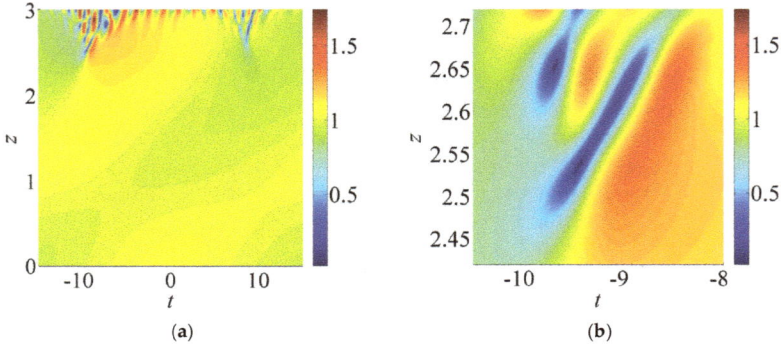

Figure 5. (a) The *q*-component of the simulated RW mode for Equation (7) with $\sigma' = 10$, $\varrho = 1$, $\delta' = 5$ and $\gamma = 2$; (b) an enlarged view of a RW-like structure in (a). The frequency *K* of the single mode disturbance is 0.2.

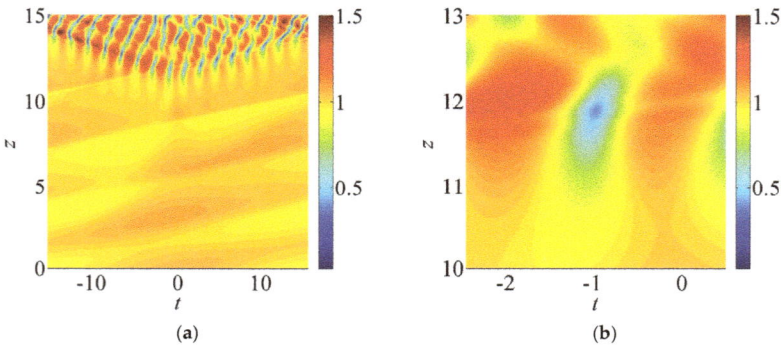

Figure 6. (a) The *q*-component of the simulated RW mode for Equation (7) with $\sigma' = 10$, $\varrho = 1$, $\delta' = 5$ and $\gamma = 0.5$; (b) an enlarged view of a RW-like structure in (a). The frequency *K* of the single mode disturbance is 0.2.

In practice, many evolution equations governing dynamical systems for laboratory and engineering settings are not integrable. Several approaches were demonstrated to be feasible as approximations or the estimations of properties of rogue waves in such non-integrable systems. For instance, analytical rogue wave solutions of the NLS equation with periodic modulated coefficients were taken as initial conditions in the numerical approximations of rogue waves for the non-integrable case with constant nonlinearity coefficient and periodic dispersion coefficient [45]. Furthermore, rogue waves from well-studied equations like the NLS equation can be utilized to study rogue waves in less thoroughly studied physical systems. With a suitable physical assumption, optical quadratic solution can be related to the solution of the NLS equation through the second-harmonic asymptotic expansion and the method of repeated substitution [46]. Hence, useful approximations of rogue waves in a quadratic medium can be obtained. It was shown that breathers and RW-like entities emerge from simulations with random initial conditions, which serve as an alternative methodology to examine rogue waves in a general nonlinear system.

In this work, an alternative method is proposed which is based on the generating mechanism of rogue waves. The method is independent of the exact rogue wave solution of the integrable system. Hence, this scheme should hopefully be quite widely applicable to general systems, and should not be restricted to the "quasi-integrable" regime where the coefficients are close to the "integrable" case. As compared to the detection of rogue waves in a chaotic field, the baseband disturbance can isolate the rogue-wave-like structures among entities such as breathers. A comprehensive study on rogue waves in a general multi-component system of coupled nonlinear Schrödinger equations will be carried out in the future. This numerical approach initiated from the insight gained from rational solutions should complement the limitation of analytical methods in the study of rogue waves.

7. Discussions and Conclusions

It was known that a multi-component Manakov system can effectively model wave propagation in a multicore optical fiber [22]. Besides temporal waveguides, such Manakov systems are relevant in other settings in optical physics too. For spatial solitons, diffraction will play the role of group velocity dispersion, and continuous variations of diffraction, nonlinearity, and gain/loss might lead to novel rogue wave patterns [47]. Similarly, Manakov soliton can arise for biased guest-host photorefractive polymer too [48]. Furthermore, reductive perturbation techniques can be employed to establish Manakov equations as approximations for propagation of electromagnetic fields along isotropic chiral metamaterials [49].

In this work, a theoretical study is performed to understand the increasing complexity of the Manakov models with larger number of components. It will be worthwhile if analytical and computational predictions here can be verified in experiments in the future. More precisely, rogue wave solutions for the defocusing three-component Manakov system with group velocity mismatch are derived by the Hirota bilinear method. The nonlinear coupling effect is highlighted, namely, the extension of existence regime of rogue waves and the enhancement of amplitude. The main focus of the work is to demonstrate the generation of rogue wave from baseband disturbance.

Recently, the onset of baseband modulation instability has been proven to be equivalent to the existence condition of rogue waves in several systems. In this work, the role of low frequency disturbance in the formation of rogue waves is further consolidated through numerical simulations. By perturbing the plane wave solution by a single mode disturbance with a small frequency, rogue wave modes with configuration similar to the analytical rogue wave solutions can be generated. This idea can be generalized to approximate rogue wave modes in non-integrable systems where most analytical methods fail. The proposed numerical methodology can greatly enrich our knowledge of rogue waves in such systems. Detailed investigations on evolution of rogue waves in general systems of coupled nonlinear Schrödinger equations without any restriction on coefficients would be conducted in the future. Moreover, modified NLS equation or the corresponding systems can be studied [50].

Secondly, the single mode wavy disturbance to a plane wave can generate breathers as well. It is well-known that the formation of breathers is closely related to MI. Depending on the MI spectrum, both symmetric and asymmetric breathers can be generated from a single mode wavy disturbance. For the case where there are multiple unstable bands, the unstable mode can generate multiple breathers which superimpose to form an asymmetric breather. On the other hand, a conventional symmetric breather is generated if there exists only one unstable band.

In conclusions, rogue wave and breather formation are closely related to the nature of the MI spectrum. Through the study of the MI spectrum supplemented with numerical simulations, more intriguing wave dynamics of general non-integrable systems can be revealed in the future. Theoretically, three-component or multi-component systems with variable coefficients and external potential can be further investigated through a similarity transformation [13]. Recently, nonlocal equations have been widely studied due to their *PT* symmetric property [51]. Many nonlocal evolution equations display remarkable similarities in comparison with intensively studied classical ones, e.g., the nonlocal NLS equation

$$iA_z + A_{tt} + AA^*(-t, z)A = 0, \tag{9}$$

which can be analyzed by direct and inverse scattering techniques [51]. Indeed, the nonlocal NLS model (Equation (9)) also possesses an infinite number of conservation laws. Thus total self-induced potential $A(t,z)A^*(-t,z)$ over the entire spatial domain is conserved, but not the usual intensity AA^*.

This feature may have implications on the size of the "elevation" and "depression" regions of the rogue waves [52]. Naturally the structure of the scattering problem and Painlevé property are slightly different from those of the classical NLS equation. These features and other extensions, e.g., coupled waveguides, higher order dispersion, and discrete models will likely constitute fruitful paths of research in the future.

Acknowledgments: Partial financial support for this project has been provided by the Research Grants Council contracts HKU711713E and HKU17200815.

Author Contributions: Kwok Wing Chow initiated the work. Hiu Ning Chan performed the analysis and simulations. Both authors participated in the interpretation of the results and the writing of the paper.

Conflicts of Interest: The authors declare no conflict of interest.

Appendix A

To analyze the shape of the rogue wave solution, consider the normalized intensity function of w defined by $F = \frac{|w|^2}{\rho^2}$. At the stationary point (0,0), the second derivatives are given by $F_{tt}|_{(0,0)} = \frac{48a^4(b^2-a^2)}{(a^2+b^2)^2}$ and $\left(F_{zz}F_{tt} - F_{zt}^2\right)|_{(0,0)} = \frac{256a^{10}(3a^2-b^2)(a^2-3b^2)}{(a^2+b^2)^4}$. Similarly, for the evolution of u and v, b is replaced by $b - \delta$ and $b + \delta$ respectively.

From the real part and imaginary part of the dispersion relation Equation (3), note that

$$\left[2\left(\delta^2 + 3\sigma\rho^2\right) + 3c\right]d^2 = c^3 + 2\left(\delta^2 + 3\sigma\rho^2\right)c^2 + \delta^4 c + 2\delta^4\sigma\rho^2 \text{ and}$$
$$d^2 = 3c^2 + 4\left(\delta^2 + 3\sigma\rho^2\right)c + \delta^4, \tag{A1}$$

where $c = a^2 - b^2$ and $d = 2ab$. It can be deduced that c is negative. Otherwise equating d^2 in both equations will result in a cubic equation in c, $8c^3 + 16\left(\delta^2 + 3\sigma\rho^2\right)c^2 + \left[2\delta^4 + 8\left(\delta^2 + 3\sigma\rho^2\right)^2\right]c + 2\delta^6 + 4\delta^4\sigma\rho^2 = 0$, which leads to a contradiction since the polynomial is positive if c is positive.

Existence of EDRW with non-zero lower bound

Without loss of generality, consider the case when both a and b are positive. If $a = b$, then it is trivial from the second order derivative test that $|v|$ attains minimum at (0,0). The case when $a > b$ will lead to a contradiction, so it remains to consider the case when $a < b$. From Equation (A1), it can be shown that $\delta^2 - 2c > \sqrt{c^2 + d^2}$ which implies that $(b + \delta)^2 - 3a^2 > 2b\delta > 0$. From the second order derivatives, it can be concluded that v is an EDRW when $\Omega_0 = a + ib$. On the other hand, u is an EDRW in the solution corresponding to the alternative root, $\Omega_0 = a - ib$. Hence, there must be a component with the configuration of an EDRW.

Moreover, the amplitude of the dark rogue wave has a lower bound of about 0.4ρ. By considering the coefficient of the quadratic term in the dispersion relation, it can be shown that $\delta > \sqrt{2}a$. With $b > a$ and $(b + \delta)^2 - 3a^2 > 2b\delta$, it is easy to obtain the bound $|v| > \frac{1}{1+\sqrt{2}}\rho \approx 0.4142\rho$. In particular, v cannot be a black rogue wave with zero minimum intensity.

Non-existence of eye-shaped bright rogue wave

Since $a^2 < b^2$, (0,0) is either a minimum point or a saddle point. Hence, w cannot be an eye-shaped bright rogue wave. Similarly, u and v cannot be an eye-shaped bright rogue wave. When $a^2 > (b + \delta)^2$, then $a^2 > b^2$ and from the above analysis $3b^2 > a^2$. It is then obvious that $3(b + \delta)^2 > a^2$ and from the sign of the Hessian, it can be concluded that (0,0) is a saddle point for $|v|$ and v cannot be an eye-shaped bright rogue wave. Furthermore, by the symmetry between u and v, it is trivial that u cannot possess such configuration either.

References

1. Kharif, C.; Pelinovsky, E.; Slunyaev, A. *Rogue Waves in the Ocean*, 1st ed.; Springer: Berlin, Germany, 2009.
2. Dysthe, K.B.; Krogstad, H.E.; Müller, P. Oceanic rogue waves. *Ann. Rev. Fluid Mech.* **2008**, *40*, 287–310. [CrossRef]

3. Adcock, T.A.A.; Taylor, P.H. The physics of anomalous ('rogue') ocean waves. *Rep. Prog. Phys.* **2014**, *77*, 105901. [CrossRef] [PubMed]
4. Cousins, W.; Sapsis, T.P. Quantification and prediction of extreme events in a one-dimensional nonlinear dispersive wave model. *Physica D* **2014**, *280–281*, 48–58. [CrossRef]
5. Onorato, M.; Residori, S.; Bortolozzo, U.; Montina, A.; Arecchi, F.T. Rogue waves and their generating mechanisms in different physical contexts. *Phys. Rep.* **2013**, *528*, 47–89. [CrossRef]
6. Shemer, L.; Alperovich, L. Peregrine breather revisited. *Phys. Fluids* **2013**, *25*, 051701. [CrossRef]
7. Grimshaw, R.H.J.; Tovbis, A. Rogue waves: analytical predictions. *Proc. R. Soc. A* **2013**, *469*, 20130094. [CrossRef]
8. Ablowitz, M.J.; Horikis, T.P. Interacting nonlinear wave envelopes and rogue wave formation in deep water. *Phys. Fluids* **2012**, *27*, 012107. [CrossRef]
9. Onorato, M.; Residori, S.; Baronio, F. *Rogue and Shock Waves in Nonlinear Dispersive Media*, 1st ed.; Springer: Basel, Switzerland, 2016.
10. Solli, D.R.; Ropers, C.; Koonath, P.; Jalali, B. Optical rogue waves. *Nature* **2007**, *450*, 1054–1058. [CrossRef] [PubMed]
11. Dudley, J.M.; Dias, F.; Erkintalo, M.; Genty, G. Instabilities, breathers and rogue waves in optics. *Nat. Photonics* **2014**, *8*, 755–764. [CrossRef]
12. Akhmediev, N.; Kibler, B.; Baronio, F.; Belić, M.; Zhong, W.-P.; Zhang, Y.; Chang, W.; Soto-Crespo, J.M.; Vouzas, P.; Grelu, P.; et al. Roadmap on optical rogue waves and extreme events. *J. Opt.* **2016**, *18*, 063001. [CrossRef]
13. Zhong, W.-P.; Belić, M.; Malomed, B.A. Rogue waves in a two-component Manakov system with variable coefficients and an external potential. *Phys. Rev. E* **2015**, *92*, 053201. [CrossRef] [PubMed]
14. Akhmediev, N.; Ankiewicz, A.; Soto-Crespo, J.M. Rogue waves and rational solutions of the nonlinear Schrödinger equation. *Phys. Rev. E* **2009**, *80*, 026601. [CrossRef] [PubMed]
15. Shrira, V.I.; Geogjaev, V.V. What makes the Peregrine soliton so special as a prototype of freak waves? *J. Eng. Math.* **2010**, *67*, 11–22. [CrossRef]
16. Chan, H.N.; Chow, K.W.; Kedziora, D.J.; Grimshaw, R.H.J.; Ding, E. Rogue wave modes for a derivative nonlinear Schrödinger model. *Phys. Rev. E* **2014**, *89*, 032914. [CrossRef] [PubMed]
17. Dhar, A.K.; Das, K.P. Fourth-order nonlinear evolution equation for two Stokes wave trains in deep water. *Phys. Fluids* **1991**, *3*, 3021–3026. [CrossRef]
18. Gramstad, O.; Trulsen, K. Fourth-order coupled nonlinear Schrödinger equations for gravity waves on deep water. *Phys. Fluids* **2011**, *23*, 062102. [CrossRef]
19. Chen, Z.; Segev, M.; Coskun, T.H.; Christodoulides, D.N.; Kivshar, Y.S. Coupled photorefractive spatial-soliton pairs. *J. Opt. Soc. Am. B* **1997**, *14*, 3066–3077. [CrossRef]
20. Agrawal, G.P. *Nonlinear Fiber Optics*, 4th ed.; Academic Press: New York, NY, USA, 2006.
21. Mecozzi, A.; Antonelli, C.; Shtaif, M. Nonlinear propagation in multi-mode fibers in the string coupling regime. *Opt. Express* **2012**, *20*, 11673–11678. [CrossRef] [PubMed]
22. Mumtaz, S.; Essiambre, R.; Agrawal, G.P. Nonlinear propagation in multimode and multicore fibers: generalization of the Manakov equations. *J. Lightwave Technol.* **2013**, *31*, 398–406. [CrossRef]
23. Guasoni, M. Generalized modulational instability in multimode fibers: wideband multimode parametric amplification. *Phys. Rev. A* **2015**, *92*, 033849. [CrossRef]
24. Wang, D.S.; Shi, Y.R.; Chow, K.W.; Yu, Z.X.; Li, X.G. Matter-wave solitons in a spin-1 Bose-Einstein condensate with time-modulated external potential and scattering lengths. *Eur. Phys. J. D* **2013**, *67*, 242. [CrossRef]
25. Baronio, F.; Degasperis, A.; Conforti, M.; Wabnitz, S. Solutions of the vector nonlinear Schrödinger equations: evidence for deterministic rogue waves. *Phys. Rev. Lett.* **2012**, *109*, 044102. [CrossRef] [PubMed]
26. Ling, L.; Guo, B.; Zhao, L.C. High-order rogue waves in vector nonlinear Schrödinger equations. *Phys. Rev. E* **2014**, *89*, 041201(R). [CrossRef] [PubMed]
27. He, J.; Guo, L.; Zhang, Y.; Chabchoub, A. Theoretical and experimental evidence of non-symmetric doubly localized rogue waves. *Proc. R. Soc. A* **2014**, *470*, 20140318. [CrossRef] [PubMed]
28. Degasperis, A.; Lombardo, S. Rational solitons of wave resonant-interaction models. *Phys. Rev. E* **2013**, *88*, 052914. [CrossRef] [PubMed]
29. Baronio, F.; Conforti, M.; Degasperis, A.; Lombardo, S.; Onorato, M.; Wabnitz, S. Vector rogue waves and baseband modulation instability in the defocusing regime. *Phys. Rev. Lett.* **2014**, *113*, 034101. [CrossRef] [PubMed]

30. Chen, S.; Soto-Crespo, J.M.; Grelu, P. Dark three-sister rogue waves in normally dispersive optical fibers with random birefringence. *Opt. Express* **2014**, *22*, 27632–27642. [CrossRef] [PubMed]

31. Li, J.H.; Chan, H.N.; Chiang, K.S.; Chow, K.W. Breathers and 'black' rogue waves of coupled nonlinear Schrödinger equations with dispersion and nonlinearity of opposite signs. *Commun. Nonlinear Sci. Numer. Simulat.* **2015**, *28*, 28–38. [CrossRef]

32. Frisquet, B.; Kibler, B.; Fatome, J.; Morin, P.; Baronio, F.; Conforti, M.; Millot, G.; Wabnitz, S. Polarization modulation instability in a Manakov fiber system. *Phys. Rev. A* **2015**, *92*, 053854. [CrossRef]

33. Frisquet, B.; Kibler, B.; Morin, P.; Baronio, F.; Conforti, M.; Millot, G.; Wabnitz, S. Optical dark rogue wave. *Sci. Rep.* **2016**, *6*, 20785. [CrossRef] [PubMed]

34. Zhao, L.C.; Liu, J. Rogue-wave solutions of a three-component coupled nonlinear Schrödinger equation. *Phys. Rev. E* **2013**, *87*, 013201. [CrossRef] [PubMed]

35. Liu, C.; Yang, Z.-Y.; Zhao, L.-C.; Yang, W.-L. Vector breathers and the inelastic interaction in a three-mode nonlinear optical fiber. *Phys. Rev. A* **2014**, *89*, 055803. [CrossRef]

36. Xu, T.; Chen, Y. Localized waves in three-component coupled nonlinear Schrödinger equation. *Chin. Phys. B* **2016**, *25*, 090201. [CrossRef]

37. Hirota, R. *The Direct Method in Soliton Theory*, 1st ed.; Cambridge University Press: Cambridge, UK, 2004.

38. Chan, H.N.; Malomed, B.A.; Chow, K.W.; Ding, E. Rogue waves for a system of coupled derivative nonlinear Schrödinger equations. *Phys. Rev. E* **2016**, *93*, 012217. [CrossRef] [PubMed]

39. Chan, H.N.; Chow, K.W. Rogue waves for an alternative system of coupled Hirota equations: structural robustness and modulation instabilities. *Stud. Appl. Math.* **2017**. [CrossRef]

40. Chen, S.; Soto-Crespo, J.M.; Grelu, P. Coexisting rogue waves within the (2+1)-component long-wave-short-wave resonance. *Phys. Rev. E* **2014**, *90*, 033203. [CrossRef] [PubMed]

41. Chan, H.N.; Ding, E.; Kedziora, D.J.; Grimshaw, R.H.J.; Chow, K.W. Rogue waves for a long wave-short wave resonance model with multiple short waves. *Nonlinear Dyn.* **2016**, *85*, 2827–2841. [CrossRef]

42. Baronio, F.; Chen, S.; Grelu, P.; Wabnitz, S.; Conforti, M. Baseband modulation instability as the origin of rogue waves. *Phys. Rev. A* **2015**, *91*, 033804. [CrossRef]

43. Yang, J. *Nonlinear Waves in Integrable and Nonintegrable Systems*, 1st ed.; SIAM: Philadelphia, PA, USA, 2010.

44. Wang, L.; Zhang, L.-L.; Zhu, Y.-J.; Qi, F.-H.; Wang, P.; Guo, R.; Li, M. Modulational instability, nonautonomous characteristics and semirational solutions for the coupled nonlinear Schrödinger equations in inhomogeneous fibers. *Commun. Nonlinear Sci. Numer. Simul.* **2016**, *40*, 216–237. [CrossRef]

45. Tiofack, C.G.L.; Coulibaly, S.; Taki, M.; De Bièvre, S.; Dujardin, G. Comb generation using multiple compression points of Peregrine rogue waves in periodically modulated nonlinear Schrödinger equations. *Phys. Rev. A* **2015**, *92*, 043837. [CrossRef]

46. Baronio, F. Akhmediev breathers and Peregrine solitary waves in a quadratic medium. *Opt. Lett.* **2017**, *42*, 1756–1759. [CrossRef] [PubMed]

47. Manikandan, K.; Senthilvelan, M.; Kraenkel, R.A. On the characterization of vector rogue waves in two-dimensional two coupled nonlinear Schrödinger equations with distributed coefficients. *Eur. Phys. J. B* **2016**, *89*, 218. [CrossRef]

48. Tan, Z.; Tian, B.; Jiang, Y.; Wang, P.; Li, M. Dynamics of the Manakov solitons in biased guest-host photorefractive polymer. *Commun. Theor. Phys.* **2013**, *60*, 150–158. [CrossRef]

49. Tsitsas, N.L.; Lakhtakia, A.; Frantzeskakis, D.J. Vector solitons in nonlinear isotropic chiral metamaterials. *J. Phys. A Math. Theor.* **2011**, *44*, 435203. [CrossRef]

50. Boscolo, S.; Peng, J.; Finot, C. Design and applications of in-cavity pulse shaping by spectral sculpturing in mode-locked fibre lasers. *Appl. Sci.* **2015**, *5*, 1379–1398. [CrossRef]

51. Ablowitz, M.J.; Musslimani, Z.H. Integrable nonlocal nonlinear Schrödinger equation. *Phys. Rev. Lett.* **2013**, *110*, 064105. [CrossRef] [PubMed]

52. Gupta, S.K.; Sarma, A.K. Peregrine rogue wave dynamics in the continuous nonlinear Schrödinger system with parity-time symmetric Kerr nonlinearity. *Commun. Nonlinear Sci. Numer. Simulat.* **2016**, *36*, 141–147. [CrossRef]

![applied sciences logo] *applied sciences*

MDPI

Article

Waveguiding Light into Silicon Oxycarbide

Faisal Ahmed Memon [1,2,*], Francesco Morichetti [1] and Andrea Melloni [1]

[1] Dipartimento di Elettronica, Informazione e Bioingegneria (DEIB), Politecnico di Milano, via Ponzio 34/5, 20133 Milan, Italy; francesco.morichetti@polimi.it (F.M.); andrea.melloni@polimi.it (A.M.)
[2] Department of Telecommunications Engineering, Mehran University of Engineering & Technology, Jamshoro 76062, Sindh, Pakistan
* Correspondence: faisalahmed.memon@polimi.it or faisal.memon@faculty.muet.edu.pk; Tel.: +39-2-2399-8978

Academic Editor: Boris Malomed
Received: 1 May 2017; Accepted: 25 May 2017; Published: 30 May 2017

Featured Application: Silicon oxycarbide is demonstrated to be a potential photonic platform for integrated optics applications.

Abstract: In this work, we demonstrate the fabrication of single mode optical waveguides in silicon oxycarbide (SiOC) with a high refractive index n = 1.578 on silica (SiO$_2$), exhibiting an index contrast of Δn = 8.2%. Silicon oxycarbide layers were deposited by reactive RF magnetron sputtering of a SiC target in a controlled process of argon and oxygen gases. The optical properties of SiOC film were measured with spectroscopic ellipsometry in the near-infrared range and the acquired refractive indices of the film exhibit anisotropy on the order of 10^{-2}. The structure of the SiOC films is investigated with atomic force microscopy (AFM) and scanning electron microscopy (SEM). The channel waveguides in SiOC are buried in SiO$_2$ (n = 1.444) and defined with UV photolithography and reactive ion etching techniques. Propagation losses of about 4 dB/cm for both TE and TM polarizations at telecommunication wavelength 1550 nm are estimated with cut-back technique. Results indicate the potential of silicon oxycarbide for guided wave applications.

Keywords: silicon oxycarbide; optical waveguides; optical materials; guided waves; integrated optics

1. Introduction

Materials offering refractive index tunability and low absorption are extremely useful for realizing optical integrated devices. In silicon nitride (SiN) compounds, like silicon oxynitride and silicon rich nitride, the refractive index can be tuned over a wide range, potentially from 1.45 to about 2, yet at the expense of higher losses [1,2]. Silicon oxycarbide (SiOC) is a versatile material that has been used in a variety of applications including Li-ion batteries [3], photoluminescence [4], electroluminescence [5], and low-k interlayer dielectric [6,7]. It has been demonstrated that the change in composition of SiOC compound can lead to a change in the refractive index n over a wide range from ~1.45 to ~3.0 under pre- and post-deposition conditions [8,9]. Structural and chemical properties of SiOC films have been extensively reported in the literature [10–13]. Recently, we have demonstrated the synthesis of SiOC films with reactive RF magnetron sputtering and showed that, under different deposition conditions, the refractive index n can be varied from 1.41 to 1.85 at wavelength λ = 1550 nm, while the extinction coefficient k is below 10^{-4} above λ = 1000 nm [14]. The low extinction coefficient k of SiOC films indicates the suitability of this material for fabrication of optical waveguides. However, to the best of our knowledge, SiOC has never been exploited as a core material for guided wave applications.

In this paper, we demonstrate the fabrication of silicon oxycarbide optical waveguides with reactive RF magnetron sputtering. Reactive sputtering is a well-established technique, relatively simple and cost effective to set up compared to conventional chemical vapor deposition (CVD). First,

SiOC films are deposited by sputtering the SiC target in the presence of argon (Ar) and oxygen (O_2) gases in a controlled process at room temperature. The prepared films are characterized with variable angle spectroscopic ellipsometry over near-infrared wavelength range. The deposited SiOC film exhibits anisotropy on the order of 10^{-2} between in-plane and out-of-plane refractive indices as acquired from the ellipsometric data. To quantify the root mean square (RMS) roughness of the deposited SiOC film on SiO_2/Si wafer, atomic force microscopy is used. RMS roughness is an important parameter that affects the performance of optical waveguides and contributes to the scattering of light. The sputtered SiOC films are then used as core layer to fabricate channel waveguides using UV photolithography and reactive ion etching (RIE) techniques. The SiOC channel waveguides are buried between two layers of SiO_2 (n = 1.444). The optical waveguides with different widths are characterized at an optical communication wavelength (1550 nm) with different lengths. Propagation losses of about 4 dB/cm for both TE and TM polarizations have been achieved indicating the suitability of sputtered SiOC for integrated optics applications.

The paper is organized as follows: Section 2 explains deposition and characterization techniques used to grow and study the properties of SiOC films (Section 2.1), and the fabrication process and the measurement setup for SiOC optical waveguides (Section 2.2). Section 3 discusses results on the characterization of SiOC films (Section 3.1) and optical waveguides (Section 3.2). Finally, conclusions are presented in Section 4.

2. Experimental Details

2.1. Layer Fabrication and Characterizations Techiques

SiOC films were deposited with reactive RF magnetron sputtering of a silicon carbide (SiC) target in presence of argon (Ar) and oxygen (O_2) gases at room temperature. The SiOC films are deposited on a 4 inches Si (100) wafer with 8 μm SiO_2 buffer layer that isolates SiOC core from high index Si substrate. To achieve the desired refractive index, an optimized sputtering recipe with the following parameters is used: rf power = 350 W, Ar flow = 60 sccm, and O_2 flow = 2.6 sccm. The sputtering was run for around three minutes to obtain a SiOC layer with a thickness of 400 nm. Further details on deposition process, obtained deposition rate of sputtering, and optical parameters of SiOC films are provided in [14].

The surface roughness of the deposited SiOC layer was determined with atomic force microscope (5600LS AFM system, Keysight, CA, USA) at five different places on the four-inch wafer. The SiOC layer was probed in contact mode over an area of 10 by 10 μm^2 at each location. Cross-sectional images of SiOC film and waveguides were obtained with scanning electron microscope (SEM LEO 1525, One Zeiss Drive, NY, USA) to investigate the morphology and verify the profile and thickness. Since SiOC is a dielectric material, a thin layer of platinum ($\approx$2 nm) was deposited on the SiOC films to avoid electrons charging effect. The high-resolution images were captured by accelerating electrons at a voltage of 20 kV and the signal was collected from in-lens detector while keeping a distance of around 6 mm between the SiOC sample and the electron gun.

The optical properties of the deposited SiOC film on SiO_2/Si wafer are measured with variable angle spectroscopic ellipsometer (VASE J.A. Woollam Inc., Lincoln, NE, USA). The ellipsometric data were acquired at multiple incidence angles (65°, 70°, and 75°) in the spectral range from 1200 to 1600 nm with a step size of 10 nm between two wavelengths. The ellipsometry measures the change in polarization state of the light reflected from the surface of the film/substrate under investigation. The measured quantities are psi (ψ) and del (Δ) which are angles defining the ratio of Fresnel reflection coefficients R_p and R_s for parallel (p) and normal (s) polarized light as [15],

$$\frac{R_p}{R_s} = \tan \psi \, exp(i\Delta) \tag{1}$$

Ellipsomteric measurements were performed taking in to account the possible depolarization originated by thickness non-uniformity and surface roughness. Since ellipsometry is an indirect technique, an optical model is necessary to obtain the optical constants (n, k) of film/substrate. A four-layer film/substrate stack including surface-roughness, SiOC film, SiO$_2$ layer, and Si substrate as shown in Figure 1a was used as a model to extract the optical constants n and k of the SiOC film. The surface roughness layer is based on Bruggeman EMA [16,17] that is AFM equivalent and provides peak-to-valley roughness. The SiOC film index was represented by Cauchy dispersion model,

$$n(\lambda) = A + \frac{B}{\lambda^2} \ (\lambda \text{ in } \mu m) \tag{2}$$

where A and B constants are the fit parameters in modelling the measured ellipsometric ψ and Δ data. The SiOC film was modeled as anisotropic layer by introducing uniaxial anisotropy in the Cauchy model. The optical axis was considered perpendicular to the surface as illustrated in Figure 1b. From the frame of reference of ellipsometry, ordinary axis n$_o$ corresponds to transverse electric (TE) polarization and extra-ordinary axis n$_e$ to transverse magnetic (TM) polarization as shown in Figure 1b. The data was fitted by the commercial WVASE32 software [17] using the optical model given in Figure 1a. The mean squared error (MSE) is the basis for validating an optical model and determining the quality of the match between the experimental and calculated data. WVASE32 software [17] uses a maximum likelihood estimator with MSE based on the Lavenberg-Marquardt regression algorithm.

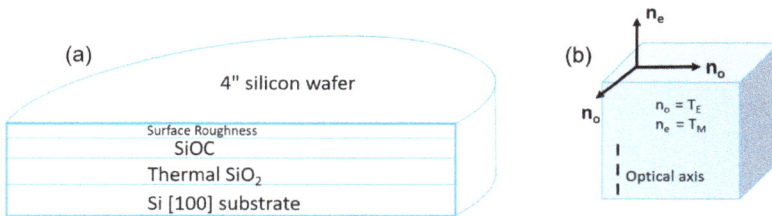

Figure 1. (**a**) Four-layer optical model employed to retrieve optical constants of SiOC film from ellipsometric data—not to scale, (**b**) schematic of SiOC anisotropic layer with optical axis perpendicular to surface and ordinary n$_o$ and extraordinary n$_e$ refractive indices.

2.2. Waveguide Fabrication and Characterizations Techniques

Before fabrication, the geometry of the SiOC waveguides was designed to obtain a single mode operation by using a commercial electromagnetic tool based on Finite Element Method (FEM). The optical waveguides were then fabricated with UV photolithography and reactive ion etching (RIE) processes as follows. To define waveguide patterns, a photoresist AZ 5214 E was spin coated on the SiOC layer and soft baked on a hot plate; then the photoresist was exposed in hard contact mode with photo mask using I-line 365 nm mercury arc lamp emitting a power of 1000 W. The photo mask includes a set of waveguides with different widths. A controlled RIE process of CHF$_3$ and O$_2$ gases mixture was used to etch channel waveguides in SiOC layer with an etch rate of 20 nm/min. The realized channel waveguides were covered with 8 µm thick SiO$_2$ (n = 1.444) deposited by a plasma enhanced chemical vapor deposition (PECVD) process (STS CVD tool) to yield symmetric optical modes.

The optical waveguides were characterized on the optical bench through butt-coupling with small-core fibers as shown in Figure 2. The schematic diagram of the waveguide measurement setup is given in Figure 2a. The light signal was coupled from a laser source operating around telecommunication wavelength λ = 1550 nm. The SiOC waveguides were aligned with small core fibers having mode field diameter MFD $\approx$ 3.6 µm using index matching oil to reduce the gap between the fiber and chip facet. The light at the output facet of SiOC chip was collected and read with optical power meter. The polarization state (TE and TM) of the light wave at the input facet of the waveguide was selected with a fiber polarization controller. The real optical waveguide measurement setup is shown in Figure 2b where a SiOC chip is aligned with small core fibers.

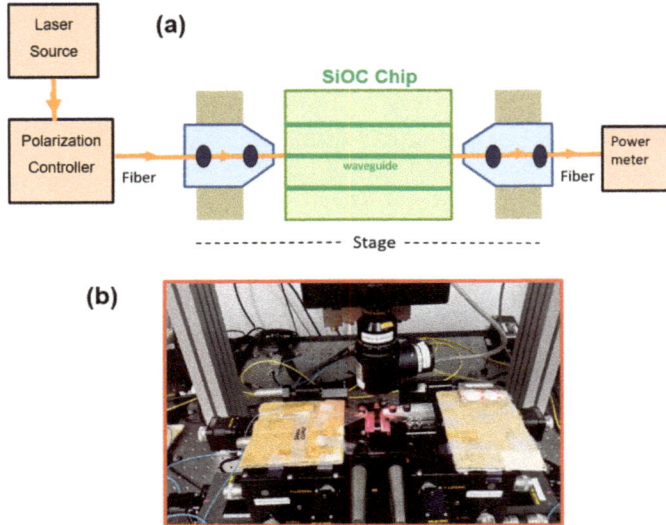

Figure 2. (a) Schematic of optical bench setup used for SiOC waveguides characterizations, (b) the measurement setup showing alignment of SiOC chip with small core optical fibers.

3. Results and Discussion

3.1. SiOC Layer Properties

SEM cross-section image of the SiOC film deposited on SiO_2/Si wafer under the optimized sputter process is displayed in Figure 3. The thickness of the SiOC layer is determined to be 400 nm. The deposited SiOC film shows dense columnar structure with pillars perpendicular to the plane of film. The columnar structure of the film provides a basis for anisotropy investigation and is taken into consideration during the optical characterization of SiOC film. The columnar structure is a typical feature in sputtered SiOC films that were deposited under different conditions, however the films were amorphous to XRD [14].

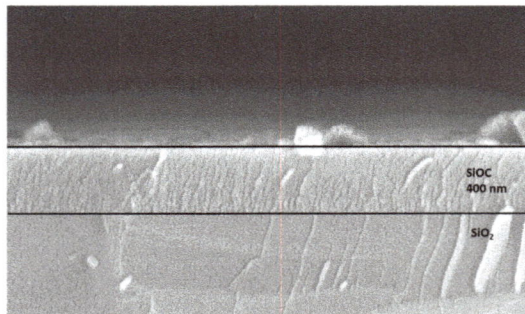

Figure 3. SEM cross-section image of SiOC deposited layer on SiO_2/Si wafer.

Figure 4a shows 3D topographical image of the SiOC film scanned with AFM in contact mode over an area of 100 μm². The scans were performed at five different locations on the four-inch wafer and the rms roughness is estimated to be 0.88 nm, sufficiently smooth for the realization of optical waveguides. In Figure 4b, 1D measured profile of the AFM image is given showing peak-to-valley roughness that is in-line with surface roughness estimated with ellipsometry.

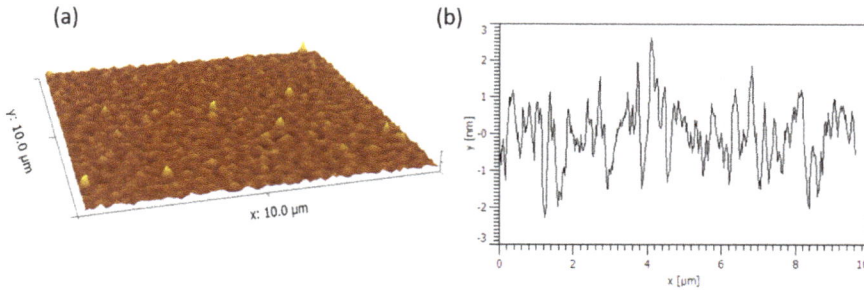

Figure 4. (**a**) 3D AFM topography scan of SiOC film surface (color scale from -5 nm to + 5 nm); (**b**) 1D measured profile of AFM image: a peak-to-valley roughness of about 2 nm is observed that agrees well with surface roughness estimated with spectroscopic ellipsometry.

The optical properties of the deposited SiOC film were characterized with variable angle spectroscopic ellipsometry in the infrared wavelength region from 1200 nm to 1600 nm. The measurement angles were 65°, 70°, and 75° which are around the Brewster's angle of Si substrate. Prior to SiOC film, the SiO_2/Si wafer was characterized with ellipsometry to obtain the thickness and accurate optical constants of SiO_2. The refractive index of SiO_2 was determined as 1.444 at $\lambda = 1550$ nm. The tabulated optical constants of crystalline Si substrate were used from the WVASE32 database [17]. Since ellipsometry is sensitive to film roughness conditions, surface roughness (SR) layer [16] was added to simulate roughness on SiOC film which improved fit and MSE. The SR layer that is the ellipsometry equivalent of AFM is estimated around 2 nm and agrees well with the AFM data. The ellispometric ψ and Δ data acquired at different angles and their respective model curves are given in Figure 5.

The four-layer anisotropic optical model discussed in Section 2 was used to fit the experimental data (green curves) and the calculated model data (red curves) matches well over the entire spectral range. Material anisotropy is expected from the columnar structure present in the sputtered SiOC film as shown in SEM image (see Figure 3). Our measurements indicate that the out-of-plane refractive index (n_e) is higher than the in-plane index (n_o) by 0.02438. These results agree with the studies reporting anisotropy in different material thin films with columnar structure where the in-plane refractive index is smaller than out-of-plane refractive index [18–20]. Therefore, the origin of the anisotropy is attributed to the columnar structure oriented normal to the plane of SiOC film. The anisotropy exhibited by SiOC film is on the order of 10^{-2}, the optical waveguides are sensitive to even this level of anisotropy because the two polarizations (TE and TM) will see a different index of refraction and propagate with different effective indices.

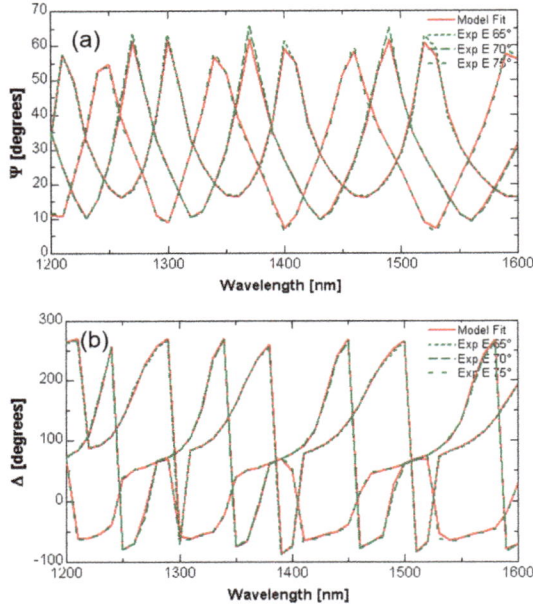

Figure 5. Ellipsometric (**a**) psi (ψ) and (**b**) del (Δ) data acquired at three angles of measurements (65°, 70°, and 75°) and their respective model curves (solid red lines) calculated with a four-layer film/substrate optical stack.

The optical constants (n, k) of the deposited SiOC film are shown in Figure 6. The extinction coefficient k is less than 10^{-4} (this value being the minimum detectable value for the ellipsometer) in agreement with previous observations [14] and is not visible in Figure 6. The ordinary axis refractive index n_o is determined to be 1.554 while it is 1.578 for extra-ordinary axis n_e at the wavelength λ = 1550 nm. The values of dispersion constants A and B of Cauchy model (Equation (2)) are 1.5528 and 4.0353×10^{-3}, respectively.

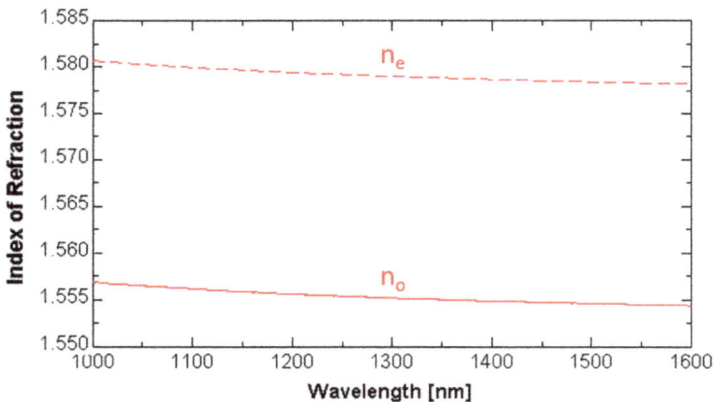

Figure 6. Ordinary (n_o) and extra-ordinary (n_e) refractive index curves of the deposited SiOC film extracted from measured ellipsometric data.

3.2. SiOC Optical Waveguides

The channel waveguides are fabricated in silicon oxycarbide (SiOC) using UV photolithography and reactive ion etching (RIE) techniques as described in Section 2.2. The channel waveguides are realized because they provide high confinement of electromagnetic light wave in two dimensions. The light confinement in the SiOC core region can be enhanced by increasing waveguide core dimensions (width and height).

After each step during the whole process, the optical chips were analyzed under optical microscope to check for the quality, defects, and residual of photoresist. The photomask used to define waveguide patterns with photolithography has a set of hundreds of waveguides with different widths ranging from 1.8 to 4.5 µm. Figure 7a shows micrograph of the SiOC waveguides (black strips) captured with optical microscope at the end of the waveguide fabrication process (after photoresist lift-off, before upper cladding deposition). The width of the waveguides increases from top to bottom (arrow direction). The waveguides definition is clear and free from defects. High resolution SEM image of one of the SiOC channel waveguides with refractive index n_e = 1.578 and dimensions of 4400 nm by 400 nm is shown in Figure 7b before the deposition of the silica upper cladding. The SiOC waveguides were buried under 8 µm thick upper SiO_2 cladding (n = 1.444) and provide an index contrast Δn of 8.2%. The index contrast between core and clad is calculated using the relation [21],

$$\Delta n = \frac{(n_{core}^2 - n_{clad}^2)}{2n_{core}^2} \qquad (3)$$

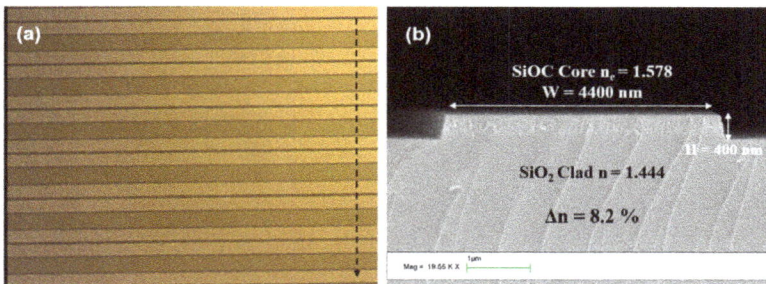

Figure 7. (**a**) Optical micrograph of straight SiOC waveguides with different widths (increasing from top to bottom with arrow) captured with optical microscope after photoresist lift-off, (**b**) high resolution SEM image of SiOC channel waveguide with dimensions (width W = 4400 nm by height H = 400 nm).

Based on the defined geometry of SiOC waveguides, electromagnetic simulations were performed to understand the mode of operation (single- or multi-mode) of the waveguides and compute the effective refractive index n_{eff} values of the propagating TE and TM modes. Figure 8a shows the fundamental TE ad TM modes of the SiOC waveguide with dimensions 4400 × 400 nm^2 (see Figure 7b) and their calculated n_{eff} values are 1.458 and 1.4602 at λ = 1550 nm. The confinement factor of the SiOC waveguides for the fundamental TE and TM modes is around 30%. In Figure 8b, n_{eff} of TE and TM modes of the SiOC waveguide are computed as a function of width (2–4.4 µm). The SiOC waveguides are single mode as no higher order TE or TM modes appear when the width is varied from 2 to 4.4 µm. The n_{eff} of fundamental TM modes is larger than TE modes due to the anisotropy exhibited by columnar structure in SiOC film that is on the order of 10^{-2}.

(a)

TE mode, neff = 1.458   TM mode, neff = 1.4602

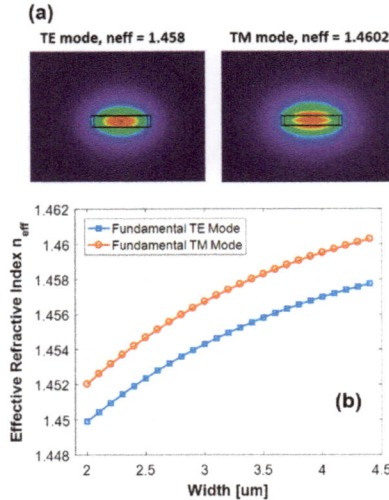

Figure 8. (a) Fundamental TE and TM modes of the SiOC waveguide (W = 4400 × H = 400 nm^2) and their respective effective indices n_{eff} computed with FEM based software, (b) Plot of effective refractive index n_{eff} of SiOC waveguide modes as a function of increasing width (2–4.4 µm), only fundamental TE and TM modes exist at λ = 1550 nm.

Figure 9 shows the experimental analyses of losses of single mode SiOC channel waveguides for different widths (2–4.4 µm) by using a cut-back technique. This technique enables to identify both the propagation losses of the waveguide and the coupling losses due to mismatch between fiber and waveguide modes overlap. As shown in Figure 9a, the propagation losses of SiOC optical waveguides having dimensions of W = 4400 nm by H = 400 nm are estimated to be 4 ± 0.5 dB/cm from the data slope versus waveguide length. The excited TE and TM polarizations (shown with black right and red left triangles, respectively) exhibit similar propagation losses. The coupling losses due to the mismatch between fiber and waveguide modes are estimated around 1.5 dB/facet, and are in good agreement with numerical simulations, which predict 80% coupling efficiency. In general, the coupling efficiency varies between fiber and waveguide modes with waveguide core size. The difference in coupling efficiency among SiOC waveguides with different core widths (between 2 µm and 4.4 µm) is 5% for both polarizations as evaluated from electromagnetic simulations. A difference of 5% means an additional loss of 0.2 dB that is within measurement error of our setup and has no appreciable effect on the assessment of waveguides losses. Furthermore, the coupling efficiency for TE and TM polarizations differing by 1% for respective waveguide width is negligible. The similar coupling efficiency is due to the anisotropy of the material that brings the n_{eff} of TE and TM modes closer to each other.

Figure 9b shows propagation loss as a function of core width. The average losses for both TE and TM polarizations are similar while there is a small increase in the propagation losses from wider (4.4 µm) to narrower (2 µm) waveguides. The confinement factor is calculated to decrease by about 8% from wider to narrower core waveguides for both TE and TM polarizations is a small number and has no impact on losses. The increase in propagation loss in narrower core waveguides can hence be attributed to the sidewall roughness that contribute to scattering of light. Moreover, from ellipsometric measurements the extinction coefficient k is undetectable which implies that SiOC material losses are low. Therefore, the origin of the waveguide losses is expected from the surface roughness and columnar structure of the sputtered SiOC film. To reduce the waveguide losses, the sputter process may further be optimized to obtain more compact SiOC films, annealing may be considered to smoothen the film structure or CVD process [22].

The results on SiOC waveguides indicate the significance of this material for guided wave applications. SiOC optical waveguides with higher refractive index (n > 1.578) are under development and will be demonstrated in a future study.

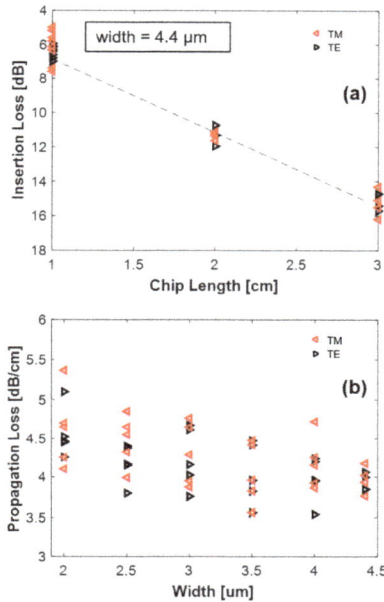

Figure 9. (**a**) Insertion loss plot of the SiOC waveguides with different chip lengths (1–3 cm), (**b**) SiOC waveguides propagation losses as a function of width (2–4.4 μm) for both TE and TM polarizations.

4. Conclusions

We have demonstrated for the first time the possibility of using silicon oxycarbide (SiOC) as a core material for guided wave applications. Detailed characterization on morphology, topography, and optical properties of sputtered SiOC film are presented. The channel waveguides are fabricated using UV photolithography and reactive ion etching processes. SiOC optical waveguides with core refractive index $n_e = 1.578$ and SiO_2 cladding provides a much higher index contrast (of about 8.2%) with respect to a conventional glass photonic platform. Moreover, SiOC offers the possibility of tuning the refractive index across a wide range, potentially from 1.45 to ~3.0, which means that index contrast as high as 38% can be achieved. Systematic investigation of SiOC waveguides with different widths (2–4.4 μm) and lengths (1–3 cm) is reported. The propagation losses of 4 ± 0.5 dB/cm for both TE and TM polarizations are presented as estimated from cut-back technique. Results indicate the potential of SiOC as a promising platform for integrated optics.

In order to reduce the losses of SiOC waveguides, the sputter process may further be optimized to achieve uniform films. Losses are perceived to originate mainly due to scattering from surface roughness and columnar structure present in the films. The annealing step may also be considered to smoothen the film structure. The study provides proof of concept for the SiOC photonic platform which may further be developed with CVD.

Higher index contrast waveguides and advanced optical devices in silicon oxycarbide are under development and will be demonstrated in future work.

Acknowledgments: Faisal Ahmed Memon acknowledges the financial support in his PhD from European Union under Erasmus Mundus 'LEADERS' project. This work was partially performed at Polifab, the micro- and nanofabrication facility of Politecnico di Milano (www.polifab.polimi.it) and the authors thank Claudio Somaschini for the valuable technical support.

Author Contributions: Faisal Ahmed Memon designed and performed the experiments for fabrication and characterizations of silicon oxycarbide thin films and optical waveguides. Faisal Ahmed Memon simulated the optical waveguides in COMSOL educational licensed (PoliMI) version (5.2) software using Wave Optics module based on finite element method (FEM). Francesco Morichetti contributed in measuring the propagation losses of the SiOC waveguides. The results were discussed by Faisal Ahmed Memon, Francesco Morichetti, and Andrea Melloni. The manuscript was written by Faisal Ahmed Memon and edited by Francesco Morichetti and Andrea Melloni.

Conflicts of Interest: The authors declare no conflict of interest.

References

1. Ng, D.K.; Wang, Q.; Wang, T.; Ng, S.K.; Toh, Y.T.; Lim, K.P.; Yang, Y.; Tan, D.T. Exploring High Refractive Index Silicon-Rich Nitride Films by Low-Temperature Inductively Coupled Plasma Chemical Vapor Deposition and Applications for Integrated Waveguides. *ACS App. Mater. Interfaces* **2015**, *7*, 21884–21889. [CrossRef] [PubMed]

2. Wörhoff, K.; Driessen, A.; Lambeck, P.V.; Hilderink, L.T.H.; Linders, P.W.C.; Popma, T.J.A. Plasma enhanced chemical vapor deposition silicon oxynitride optimized for application in integrated optics. *Sens. Actuators A Phys.* **1999**, *74*, 9–12. [CrossRef]

3. Pradeep, V.S.; Graczyk-Zajac, M.; Riedel, R.; Soraru, G.D. New Insights in to the Lithium Storage Mechanism in Polymer Derived SiOC Anode Materials. *Electrochim. Acta* **2014**, *119*, 78–85. [CrossRef]

4. Peng, Y.; Zhou, J.; Zheng, X.; Zhao, B.; Tan, X. Structure and photoluminescence properties of silicon oxycarbide thin films deposited by the rf reactive sputtering. *Int. J. Mod. Phys. B* **2011**, *25*, 2983. [CrossRef]

5. Ding, Y.; Shirai, H.; He, D. White light emission and electrical properties of silicon oxycarbide-based metal-oxide-semiconductor diode. *Thin Solid Films* **2011**, *519*, 2513–2515. [CrossRef]

6. Kim, H.J.; Shao, Q.; Kim, Y.-H. Characterization of low-dielectric-constant SiOC thin films deposited by PECVD for interlayer dielectrics of multilevel interconnection. *Surf. Coat. Technol.* **2003**, *171*, 39–45. [CrossRef]

7. Wang, M.R.; Rusli; Xie, J.L.; Babu, N.; Li, C.Y.; Rakesh, K. Study of oxygen influences on carbon doped silicon oxide low k thin films deposited by plasma enhanced chemical vapor deposition. *J. Appl. Phys.* **2004**, *96*, 829–834. [CrossRef]

8. Ryan, J.V.; Pantano, C.G. Synthesis and characterization of inorganic silicon oxycarbide glass thin films by reactive rf-magnetron sputtering. *J. Vac. Sci. Technol. A* **2007**, *25*, 153–159. [CrossRef]

9. Gallis, S.; Nikas, V.; Huang, M.; Eisenbraun, E.; Kaloyeros, A.E. Comparative study of the effects of thermal treatment on the optical properties of hydrogenated amorphous silicon-oxycarbide. *J. Appl. Phys.* **2007**, *102*, 24302. [CrossRef]

10. Pantano, C.G.; Singh, A.K.; Zhang, H. Silicon Oxycarbide Glasses. *J. Sol-Gel. Sci. Technol.* **1999**, *14*, 7–25. [CrossRef]

11. Renlund, G.; Prochazka, S.; Doremus, R. Silicon oxycarbide glasses: Part I. Preparation and chemistry. *J. Mat. Res.* **1991**, *6*, 2716–2722. [CrossRef]

12. Renlund, G.; Prochazka, S.; Doremus, R. Silicon oxycarbide glasses: Part II. Structure and properties. *J. Mater. Res.* **1991**, *6*, 2723–2734. [CrossRef]

13. Sorarù, G.D.; D'Andrea, G.; Glisenti, A. XPS characterization of gel-derived silicon oxycarbide glasses. *Mater. Lett.* **1996**, *27*, 1–5. [CrossRef]

14. Memon, F.A.; Morichetti, F.; Abro, M.I.; Iseni, G.; Somaschini, C.; Aftab, U.; Melloni, A. Synthesis, Characterization and Optical Constants of Silicon Oxycarbide. *EPJ Web Conf.* **2017**, *139*. [CrossRef]

15. Fujiwara, H. *Spectroscopic Ellipsometry: Principles and Applications*; Wiley: West Sussex, UK, 2007.

16. Bruggeman, D.A.G. Berechnung verschiedener physikalischer Konstanten von heterogenen Substanzen I. Dielektrizitätskonstanten und Leitfähigkeiten der Mischkörper aus isotropen Substanzen. *Ann. Phys.* **1935**, *24*, 636–679. [CrossRef]

17. J.A. Woollam Co., Inc. *Guide to Using WVASE® Spectroscopic Ellipsometry Data Acquisition and Analysis Software;* J.A. Woollam Co., Inc.: Lincoln, NE, USA, 1994–2012.

18. Valyukh, I.; Green, S.; Arwin, H.; Niklasson, G.A.; Wäckelgård, E.; Granqvist, C.G. Spectroscopic ellipsometry characterization of electrochromic tungsten oxide and nickel oxide thin films made by sputter deposition. *Sol. Energy Mater. Sol. Cells* **2010**, *94*, 724–732. [CrossRef]

19. Liu, T.; Henderson, C.L.; Samuels, R. Quantitative characterization of the optical properties of absorbing polymer films: Comparative investigation of the internal reflection intensity analysis method. *J. Polym. Sci. B Polym. Phys.* **2003**, *41*, 842–855. [CrossRef]

20. Prest, W.M., Jr.; Luca, D.J. The origin of the optical anisotropy of solvent cast polymeric films. *J. Appl. Phys.* **1979**, *50*, 6067. [CrossRef]

21. Senior, J.M. *Optical Fiber Communications: Principles and Practice*, 3rd ed.; Pearson Education Limited: Edinburgh, UK, 2009; pp. 19–20.

22. Mandracci, P.; Frascella, F.; Rizzo, R.; Virga, A.; Rivolo, P.; Descrovi, E.; Giorgis, F. Optical and structural properties of amorphous silicon-nitrides and silicon-oxycarbides: Application of multilayer structures for the coupling of Bloch Surface Waves. *J. Non-Cryst. Solids* **2016**, *453*, 113–117. [CrossRef]

![applied sciences logo]

applied
sciences

MDPI

Article

Photon Propagation through Linearly Active Dimers

José Delfino Huerta Morales [1,†] and Blas Manuel Rodríguez-Lara [1,2,*,†]

[1]   Instituto Nacional de Astrofísica, Óptica y Electrónica, Calle Luis Enrique Erro No. 1, Sta. Ma. Tonantzintla, Puebla CP 72840, Mexico; jd_huerta@inaoep.mx

[2]   Photonics and Mathematical Optics Group, Tecnológico de Monterrey, Monterrey 64849, Mexico

*   Correspondence: bmlara@itesm.mx; Tel.: +52-81-8358-2000 (ext. 4640)

†   These authors contributed equally to this work.

Academic Editor: Boris Malomed

Received: 11 March 2017; Accepted: 22 May 2017; Published: 7 June 2017

Abstract: We provide an analytic propagator for non-Hermitian dimers showing linear gain or losses in the quantum regime. In particular, we focus on experimentally feasible realizations of the $\mathcal{PT}$-symmetric dimer and provide their mean photon number and second order two-point correlation. We study the propagation of vacuum, single photon spatially-separable, and two-photon spatially-entangled states. We show that each configuration produces a particular signature that might signal their possible uses as photon switches, semi-classical intensity-tunable sources, or spatially entangled sources to mention a few possible applications.

Keywords: $\mathcal{PT}$-symmetric dimer; non-Hermitian dimer; quantum linear dimer; optical waveguide couplers; photon sampling

1. Introduction

Propagation through classical $\mathcal{PT}$-symmetric optical systems has been extensively studied; cf. Ref. [1] and references therein from the initial description of linear loses in directional couplers with a more complex non-Hermitian symmetry [2], the quantum-like description of classical planar waveguides [3], to all the contemporaneous work derived from the seminal introduction of optical $\mathcal{PT}$-symmetric structures [4]. It is well known that the propagation of electromagnetic field through a linearly active two-waveguide coupler can be described by the classical $\mathcal{PT}$-symmetric dimer,

$$-i\frac{d}{dz}\begin{pmatrix}\mathcal{E}_1(z)\\\mathcal{E}_2(z)\end{pmatrix}=\begin{pmatrix}i\gamma & g\\ g & -i\gamma\end{pmatrix}\begin{pmatrix}\mathcal{E}_1(z)\\\mathcal{E}_2(z)\end{pmatrix},\tag{1}$$

where the effective evanescent coupling between the two single waveguide field modes is given by the real parameter g and the effective gain and loss by the real positive parameter γ. Such a system can be realized experimentally by balanced gain and loss in the waveguides, but it is also possible to realize it with gain-gain, loss-loss, passive-gain [5], and passive-loss [6,7] waveguide, microcavity rings [8,9], and electric cirtcuits [10] setups.

In the quantum regime [11–14], the importance of adequately modeling media with linear gain or loss has been brought forward recently [15,16]. In this regime, linear media induces quantum fluctuations, such that the ideal $\mathcal{PT}$-symmetric optical dimer dynamics is effectively described by quantum Langevin equations [15],

$$-i\frac{d}{dz}\begin{pmatrix}\hat{a}_1(z)\\\hat{a}_2(z)\end{pmatrix}=\begin{pmatrix}i\gamma & g\\ g & -i\gamma\end{pmatrix}\begin{pmatrix}\hat{a}_1(z)\\\hat{a}_2(z)\end{pmatrix}+\begin{pmatrix}\hat{f}_1(z)\\\hat{f}_2(z)\end{pmatrix},\tag{2}$$

in terms of the effective balanced gain and loss parameter, γ, the effective mode coupling, g, between the two modes of the optical resonators, described by the annihilation operators $\hat{a}_1(z)$ and $\hat{a}_2(z)$, and the delta-correlated Langevin forces introduced by the gain and loss media,

$$
\begin{aligned}
\langle \hat{f}_1^{\dagger}(z)\hat{f}_1(\zeta) \rangle &= 2\gamma\delta(z-\zeta), \\
\langle \hat{f}_2(z)\hat{f}_2^{\dagger}(\zeta) \rangle &= 2\gamma\delta(z-\zeta),
\end{aligned}
\tag{3}
$$

in that order. These Gaussian fluctuations modify the well-known dynamics produced by the classical optical $\mathcal{PT}$-symmetric dimer and generate second order correlations in the ideal quantum optical $\mathcal{PT}$-symmetric dimer [15], Figure 1. Here, we want to discuss the different types of correlations that might arise from feasible experimental realizations of the quantum $\mathcal{PT}$-symmetric dimer beyond the balanced gain-loss setup. In the following section, we will introduce the quantum model for a generalized linear active dimer, a non-Hermitian quantum optical dimer, and provide its propagation solution. Then, we will discuss the dynamics of spontaneous photon generation as well as photon bunching and anti-bunching in the different configurations. We will study single-photon propagation with spatially separable and two-photon propagation with entangled states through the mean photon number and second order two-point correlations. Finally, we will close with a brief conclusion and discuss how mean photon propagation and second order spatial correlations might provide insight to their application in photon switching or as semi(non)-classical light sources.

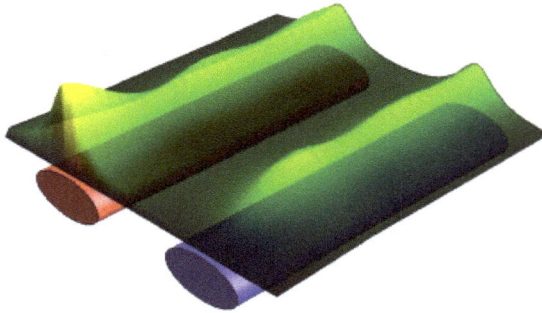

Figure 1. Schematic showing the renormalized light intensity arising from the spontaneous generation of photons through two coupled waveguides in the $\mathcal{PT}$-symmetry regime with balanced gain-loss configuration.

2. Quantum Model and Configurations

In the laboratory, we can think about a more general realization of the effective quantum $\mathcal{PT}$-symmetric dimer provided by the quantum non-Hermitian dimer,

$$
-i\frac{d}{dz}\begin{pmatrix} \hat{a}_1(z) \\ \hat{a}_2(z) \end{pmatrix} = \begin{pmatrix} n_1 & g \\ g & n_2 \end{pmatrix}\begin{pmatrix} \hat{a}_1(z) \\ \hat{a}_2(z) \end{pmatrix} + \begin{pmatrix} \hat{f}_1(z) \\ \hat{f}_2(z) \end{pmatrix}.
\tag{4}
$$

Again, the field annihilation operators in the first and second waveguides are given by the operators $\hat{a}_1(z)$ and $\hat{a}_2(z)$, in that order. The effective refractive indices of the optical waveguides are given by the complex numbers n_1 and n_2, the effective coupling between the optical modes is given by the real positive parameter g. Finally, the Gaussian fluctuations, due to the active linear media, are described by the Langevin forces $\hat{f}_1(z)$ and $\hat{f}_2(z)$, such that the only nonzero mean values involving them are their second order correlations [17],

$$
\begin{aligned}
\langle \hat{f}_j^\dagger(z) \hat{f}_j(z') \rangle &= 2\Im(n_j)\delta(z - z'), \quad \text{for gain media,} \\
\langle \hat{f}_j(z) \hat{f}_j^\dagger(z') \rangle &= 2\Im(n_j)\delta(z - z'), \quad \text{for loss media.}
\end{aligned}
\tag{5}
$$

It is straightforward to write a formal propagator for this quantum optical system,

$$
\begin{pmatrix} \hat{a}_1(\zeta) \\ \hat{a}_2(\zeta) \end{pmatrix} = e^{in_0\zeta} \begin{pmatrix} \hat{b}_1(\zeta) \\ \hat{b}_2(\zeta) \end{pmatrix},
\tag{6}
$$

where we have factorized an average refractive index, $n_0 = (n_1 + n_2)/(2g)$, that introduces a common phase, $\Re(n_0)$, and a scaling factor due to gain or loss, $\Im(n_0)$. We have also scaled the propagation variable by the effective coupling of the dimer, $\zeta = gz$. Note that the average refractive index cannot be zero for experimental realizations. It becomes a pure real number, a phase factor, for passive materials, and a pure imaginary number, a scaling factor, for identical media with balanced gain and loss. The second term in the propagator,

$$
\begin{pmatrix} \hat{b}_1(\zeta) \\ \hat{b}_2(\zeta) \end{pmatrix} = e^{i\hat{H}\zeta} \begin{pmatrix} \hat{b}_1(0) \\ \hat{b}_2(0) \end{pmatrix} + \int_0^\zeta e^{i\hat{H}(\zeta - t)} e^{-in_0 t} \begin{pmatrix} \hat{f}_1(t) \\ \hat{f}_2(t) \end{pmatrix} dt,
\tag{7}
$$

provides us with the effective dynamics of the system. Here, we need use the scaling property of Dirac delta, $\delta(\zeta) = \delta(z)/|g|$, and have defined a complex effective refractive index $n = (n_1 - n_2)/(2g)$ that provides us with an auxiliary effective non-Hermitian matrix,

$$
\hat{H} = \begin{pmatrix} n & 1 \\ 1 & -n \end{pmatrix}.
\tag{8}
$$

Note that the case of waveguides with identical real part of their effective refractive indices, $\Re(n_1) = \Re(n_2)$, yields a purely imaginary effective refractive index that we can rename as $n = i\gamma$ in order to recover the standard quantum $\mathcal{PT}$-symmetric dimer [15]. Again, let us stress that the dynamics introduced by a more realistic model of linearly active media, where configurations beyond balanced gain-loss are easily obtained, will include a phase and scaling factor proportional to the average refractive index that are not taken into consideration in the ideal $\mathcal{PT}$-symmetric dimer configuration.

The coupling matrix exponential can be easily calculated following an approach similar to that used in the classical $\mathcal{PT}$-symmetric dimer [1],

$$
\begin{aligned}
\hat{U}(\zeta) &= e^{i\hat{H}\zeta}, \\
&= \hat{1} \cos(\Omega\zeta) + \hat{H}\, \zeta \, \mathrm{sinc}(\Omega\zeta), \quad \Omega \in \mathbb{C}.
\end{aligned}
\tag{9}
$$

Here, the symbol $\hat{1}$ stands for the two by two identity matrix, we have used the cardinal sine function $\mathrm{sinc}(x) = \sin(x)/x$, and the complex dispersion relation is given by the following expression,

$$
\Omega = \sqrt{1 + n^2}.
\tag{10}
$$

Note that this analytic propagator can describe any non-Hermitian dimer and, in the special case of purely imaginary auxiliary refractive index, $n = i\gamma$, we recover the propagator for the standard classical $\mathcal{PT}$-symmetric dimer [18],

$$
\hat{U}(\zeta) = \begin{cases} \hat{\mathbb{1}}\cos\left(\Omega\zeta\right) + i\,\hat{H}\,\zeta\,\text{sinc}\left(\Omega\zeta\right) & \gamma < 1, \\ \hat{\mathbb{1}} + i\,\zeta\,\hat{H}, & \gamma = 1, \\ \hat{\mathbb{1}}\cosh\left(|\Omega|\zeta\right) + i\,\hat{H}\,\zeta\,\text{sinhc}\left(|\Omega|\zeta\right), & \gamma > 1. \end{cases} \tag{11}
$$

Experimentally, the effective $\mathcal{PT}$-symmetric dimer can be reached via different configurations: balanced gain-loss dimer, $n_1 = n_R - in_I$ and $n_2 = n_R + in_I$, in that order, such that $n_0 = n_R/g$ and $\gamma = -n_I/g$; gain-gain dimer, $n_1 = n_R - in_{I,1}$ and $n_2 = n_R - in_{I,2}$ such that $n_0 = [2n_R - i(n_{I,1} + n_{I,2})]/(2g)$ and $\gamma = -(n_{I,1} - n_{I,2})/(2g)$; passive-gain dimer, $n_1 = n_R$ and $n_2 = n_R - in_I$, in that order, such that $n_0 = (2n_R - in_I)/(2g)$ and $\gamma = n_I/(2g)$; passive-loss dimer, $n_1 = n_R$ and $n_2 = n_R + in_I$, in that order, such that $n_0 = (2n_R + in_I)/(2g)$ and $\gamma = -n_I/(2g)$; and loss-loss dimer, $n_1 = n_R + in_{I,1}$ and $n_2 = n_R + in_{I,2}$ such that $n_0 = [2n_R + i(n_{I,1} + n_{I,2})]/(2g)$ and $\gamma = (n_{I,1} - n_{I,2})/(2g)$; for all of these configurations, we have assumed $n_R, n_I, n_{I,j} > 0$. The formal solution presented in this section allows us to explore the propagation properties of all of these experimentally feasible configurations. Table 1 shows the values of these parameters for the different configurations delivering an effective $\mathcal{PT}$-symmetric dimer; and we have added the imaginary part of the effective bias refractive index, $\beta = (n_{I,1} + n_{I,2})/(2g) = -\Im(n_0)$, as it will be useful in the following sections.

Table 1. A summary of the parameters involved in the different feasible experimental realizations of the $\mathcal{PT}$-symmetric dimer.

Realization	n_1	n_2	n	n_0	γ	β
Gain-loss	$n_R - in_I$	$n_R + in_I$	$-i\frac{n_I}{g}$	$\frac{n_R}{g}$	$-\frac{n_I}{g}$	0
Gain-gain	$n_R - in_{I,1}$	$n_R - in_{I,2}$	$i\frac{-n_{I,1}+n_{I,2}}{2g}$	$\frac{n_R}{g} - i\frac{n_{I,1}+n_{I,2}}{2g}$	$\frac{-n_{I,1}+n_{I,2}}{2g}$	$\frac{n_{I,1}+n_{I,2}}{2g}$
Gain-passive	$n_R - in_I$	n_R	$-i\frac{n_I}{2g}$	$\frac{n_R}{g} + i\frac{n_I}{2g}$	$-\frac{n_I}{2g}$	$\frac{n_I}{2g}$
Passive-loss	n_R	$n_R + in_I$	$-i\frac{n_I}{2g}$	$\frac{n_R}{g} + i\frac{n_I}{2g}$	$-\frac{n_I}{2g}$	$-\frac{n_I}{2g}$
Loss-loss	$n_R + in_{I,1}$	$n_R + in_{I,2}$	$i\frac{n_{I,1}-n_{I,2}}{2g}$	$\frac{n_R}{g} + i\frac{n_{I,1}+n_{I,2}}{2g}$	$\frac{n_{I,1}-n_{I,2}}{2g}$	$\frac{-n_{I,1}-n_{I,2}}{2g}$

3. Spontaneous Generation of Photons

The first signature that differentiates a quantum from a classical $\mathcal{PT}$-symmetric dimer is the spontaneous generation of photons due to the gain medium in presence of vacuum input fields,

$$
\begin{aligned} n_j^{(00)}(\zeta) &= \langle 0,0|\hat{a}_j^\dagger(\zeta)\hat{a}_j(\zeta)|0,0\rangle, \\ &= e^{2\beta\zeta}\langle 0,0|\hat{b}_j^\dagger(\zeta)\hat{b}_j(\zeta)|0,0\rangle. \end{aligned} \tag{12}
$$

In other words, vacuum fluctuations are enough to make the linear active media spontaneously generate photons [17]; something that is lacking in the classical model. The signatures available through the spontaneous generation of photos in diverse configurations are provided in the following for the standard balanced **gain-loss** dimer,

$$
\begin{aligned} n_1^{(00)}(\zeta) &= -2\gamma\int_0^\zeta |\hat{U}_{11}(t)|^2 dt, \\ n_2^{(00)}(\zeta) &= -2\gamma\int_0^\zeta |\hat{U}_{21}(t)|^2 dt, \end{aligned} \tag{13}
$$

where the (i,j)-th component of the two by two propagation matrix has been written as $\hat{U}_{ij}(\zeta)$, and it is important to note that these integrals can be solved analytically but yield expressions too long to write here. The **gain-gain** dimer yields the following expressions,

$$
\begin{aligned}
n_1^{(00)}(\zeta) &= 2(\beta - \gamma) \int_0^\zeta \left|\hat{U}_{11}(t)\right|^2 e^{2\beta t} dt + 2(\beta + \gamma) \int_0^\zeta \left|\hat{U}_{12}(t)\right|^2 e^{2\beta t} dt, \\
n_2^{(00)}(\zeta) &= 2(\beta - \gamma) \int_0^\zeta \left|\hat{U}_{21}(t)\right|^2 e^{2\beta t} dt + 2(\beta + \gamma) \int_0^\zeta \left|\hat{U}_{22}(t)\right|^2 e^{2\beta t} dt.
\end{aligned}
\tag{14}
$$

For the **gain-passive** dimer, we can write the spontaneous generation as:

$$
\begin{aligned}
n_1^{(00)}(\zeta) &= -4\gamma \int_0^\zeta \left|\hat{U}_{11}(t)\right|^2 e^{-2\gamma t} dt, \\
n_2^{(00)}(\zeta) &= -4\gamma \int_0^\zeta \left|\hat{U}_{21}(t)\right|^2 e^{-2\gamma t} dt,
\end{aligned}
\tag{15}
$$

and, obviously, there is not spontaneous generation in the **passive-loss** and **loss-loss** dimer,

$$
n_1^{(00)}(\zeta) = n_2^{(00)}(\zeta) = 0.
\tag{16}
$$

While the expressions for the spontaneous generation are complicated, it is straightforward to realize from the analytic expressions that they will present different signatures through propagation in the dimer.

The signatures from the different configurations can be seen in Figure 2, where we show an instantaneously renormalized spontaneous generation of photons,

$$
\tilde{n}_j^{(00)}(\zeta) = \frac{n_j^{(00)}(\zeta)}{n_1^{(00)}(\zeta) + n_2^{(00)}(\zeta)}, \qquad \zeta > 0,
\tag{17}
$$

and avoid the position $\zeta = 0$ due to the divergence induced by the initial vacuum field. The rows in Figure 2 present the spontaneous emission in the balanced gain-loss, gain-gain, gain-passive configurations, from top to bottom, and the columns show results in the $\mathcal{PT}$-symmetric, Kato exceptional point, and broken symmetry regimes, from left to right. The spontaneous emission in the ideal $\mathcal{PT}$-symmetric is shown in Figure 2a–c. Note that the oscillations in the $\mathcal{PT}$-symmetric regime appear earlier in the propagation for the gain-gain, Figure 2d, and gain-passive configurations, Figure 2g. In the Kato exceptional point, the spontaneous emission is equivalent in the ideal dimer, Figure 2b, and the gain-passive configurations, Figure 2h, and follows a slightly different initial distribution in the gain-gain case Figure 2e. Finally, something similar happens in the broken symmetry regime, the distinction between spontaneous emission in the waveguides for the ideal, Figure 2c, and gain-passive configurations, Figure 2i, is almost null and follows a slightly different initial distribution for the gain-gain case, Figure 2f. Note that the spontaneous generation is asymptotically equal in both waveguides when the dimer is at the Kato exceptional point for any configuration with linear gain, Figure 2b,e,h. In addition, in the broken symmetry regime, the asymptotic value for the renormalized spontaneous generation is the same for the different configurations including linear gain, Figure 2c,f,i, and converges to the same value than the classical $\mathcal{PT}$-symmetric dimer [1,18],

$$
\begin{aligned}
\lim_{\zeta \to \infty} \tilde{n}_1^{(00)}(\zeta) &= \frac{1}{2\gamma} \left(\gamma + \sqrt{\gamma^2 - 1}\right)^{-1}, \\
\lim_{\zeta \to \infty} \tilde{n}_2^{(00)}(\zeta) &= \frac{1}{2\gamma} \left(\gamma + \sqrt{\gamma^2 - 1}\right).
\end{aligned}
\tag{18}
$$

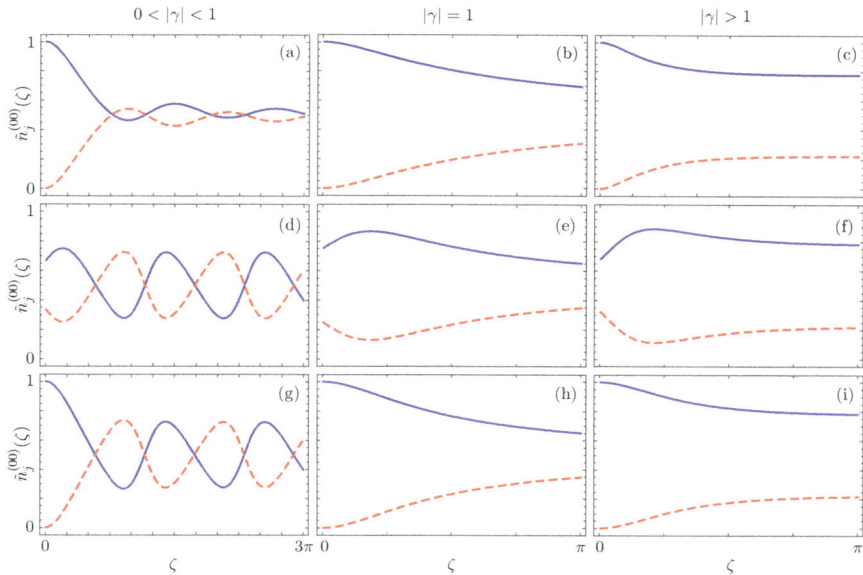

Figure 2. Instantaneously renormalized spontaneous generation, $\tilde{n}_j^{(00)}(\zeta)$, along different realizations of the effective $\mathcal{PT}$-symmetric dimer. The first row, (**a–c**), shows balanced gain-loss, second row, (**d–f**), shows gain-gain, and third row, (**g–i**), shows gain-passive configurations in the $\mathcal{PT}$-symmetric regime, the first column with $|\gamma| = 0.5$, the Kato point with $|\gamma| = 1$, the second column, and broken symmetry regime, and the third column with $|\gamma| = 1.2$. Values for the first and second waveguides are shown with a solid blue and a dashed red lines, in that order. Note the oscillatory behavior of the spontaneous generation inside and its asymptotic behavior outside the $\mathcal{PT}$-symmetric regime.

4. Photon Bunching in Spontaneous Generation

We can also look at the signatures provided by the probability to detect simultaneously one photon at each of the waveguides output in the different configurations. This can be written in the following form for the spontaneous generation of photons,

$$
\begin{aligned}
\langle 0,0|\hat{a}_1^\dagger(\zeta)\,\hat{a}_2^\dagger(\zeta)\,\hat{a}_1(\zeta)\,\hat{a}_2(\zeta)|0,0\rangle &= \langle 0,0|\hat{a}_1^\dagger(\zeta)\,\hat{a}_1(\zeta)|0,0\rangle\,\langle 0,0|\hat{a}_2^\dagger(\zeta)\,\hat{a}_2(\zeta)|0,0\rangle \\
&+ \langle 0,0|\hat{a}_1^\dagger(\zeta)\,\hat{a}_2(\zeta)|0,0\rangle\,\langle 0,0|\hat{a}_2^\dagger(\zeta)\,\hat{a}_1(\zeta)|0,0\rangle \\
&= n_1^{(00)}(\zeta)n_2^{(00)}(\zeta) + \left|n_{12}^{(00)}(\zeta)\right|^2,
\end{aligned}
\tag{19}
$$

where he have used the Gaussian nature of Langevin forces [15] to simplify this probability using the expressions derived in the last section and defining the following first order two-point correlation,

$$
\begin{aligned}
n_{12}^{(00)}(\zeta) &= \langle 0,0|\hat{a}_1^\dagger(\zeta)\hat{a}_2(\zeta)|0,0\rangle, \\
&= e^{2\beta\zeta}\langle 0,0|\hat{b}_1^\dagger(\zeta)\hat{b}_2(\zeta)|0,0\rangle.
\end{aligned}
\tag{20}
$$

Note that the detection probability will always be positive independently of the form taken by the first order two-point correlation for the different configurations: **balanced gain-loss** dimer,

$$
n_{12}^{(00)}(\zeta) = -2\gamma \int_0^\zeta \hat{U}_{11}^*(t)\,\hat{U}_{21}(t)\,dt,
\tag{21}
$$

gain-gain dimer,

$$n_{12}^{(00)}(\zeta) = 2(\beta - \gamma)\int_0^\zeta \hat{U}_{11}^*(t)\,\hat{U}_{21}(t)e^{2\beta t}dt + 2(\beta + \gamma)\int_0^\zeta \hat{U}_{12}^*(t)\,\hat{U}_{22}(t)e^{2\beta t}dt, \qquad (22)$$

gain-passive dimer,

$$n_{12}^{(00)}(\zeta) = -4\gamma\int_0^\zeta \hat{U}_{11}^*(t)\,\hat{U}_{21}(t)e^{-2\gamma t}dt, \qquad (23)$$

and, obviously, for the **passive-loss** and **loss-loss** dimer,

$$n_{12}^{(00)}(\zeta) = 0. \qquad (24)$$

Again, we have to be careful to consider the appropriate parameters γ and β for each configuration summarized in Table 1.

In order to visualize the information, we can use a quantity similar to Mandel Q-parameter [19],

$$\begin{aligned} q^{(00)}(\zeta) &= g_2^{(00)}(\zeta) - 1, \\ &= \frac{|n_{12}^{(00)}(\zeta)|^2}{n_1^{(00)}(\zeta)n_2^{(00)}(\zeta)}, \qquad \zeta > 0, \end{aligned} \qquad (25)$$

given in terms of the second order two-point correlation function,

$$g_2^{(00)}(\zeta) = 1 + \frac{|n_{12}^{(00)}(\zeta)|^2}{n_1^{(00)}(\zeta)n_2^{(00)}(\zeta)}, \qquad \zeta > 0, \qquad (26)$$

that provides us with the probability of simultaneously detecting a photon in each waveguide output. The values of this two-point parameter are always positive, $q^{(00)}(\zeta) \geq 0$, thus, the different configurations only show photon bunching. Figure 3 shows the photon bunching signatures obtained with the different dimer configurations discussed before. The ordering is the same than in Figure 2, rows show the balanced gain-loss, gain-gain, and gain-passive dimer configurations, in that order, and columns show the symmetric, exceptional and broken symmetry regimes from left to right. We can immediately see that the signatures in the $\mathcal{PT}$-symmetric regime, Figure 3a,d,g, present an oscillatory behavior, while those in the Kato exceptional point, Figure 3b,e,h, and the broken $\mathcal{PT}$-symmetry regimes, and Figure 3c,f,i, saturate to the unit. The dimer configuration also influences the photon bunching signature, the balanced gain-loss dimer presents oscillations that do not approach zero, Figure 3a, while the gain-gain and the gain-passive dimers present oscillations that reach zero, Figure 3d,g. In addition, the balanced gain-loss dimer saturates in a manner similar to that of the gain-passive dimer starting from a nonzero value, Figure 3b,h as well as Figure 3c,i, while the gain-gain dimer starts from zero and saturates faster, Figure 3e,f.

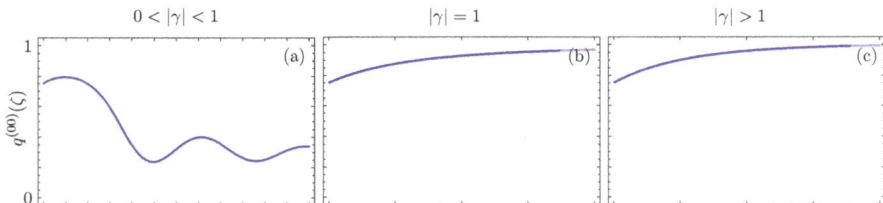

Figure 3. *Cont.*

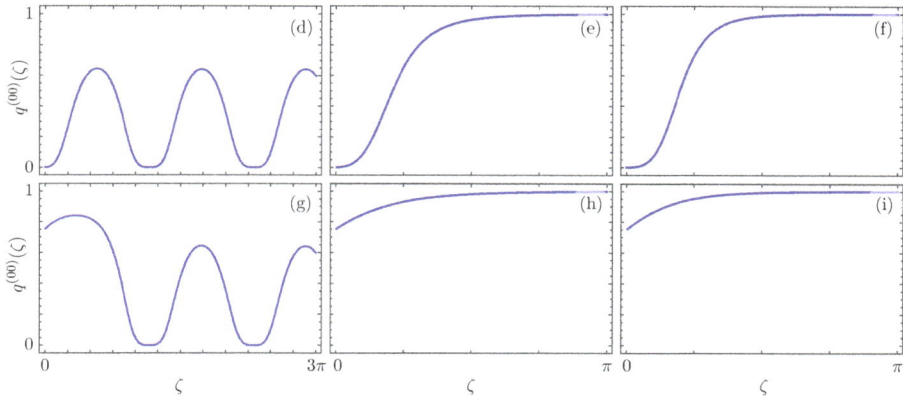

Figure 3. Photon bunching shown in terms of the $q^{(00)}(\zeta)$ parameter for different realizations of the effective $\mathcal{PT}$-symmetric dimer. The first row, (**a–c**), shows balanced gain-loss, second row, (**d–f**), shows gain-gain, and third row, (**g–i**), shows gain-passive configurations in the $\mathcal{PT}$-symmetric regime, the first column with $|\gamma| = 0.5$, the Kato point with $|\gamma| = 1$, the second column, and broken symmetry regime, and the third column with $|\gamma| = 1.2$.

5. Photon Propagation

In general, we can study the propagation of any given initial photon state through our dimer,

$$
\begin{aligned}
n_j^{(\psi_0)}(\zeta) &= \langle\psi_0|\hat{a}_j^\dagger(\zeta)\hat{a}_j(\zeta)|\psi_0\rangle, \\
&= e^{2\beta\zeta}\langle\psi_0|\hat{b}_j^\dagger(\zeta)\hat{b}_j(\zeta)|\psi_0\rangle,
\end{aligned}
\tag{27}
$$

and realize that the mean photon number in each waveguide,

$$
\begin{aligned}
n_1^{(\psi_0)}(\zeta) &= e^{2\beta\zeta}|\hat{U}_{11}(\zeta)|^2\langle\psi_0|\hat{a}_1^\dagger(0)\hat{a}_1(0)|\psi_0\rangle + e^{2\beta\zeta}\hat{U}_{11}^*(\zeta)\hat{U}_{12}(\zeta)\langle\psi_0|\hat{a}_1^\dagger(0)\hat{a}_2(0)|\psi_0\rangle \\
&+ e^{2\beta\zeta}\hat{U}_{12}^*(\zeta)\hat{U}_{11}(\zeta)\langle\psi_0|\hat{a}_2^\dagger(0)\hat{a}_1(0)|\psi_0\rangle + e^{2\beta\zeta}|\hat{U}_{12}(\zeta)|^2\langle\psi_0|\hat{a}_2^\dagger(0)\hat{a}_2(0)|\psi_0\rangle \\
&+ e^{2\beta\zeta}\int_0^\zeta\int_0^\zeta \hat{U}_{11}^*(\zeta-t')\hat{U}_{11}(\zeta-t)e^{in_0^*t'}e^{-in_0t}\left\langle\psi_0|\hat{f}_1^\dagger(t')\hat{f}_1(t)|\psi_0\right\rangle dt'dt \\
&+ e^{2\beta\zeta}\int_0^\zeta\int_0^\zeta \hat{U}_{12}^*(\zeta-t')\hat{U}_{12}(\zeta-t)e^{in_0^*t'}e^{-in_0t}\left\langle\psi_0|\hat{f}_2^\dagger(t')\hat{f}_2(t)|\psi_0\right\rangle dt'dt, \\
n_2^{(\psi_0)}(\zeta) &= e^{2\beta\zeta}|\hat{U}_{21}(\zeta)|^2\langle\psi_0|\hat{a}_1^\dagger(0)\hat{a}_1(0)|\psi_0\rangle + e^{2\beta\zeta}\hat{U}_{21}^*(\zeta)\hat{U}_{22}(\zeta)\langle\psi_0|\hat{a}_1^\dagger(0)\hat{a}_2(0)|\psi_0\rangle \\
&+ e^{2\beta\zeta}\hat{U}_{22}^*(\zeta)\hat{U}_{21}(\zeta)\langle\psi_0|\hat{a}_2^\dagger(0)\hat{a}_1(0)|\psi_0\rangle + e^{2\beta\zeta}|\hat{U}_{22}(\zeta)|^2\langle\psi_0|\hat{a}_2^\dagger(0)\hat{a}_2(0)|\psi_0\rangle \\
&+ e^{2\beta\zeta}\int_0^\zeta\int_0^\zeta \hat{U}_{21}^*(\zeta-t')\hat{U}_{21}(\zeta-t)e^{in_0^*t'}e^{-in_0t}\left\langle\psi_0|\hat{f}_1^\dagger(t')\hat{f}_1(t)|\psi_0\right\rangle dt'dt \\
&+ e^{2\beta\zeta}\int_0^\zeta\int_0^\zeta \hat{U}_{22}^*(\zeta-t')\hat{U}_{22}(\zeta-t)e^{in_0^*t'}e^{-in_0t}\left\langle\psi_0|\hat{f}_2^\dagger(t')\hat{f}_2(t)|\psi_0\right\rangle dt'dt,
\end{aligned}
\tag{28}
$$

will have a component related to spontaneous generation, those terms with the integrals, and another to stimulated generation, the rest. As a practical example, let us use as the initial state a single photon impinging the first waveguide, $|\psi_0\rangle = |10\rangle$. Again, we can calculate the mean photon number in the **balanced gain-loss** dimer,

$$
\begin{aligned}
n_1^{(10)}(\zeta) &= |\hat{U}_{11}(\zeta)|^2 - 2\gamma\int_0^\zeta|\hat{U}_{11}(t)|^2 dt, \\
n_2^{(10)}(\zeta) &= |\hat{U}_{21}(\zeta)|^2 - 2\gamma\int_0^\zeta|\hat{U}_{21}(t)|^2 dt,
\end{aligned}
\tag{29}
$$

and realize that the spontaneous generation terms are identical to those in the spontaneous generation in Equation (13). The same occurs to all the dimer configurations. In the **gain-gain** dimer,

$$
\begin{aligned}
n_1^{(10)}(\zeta) &= e^{2\beta\zeta}\left|\hat{U}_{11}(\zeta)\right|^2 + 2(\beta-\gamma)\int_0^\zeta \left|\hat{U}_{11}(t)\right|^2 e^{2\beta t}dt + 2(\beta+\gamma)\int_0^\zeta \left|\hat{U}_{12}(t)\right|^2 e^{2\beta t}dt, \\
n_2^{(10)}(\zeta) &= e^{2\beta\zeta}\left|\hat{U}_{21}(\zeta)\right|^2 + 2(\beta-\gamma)\int_0^\zeta \left|\hat{U}_{21}(t)\right|^2 e^{2\beta t}dt + 2(\beta+\gamma)\int_0^\zeta \left|\hat{U}_{22}(t)\right|^2 e^{2\beta t}dt,
\end{aligned}
\tag{30}
$$

we recover the spontaneous generation terms from Equation (14). In the **gain-passive** dimer,

$$
\begin{aligned}
n_1^{(10)}(\zeta) &= e^{2\beta\zeta}\left|\hat{U}_{11}(\zeta)\right|^2 - 4\gamma\int_0^\zeta \left|\hat{U}_{11}(t)\right|^2 e^{-2\gamma t}dt, \\
n_2^{(10)}(\zeta) &= e^{2\beta\zeta}\left|\hat{U}_{21}(\zeta)\right|^2 - 4\gamma\int_0^\zeta \left|\hat{U}_{21}(t)\right|^2 e^{-2\gamma t}dt,
\end{aligned}
\tag{31}
$$

we recover those from the spontaneous generation in Equation (15). Finally, in the **passive-loss** and **loss-loss** dimers, the intensity only depends on the initial state and decays due to the nature of the auxiliary β parameter,

$$
\begin{aligned}
n_1^{(10)}(\zeta) &= e^{2\beta\zeta}\left|\hat{U}_{11}(\zeta)\right|^2, \\
n_2^{(10)}(\zeta) &= e^{2\beta\zeta}\left|\hat{U}_{21}(\zeta)\right|^2.
\end{aligned}
\tag{32}
$$

In these expressions, it is easier to identify that the spontaneous generation component is identical to the one in the vacuum propagation case and the stimulated component, in the specific case of single photon propagation, will be the same than in the classical dimer, as expected.

Figure 4 shows the renormalized mean photon number for the propagation of a single photon in the first waveguide,

$$
\tilde{n}_j^{(10)}(\zeta) = \frac{n_j^{(10)}(\zeta)}{n_1^{(10)}(\zeta) + n_2^{(10)}(\zeta)}.
\tag{33}
$$

Note that the initial state is not excluded, as in the spontaneous generation case, because now there will always be a nonzero probability that the dimer will have a photon propagating through it. Now, as the renormalized mean photon number will have a spontaneous and stimulated component, it is possible to see that the strongest differences in the $\mathcal{PT}$-symmetric regime will occur at small propagation distances, and the oscillation frequency will be larger than in the spontaneous generation case, Figure 4a,d,g. In the Kato exceptional point, Figure 4b,e,h, and the broken symmetry regime, Figure 4c,f,i, the same will happen. The strongest deviation from the spontaneous generation signature will occur for small propagation distances and it will take slightly longer propagation distances to reach an asymptotic limit identical in value to that of the spontaneous generation,

$$
\begin{aligned}
\lim_{\zeta\to\infty}\tilde{n}_1^{(10)}(\zeta) &= \tfrac{1}{2\gamma}\left(\gamma+\sqrt{\gamma^2-1}\right)^{-1}, \\
\lim_{\zeta\to\infty}\tilde{n}_2^{(10)}(\zeta) &= \tfrac{1}{2\gamma}\left(\gamma+\sqrt{\gamma^2-1}\right).
\end{aligned}
\tag{34}
$$

Finally, it is possible to follow the renormalized mean photon number for the passive-loss and loss-loss dimer which will have identical signatures in all regimes, Figure 4j,k,l, and is able to provide photon localization at any of the waveguides, Figure 4j.

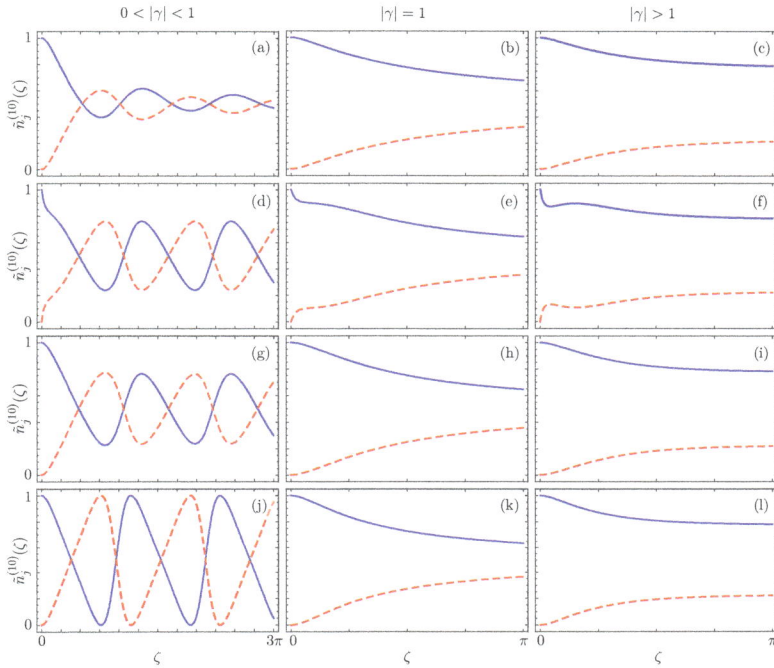

Figure 4. Instantaneously renormalized mean photon number, $\hat{n}_j^{(10)}(\zeta)$, along different realizations of the effective $\mathcal{PT}$-symmetric dimer. The first row, (**a**–**c**), shows balanced gain-loss, the second row, (**d**–**f**), shows gain-gain, the third row, (**g**–**i**), shows gain-passive, and the fourth row, (**j**–**l**), shows both passive-loss and loss-loss configurations in the $\mathcal{PT}$-symmetric regime, the first column with $|\gamma| = 0.5$, the Kato point with $|\gamma| = 1$, the second column, and the broken symmetry regime, and the third column with $|\gamma| = 1.2$. Values for the first and second waveguides are shown with a solid blue and a dashed red lines, in that order.

6. Photon Bunching and Anti-Bunching in Photon Propagation

We can also study the effect of photon propagation on second order two-point correlations as we did for spontaneous generation. In order to study a different state, we will consider as initial state a *N00N* state, $|N00N\rangle = (|N0\rangle + |0N\rangle) / \sqrt{2}$, in order to see negative values of the two-point Mandel parameter at least for the initial state,

$$q^{(N00N)}(\zeta) \quad = \quad \frac{n_{1212}^{(N00N)}(\zeta)}{n_1^{(N00N)}(\zeta) n_2^{(N00N)}(\zeta)} - 1, \tag{35}$$

where we have defined the following second order two-point correlation function,

$$n_{1212}^{(N00N)}(\zeta) = e^{4\beta\zeta} \left\langle N00N | \hat{b}_1^\dagger(\zeta)\, \hat{b}_2^\dagger(\zeta)\, \hat{b}_1(\zeta)\, \hat{b}_2(\zeta)\, | N00N \right\rangle, \tag{36}$$

and we take the mean photon numbers as defined in Equation (27).

For the two-photon *N00N* state, $N = 2$, the mean photon numbers at the waveguides are provided by the following expressions,

$$
\begin{aligned}
n_1^{(2002)}(\zeta) &= e^{2\beta\zeta} \left| \hat{U}_{11}(\zeta) \right|^2 + e^{2\beta\zeta} \left| \hat{U}_{12}(\zeta) \right|^2 + n_1^{(00)}(\zeta), \\
n_2^{(2002)}(\zeta) &= e^{2\beta\zeta} \left| \hat{U}_{21}(\zeta) \right|^2 + e^{2\beta\zeta} \left| \hat{U}_{22}(\zeta) \right|^2 + n_2^{(00)}(\zeta),
\end{aligned}
\tag{37}
$$

where the first and second terms corresponds to the stimulated generation, and the third term is related to the spontaneous generation and can be recovered from Section 3 for each and every configuration. The general expression for the two-point correlation is summarized in the following,

$$
\begin{aligned}
n_{1212}^{(2002)}(\zeta) &= e^{4\beta\zeta} |\hat{U}_{11}\hat{U}_{21} + \hat{U}_{12}\hat{U}_{22}|^2 + n_1^{(00)}(\zeta) n_2^{(00)}(\zeta) + |n_{12}^{(00)}(\zeta)|^2 \\
&+ e^{2\beta\zeta} \left[n_1^{(00)}(\zeta) \left(|\hat{U}_{21}|^2 + |\hat{U}_{22}|^2 \right) + n_2^{(00)}(\zeta) \left(|\hat{U}_{11}|^2 + |\hat{U}_{12}|^2 \right) \right] \\
&+ 2\, e^{2\beta\zeta} \, \Re \left[n_{12}^{(00)}(\zeta) \left(\hat{U}_{21}^* \hat{U}_{11} + \hat{U}_{22}^* \hat{U}_{12} \right) \right].
\end{aligned}
\tag{38}
$$

For the **balanced gain-loss** dimer, we use the spontaneous generation terms, $n_j^{(00)}(\zeta)$, provided by Equation (13), and the first order two-point correlation, $n_{12}^{(00)}(\zeta)$, from Equation (21). In the **gain-gain** dimer, the expressions for $n_j^{(00)}(\zeta)$ and $n_{12}^{(00)}(\zeta)$ are given by Equations (14) and (22), in that order. In the **gain-passive** dimer, we only need the definitions provided by Equations (15) and (23) for the spontaneous generation and the first order two-point correlation, respectively. Finally, for **passive-loss** and **loss-loss** dimers, the spontaneous generation is null, $n_j^{(00)}(\zeta) = n_{12}^{(00)}(\zeta) = 0$, such that

$$
n_{1212}^{(2002)}(\zeta) = e^{4\beta\zeta} |\hat{U}_{11}\hat{U}_{21} + \hat{U}_{12}\hat{U}_{22}|^2.
\tag{39}
$$

Obviously, the adequate parameter β from Table 1 should be used for each configuration. These expressions become complicated enough that we must rely on a figure-based analysis.

Figure 5 shows the two-point Mandel parameter for the different dimer configurations in the $\mathcal{PT}$-symmetric, the Kato exceptional point, and broken $\mathcal{PT}$-symmetry regimes. Now, the initial state shows anti-bunching, a negative value of the two-point Mandel parameter, due to its delocalization of the two-photon state. For the balanced gain-loss, the initial state propagates and, after a critical propagation distance, presents bunching in the $\mathcal{PT}$-symmetric, Figure 5a, the Kato exceptional point, Figure 5b, broken symmetry, Figure 5c, and regimes. The gain-gain and gain-passive dimers show a similar, more interesting behavior where the propagated state oscillates between an anti-bunched and bunched state in the $\mathcal{PT}$-symmetric regime, Figure 5d,g, and the transition from an anti-bunched to a bunched state in the Kato exceptional point, Figure 5e,h, and broken symmetry regimes, Figure 5f,i. Finally, in both the passive-loss and loss-loss dimers, the probability of losing photons makes the propagated state anti-bunched in all regimes, Figure 5j–l.

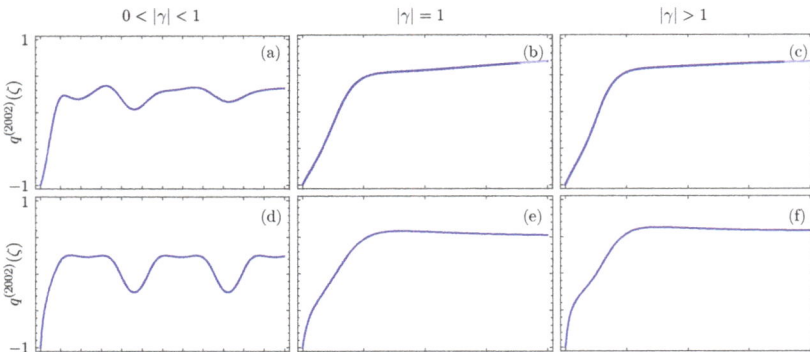

Figure 5. *Cont.*

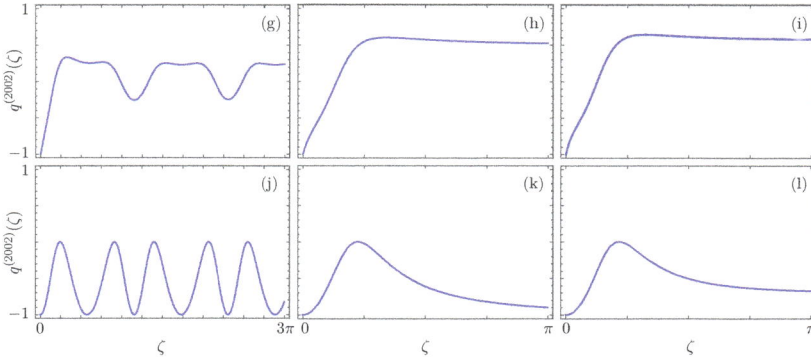

Figure 5. Photon bunching and anti-bunching shown in terms of the $q^{(2002)}(\zeta)$ parameter for different realizations of the effective $\mathcal{PT}$-symmetric dimer. The first row, (**a**–**c**), shows balanced gain-loss, the second row, (**d**–**f**), shows gain-gain, the third row, (**g**–**i**), shows gain-passive, and the fourth row, (**j**–**l**), shows both passive-loss and loss-loss configurations in the $\mathcal{PT}$-symmetric regime, the first column with $|\gamma| = 0.5$, the Kato point with $|\gamma| = 1$, the second column, and the broken symmetry regime, the third column with $|\gamma| = 1.2$.

7. Conclusions

We have calculated the propagation of photon states through non-Hermitian linear dimers with Gaussian gain and losses. As a practical example, we studied propagation in the different experimentally feasible configurations of the $\mathcal{PT}$-symmetric dimer and show that each and every configuration presents a different signature in the propagation of vacuum, single and two-photon states. These signatures go beyond the mean photon number at the waveguides, which can be split into spontaneous and stimulated generation components, and can be found, for example, in the second order two-point correlation of the photon state propagating through the dimers.

First, let us focus on the mean photon number signatures. They show that the propagation of vacuum and single photons through dimers in the $\mathcal{PT}$-symmetry regime might provide us with directional coupler devices controllable by the propagation length. Furthermore, devices that provide full single photon switching can only be designed using dimers in the passive-loss and loss-loss configuration. Outside the $\mathcal{PT}$-symmetric regime, the asymptotic stability of the dimers in any given configuration suggest their use as symmetric intensity sources at the Kato exceptional point and, in the broken symmetry regime, they might provide asymmetric intensity sources where the intensities ratio can be controlled by the linear properties of the material.

On the other hand, second order two-point correlation signatures can help us choose configurations depending on the type of state needed for a particular application. For example, if spatially separable states are needed, choosing a configuration showing photon bunching comes naturally. Spatially entangled states are provided by passive-loss and loss-loss configurations where the initial two-photon state is delocalized in the waveguides.

We want to note that our description of linear media is far from complete. Real world linear materials saturate; this induces further restrictions on the model that we do not consider in this manuscript and could provide further avenues of research. Furthermore, the addition of new interactions, like introducing a two-level system to create more complex hybrid devices [20], requires a first principle reformulation in order to recover adequate effective models.

Acknowledgments: José Delfino Huerta Morales acknowledges financial support from CONACYT #294921 PhD grant.

Conflicts of Interest: The authors declare no conflict of interest.

Abbreviations

The following abbreviations are used in this manuscript:

MDPI	Multidisciplinary Digital Publishing Institute.
$\mathcal{PT}$	Parity-Time.
$\Re(\alpha)$ and $\Im(\alpha)$	Real and imaginary parts of a complex number α, in that order.
CONACYT	Consejo Nacional de Ciencia y Tecnología

References

1. Huerta Morales, J.D.; Guerrero, J.; Lopez-Aguayo, S.; Rodríguez-Lara, B.M. Revisiting the optical $\mathcal{PT}$-symmetric dimer. *Symmetry* **2016**, *8*, 83, doi:10.3390/sym8090083.
2. Somekh, S.; Garmire, E.; Yariv, A.; Garvin, H.L.; Hunsperger, R.G. Channel optical waveguide directional couplers. *Appl. Phys. Lett.* **1973**, *22*, 46–47.
3. Ruschhaupt, A.; Delgado, F.; Muga, J.G. Physical realization of $\mathcal{PT}$-symmetric potential scattering in a planar slab waveguide. *J. Phys. A Math. Gen.* **2005**, *38*, L171–L176.
4. El-Ganainy, R.; Makris, K.G.; Christodoulides, D.N.; Musslimani, Z.H. Theory of coupled optical $\mathcal{PT}$-symmetric structures. *Opt. Lett.* **2007**, *32*, 2632–2634.
5. Rüter, C.E.; Makris, K.G.; El-Ganainy, R.; Christodoulides, D.N.; Kip, D. Observation of parity-time symmetry in optics. *Nat. Phys.* **2010**, *6*, 192–195.
6. Guo, A.; Salamo, G.J.; Duchesne, D.; Morandotti, R.; Volatier-Ravat, M.; Aimez, V.; Siviloglou, G.A.; Christodoulides, D.N. Observation of $\mathcal{PT}$-symmetry breaking in complex optical potentials. *Phys. Rev. Lett.* **2009**, *103*, 093902, doi:10.1103/PhysRevLett.103.093902.
7. Ornigotti, M.; Szameit, A. Quasi $\mathcal{PT}$-symmetry in passive photonic lattices. *J. Opt.* **2014**, *16*, 065501, doi:10.1088/2040-8978/16/6/065501.
8. Peng, B.; Ozdemir, S.K.; Lei, F.C.; Monifi, F.; Gianfreda, M.; Long, G.L.; Fan, S.H.; Nori, F.; Bender, C.M.; Yang, L. Parity-time-symmetric whispering-gallery microcavities. *Nat. Phys.* **2014**, *10*, 394–398.
9. Hodaei, H.; Miri, M.A.; Heinrich, M.; Christodoulides, D.N.; Khajavikhan, M. Parity-time-symmetric microring lasers. *Science* **2014**, *346*, 975–978.
10. Schindler, J.; Lin, Z.; Lee, J.M.; Ramezani, H.; Ellis, F.M.; Kottos, T. $\mathcal{PT}$-symmetric electronics. *J. Phys A Math. Theor.* **2012**, *45*, 444029, doi:10.1088/1751-8113/45/44/444029.
11. Politi, A.; Cryan, M.J.; Rarity, J.G.; Yu, S.; O'Brien, J.L. Silica-on-Silicon waveguide quantum circuits. *Science* **2008**, *320*, 646–649.
12. Bromberg, Y.; Lahini, Y.; Morandotti, R.; Silberberg, Y. Quantum and Classical Correlations in Waveguide Lattices. *Phys. Rev. Lett.* **2009**, *102*, 253904, doi:10.1103/PhysRevLett.102.253904.
13. Peruzzo, A.; Lobino, M.; Matthews, J.C.F.; Matsuda, N.; Politi, A.; Poulios, K.; Zhou, X.Q.; Lahini, Y.; Ismail, N.; Wörhoff, K.; et al. Quantum Walks of Correlated Photons. *Science* **2010**, *329*, 1500–1503.
14. Joglekar, Y.N.; Thompson, C.; Scott, D.D.; Vemuri, G. Optical waveguide arrays: Quantum effects and $\mathcal{PT}$ symmetry breaking. *Eur. Phys. J. Appl. Phys.* **2013**, *63*, 30001, doi:10.1051/epjap/2013130240.
15. Agarwal, G.S.; Qu, K. Spontaneous generation of photons in transmission of quantum fields in $\mathcal{PT}$-symmetric optical systems. *Phys. Rev. A* **2012**, *85*, 031802, doi:10.1103/PhysRevA.85.031802.
16. Gräfe, M.; Heilmann, R.; Keil, R.; Eichelkraut, T.; Heinrich, M.; Nolte, S.; Szameit, A. Correlations of indistinguishable particles in non-Hermitian lattices. *New J. Phys.* **2013**, *15*, 033008, doi:10.1088/1367-2630/15/3/033008.
17. Scully, M.O.; Zubairy, M.S. *Quantum Optics*; Cambridge University Press: Cambridge, UK, 2001.
18. Rodríguez-Lara, B.M.; Guerrero, J. Optical finite representation of the Lorentz group. *Opt. Lett.* **2015**, *40*, 5682–5685.
19. Mandel, L. Sub-Poissonian photon statistics in resonance fluorescence. *Opt. Lett.* **1979**, *4*, 205–207.
20. Lepert, G.; Trupke, M.; Hartmann, M.J.; Plenio, M.B.; Hinds, E.A. Arrays of waveguide-coupled optical cavities that interact strongly with atoms. *New. J. Phys.* **2011**, *13*, 113002, doi:10.1088/1367-2630/13/11/113002.

applied
sciences

MDPI

Article

Demonstration of High-Speed Optical Transmission at 2 μm in Titanium Dioxide Waveguides

Manon Lamy, Christophe Finot, Julien Fatome, Juan Arocas, Jean-Claude Weeber and Kamal Hammani *

Laboratoire Interdisciplinaire Carnot de Bourgogne (ICB), UMR 6303 CNRS–Université de Bourgogne Franche-Comté, 9 Avenue Alain Savary, BP 47870, 21078 Dijon CEDEX, France; manon.lamy@u-bourgogne.fr (M.L.); cfinot@u-bourgogne.fr (C.F.); julien.fatome@u-bourgogne.fr (J.F.); juan.arocas@u-bourgogne.fr (J.A.); jean-claude.weeber@u-bourgogne.fr (J.-C.W.)
* Correspondence: kamal.hammani@u-bourgogne.fr

Academic Editor: Boris Malomed
Received: 16 May 2017; Accepted: 15 June 2017; Published: 17 June 2017

Abstract: We demonstrate the transmission of a 10-Gbit/s optical data signal in the 2 μm waveband into titanium dioxide waveguides. Error-free transmissions have been experimentally achieved taking advantage of a 23-dB insertion loss fiber-to-fiber grating-based injection test-bed platform.

Keywords: integrated optics; optical gratings; titanium dioxide; optical communications at 2 μm wavelengths

1. Introduction

Nowadays, optical communication traffic continuously increases, inexorably approaching a "capacity crunch" [1–3]: the conventional C-band around 1.55 μm will not be sufficient anymore and alternative approaches have to be adopted [4]. Recently, the 2 μm spectral region has been suggested as a new transmission window [5], benefiting from the emergence of thulium-doped fiber amplifiers (TDFA) with broadband and high gain spanning from 1900 to 2100 nm [6]. This has stimulated studies of dedicated photonic components such as InP-based modulators [7,8] or arrayed waveguide gratings [9]. High bit rate communications over distances exceeding one hundred meters have already been successfully demonstrated [10–13] in low-loss hollow core bandgap photonic fibers designed to present minimal losses around 2000 nm [14] or in solid-core single mode fibers [15].

Optical transmissions over much shorter distances (typically a few hundred micrometers) also deserve interest in the context of on-board connections and photonic routing operations. Therefore, this open issue requires further experimental investigations to evaluate the potential of various materials transparent in this new spectral band. Optimally, such a material should be also transparent in other telecommunication bands ranging from O-band (1310 nm) to C-band (1550 nm) and eventually also in the 850-nm band. Regarding the recent research, a natural choice could be silicon nitride (Si_3N_4), which has already stimulated many works in the visible but also in the mid-infrared range, mainly in the context of frequency combs [16]. In this new contribution, we are interested in another material which remains to date relatively unexplored: titanium dioxide (TiO_2). This cost-efficient material is indeed transparent from visible to mid-infrared wavelengths [17] and can be considered as a complementary metal-oxide semiconductor (CMOS) compatible material [18]. Compared to Si_3N_4, TiO_2 presents several advantages. Among them, it presents lower stress constraints for thicknesses beyond 250 nm and has an easier deposition process possible at lower temperatures. Note also that it exhibits a higher linear refractive index, which is critical for a stronger confinement [19]. Up to now, detailed studies have taken advantage of its negative thermo-optic coefficient [18,20] and its transparency in the visible range [21,22], or have reported on its linear and nonlinear properties in the C-band [19,23,24]. In this work, we experimentally explore, for the first time, TiO_2 as an efficient medium for a photonic component operating at 2 μm. After describing the design and fabrication of our TiO_2 waveguides, we detail the experimental setup under use and validate the device for error-free transmission of a 10 Gbit/s on-off keying signal for both a subwavelength single mode waveguide and a multimode waveguide.

2. Design and Fabrication of the Photonic Structure

2.1. Design and Fabrication of the Photonic Structure

Whatever the platform involved for the design, a critical issue in integrated photonics is always how to efficiently couple the light into the device. In a dedicated article [25], we recently investigated and experimentally validated a new kind of metal grating that is embedded directly within the dielectric layer instead of being deposited on it [26]. Thus, we were able to efficiently couple a 1.55 μm signal into a TiO_2 photonic waveguide. Here, a similar design is exploited, as depicted in Figure 1a. More precisely, the structure consists of a metal (Au) grating embedded between two TiO_2 layers on a glass substrate.

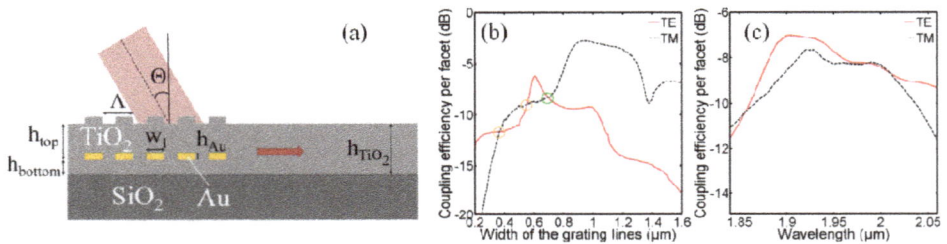

Figure 1. (a) Embedded metal gratings in TiO_2 layout; (b) Numerical simulations of the coupling efficiency for one facet as a function of the width w_1 of the grating lines for the following parameters: h_{bottom} = 70 nm, h_{Au} = 5 nm, h_{top} = 234 nm, Λ = 1900 nm at the central wavelength of our laser source (1.98 μm). The results for the transverse electric (TE) mode are compared to those for the transverse magnetic (TM) mode. The circles highlight the values where the efficiencies of both modes are equal. The green circle corresponds to the best value; (c) Corresponding coupling efficiency per facet for the TE and TM modes as a function of the wavelength (for w_1 = 700 nm—green circle on panel (b)).

Using a commercial finite element-based software (Comsol Multiphysics), we can optimize the parameters of the design to obtain the best coupling around the 2 μm wavelength. We have taken into account the fabrication and experimental setup constraints, leading to fixed values for the bottom layer height h_{bottom} = 70 nm and of the incident angle Θ = 30°. Thus, using a Monte Carlo algorithm varying (h_{Au}, h_{top}, Λ, w_1) in a 4D parameter space, the coupling efficiency could reach 52% for transverse magnetic (TM) mode when the gold height h_{Au} = 57 nm, the top layer height h_{top} = 234 nm, the period Λ = 1904 nm, and the width of grating lines w_1 = 980 nm. However, with these parameters, the coupling efficiency for the transverse electric (TE) mode drops down to 10%. Similarly, the optimization of parameters for the TE mode gives an efficiency reaching up to 36% for h_{Au} = 45 nm, h_{top} = 234 nm, Λ = 1679 nm, and w_1 = 356 nm, whereas those parameters give a poor efficiency of 1% for the TM mode.

However, contrary to Reference [25], our aim is not to reach the best coupling efficiency. Here, we tried to find numerically the geometric parameters that allow a fair coupling efficiency in both the TE and TM modes for a slab. From the previous optimization, the top layer should be 234 nm and, given that h_{bottom} = 70 nm, the total TiO_2 thickness considered is 304 nm. The thickness of the gold lines h_{Au} is chosen to be 56 nm, while the period Λ = 1900 nm. To adjust the efficiency of the TM mode compared to the TE mode, we adjust the filling factor by sequentially varying the width w_1, as shown in Figure 1b. It appears that three values of w_1 allow a similar coupling for the TE and TM modes, but the width of the grating lines that gives the best efficiency (-8.5 dB) is around 700 nm.

2.2. Modal Analysis

Here, we focus our attention on two strip waveguides (Figure 2a): an 8.0-µm wide waveguide and a subwavelength 1.6-µm wide waveguide. The 8-µm wide waveguide is clearly multimode with nine TE modes and three TM modes. Considering a refractive index of 1.44 for the glass substrate and 2.41 for the TiO$_2$ layer [17], Figure 2b shows that the first TE and TM modes in the multimode (MM) waveguide have an effective index of 1.87 and 1.49, respectively. Note that the TM mode is clearly much less confined than the TE mode, as can be seen in Figure 2(b2), but over the length considered, the lossy behavior of the TM mode has very low impact, as demonstrated in the following sections. The subwavelength 1.6-µm wide waveguide is single-mode (SM) with an effective index of the TE mode around 1.78, shown in Figure 2c.

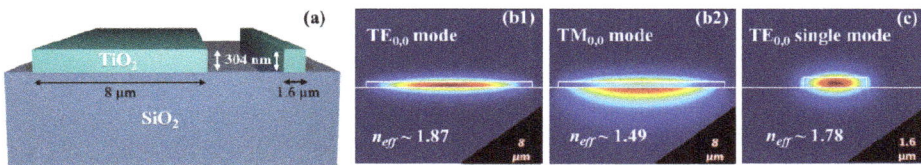

Figure 2. (**a**) Sketch of the cross-section of the 304-nm strip waveguides. Corresponding mode profile of the electric field and associated effective index for (**b**) a multimode waveguide (width, 8 µm) and (**c**) a subwavelength waveguide (width, 1.6 µm). Subplot (**b1**) corresponds to the fundamental TE mode, whereas subplot (**b2**) is related to fundamental TM mode.

2.3. Fabrication

The fabrication process relies on traditional techniques. Titanium dioxide layers are deposited on a glass substrate by reactive direct current (DC) magnetron sputtering of a 99.9% pure titanium target under argon and oxygen control atmosphere. Electron-beam lithography, followed by thermal gold evaporation (here, 3 nm of chromium are used as an adhesion layer) and a lift-off process is used to fabricate the gold gratings. Then the top layer of TiO$_2$ is deposited, followed by overlay electron-beam lithography. After metallic mask evaporation, reactive ion etching is performed to make the waveguides. Finally, wet etching removes the mask to obtain the final device to be tested.

As mentioned previously, we fabricated two structures of interest (Figure 3): an MM waveguide (with a width of 8 µm) and an SM waveguide (slightly overexposed leading to a width of 1.65 µm).

These waveguides have the same length fixed at 575 µm but, for the SM waveguide, two 85-µm long tapers (with a maximal width of 30 µm) are used at the input and output. The parameters measured on the fabricated device differ slightly from the targeted one: we measured a 304-nm total thicknesss with h_{bottom} = 69 nm and h_{top} = 235 nm. Moreover, the waveguide width is actually 1.65 µm, which induces the existence of the TE$_{0,1}$ with an effective index close to 1.47. For each structure, two widths of grating lines (related to filling factor) have been fabricated. The two targeted widths were the best values for which TE and TM were equal in Figure 1b (i.e., 550 and 700 nm). However, due to overexposure, we measured both widths to be 630 and 860 nm.

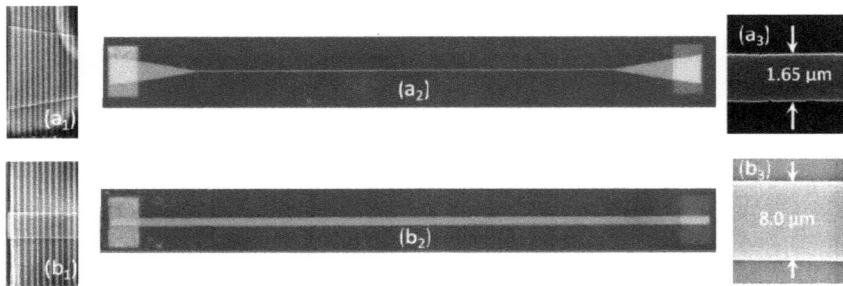

Figure 3. Images of (**a**) a subwavelength waveguide and (**b**) a large waveguide considered as a slab. Insets 1 show scanning electron microscopy (SEM) pictures of embedded metal grating on one end of the waveguides whereas insets 2 show optical images of the waveguides. Contrary to the slab, the subwavelength waveguide is equipped with tapers. Insets 3 correspond to SEM pictures of the subwavelength that has a width of 1.65 µm (**a3**), and the slab which is actually 8.0 µm wide (**b3**).

2.4. *Test of the Device*

We then characterized the coupling efficiency of the waveguides with a setup similar as in Reference [25]. Two focusers oriented with an angle of 30° were used to inject and collect the signal at the input and output gratings, respectively. The injection of light emitted by an amplified spontaneous emission (ASE) source spanning from 1900 to 2050 nm was adjusted (thanks to two cameras operating in the visible and 2-µm ranges (model Xeva 2.35-320, Xenics nv, Leuven, Belgium). After transmission in the waveguide and decoupling by the grating, the output light was collected by a lensed-fiber focuser and recorded with an optical spectrum analyzer (OSA) (model AQ6375B, Yokogawa Electric Corporation, Musashino, Japan). Figure 4 shows the total insertion losses as a function of the injected wavelength for the two waveguides under test. Note that those transmission spectra are normalized with respect to a reference spectrum obtained by optimizing the light collected by the output focuser after specular reflection of the incident light onto a gold mirror [25]. It is noted that, in Figure 4, oscillations of high amplitude and short period are particularly marked for the large waveguide. These spectral oscillations are attributed to a Fabry–Perot effect, the spectral range between two maxima being in agreement with what can be expected from the roundtrip distance. The experimental loss should be twice as large as the value expected from our numerical simulation (Figure 1) performed for one facet, given that we neglect the propagation loss. Then, the discrepancy between the numerical simulations and the maximum experimental efficiency is about 3 dB, mainly attributed to the strong dependency on the width of the grating lines and also probably due to fabrication issues (in particular related to the roughness). Even though, for the MM waveguide configuration, the TE mode was expected to be as efficient as the TM mode for a grating line width of 700 nm, here it was the case for $w_1 = 630$ nm. Regarding the TE SM waveguide, as expected, the coupling efficiency is similar to the one obtained for the large waveguide. Let us once again recall that a (6 dB) better efficiency can be reached for optimized parameters, but our goal here was to establish a fair comparison between the MM and SM components.

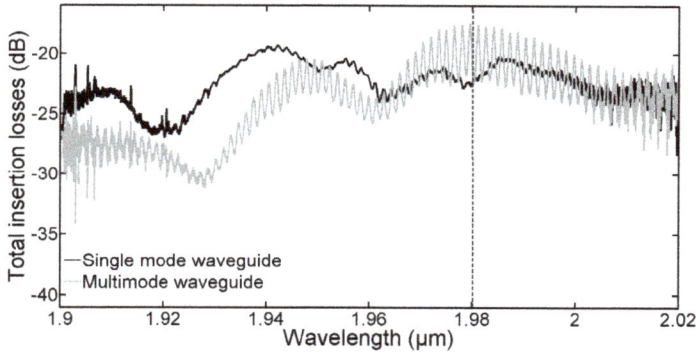

Figure 4. Total insertion losses as a function of the injected wavelength for an MM waveguide (grey curve) and an SM waveguide (black curve) obtained with the amplified spontaneous emission (ASE) source (for $w_1 = 630$ nm). The oscillations of short period on the grey curve are due to a Perot–Fabry effect. The grey dashed line corresponds to the central wavelength of the laser source used in the next section.

3. Validation of the Transmission of a 10 Gbit/s Signal

In order to demonstrate the suitability of our TiO_2 devices for 2-µm optical communications, we implemented the experimental setup detailed in Figure 5, based on 2-µm commercially available devices. The transmitter (TX) was based on a laser diode centered at 1980 nm and was intensity modulated by means of a commercial Niobate–Lithium modulator (model MX2000-LN-10, iXblue Photonics, Besançon, France). The Non-Return-to-Zero On-Off-Keying signal under test was a $2^{31}-1$ pseudorandom bit sequence (PRBS) at 10 Gbit/s. Since the SM waveguide is polarization-sensitive, a polarization controller was used after the intensity modulator. Then, a thulium-doped fiber amplifier (TDFA) was used before a 90/10 coupler, allowing us to monitor the power injected into the waveguide. Note that the variable optical attenuator is usually implemented just in front of the receiver, but due to limited sensitivity of the available power meters working at 2 µm, it was more convenient to insert it before the waveguide.

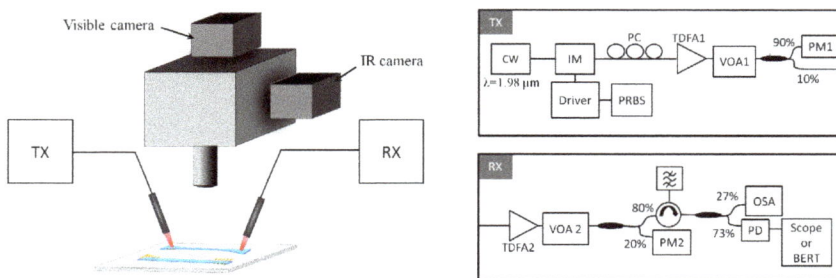

Figure 5. Experimental setup for a 10-Gbit/s 2-µm optical transmission. CW: continuous wave; IM: intensity modulator; PC: polarization controller; PRBS: pseudorandom binary sequence; TDFA: thulium-doped fiber amplifier; VOA: variable optical attenuator; PM: power meter; OSA: optical spectrum analyzer; PD: photodiode; BERT: bit error rate tester (model MU181040A, Anritsu Corporation, Atsugi-shi, Japan).

The receiver (RX) was based on a second TDFA. This TDFA was set to work with a constant gain instead of a constant output power, as usually used in C-band. Therefore, a variable optical attenuator was implemented at its output to ensure that the photodiode operated at a constant power level. An optical bandpass filter (0.64 nm bandwidth) (model OETFG-100, O/E Land Inc., LaSalle, QC, Canada) was also inserted at the output of the system, in order to limit the accumulation of amplified spontaneous emission from the TDFAs. The signal was finally analyzed both in the spectral and temporal domains. An optical spectrum analyzer was used to measure the optical signal-to-noise ratio (OSNR) of the received signal (using the usual 0.1 nm noise bandwidth), whereas a photodiode (12.5 GHz electrical bandwidth) (model 818-BB-51F, Newport corporation, Irvine, CA, USA) enabled bit error rate measurements in addition to recording of the output eye diagram on a high-speed sampling oscilloscope.

Figure 6 summarizes the results obtained at 1980 nm. Error-free operation can be achieved for back-to-back measurements as well as in the presence of the TiO_2 waveguides. Examples of the corresponding eye-diagrams are provided in panels (a): in both configurations, a widely open eye can be recorded and the insertion of the waveguides under test does not induce any visible degradations of the transmission.

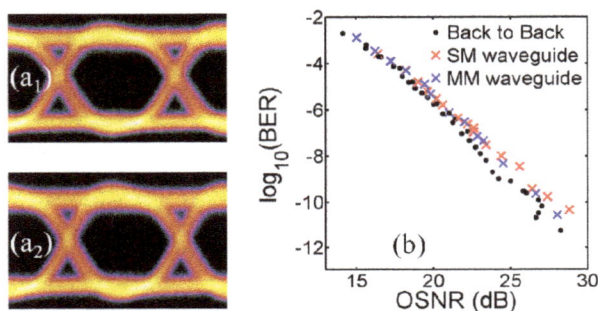

Figure 6. (**a**) Eye diagrams for the back-to-back configuration and after transmission in the 1.65 μm wide waveguide (panels 1 and 2 respectively). In both cases, eyes diagrams were recorded for error-free measurements; (**b**) Bit-Error-Rate (BER) as a function of optical signal-to-noise ratio (OSNR) for the two previously described waveguides. On the graph, the black points of measurements are associated with back-to-back configuration. The crosses are used for BER measurements for TiO_2 devices (red for the 1.65-μm SM wide waveguide and blue for the 8-μm wide MM waveguide).

The quality of the transmission through the TiO_2 waveguide was more quantitatively evaluated through systematic measurements of the Bit-Error-Rate according to the OSNR on the receiver. The results obtained for the various configurations are summarized in Figure 6b. From a general point of view, the global trends are very similar, with very moderate penalty (~1.5 dB) obtained after transmission through the waveguides and compared to the back-to-back configuration for BER lower than 10^{-8}. Moreover, no significant difference was observed between the SM and MM waveguides. Therefore, for the length of propagation under consideration, the multimode nature of the waveguide does not seem to impair the transmission quality.

Before concluding, we would like to emphasize here the fact that such measurements are not as straightforward as in the C-band. Indeed, the 2-μm devices have not reached the same level of maturity and it is still difficult to find such devices commercially available. This experiment shows that despite the lack of optimized devices, error-free transmission is possible, confirming that this new waveband is definitely an effective alternative, especially if new solutions appear in the next few years.

4. Conclusions

To conclude (thanks to embedded metal gratings) we have been able to efficiently couple an incident light beam into a TiO_2 waveguide in the 2-μm spectral range. This particular design allows us to demonstrate, for the first time, an error-free transmission of a 2-μm optical data stream at 10 Gbit/s in a 575-μm long TiO_2 waveguide. A full set of BER measurements has been performed with a fair comparison between a single-mode 1.65-μm wide waveguide and a multimode 8-μm wide waveguide insensitive to the polarization. No significant difference was then observed. With the future progress of the emitter/receiver stages, we believe that the present component will also be able to handle higher transmission speeds.

This study paves the way to integrated photonics at 2 μm, and introduces titanium dioxide as a serious candidate for photonics from the visible to the mid-infrared range. With technological progress and the maturity gain that can be expected in the near future for fiber and optoelectronics solutions operating around 2 μm, we are confident in the possibility of involving longer TiO_2 waveguides up to a few centimeters in length.

Acknowledgments: This work is financially supported by PARI PHOTCOM Région Bourgogne, by Carnot Arts Institute (PICASSO 2.0 project), by the Institut Universitaire de France, and by FEDER-FSE Bourgogne 2014/2020. The research work has benefited from the PICASSO experimental platform of the University of Burgundy. We thank M. G. Nielsen for advices and fruitful discussions on the grating design issues.

Author Contributions: Manon Lamy, Kamal Hammani, Christophe Finot and Julien Fatome conceived, designed and performed the experiment of optical transmission; Manon Lamy, Kamal Hammani, Juan Arocas and Jean-Claude Weeber designed and fabricated the optical waveguides and the associated gratings. Manon Lamy, Kamal Hammani and Jean-Claude Weeber took part in the numerical modelling and interpretation. All the authors contribute to the analysis of the results and to the writing of the manuscript. Kamal Hammani and Christophe Finot supervised the project.

Conflicts of Interest: The authors declare no conflict of interest.

References

1. Richardson, D.J. Filling the light pipe. *Science* **2010**, *330*, 327–328. [CrossRef] [PubMed]
2. Ellis, A.D.; Suibhne, N.M.; Saad, D.; Payne, D.N. Communication networks beyond the capacity crunch. *Philos. Trans. R. Soc. A* **2016**, *374*. [CrossRef] [PubMed]
3. Desurvire, E.B. Capacity demand and technology challenges for lightwave systems in the next two decades. *J. Lightwave Technol.* **2006**, *24*, 4697–4710. [CrossRef]
4. Richardson, D.J. New optical fibres for high-capacity optical communications. *Philos. Trans. R. Soc. A* **2016**, *374*. [CrossRef] [PubMed]
5. Kavanagh, N.; Sadiq, M.; Shortiss, K.; Zhang, H.; Thomas, K.; Gocalinska, A.; Zhao, Y.; Pelucchi, E.; Brien, P.O.; Peters, F.H.; et al. Exploring a new transmission window for telecommunications in the 2 μm waveband. In Proceedings of the 18th International Conference on Transparent Optical Networks (ICTON 2016), Trento, Italy, 10–14 July 2016; pp. 1–4.
6. Li, Z.; Heidt, A.M.; Daniel, J.M.O.; Jung, Y.; Alam, S.U.; Richardson, D.J. Thulium-doped fiber amplifier for optical communications at 2 μm. *Opt. Express* **2013**, *21*, 9289–9297. [CrossRef] [PubMed]
7. Ye, N.; Gleeson, M.R.; Sadiq, M.U.; Roycroft, B.; Robert, C.; Yang, H.; Zhang, H.; Morrissey, P.E.; Mac Suibhne, N.; Thomas, K.; et al. InP-based active and passive components for communication systems at 2 μm. *J. Lightwave Technol.* **2015**, *33*, 971–975. [CrossRef]
8. Sadiq, M.U.; Gleeson, M.R.; Ye, N.; O'Callaghan, J.; Morrissey, P.; Zhang, H.Y.; Thomas, K.; Gocalinska, A.; Pelucchi, E.; Gunning, F.C.G.; et al. 10 Gb/s InP-based Mach–Zehnder modulator for operation at 2 μm wavelengths. *Opt. Express* **2015**, *23*, 10905–10913. [CrossRef] [PubMed]
9. Zhang, H.; Gleeson, M.; Ye, N.; Pavarelli, N.; Ouyang, X.; Zhao, J.; Kavanagh, N.; Robert, C.; Yang, H.; Morrissey, P.E.; et al. Dense WDM transmission at 2 μm enabled by an arrayed waveguide grating. *Opt. Lett.* **2015**, *40*, 3308–3311. [CrossRef] [PubMed]
10. Petrovich, M.N.; Poletti, F.; Wooler, J.P.; Heidt, A.M.; Baddela, N.K.; Li, Z.; Gray, D.R.; Slavík, R.; Parmigiani, F.; Wheeler, N.V.; et al. Demonstration of amplified data transmission at 2 μm in a low-loss wide bandwidth hollow core photonic bandgap fiber. *Opt. Express* **2013**, *21*, 28559–28569. [CrossRef] [PubMed]

11. Zhang, H.; Kavanagh, N.; Li, Z.; Zhao, J.; Ye, N.; Chen, Y.; Wheeler, N.V.; Wooler, J.P.; Hayes, J.R.; Sandoghchi, S.R.; et al. 100 Gbit/s WDM transmission at 2 μm: Transmission studies in both low-loss hollow core photonic bandgap fiber and solid core fiber. *Opt. Express* **2015**, *23*, 4946–4951. [CrossRef] [PubMed]

12. Hayes, J.R.; Sandoghchi, S.R.; Bradley, T.D.; Liu, Z.; Slavik, R.; Gouveia, M.A.; Wheeler, N.V.; Jasion, G.T.; Chen, Y.; Numkam-Fokoua, E.; et al. Antiresonant hollow core fiber with octave spanning bandwidth for short haul data communications. In Proceedings of the OFC Conference Postdeadline Papers: p Th5A.3, Anaheim, CA, USA, 20 March 2016.

13. Liu, Z.; Chen, Y.; Li, Z.; Kelly, B.; Phelan, R.; O'Carroll, J.; Bradley, T.; Wooler, J.P.; Wheeler, N.V.; Heidt, A.M.; et al. High-capacity directly modulated optical transmitter for 2 μm spectral region. *J. Lightwave Technol.* **2015**, *33*, 1373–1379. [CrossRef]

14. Roberts, P.J.; Couny, F.; Sabert, H.; Mangan, B.J.; Williams, D.P.; Farr, L.; Mason, M.W.; Tomlinson, A.; Birks, T.A.; Knight, J.C.; et al. Ultimate low loss of hollow-core photonic crystal fibres. *Opt. Express* **2005**, *13*, 236–244. [CrossRef] [PubMed]

15. Xu, K.; Sun, L.; Xie, Y.; Song, Q.; Du, J.; He, Z. Transmission of IM/DD signals at 2 μm wavelength using PAM and CAP. *IEEE Photonics J.* **2016**, *8*, 1–7. [CrossRef]

16. Luke, K.; Okawachi, Y.; Lamont, M.R.E.; Gaeta, A.L.; Lipson, M. Broadband mid-infrared frequency comb generation in a Si_3N_4 microresonator. *Opt. Lett.* **2015**, *40*, 4823–4826. [CrossRef] [PubMed]

17. Kischkat, J.; Peters, S.; Gruska, B.; Semtsiv, M.; Chashnikova, M.; Klinkmüller, M.; Fedosenko, O.; Machulik, S.; Aleksandrova, A.; Monastyrskyi, G.; et al. Mid-infrared optical properties of thin films of aluminum oxide, titanium dioxide, silicon dioxide, aluminum nitride, and silicon nitride. *Appl. Opt.* **2012**, *51*, 6789–6798. [CrossRef] [PubMed]

18. Shang, K.; Djordjevic, S.S.; Li, J.; Liao, L.; Basak, J.; Liu, H.-F.; Yoo, S.J.B. Cmos-compatible titanium dioxide deposition for athermalization of silicon photonic waveguides. In Proceedings of the Optical Society of America CLEO: 2013, San Jose, CA, USA, 9 June 2013.

19. Bradley, J.D.B.; Evans, C.C.; Choy, J.T.; Reshef, O.; Deotare, P.B.; Parsy, F.; Phillips, K.C.; Lončar, M.; Mazur, E. Submicrometer-wide amorphous and polycrystalline anatase TiO_2 waveguides for microphotonic devices. *Opt. Express* **2012**, *20*, 23821–23831. [CrossRef] [PubMed]

20. Reshef, O.; Shtyrkova, K.; Moebius, M.G.; Griesse-Nascimento, S.; Spector, S.; Evans, C.C.; Ippen, E.; Mazur, E. Polycrystalline anatase titanium dioxide microring resonators with negative thermo-optic coefficient. *J. Opt. Soc. Am. B* **2015**, *32*, 2288–2293. [CrossRef]

21. Choy, J.T.; Bradley, J.D.B.; Deotare, P.B.; Burgess, I.B.; Evans, C.C.; Mazur, E.; Lončar, M. Integrated tio2 resonators for visible photonics. *Opt. Lett.* **2012**, *37*, 539–541. [CrossRef] [PubMed]

22. Häyrinen, M.; Roussey, M.; Säynätjoki, A.; Kuittinen, M.; Honkanen, S. Titanium dioxide slot waveguides for visible wavelengths. *Appl. Opt.* **2015**, *54*, 2653–2657. [CrossRef] [PubMed]

23. Evans, C.C.; Liu, C.; Suntivich, J. Low-loss titanium dioxide waveguides and resonators using a dielectric lift-off fabrication process. *Opt. Express* **2015**, *23*, 11160–11169. [CrossRef] [PubMed]

24. Evans, C.C.; Shtyrkova, K.; Bradley, J.D.B.; Reshef, O.; Ippen, E.; Mazur, E. Spectral broadening in anatase titanium dioxide waveguides at telecommunication and near-visible wavelengths. *Opt. Express* **2013**, *21*, 18582–18591. [CrossRef] [PubMed]

25. Lamy, M.; Hammani, K.; Arocas, J.; Finot, C.; Weeber, J. Broadband Metal Grating Couplers Embedded in Titanium Dioxide Waveguides. 2017. Available online: https://hal.archives-ouvertes.fr/hal-01503970 (accessed on 9 June 2017).

26. Scheerlinck, S.; Schrauwen, J.; Van Laere, F.; Taillaert, D.; Van Thourhout, D.; Baets, R. Efficient, broadband and compact metal grating couplers for silicon-on-insulator waveguides. *Opt. Express* **2007**, *15*, 9625–9630. [CrossRef] [PubMed]

applied sciences

MDPI

Review

Soliton Content of Fiber-Optic Light Pulses

Fedor Mitschke *,†, Christoph Mahnke † and Alexander Hause †

Institut für Physik, Universität Rostock, A.-Einstein-Str. 23, 18059 Rostock, Germany;
christoph.mahnke@uni-rostock.de (C.M.); alexander.hause2@uni-rostock.de (A.H.)
* Correspondence: fedor.mitschke@uni-rostock.de; Tel.: +49-381-498-6820
† These authors contributed equally to this work.

Academic Editors: Boris Malomed and Totaro Imasaka
Received: 11 May 2017; Accepted: 11 June 2017; Published: 19 June 2017

Abstract: This is a review of fiber-optic soliton propagation and of methods to determine the soliton content in a pulse, group of pulses or a similar structure. Of central importance is the nonlinear Schrödinger equation, an integrable equation that possesses soliton solutions, among others. Several extensions and generalizations of this equation are customary to better approximate real-world systems, but this comes at the expense of losing integrability. Depending on the experimental situation under discussion, a variety of pulse shapes or pulse groups can arise. In each case, the structure will contain one or several solitons plus small amplitude radiation. Direct scattering transform, also known as nonlinear Fourier transform, serves to quantify the soliton content in a given pulse structure, but it relies on integrability. Soliton radiation beat analysis does not suffer from this restriction, but has other limitations. The relative advantages and disadvantages of the methods are compared.

Keywords: direct scattering transform; fiber optics; inverse scattering transform; nonlinear Fourier transform; nonlinear Schrödinger equation; soliton; soliton radiation beat analysis

1. Introduction

Light pulses in optical fibers carry the bulk of all telecommunication today. In comparison to electrical pulses in cables, fiber-optic transmission is vastly superior, mostly due to two fundamental advantages: light in fibers suffers extremely low power loss, and fibers provide an extremely wide bandwidth. For the loss, a rough number is 0.2 dB/km. This implies that even after 100 km, still, 1% of the launch power is left, which is unrivaled. The available bandwidth can be estimated as 30 THz (maybe 50 THz if one accepts slightly higher loss), which is several orders of magnitude better than anything that can be achieved with electronics.

Further comparison of optical and electrical transmission brings us to an imperfection shared by both: in either type of conduit, group velocity depends on frequency; as any signal requires a certain bandwidth, transmitted signals suffer from distortion due to group velocity dispersion. This is a linear distortion, which means that it can be perfectly compensated by adding an element of opposite dispersion.

In one respect, however, optical fibers are very different from electrical cables. The response of glass to light is nonlinear, whereas that of copper to current is not (within reasonable limits). Therefore, the physics of data transmission with light pulses through optical fibers is fundamentally different in that it involves a nonlinear response of the material to the signal [1,2]. Nonlinear distortions are not as easily compensated as linear ones. This has led to a widely-held conception that optical power in fibers must always be kept low enough so that nonlinear distortions are avoided.

To this argument, one can give two responses: One, nonlinearity can actually be put to good use, even be used to an advantage, when the concept of solitons is adopted. It has already been demonstrated even in a commercial setting. Two, the ever-growing demand for data-carrying capacity brings us close to a capacity crunch, unless means are found in just a few years to improve technology. It may become necessary to look at solitonic formats again.

2. The Nonlinear Schrödinger Equation and Some of Its Solutions

The glass of optical fiber is a dispersive material. With the index being a function of wavelength or optical frequency, i.e., $n = n(\omega)$, the group velocity is also frequency dependent. The glass is also a nonlinear material. This chiefly means that the refractive index follows the instantaneous intensity, i.e., $n = n(I)$. Across the duration of a short pulse of light, the index experiences a slight modulation, which gives rise to intensity-dependent phase shifts known as self-phase modulation [3].

A propagation equation for light pulses in fibers must contain these two influences, at least in leading order. The equation that fits this description is known as the nonlinear Schrödinger equation (NLSE). It plays an absolutely central role in all of the nonlinear fiber optics. It describes the evolution of the pulse amplitude envelope (the fast oscillation at some central optical frequency ω_0 is removed) in a frame of reference comoving with the traveling light, i.e., with the group velocity at ω_0. Written in physical quantities, it is given by:

$$i\frac{\partial}{\partial z}A - \frac{\beta_2}{2}\frac{\partial^2}{\partial t^2}A + \gamma|A|^2A = 0 \quad . \tag{1}$$

Here, $A = A(z,t)$ is the complex amplitude with z the position along the fiber and t the retarded time. β_2 is the coefficient of group velocity dispersion, and γ is the coefficient of nonlinearity. Depending on the situation, it may become necessary to add corrective terms for loss, higher-order dispersion or other nonlinear effects, as we shall see in Section 3.1. The equation received its name due to its formal similarity to the Schrödinger equation of quantum mechanics fame: The quantum mechanical version typically describes how a wave function spreads out in space as time goes by; here, the equation describes how a short pulse gets broadened temporally as it propagates down the fiber. Therefore, time and space coordinates switch roles. The potential, in the nonlinear case, comes from the self-phase modulation, which is captured in the nonlinear term.

The effort to find solutions to this equation benefited from the fact that the NLSE is integrable. Integrable equations have very special mathematical properties like an infinite number of preserved quantities. A first solution was found by Vladimir E. Zakharov and Aleksei B. Shabat in 1972 [4] using the inverse scattering technique, developed by Gardner et al. in 1967 [5] (see Section 4.1). It exists for a particular combination of algebraic signs of the two parameters; with γ always positive, one would have the correct signs for $\beta_2 < 0$, i.e., for what is known as anomalous dispersion. Two years later, Junkichi Satsuma and Nubuo Yajima [6] treated the pertaining initial-value problem. Between them, these two papers already form a solid base of the understanding of solitons. In the time between those two, Akira Hasegawa and Frederick Tappert suggested that the NLSE was applicable to pulse propagation in optical fibers, so that fiber-optic solitons would exist [7]. At that time, optical fibers were a novelty and not yet technologically matured. Attempts to transmit light pulses for data transmission suffered from dispersive distortions. It was therefore a bold proposal that the use of solitons as signaling pulses might overcome dispersive broadening, in other words that nonlinearity could be exploited to cancel distortions arising from a linear mechanism.

Of course, the proposal required as a very first step that the existence of fiber-optic solitons be corroborated experimentally. In the anomalously dispersive regime, i.e., at wavelengths longer than ca. 1.3 μm, they were still too lossy for any meaningful test. By 1980, this situation had improved, and Linn F. Mollenauer, Roger H. Stolen and James P. Gordon could experimentally verify the existence of solitons [8].

2.1. Soliton Solutions

2.1.1. The Fundamental Soliton

In the anomalous dispersion regime, the NLSE Equation (1) supports a stable solution known as the fundamental soliton:

$$
\begin{aligned}
A(z,t) &= \sqrt{P_0}\ \mathrm{sech}\left(\frac{(t-t_s)-\Omega\beta_2(z-z_s)}{T_0}\right) \\
&\times \exp\left[i\frac{\gamma P_0}{2}(z-z_s)-i\Omega\left((t-t_s)-\frac{1}{2}\Omega\beta_2(z-z_s)\right)+i\varphi_s\right] \quad .
\end{aligned}
\tag{2}
$$

Here, P_0 is the peak power; T_0 is the pulse duration; Ω is the deviation of the soliton's center optical frequency from that of the frame of reference, ω_0. t_s, z_s and φ_s are the initial values of center time, start position and phase offset, respectively. As long as a single soliton is considered, frequency offset and initial values may be set to zero without loss of generality so that a considerably simplified version:

$$
A(z,t) = \sqrt{P_0}\ \mathrm{sech}\left(\frac{t}{T_0}\right)\exp\left[i\frac{\gamma P_0}{2}z\right]
\tag{3}
$$

remains. In either case, amplitude and duration are coupled and must fulfil:

$$
P_0 T_0^2 = \frac{|\beta_2|}{\gamma} \quad .
\tag{4}
$$

The propagation of a fundamental soliton with $\Omega = t_s = z_s = \varphi_s = 0$ is shown in Figure 1. As the temporal evolution (sech term) does not contain z, the temporal shape does not change during propagation. Without nonlinearity, the shape would change appreciably after one dispersion length $L_D = T_0^2/|\beta_2|$. The other relevant length scale is the nonlinearity length $L_{NL} = 1/(\gamma P_0)$ after which self-phase modulation becomes appreciable. As Equation (4) shows, for the soliton $L_D = L_{NL}$. This equality expresses the balance between both.

Among the conserved quantities of the integrable equation are the soliton energy:

$$
E_{sol} = 2P_0 T_0
\tag{5}
$$

and the soliton center frequency Ω. As will be described below (Section 4.1), these two values are found from inverse scattering in the form of imaginary and real parts of a complex eigenvalue. We point out that in a dispersive medium, a nonzero frequency translates to a relative motion, so that the frequency is usually referred to as the velocity. The number of solitons is also preserved, even in cases when there is more than one.

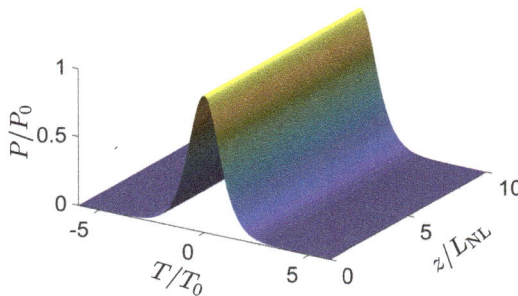

Figure 1. Evolution of a fundamental nonlinear Schrödinger equation (NLSE) soliton.

2.1.2. Radiation and Higher-Order Solitons

To create a soliton, one can launch a sech-shaped pulse as in Equation (3) for $z = 0$ into a fiber with parameters conforming to Equation (4). If, however, the parameters are only approximately right, we can do as follows: In the spirit of [6] and as also detailed in [9] (but they call A what we call N as in [1,2,10]), we write the initial condition as:

$$A(0,t) = N\sqrt{P_0}\ \text{sech}\left(\frac{t}{T_0}\right) \quad .$$
(6)

where the pulse is scaled by a factor $N \geq 0$, $N \in \mathbb{R}$ called the soliton order. The energy is then:

$$E = N^2 E_{\text{sol}} \quad .$$
(7)

Of course, at $N = 1$, there is the fundamental soliton, but at non-integer N, there is also a part of the pulse energy that is not part of the soliton energy. This balance is a linear wave; in the soliton context, it is called radiation. Linear waves are subject to dispersive broadening, so that after a sufficiently long propagation, radiation will be dispersed away from the soliton, and the soliton itself emerges. Until that happens, the different phase evolution of soliton and radiation creates a beat note, visible as a slowly-decaying beat pattern. This is shown for $N = 1.2$ in Figure 2 (left), where lengths are scaled to $z_0 = (\pi/2)\,L_D$. Evaluation of the beat pattern can yield soliton parameters, see Section 4.2.

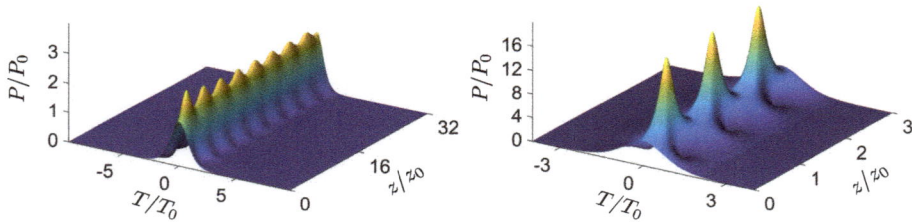

Figure 2. (Left) If a pulse with soliton order $N = 1.2$ is launched, a slowly decaying beat with radiation lets the soliton emerge gradually; (right) at $N = 2$, one generates a higher-order soliton, a structure with periodic shape oscillation.

There is a threshold value of N to generate a soliton. As long as $N < 1/2$, all energy goes into radiation. At $N = 1$, the radiative part dips to zero. At $N = 3/2$, it reaches the same energy as at $N = 1/2$ again, and that suffices to generate a second soliton. At $N = 2$, there is a combination of two solitons without any radiative part. This structure, for which an explicit expression was already given in [6], is called an $N = 2$ soliton (Figure 2 right). More generally speaking, for any $N \in \mathbb{N}$, $N > 1$, one has a so-called higher-order soliton of order N in the pure form, i.e., without radiation. An explicit expression for the $N = 3$ case solitons was found in [11]. The power profile of all higher-order solitons oscillates with period z_0. For all non-integer values of N, there is a radiative contribution.

Figure 3 gives an overview of this situation. In its upper part, the pulse energy according to Equation (7), shown in units of E_{sol}, is represented by the dashed parabola. Beginning at $N = 1/2$, the first soliton appears. Its energy then rises linearly and passes through unity at $N = 1$. Similarly, at other half-integer values of N, more solitons begin, each of them contributing energy to a cumulative value shown in blue. The latter is tangent to the parabola at integer N values; here, the individual soliton energies are also integer in E_{sol} units. They take the first N values of the sequence $1, 3, 5, 7, \ldots$, which add up to N^2. For all non-integer N, there is a gap between the cumulative solitonic and the total energy, which represents the radiative part of the energy. Radiation energy is shown in the figure's lower part on an inverted, magnified scale (green curve). Whenever, at some half-integer N, it reaches $E_{\text{rad}} = 1/4\,E_{\text{sol}}$, the next soliton is created. Always the integer number closest to N determines the number of solitons involved, while the fractional part determines the amount of radiation created along with them.

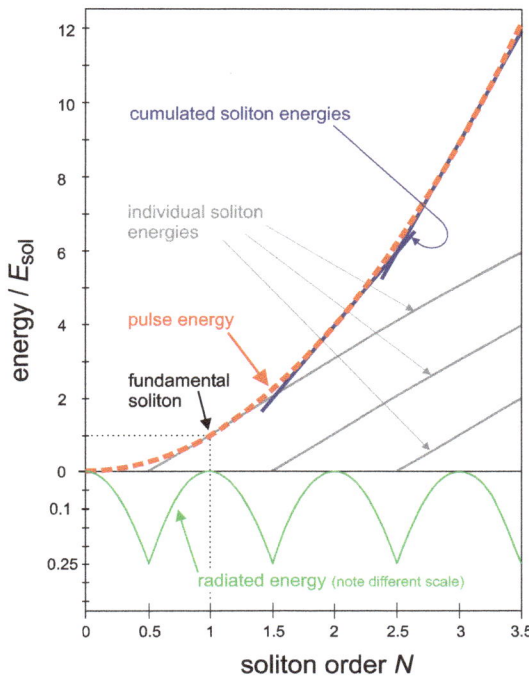

Figure 3. Soliton and radiation energies as a function of soliton order N. Energy is normalized to units of fundamental soliton E_{sol} of Equation (5). The fundamental soliton is at the black dotted markers. Individual soliton energies (grey lines) add up to a cumulative value (piecewise linear blue trace); the latter approximates the total pulse energy (red dashed parabola). The difference between both is shown on an expanded, inverted scale in the lower part (green curve). From [2].

2.1.3. Soliton Interaction

When two or more pulses are launched in rapid succession, each of them is affected by the presence of the others due to the index nonlinearity. In an integrable system, they cannot merge (the number is preserved), but they can be set in relative motion as if there were interaction forces. Gordon investigated the soliton-soliton interaction [12] and found that an effective force would decay exponentially with growing separation between pulses and vary sinusoidally with their relative phase. This means that there can be both attraction (in phase) or repulsion (opposite phase). This prediction was tested experimentally in [13] and was found to be fully correct, with the only caveat that in the attractive case, the first close encounter (collision) leads to processes of higher order not captured in the NLSE.

Note that this view of interaction forces between particles is metaphorical and justifiable only if both are well separated. In a nonlinear system, the superposition principle does not hold; therefore, two sech-shaped pulses in close proximity to each other are not really two identical, but overlapping solitons, but form a somewhat more involved two-soliton compound [14]. For that, energy and velocity are preserved quantities.

2.2. Breather Solutions

There is a different family of solutions of the NLSE, which is best explained by starting with the continuous wave (cw) case. The NLSE is solved by the cw ansatz:

$$A = \sqrt{P_0}\, e^{i\gamma P_0 z} \tag{8}$$

with constant power P_0. It turns out that this solution is stable for normal ($\beta_2 > 0$) and unstable for anomalous dispersion ($\beta_2 < 0$). The instability implies growth of small perturbations of the cw; this is known as modulation instability [1]. The frequency of maximum gain is offset by $\pm\omega_{\max}$ from the carrier frequency, with:

$$\omega_{\max} = \sqrt{\frac{2\gamma P_0}{|\beta_2|}} \quad . \tag{9}$$

If the cw is perturbed by white noise, there is always energy in these Fourier components, and a ripple with frequency $\omega_{\max}$ will begin to grow with gain:

$$g_{\max} = 2\gamma P_0 \quad . \tag{10}$$

The further fate was widely appreciated only a couple of years ago even though the mathematical results had all been worked out as early as 1986 [15]. There is a family of solutions of the NLSE characterized by an infinitely extended cw background and some modulation on top. Its general form can be written as [16]:

$$A(a,z,t) = \sqrt{P_0}\, \left[1 + M(a,z,t)\right] \exp(i\gamma P_0 z) \quad . \tag{11}$$

where $M(a,z,t)$ describes the modulation term. In the case of what is now known as the Akhmediev breather, it takes the form:

$$M(a,z,t) = \frac{2(1-2a)\cosh\left[b(a)\gamma P_0 z\right] + ib(a)\sinh\left[b(a)\gamma P_0 z\right]}{\sqrt{2a}\,\cos\left[\omega_{\mathrm{mod}}t\right] - \cosh\left[b(a)\gamma P_0 z\right]} \quad . \tag{12}$$

with parameters $0 < a < 1/2$ and $b(a) = \sqrt{8a - 16a^2}$. The modulation frequency is $\omega_{\mathrm{mod}} = \omega_{\mathrm{c}}\sqrt{1-2a}$, and:

$$\omega_{\mathrm{c}} = \sqrt{\frac{4\gamma P_0}{|\beta_2|}} \geq |\omega_{\mathrm{mod}}| \tag{13}$$

is the frequency range where gain is possible. Gain maximum occurs at $a = 1/4$ with $\omega_{\max} = \omega_{\mathrm{mod}}/\sqrt{2}$, and Equations (9) and (10) are recovered. This particular case is illustrated in Figure 4. From a cw background, the modulation grows, until at $z = 0$, it reaches a maximum amplitude. Here, one has a periodic train of pulses. During further evolution, the modulation decays until for $z \to \infty$, the cw is recovered.

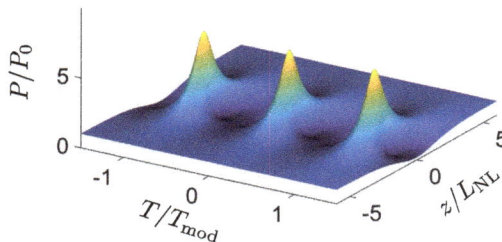

Figure 4. Evolution of an Akhmediev breather at $a = 1/4$. $T_{\mathrm{mod}} = 2\pi/\omega_{\mathrm{mod}}$.

The first experiments to demonstrate the Akhmediev breather in optical fibers were described in [17] and in more detail in [18]. The process was initiated either from random perturbation [17] or from a suitable weak modulation as suggested in [19] and performed experimentally in [18].

If we let $a \to 1/2$, this implies that $\omega_{mod} \to 0$, and the temporal separation between the pulses in the train diverges. The modulation then takes the form:

$$M(z,t) = -\frac{4(1 + i2\gamma P_0 z)}{1 + \omega_c^2 t^2 + 4(\gamma P_0 z)^2} \quad .$$ (14)

This is a structure that is localized in both time and space, known as the Peregrine soliton [20]. It is shown in Figure 5 (left); a first experimental demonstration was given in [21]. A further member of this family of NLSE solutions was studied in [22] and is known as the Kuznetsov–Ma soliton. Mathematically, that solution is described by the modulation term:

$$M(a,z,t) = \frac{2(1 - 2a)\cos\left[|b(a)|\gamma P_0 z\right] - i|b(a)|\sin\left[|b(a)|\gamma P_0 z\right]}{\sqrt{2a}\,\cosh\left[|\omega_m|t\right] - \cos\left[|b(a)|\gamma P_0 z\right]}$$ (15)

with a parameter range of $a > 1/2$. It consists of a periodically-recurring peak; see Figure 5 (right).

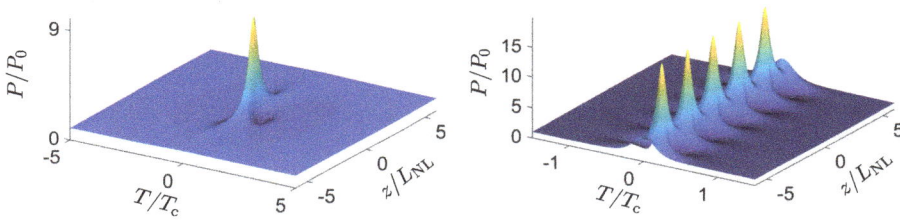

Figure 5. Evolution of a Peregrine soliton (left) and a Kuznetsov–Ma soliton (right). $T_c = 2\pi/\omega_c$.

3. Extensions to the Nonlinear Schrödinger Equation

3.1. Generalized Nonlinear Schrödinger Equation

The NLSE neglects linear loss, approximates dispersion by the leading group velocity dispersion term and ignores nonlinear effects other than the intensity dependence of the refractive index. Depending on the situation and/or the required accuracy of the calculated results, terms must be added to include what is not represented. This leads to a generalized NLSE, which may take the form:

$$i\frac{\partial}{\partial z}A - \frac{\beta_2}{2}\frac{\partial^2}{\partial t^2}A - \underbrace{i\frac{\beta_3}{6}\frac{\partial^3}{\partial t^3}A + \frac{\beta_4}{24}\frac{\partial^4}{\partial t^4}A \ldots}_{\text{higher-order dispersion}} + \gamma|A|^2 A$$

$$+ \underbrace{i\frac{\gamma}{\omega_0}\frac{\partial}{\partial t}\left(|A|^2 A\right)}_{\text{self-steepening}} - \underbrace{T_R \gamma A\frac{\partial}{\partial t}|A|^2}_{\text{Raman}} + \underbrace{i\frac{\alpha}{2}A}_{\text{loss}} = 0 \quad .$$ (16)

As the generalized NLSE is not integrable, it usually must be treated numerically. The impact of the new terms is as follows:

The dispersion curve can be written as a Taylor expansion around the carrier frequency ω_0; the NLSE is written so that the β_2 term appears. All higher-order terms (shown here up to the fourth order) provide corrections that become relevant when either the pulse spectrum gets quite wide or when the carrier frequency is close to a zero-dispersion point (wavelength where β_2 passes through zero) of the fiber.

The 'self-steepening' term [1,23] results from the intensity dependence of the group velocity and leads to an asymmetry in the pulse shape. It can produce shifts in spectral and temporal positions even in the absence of the Raman term.

The Raman effect leads to a redistribution of spectral power to the advantage of the low-frequency slope of the pulse spectrum. A first study of the Raman gain in fibers was done by R. H. Stolen et al. [24]. The gain curve peaks at a frequency offset of ≈13 THz, but is nonzero all the way down to zero offset. As a result, the spectral 'center of mass' of a pulse is shifted to lower frequencies. This 'redshift' scales with the inverse fourth power of the pulse width [25]. The term expressing this in the NLSE is shown here in a simplified version, which contains the Raman response time T_R. This model is valid for pulses being longer than ≈100 fs; more accurate models are described in [26–28].

The loss term contains Beer's loss coefficient α; the impact of the loss can be characterized by the characteristic loss length $L_\alpha = 1/\alpha$. As long as the loss is weak, i.e., $L_\alpha \gg L_D, L_{NL}$, the equilibrium of dispersive and nonlinear effects that characterizes the soliton is only mildly perturbed. In that situation, the soliton can rearrange its shape: With the peak power drooping and the width increasing slightly, Equation (4) can still be approximately fulfilled even when the energy is reduced [29,30].

3.2. Dispersion-Managed Fibers

As a further complication in a realistic assessment of pulse propagation in fibers, one must acknowledge that in the telecommunications field, a special type of fiber has been preferred for 20 years now in which β_2 is not a constant. Originally, an attempt was made to cancel dispersive effects by concatenating segments of fiber with alternatingly positive and negative group velocity dispersion. As it turned out, a perfect cancellation (path average $\bar\beta_2 = 0$)—possible at a single wavelength only anyway—did not work as well as if a small negative value were left. This is on account of the fiber's nonlinearity.

Figure 6 illustrates the structure of a dispersion-managed fiber (DM fiber). The dispersion map is described by the spatial period $L_{map} = L^+ + L^-$ with which the alternating pattern is repeated [31]. Strictly speaking, the value of the nonlinearity coefficient γ will also alternate from one fiber segment to the other.

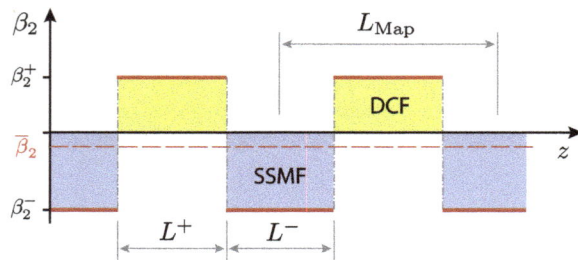

Figure 6. Dispersion-managed fiber consists of alternatingly normally (dispersion compensating fiber (DCF)) and anomalously dispersive fiber (standard single mode fiber (SSMF)) segments. The dispersion parameter β_2 alternates between values of β_2^+ and β_2^-; the lengths of segments are L^+ and L^-, respectively. The dispersion map period is $L_{map} = L^+ + L^-$.

It is not immediately clear whether in a dispersion-managed (DM) fiber something like solitons can exist, as they are solutions of the NLSE for anomalous dispersion only. However, as it turned out, a type of pulse exists that is stabilized by nonlinearity [32–37]. These pulses were then called dispersion-managed (DM) solitons. Their shape is neither of the usual unchirped sech type, nor constant. Rather, it breathes over a full dispersion period so that after a full period, the shape in both the amplitude and phase profile is restored. An example of a DM soliton is shown in Figure 7.

Mathematically, dispersion management is captured by making fiber parameters z-dependent:

$$i\frac{\partial}{\partial z}A - \frac{\beta_2(z)}{2}\frac{\partial^2}{\partial t^2}A + \gamma(z)|A|^2 A = 0 \quad . \tag{17}$$

This modifies the NLSE into a non-integrable form so that it does not support solitons in the usual sense. Depending on the dispersion modulation strength (essentially, $\beta_2^+ - \beta_2^-$), the DM soliton shape may be nearly sech-shaped or closer to a Gaussian [38]. In the strong modulation limit DM, solitons exhibit oscillating tails [38,39]. There is no closed analytical expression for it, but good approximations to its shape can be found by numerical procedures like that given in [40]. During propagation, a DM soliton suffers from continuous radiative loss, which may be weak, but nevertheless must lead to its eventual decay. Fortunately, for applications, the decay distance typically exceeds any practically relevant fiber length.

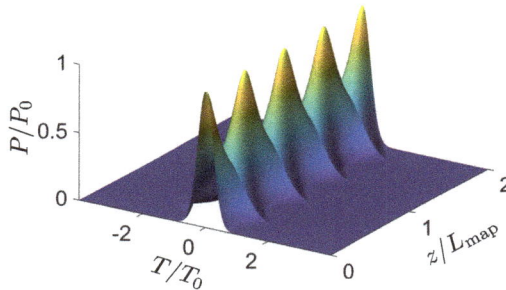

Figure 7. Evolution of a dispersion-managed soliton over two dispersion map periods.

It was eventually understood that dispersion-management provides advantages over a constant-dispersion fiber. It suppresses four-wave mixing, a process occurring in a transmission system with many wavelength channels filled simultaneously with independent data streams, where it leads to cross-talk and, thus, errors. An extensive review of dispersion-management is given in [31].

3.3. Cases Involving Several Solitons

We mention a few more cases in which solitons or their compounds play a role and which have received attention in the research literature.

3.3.1. Soliton Molecules

In DM fibers, stable compounds of solitons exist, which have been called soliton molecules [41]. The alternating dispersion creates a rapid to-and-fro of the chirp of DM solitons and, thus, a varying mutual force. If two DM solitons are in a particular mutual separation, the net force is zero, and an equilibrium is obtained [42,43]. For larger separations, there is attraction, for smaller, repulsion: it is a stable equilibrium. Soliton molecules also exist for more than two DM solitons and have been discussed as information carriers in data transmission [44,45]. As it turns out, there can be several equilibrium positions [46].

3.3.2. Soliton Gas and Crystal

Another situation in which soliton propagation is subject to periodic perturbation was described in a series of experiments about a synchronously-driven fiber resonator [47] (using non-DM fiber). Free propagation in the fiber alternated with interference at an input coupler. As a result, relatively long input pulses organized themselves into a pattern that contained a multitude of short pulses;

each of those fulfilled the soliton condition Equation (4). The pattern could be either chaotic (soliton gas) [48] or periodic (called soliton crystal) [49], depending on the drive power.

Similar phenomena have also been reported from fiber lasers [50–52], which, however, are not described by the NLSE, as that does not consider nonlinear gain and loss. An equation including these effects is the cubic-quintic complex Ginzburg–Landau equation. One finds that there are soliton-like solutions of this equation that are called dissipative solitons [53–55].

3.3.3. Supercontinuum Generation

For some metrological purposes like coherence tomography, wide bandwidth light with good focusability is required. Thermal light sources cannot deliver adequate power due to both their insufficient spatial coherence and the limitations set by Planck's radiation law. One can, however, generate broadband light in a fiber by exploiting the nonlinearity to create new frequency components. Experiments involve, in most cases, a powerful mode-locked laser and a piece of highly nonlinear fiber. Either a modulation instability according to an Akhmediev breather scenario evolves, but the structure then breaks apart, or the input pulse represents a high-order ($N \gg 1$) soliton, which undergoes fission [56]. In either event, the initial light signal breaks up into an arrangement of several pulses, many of which have soliton characteristics, and radiation. This arrangement has similarities to the soliton gas described above. Through the action of the Raman effect, four-wave mixing and cross-phase modulation combined with complex phase matching properties in the fiber's dispersion curve (see, e.g., [57]), the light quickly develops a spectrum that may easily be one octave wide and often much more. Such a structure is referred to as an optical supercontinuum and has found much research interest; commercial supercontinuum sources are being offered. A good review is [58].

A corollary to this is the phenomenon of optical rogue waves. Rogue waves of the ocean have been identified as a real phenomenon, not a myth. Unusually, large single waves can occasionally appear without warning, possibly wreaking havoc on ships, and disappear again just as suddenly. It was first discussed in [59] that in the context of optical supercontinuum generation, single spikes of extreme power can occur every once in a while. There is an ongoing discussion about the mechanisms [60].

4. Methods to Verify Soliton Content

The inverse scattering technique allows one to find solutions to integrable nonlinear wave equations. However, in practice, one quite often is faced with the reversed task of determining the soliton content of a given signal. The task is relatively straightforward when the signal consists of isolated bell-shaped pulses. One can check whether the power profile approximates a $\mathrm{sech}^2(t)$ shape and whether the chirp is close to zero. Then, duration and peak power can be checked with Equation (4). This way, an $N = 1$ soliton can be directly identified, and for pulses with $N \neq 1$, the local value of N can be found. However, the applicability of this direct method is limited to simple cases. It is of no use for waveforms, as they may be found in more complex situations like soliton gas or optical supercontinuum; see Section 4.1.3.

We will therefore briefly describe more advanced techniques. The direct scattering method is often considered as the benchmark for the determination of soliton parameters. However, it has its limitations, chief among which is that it relies on the integrability of the system. An alternative is soliton radiation beat analysis, which does not have that same restriction, but the price to be paid for that is that (i) more input data are required and that (ii) in complex cases, the obtained pattern may get too complicated to interpret.

4.1. The Inverse and Direct Scattering Transform

4.1.1. The Method

The inverse scattering transform (IST) is a technique that can be used to solve certain nonlinear partial differential equations (PDE) like the Korteweg–de Vries equation, the sine-Gordon equation or the nonlinear Schrödinger equation. The basic concepts of this method were presented in 1967 in a groundbreaking paper by Gardner and coworkers dealing with the Korteweg–de Vries equation [5]. Shortly after, Lax showed how to apply the method also to other nonlinear PDEs when certain conditions are met [61]. The IST was successfully adapted to the NLSE by Zakharov and Shabat in 1972 [4].

The method of IST is based on transformations between the time-space domain and a nonlinear spectral domain, in which the nonlinear evolution of some given input field reduces to a linear problem. In the spectral domain, a given field is represented by its scattering spectrum, which is calculated by the so-called direct scattering transform (DST). The inverse operation is called the inverse scattering transform. As these transformations can be regarded as extensions to the well-known Fourier transform, they are also known as nonlinear Fourier transform and inverse nonlinear Fourier transform, respectively. Using the nonlinear Fourier transform, one can get insight into the linear and nonlinear components of a given field by analyzing its scattering spectrum. Generally, it consists of a continuous part, which represents the small amplitude radiation, and a discrete part, which can be attributed to the solitons contained.

In recent years, several groups took up the idea of using multi-soliton pulses for nonlinear optical data transmission, which originally was suggested by Hasegawa and Nyu [62]. In the course of this renewed attention, several nonlinear transmission schemes and techniques for the nonlinear Fourier transform and its inverse transform have been (and still are) developed to overcome the linear capacity limits of fiber-optical transmission lines. An overview about the latest developments can be found in [63].

DST is a great tool to analyze given pulses, but its application is somewhat involved. The DST has been applied analytically to only a few simple pulse shapes like sech-shaped [6] or rectangular pulses [64]. Even for such a standard shape as a Gaussian,

$$A(t) = A_0 \sqrt{P_0} \exp\left[-\frac{1}{2}\left(\frac{t}{T_0}\right)^2\right] \quad , \tag{18}$$

no analytical result is known to the authors' best knowledge. Instead, approximate solutions can be found numerically. As an example, when setting $A_0 = 3$ and choosing P_0 and T_0 according to Equation (4), DST reveals that the pulse consists of two solitons and a small amplitude radiation part of nontrivial spectral shape (see Figure 8).

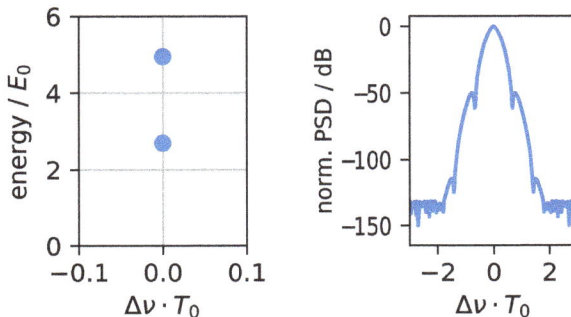

Figure 8. Nonlinear spectrum calculated for a Gaussian pulse shape (Equation (18)) with $A_0 = 3$ and P_0, T_0 fulfilling Equation (4). (Left) The discrete spectrum, consisting of two solitons of the same center frequency. The reference energy here is chosen as $E_0 = 2P_0T_0$. (Right) The power spectral density (PSD) of the linear radiation part.

Various schemes for numerical calculations of the DST are available [63,65–68]. They all solve the Zakharov–Shabat eigenvalue problem of a discretized input field with M sample points and determine the nonlinear spectrum. One can distinguish between matrix methods that calculate the eigenvalues directly from a $M \times M$ matrix and search methods in which the eigenvalues are found in an iterative manner. For a recent discussion, see, e.g., [63,66–68].

In the following, we will sketch briefly how to calculate the scattering spectrum using a search method. Using the formulation of Ablowitz et al. [69], one finds that solving the dimensionless NLSE:

$$\frac{\partial q}{\partial z} = \frac{i}{2}\frac{\partial^2 q}{\partial t^2} + i|q|^2 q \tag{19}$$

is equivalent to the integration of the system of equations:

$$\frac{\partial \Phi}{\partial t} + T\Phi = 0 \quad, \tag{20}$$

$$\frac{\partial \Phi}{\partial z} + Z\Phi = 0 \quad. \tag{21}$$

T and Z are operators (matrices) that determine the temporal and spatial evolution of some auxiliary function:

$$\Phi = \begin{pmatrix} \phi_1 \\ \phi_2 \end{pmatrix} \tag{22}$$

where ϕ_1 and ϕ_2 are called Jost functions. Explicitly, the operator T can be written as:

$$T = \begin{pmatrix} -i\lambda & q \\ -q^* & i\lambda \end{pmatrix} \quad. \tag{23}$$

Calculation of the DST can be done by integrating Equation (20) with respect to time with different test values of the eigenvalue candidate $\lambda = \zeta$. The initial condition can be chosen as:

$$\Phi(z_0, t, \zeta) = \begin{pmatrix} 1 \\ 0 \end{pmatrix} \exp(-i\zeta t) \quad \text{with} \quad t \to -\infty \quad, \tag{24}$$

when q is localized in the sense that for $t \to \pm\infty$, $q \to 0$ at an exponential rate [65,66]. As a result, one obtains the scattering coefficients $a(\zeta)$ and $b(\zeta)$ from $\Phi(z_0, t, \zeta)$:

$$a(z_0, \zeta) = \phi_1(z_0, t, \zeta) \exp(i\zeta t) \quad \text{with} \quad t \to \infty \quad, \tag{25}$$
$$b(z_0, \zeta) = \phi_2(z_0, t, \zeta) \exp(-i\zeta t) \quad \text{with} \quad t \to \infty \quad. \tag{26}$$

Additionally, the derivatives of a and b with respect to ζ can also be calculated by extending the integration scheme to include $\Phi' = \partial\Phi/\partial\zeta$.

In the scattering spectrum for anomalous dispersion, solitons are present in the form of discrete eigenvalues ζ_k with $a(\zeta_k) = 0$ and $\text{Im}(\zeta_k) > 0$. $\text{Re}(\zeta_k)$ represents the frequency, and $2\,\text{Im}(\zeta_k)$ corresponds to the amplitude of the soliton; both values are preserved during propagation. The continuous part of the spectrum (radiation) can be calculated from the values of $a(\zeta)$, with $\text{Re}(\zeta) = 0$. The spectral power density $|\mathcal{F}|^2$ is then obtained as:

$$|\mathcal{F}|^2 = -\frac{1}{\pi}\ln(|a(\zeta, z)|). \tag{27}$$

4.1.2. Numerical Restrictions

We have applied numerical DST to a case for which exact analytical results are known, i.e., the fundamental soliton Equation (3), to assess inaccuracies of the numerical procedure. They arise from various sources; we identify three main error sources.

Truncation error: At the edges of the computational time window of finite width T_W, the pulse wings are truncated. A fraction E_{trunc} of the energy is not represented in the numerical field:

$$E_{trunc} = 2 \int_{T_W/2}^{\infty} \hat{P} \operatorname{sech}^2 \left(\frac{T}{T_0} \right) dT = E_{tot} \left(1 - \tanh \left(\frac{T_W/2}{T_0} \right) \right) \quad . \tag{28}$$

Sampling error: Sampling error arises from coarse sampling when the temporal discretization step is too wide for the pulse duration.

Floating point error: Computers store numbers in some format with a limited number of digits; 'double precision' usually offers 15 digits plus the exponent. After a large number of floating point operations, numerical errors accumulate.

To disentangle these contributions, we varied the number of sampling points and the width of the time window for a given pulse width; the results are shown in Figure 9. To the right, the computational window is sufficiently wide ($T_W \gg T_0$). Then, the three curves shown are for discretization step numbers as indicated by the labels. For 2^{12} points, sampling is increasingly coarse when these points are spread over a wider window, and the error rises. Larger numbers of points improve the situation quadratically because the error scales with the square of the sampling point separation. To the left, truncation error dominates; the prediction of Equation (28), plotted as a solid curve, fits well. Floating point error becomes relevant only where these two contributions are small and where the number of calculation steps is large; in our example, it appears as 'noise' in the central section of the curve for 2^{22} points. We convinced ourselves that the transfer-matrix (TM) method presented in [65] yields the same results.

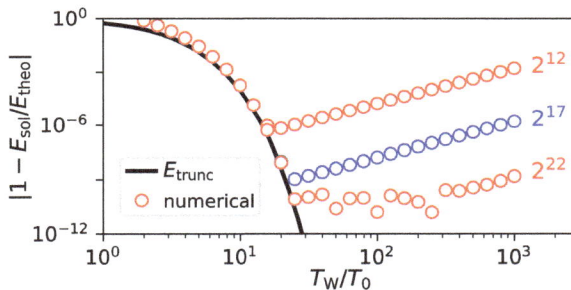

Figure 9. Direct scattering analysis of an $N = 1$ soliton with different time window widths T_W/T_0. The difference between the soliton energy found by direct scattering transform (DST) and the analytical value is shown. Labels on the right indicate the number of sample points used. Solid curve: expected truncation error, Equation (28) of the input pulse.

4.1.3. Applications and Limitations of DST

The DST method as described so far is quite generally applicable to arbitrary pulse profiles provided two conditions are met: One is integrability, which is not fulfilled for many structures in the generalized NLSE in Section 3.1 and for DM systems in Section 3.2. The other is that for time to infinity, the structure must decay to zero sufficiently fast [65,66]. This is not the case for the structures described in Section 2.2.

If integrability is violated only weakly either by loss or through the Raman effect, the concept of solitons remains valid in an approximate sense as pulses that maintain an (near-) equilibrium between dispersive and nonlinear influences. It might be more accurate to speak then of solitary pulses. Strictly speaking, though, all preserved quantities are no more. Perturbation theory has shown [29,30] that solitons rearrange their shape in terms of width and peak power; in the process, they acquire a new energy value, as well. In the sense that this may still be interpreted as solitons, some researchers have adopted the following strategy: In a mildly non-integrable system with loss [70] or Raman effect [71–73], the pulse structure is propagated numerically up to some position z_i, where DST is then applied to find what one may call 'local eigenvalues'. Then, one lets z_i slide and repeats DST, in order to find the position dependence, or spatial evolution, of the eigenvalues. An example of such a case is shown in Figure 10 where the Raman effect, third-order dispersion and loss are treated separately. The most conspicuous effects are the energy reduction by the loss term and the frequency downshift by the Raman effect.

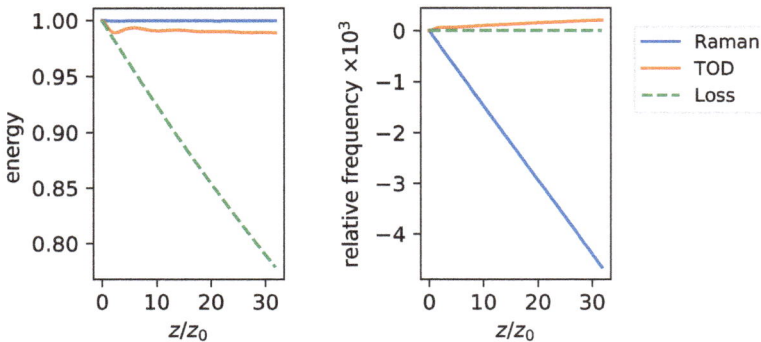

Figure 10. Energy and frequency eigenvalues of a fundamental soliton obtained using the DST method. The soliton was perturbed by Raman self-frequency shift (blue, solid, Raman response time: $T_R = 0.03\,T_0$), third-order dispersion (orange, solid, TODparameter: $\beta_3 = 0.5\,T_0\beta_2$), and linear loss (green, dashed, loss length: $L_\alpha = 200\,L_D$).

When it comes to complex situations like in the context of optical supercontinuum generation where many interactions and collisions of several pulses take place, energy can be transferred between pulses, and more changes to local eigenvalues will occur. In any event, one has to interpret the results produced in this way with caution.

If one considers infinitely-extended structures, one first needs to acknowledge that they have infinite energy. One possible approach is to start with a finite piece ('window') of the infinite structure and determine eigenvalues, then increment the window width and follow the eigenvalues; finally, let the window width tend to infinity [74]. In the limit, one obtains an infinite number of eigenvalues, but with a finite density (number of eigenvalues per unit time). These eigenvalues cover a continuum of energies from zero up to some maximum; if one considers the simplest case of an unmodulated continuous wave, this maximum depends only on its power P_0 and is given by $E_{max} = 2\sqrt{|\beta_2|\,P_0/\gamma}$. With respect to the interpretation of these eigenvalues as signatures of solitons, the following was shown in [74]: Consider a finite window, which contains m eigenvalues, and allow it to propagate. On account of integrability, it contains a constant number of m solitons. If one then applies a perturbation, e.g., by Raman effect, precisely m separate pulses emerge; once they are sufficiently separated from each other after some propagation, it can be checked that each fulfills the soliton criterion Equation (4).

The other approach to infinitely extended fields is to consider periodic problems, like the Akhmediev breather. Adaptations of IST to periodic boundary conditions (periodic IST, also called finite gap theory) have been given in [75–77] where the obtained eigenvalues are not necessarily interpreted as solitons, but rather as the representation of linear and nonlinear modes [78–80]. Like the nonperiodic IST, this periodic case gained renewed interest in recent years. Different theoretical and numerical schemes have been developed with a view toward the application of nonlinear data transmission [63,66,67,81–83]. A variant of the periodic DST based on the artificial periodization of field components was used to analyze rogue events and numerically to find the eigenvalue spectra of the breather solutions of Section 2.2 [84]. Another approach to characterize rogue events via DST was taken in [85,86].

We have here concentrated on pulse-like initial conditions. Recently, the scope was extended to integrable turbulence [87] with a view towards understanding rogue waves. Initial conditions for the NLSE were generated by superimposing random perturbations in both amplitude and phase on a continuous wave; the resulting field was characterized by its eigenvalue spectrum.

4.2. Soliton Radiation Beat Analysis

Soliton radiation beat analysis (SRBA) is a numerical method to determine the soliton content of a pulse in an optical fiber even if the system is not integrable. In contrast to DST, which requires the pulse profile at a single position in the fiber, SRBA requires knowledge of the evolution over a certain length section of the fiber. This is easily understood: In integrable systems, soliton parameters are invariant so that if they are known in one place, they are known everywhere. SRBA can follow the evolution of soliton parameters in a non-integrable setting; the price to pay is that representative information about it must be introduced into the calculation.

As early as in 1974, it was noted [6] that the envelope of a pulse not exactly conforming to a soliton undergoes oscillations. Such a pulse contains a soliton and some radiation; the oscillation arises as a beating between both. The radiation evolves linearly; the soliton contains a power-dependent phase term as shown in Equation (3). The frequency of the beat note therefore allows a direct conclusion about the soliton energy.

The SRBA method was first introduced in [88]. The propagation of a sech pulse with the initial form $A(z = 0, t) = N\sqrt{P_0}\,\text{sech}(t/T_0)$ was simulated numerically. This was repeated while incrementing the soliton parameter N. As the oscillation of the peak power is damped due to dispersive spreading of the radiative part, it was found advantageous to evaluate the peak spectral power, which undergoes undamped oscillation. A windowed Fourier transform was applied to the spatial evolution of the power at the center frequency $|\tilde{A}(\Omega = 0, z)|^2$. This produces a spectrogram of spatial frequencies called the soliton radiation beat pattern. An example is shown in Figure 11 (left).

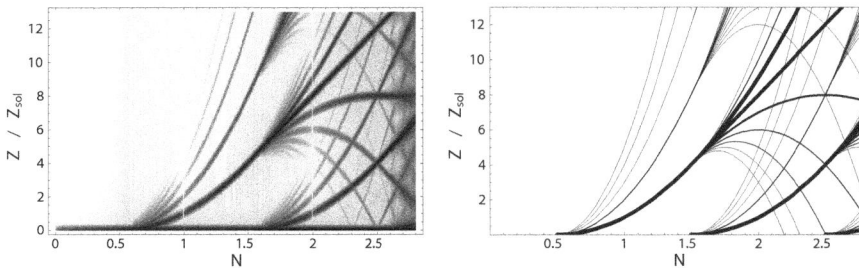

Figure 11. (Left) Soliton radiation beat analysis (SRBA) chart from repeated numerical simulations with increasing soliton number N; (right) corresponding predictions of beat signals and its overtones (from [88]).

As N is increased, a soliton radiation beat first appears at its expected threshold near $N = 1/2$. The figure shows that this beat note also has overtones. Its amplitude (grey scale) vanishes at integer N (apparent 'white stripes'), which is of course because there, the radiation amplitude vanishes; the beat amplitude is proportional to the product of both amplitudes.

If there is more than one soliton involved, the beat pattern will contain the frequencies of beats between all pairs of solitons, as well as that of each soliton beating with radiation. In Figure 11 (left), traces pertaining to the second soliton begin at $N = 3/2$ as expected and have amplitude nulls at $N = 2$. Difference frequencies are visible as downward-bending curves. In the right panel of Figure 11, analytical results for beat frequencies are plotted for comparison. Shown are all calculated frequencies, up to four overtones, and all combination frequencies of this multitude. It is obvious that all SRBA traces of the left panel can be identified in this way. SRBA provides the additional information that there are notches in the amplitudes of all traces involving radiation at integer N values.

This shows that SRBA reproduces all known features of the soliton energies as known analytically from IST. In [89], this method was shown to work for chirped Gaussian pulses in the NLSE. Going beyond the NLSE to non-integrable situations is also possible as SRBA makes no assumption about integrability. A lossy fiber was treated in [90]. It was shown that a mild, 'adiabatic' loss, if it persists for a long fiber length, eventually turns diabatic because the relevant length scales L_D and L_{NL} also change in the process. Once the loss is diabatic, the soliton decays. In that situation, however, the applicability of DST even in the approximate sense as described above in Section 4.1.3 is ultimately lost, and interpretations based on that approach [70] are doubtful and at variance to the SRBA results.

A further important example of the application of SRBA is the determination of the dispersion-managed soliton content. This case is just not accessible for DST, not even in the sense of 'local' analysis as in Section 4.1.3. Other than SRBA, no other numerical method to analyze DM structures is known. In [88], a suitably-scaled Gaussian pulse was used as a good approximation to an ideal DM soliton, and an SRBA pattern was obtained. Both the modification of the soliton energy and the characteristic minima of soliton radiation beat traces were found.

In subsequent research, SRBA was also applied to analyze Akhmediev breathers and a sinusoidally modulated cw [91], as well as soliton crystals [92]. One can assume that in principle, the SRBA method can be applied to all non-integrable systems where a nonlinear phase shift appears. Potential applications are soliton molecules (see Section 3.3.1) or dissipative solitons (see Section 3.3.2).

SRBA as described so far relies on the evaluation of the central spectral power $\widetilde{A}^2(\Omega = 0, z)$. In a generalization of the technique, the consideration is extended to the full spectrum, $\widetilde{A}^2(\Omega, z)$ [93]. In this variant, the technique also yields information about soliton velocities. Figure 12 shows an example for the case of an $N = 2$ soliton.

The pattern was generated using a value of $N = 2.01$, so that there is always a little bit of radiation copropagating with the second-order soliton, and a beat note is generated. Parabola-like curves are due to radiation; horizontal flat lines touching the parabolas are due to solitons. Once the traces are identified, the soliton energies can be obtained from the spatial frequency of the nonlinear phase rotation. To illustrate the interpretation of such data, we point out that the two solitons appear with spatial frequencies in the ratio of 9:1. According to Equation (3), this means that the peak powers are $P_0(S_1) = 9P_0(S_2)$. Equation (4) then dictates that $T_0(S_1) = 1/3\,T_0(S_2)$. The ratio of energies obtained from Equation (5) is then 3:1, precisely as expected from Figure 3. Any frequency shifts would appear as horizontal displacements; in this particular example, there are none. There is no scan of N here; this extended procedure has the advantage that only a single propagation simulation is necessary.

A particular difficulty with DM solitons is that their shape undergoes continuous changes. At various positions within L_{map}, the shape varies considerably, but the pulse shape repeats after L_{map} (it is 'stroboscopically' stable). Even then, when the power is modified, the shape varies from 'nearly sech' to 'nearly Gaussian', and depending on the dispersion modulation depth, oscillating tails may appear. Therefore, extra measures need to be taken when the energy is scanned for an SRBA pattern calculation. For the data in the left panel of Figure 13, at 0.5 pJ increments of energy, the best approximation to the DM soliton shape was found using Nijhof's method [40]. If the pulse shape were exact, there would be no beat note with radiation. This is why in 0.5 pJ increments, there seem to be white lines in the figure. Above and below each such step, the energy was scanned by simple scaling of the pulse shape to ±0.25 pJ. The slight deviation from the best pulse shape suffices to produce beat notes, and a full survey of their positions is obtained. For better clarity, the data from this calculation are transferred to the right panel of Figure 13, and different line thicknesses are standardized into three classes in the process. The data show that a DM soliton consists of several constituent solitons; the pertaining traces are labeled S_1–S_3.

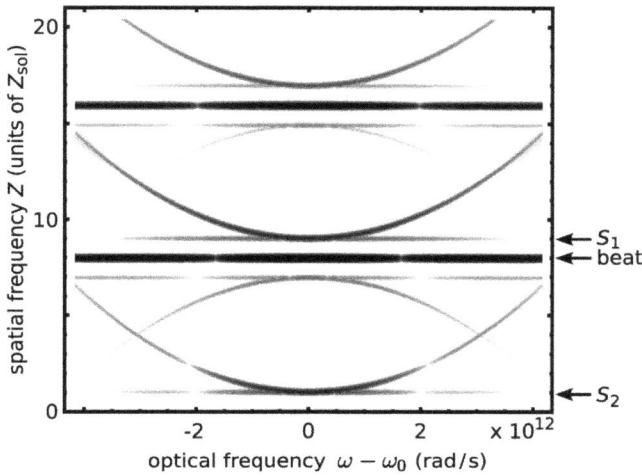

Figure 12. Full-frequency SRBA chart to determine the soliton content of an $N = 2$ soliton. Grey scale corresponds to the log of the Fourier transform of the spatial evolution of the spectral power density. Arrows labeled S_1 and S_2 mark traces pertaining to the first and second soliton; their beat note is also highlighted (from [93]).

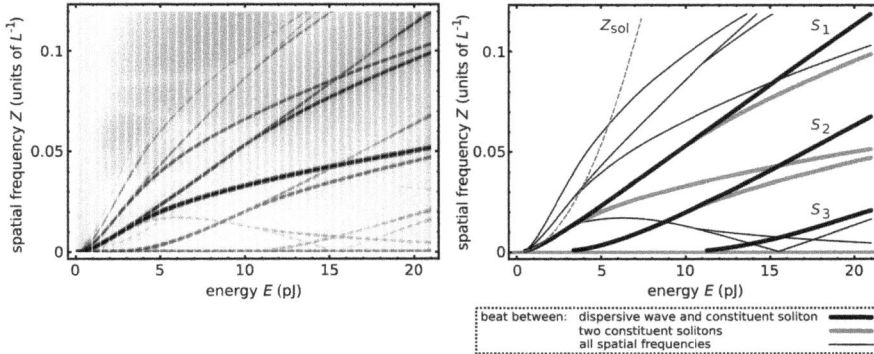

Figure 13. (Left) SRBA chart for an energy scan of a DM soliton. For an explanation of the procedure, and the apparent white vertical stripes in particular, see the text. (Right) Extract of the left panel, with standardized line types. The fundamentals pertaining to three constituent solitons are labeled S_1–S_3. The dashed line labeled 'sol' is for reference; it represents the spatial frequency for a non-DM soliton in a fiber with $\beta_2 = \bar{\beta}_2$ (from [93]).

A comparison of DST and SRBA reveals the advantages and weaknesses of both. SRBA requires more input data. These can only be available from numerical simulations; it is hard to imagine how suitable data could be obtained from an experiment. On the other hand, SRBA can treat cases that are inaccessible to DST. Its limit is reached when for a growing number of solitons involved, the number of traces becomes too large to disentangle them all. DST can be applied to many-soliton cases [94], but only in strictly integrable systems, or with the limitations in 'nearly integrable' systems as pointed out in Section 4.1.3.

5. Conclusions

Solitons in optical fibers are pulses of light that maintain a balance between linear and nonlinear distortions. Real-world systems are rarely, if ever, of the integrable kind. It is therefore appropriate to apply the word 'soliton' also to pulses for which the balance is maintained at least approximately; to do so is now common usage. However, one occasionally sees research papers in which the word is used in a very loose sense, with no effort being made to actually check whether the balance is fulfilled.

In this review, we outline the concept of fiber-optic solitons, with a particular view toward methods of verification of the solitonic nature of a given pulse, or group of pulses, or a similar structure. Soliton content has been discussed in the context of structures like soliton molecules, soliton gas and crystal, Akhmediev breather, optical supercontinuum and rogue waves. The simplest method is to check pulse parameters with Equation (4), which directly expresses an equilibrium condition. Unfortunately, this straightforward test is only applicable in very simple situations when there are clean, well-isolated pulses, and it does not fully quantify the soliton content. Grown from inverse scattering theory, the first method to analytically find the soliton solution of the NLSE is direct scattering transform. It basically tests whether the structure under test contains pulses that hang together as entities even when they collide, the property that gave solitons their 'particle-like' name [95]. An alternative method is soliton radiation beat analysis, which basically tracks the nonlinear phase rotation, which is the signature of solitons. We discuss some issues of practicality and accuracy and compare the ranges of applicability of both DST and SRBA.

There is no single method that covers all cases; indeed, in some relevant cases, no available method is fully satisfactory. In the complex evolution of supercontinuum, for example, the method is often adapted ad hoc to the situation, for want of alternatives. In some interesting cases, soliton content analysis has not been pursued yet, to our knowledge. Dissipative solitons are subject to yet another condition beyond Equation (4), but should be accessible for SRBA. This article is meant to provide the reader with a survey to what is currently known, to foster future research.

Author Contributions: F.M., C.M. and A.H.have contributed equally to this manuscript.

Conflicts of Interest: The authors declare no conflict of interest.

Abbreviations

The following abbreviations are used in this manuscript:

cw	continuous wave
DM	dispersion-management
DST	direct scattering transform
IST	inverse scattering transform
NLSE	nonlinear Schrödinger equation
SRBA	soliton radiation beat analysis

References

1. Agrawal, G.P. *Nonlinear Fiber Optics*, 5th ed.; Academic Press: Oxford, UK, 2013.
2. Mitschke, F. *Fiber Optics. Physics and Technology*, 2nd ed.; Springer: Heidelberg, Germany, 2016.
3. Stolen, R.H.; Lin, C. Self-phase-modulation in silica optical fibers. *Phys. Rev. A* **1978**, *17*, 1448–1453.
4. Zakharov, V.E.; Shabat, A.B. Exact Theory of Two-Dimensional Self-Focusing and One-Dimensional Self-Modulation of Waves in Nonlinear Media. *Sov. Phys. JETP* **1972**, *34*, 62–69.
5. Gardner, C.S.; Greene, J.M.; Kruskal, M.D.; Miura, R.M. Method for Solving the Korteweg-de Vries Equation. *Phys. Rev. Lett.* **1967**, *19*, 1095–1097.

6. Satsuma, J.; Yajima, N. Initial Value Problem of One-Dimensional Self-Modulation of Nonlinear Waves in Dispersive Media. *Prog. Theor. Phys. Suppl.* **1974**, *55*, 284–306.
7. Hasegawa, A.; Tappert, F. Transmission of stationary nonlinear optical pulses in dispersive dielectric fibers. I. Anomalous dispersion. *Appl. Phys. Lett.* **1973**, *23*, 142–144.
8. Mollenauer, L.F.; Stolen, R.H.; Gordon, J.P. Experimental Observation of Picosecond Pulse Narrowing and Solitons in Optical Fibers. *Phys. Rev. Lett.* **1980**, *45*, 1095–1098.
9. Yang, J. *Nonlinear Waves in Integrable and Nonintegrable Systems*; Society for Industrial and Applied Mathematics (SIAM): Philadelphia, PA, USA, 2010.
10. Kivshar, Y.S.; Agrawal, G.P. *Optical Solitons. From Fibers to Photonic Crystals*; Academic Press: London, UK, 2003.
11. Schrader, D. Explicit calculation of N-Soliton Solutions of the nonlinear Schrödinger equation. *IEEE J. Quantum Electron.* **1995**, *31*, 2221–2225.
12. Gordon, J.P. Interaction forces among solitons in optical fibers. *Opt. Lett.* **1983**, *8*, 596–598.
13. Mitschke, F.; Mollenauer, L.F. Experimental observation of interaction forces between solitons in optical fibers. *Opt. Lett.* **1987**, *12*, 355–357.
14. Desem, C.; Chu, P.L. Reducing soliton interaction in single-mode optical fibres. *IEEE Proc.* **1987**, *134*, 145–151.
15. Akhmediev, N.N.; Korneev, V.I. Modulation instability and periodic solutions of the nonlinear Schrödinger equation. *Theor. Math. Phys.* **1986**, *69*, 1089–1093.
16. Mahnke, C.; Mitschke, F. Ultrashort Light Pulses Generated from Modulation Instability: Background Removal and Soliton Content. *Appl. Phys. B* **2013**, *116*, 15–20.
17. Dudley, J.M.; Genty, G.; Dias, F.; Kibler, B.; Akhmediev, N. Modulation instability, Akhmediev Breathers and continuous wave supercontinuum generation. *Opt. Express* **2009**, *17*, 21497–21508.
18. Hammani, K.; Wetzel, B.; Kibler, B.; Fatome, J.; Finot, C.; Millot, G.; Akhmediev, N.; Dudley, J.M. Spectral dynamics of modulation instability described using Akhmediev breather theory. *Opt. Lett.* **2011**, *36*, 2140–2142.
19. Hasegawa, A. Generation of a train of soliton pulses by induced modulational instability in optical fibers. *Opt. Lett.* **1984**, *9*, 288–290.
20. Peregrine, D.H. Water waves, nonlinear Schrödinger equations and their solutions. *J. Aust. Math. Soc. B* **1983**, *25*, 16–43.
21. Kibler, B.; Fatome, J.; Finot, C.; Dias, F.; Genty, G.; Akhmediev, N.; Dudley, J.M. The Peregrine soliton in nonlinear fiber optics. *Nat. Phys.* **2010**, *6*, 790–795.
22. Kibler, B.; Fatome, J.; Finot, C.; Millot, G.; Genty, G.; Wetzel, B.; Akhmediev, N.; Dias, F.; Dudley, J.M. Observation of the Kuznetsov-Ma soliton dynamics in optical fibre. *Sci. Rep.* **2012**, *2*, 463.
23. Anderson, D.; Lisak, M. Nonlinear asymmetric self-phase modulation and self-steepening of pulses in long optical waveguides. *Phys. Rev. A* **1983**, *27*, 1393–1398.
24. Stolen, R.H.; Lee, C.; Jain, R.K. Development of the stimulated Raman spectrum in single-mode silica fibers. *J. Opt. Soc. Am. B* **1984**, *1*, 652–657.
25. Gordon, J.P. Theory of the soliton self frequency shift. *Opt. Lett.* **1986**, *11*, 662–664.
26. Stolen, R.H.; Gordon, J.P.; Tomlinson, W.J.; Haus, H.A. Raman response function of silica-core fibers. *J. Opt. Soc. Am. B* **1989**, *6*, 1159–1166.
27. Mamyshev, P.V.; Chernikov, S.V. Ultrashort-pulse propagation in optical fibers. *Opt. Lett.* **1990**, *15*, 1076–1078.
28. Lin, Q.; Agrawal, G.P. Raman response function for silica fibers. *Opt. Lett.* **2006**, *31*, 3086–3088.
29. Kaup, D.J. A perturbation expansion for the Zakharov–Shabat inverse scattering transform. *SIAM J. Appl. Math.* **1976**, *31*, 121–133.
30. Hasegawa, A.; Kodama, Y. Amplification and reshaping of optical solitons in a glass fiber—I. *Opt. Lett.* **1982**, *7*, 285–287.
31. Turitsyn, S.K.; Bale, B.G.; Fedoruk, M.P. Dispersion-managed solitons in fibre systems and lasers. *Phys. Rep.* **2012**, *521*, 135–203.
32. Smith, N.J.; Knox, F.M.; Doran, N.J.; Blow, K.J.; Bennion, I. Enhanced power solitons in optical fibres with periodic dispersion management. *Electron. Lett.* **1996**, *32*, 54–55.
33. Nijhof, J.H.B.; Doran, N.J.; Forysiak, W.; Knox, F.M. Stable soliton-like propagation in dispersion managed systems with net anomalous, zero and normal dispersion. *Electron. Lett.* **1997**, *33*, 1726–1727.
34. Chen, Y.; Haus, H.A. Dispersion-managed solitons with net positive dispersion. *Opt. Lett.* **1998**, *23*, 1013–1015.

35. Turytsin, S.K.; Shapiro, E.G. Dispersion-managed solitons in optical amplifier transmission systems with zero average dispersion. *Opt. Lett.* **1998**, *23*, 682–684.
36. Kutz, J.N.; Evangelides, S.G. Dispersion-managed breathers with average normal dispersion. *Opt. Lett.* **1998**, *23*, 685–687.
37. Grigoryan, V.S.; Menyuk, C.R. Dispersion-managed solitons at normal average dispersion. *Opt. Lett.* **1998**, *23*, 609–611.
38. Turitsyn, S.; Shapiro, E.; Medvedev, S.; Fedoruk, M.P.; Mezentsev, V. Physics and mathematics of dispersion managed optical solitons. *C. R. Phys.* **2003**, *4*, 145–161.
39. Lushnikov, P.M. Oscillating tails of a dispersion-managed soliton. *J. Opt. Soc. Am. B* **2004**, *21*, 1913–1918.
40. Nijhof, J.H.B.; Forysiak, W.; Doran, N.J. The averaging method for finding exactly periodic dispersion-managed solitons. *IEEE J. Sel. Top. Quantum Electron.* **2000**, *5*, 330–336.
41. Stratmann, M.; Pagel, T.; Mitschke, F. Experimental Observation of Temporal Soliton Molecules. *Phys. Rev. Lett.* **2005**, *95*, 143902.
42. Hause, A.; Hartwig, H.; Seifert, B.; Stolz, H.; Böhm, M.; Mitschke, F. Phase structure of soliton molecules. *Phys. Rev. A* **2007**, *75*, 063836.
43. Hause, A.; Hartwig, H.; Böhm, M.; Mitschke, F. Binding mechanism of temporal soliton molecules. *Phys. Rev. A* **2008**, *78*, 063817.
44. Rohrmann, P.; Hause, A.; Mitschke, F. Solitons Beyond Binary: Possibility of Fibre-Optic Transmission of Two Bits per Clock Period. *Sci. Rep.* **2012**, *2*, 866.
45. Rohrmann, P.; Hause, A.; Mitschke, F. Two-soliton and three-soliton molecules in optical fibers. *Phys. Rev. A* **2013**, *87*, 043834.
46. Hause, A.; Mitschke, F. Higher-order equilibria of temporal soliton molecules in dispersion-managed fibers. *Phys. Rev. A* **2013**, *88*, 063843.
47. Steinmeyer, G.; Buchholz, A.; Hänsel, M.; Heuer, M.; Schwache, A.; Mitschke, F. Dynamical pulse shaping in a nonlinear resonator. *Phys. Rev. A* **1995**, *52*, 830–838.
48. Schwache, A.; Mitschke, F. Properties of an optical soliton gas. *Phys. Rev. E* **1997**, *55*, 7720–7725.
49. Malomed, B.A.; Schwache, A.; Mitschke, F. Soliton lattice and gas in passive fiber-ring resonators. *Fiber Integr. Opt.* **1998**, *17*, 267–277.
50. Amrani, F.; Haboucha, A.; Salhi, M.; Leblond, H.; Komarov, A.; Sanchez, F. Dissipative solitons compounds in a fiber laser. Analogy with the states of the matter. *Appl. Phys. B* **2010**, *99*, 107–114.
51. Grelu, P.; Soto-Crespo, J.M. Temporal soliton "molecules" in mode-locked lasers: Collisions, pulsations, and vibrations. In *Dissipative Solitons: From Optics to Biology and Medicine*; Lecture Notes in Physics; Akhmediev, N., Ankiewicz, A., Eds.; Springer: Berlin/Heidelberg, Germany, 2008; Volume 751, pp. 137–173.
52. Chouli, S.; Grelu, P. Soliton rains in a fiber laser: An experimental study. *Phys. Rev. A* **2010**, *81*, 063829.
53. Soto-Crespo, J.M.; Akhmediev, N.N.; Afanasjev, V.V. Stability of the pulselike solutions of the quintic complex Ginzburg–Landau equation. *J. Opt. Soc. Am. B* **1996**, *13*, 1439–1449.
54. Grelu, P.; Akhmediev, N.N. Dissipative solitons for mode-locked lasers. *Nat. Photonics* **2012**, *6*, 84–92.
55. Akhmediev, N.; Ankiewicz, A. (Eds.) *Dissipative Solitons: From Optics to Biology and Medicine*; Lecture Notes in Physics; Springer: Berlin/Heidelberg, Germany, 2008; Volume 751.
56. Husakou, A.V.; Herrmann, J. Supercontinuum generation of higher-order solitons by fission in photonic crystal fibers. *Phys. Rev. Lett.* **2001**, *87*, 203901.
57. Roy, S.; Bhadra, S.K.; Agrawal, G.P. Perturbation of higher-order solitons by fourth-order dispersion in optical fibers. *Opt. Commun.* **2009**, *282*, 3798–3803.
58. Dudley, J.M.; Genty, G.; Coen, S. Supercontinuum generation in photonic crystal fiber. *Rev. Mod. Phys.* **2006**, *78*, 1135–1184.
59. Solli, D.R.; Ropers, C.; Koonath, P.; Jalali, B. Optical Rogue Waves. *Nature* **2007**, *450*, 1054–1057.
60. Akhmediev, N.; Kibler, B.; Baronio, F.; Belić, M.; Zhong, W.P.; Zhang, Y.; Chang, W.; Soto-Crespo, J.M.; Vouzas, P.; Grelu, P.; et al. Roadmap on optical rogue waves and extreme events. *J. Opt.* **2016**, *18*, 063001.
61. Lax, P.D. Integrals of nonlinear equations of evolution and solitary waves. *Commun. Pure Appl. Math.* **1968**, *21*, 467–490.
62. Hasegawa, A.; Nyu, T. Eigenvalue communication. *J. Lightwave Technol.* **1993**, *11*, 395–399.

63. Turitsyn, S.K.; Prilepsky, J.E.; Le, S.T.; Wahls, S.; Frumin, L.L.; Kamalian, M.; Derevyanko, S.A. Nonlinear Fourier transform for optical data processing and transmission: Advances and perspectives. *Optica* **2017**, *4*, 307–322.

64. Manakov, S.V. Nonlinear Fraunhofer diffraction. *Zh. Eksp. Teor. Fiz.* **1973**, *65*, 1392–1398.

65. Boffetta, G.; Osborne, A. Computation of the Direct Scattering Transform for the Nonlinear Schroedinger Equation. *J. Comput. Phys.* **1992**, *102*, 252–264.

66. Kamalian, M.; Prilepsky, J.E.; Le, S.T.; Turitsyn, S.K. Periodic nonlinear Fourier transform for fiber-optic communications, Part I: Theory and numerical methods. *Opt. Express* **2016**, *24*, 18353–18369.

67. Kamalian, M.; Prilepsky, J.E.; Le, S.T.; Turitsyn, S.K. Periodic nonlinear Fourier transform for fiber-optic communications, Part II: Eigenvalue communication. *Opt. Express* **2016**, *24*, 18370–18381.

68. Yousefi, M.I.; Kschischang, F.R. Information transmission using the nonlinear Fourier transform, Part II: Numerical methods. *IEEE Trans. Inf. Theory* **2014**, *60*, 4329–4345.

69. Ablowitz, M.J.; Kaup, D.J.; Newell, A.C.; Segur, H. Nonlinear-Evolution Equations of Physical Significance. *Phys. Rev. Lett.* **1973**, *31*, 125–127.

70. Prilepsky, J.E.; Derevyanko, S.A. Breakup of a multisoliton state of the linearly damped nonlinear Schrödinger equation. *Phys. Rev. E* **2007**, *75*, 036616.

71. Kodama, Y.; Hasegawa, A. Nonlinear pulse propagation in a monomode dielectric guide. *IEEE J. Quantum Electron.* **1987**, *23*, 510–524.

72. Tai, K.; Bekki, N.; Hasegawa, A. Fission of optical solitons induced by stimulated Raman effect. *Opt. Lett.* **1988**, *13*, 392–394.

73. Gölles, M.; Uzunov, I.M.; Lederer, F. Break up of N-soliton bound states due to intrapulse Raman scattering and third-order dispersion—An eigenvalue analysis. *Phys. Lett. A* **1997**, *231*, 195–200.

74. Mahnke, C.; Mitschke, F. Possibility of an Akhmediev breather decaying into solitons. *Phys. Rev. A* **2012**, *82*, 033808.

75. Its, A.R.; Matveev, V.B. Schrödinger operators with finite-gap spectrum and N-soliton solutions of the Korteweg—de Vries equation. *Theor. Math. Phys.* **1975**, *23*, 51–68.

76. Kawata, T.; Inoue, H. Inverse scattering method for the nonlinear evolution equations under nonvanishing conditions. *J. Phys. Soc. Jpn.* **1978**, *44*, 1722–1729.

77. Ma, Y.C.; Ablowitz, M.J. The periodic cubic Schrodinger equation. *Stud. Appl. Math.* **1981**, *65*, 113–158.

78. Ma, Y.C. The Perturbed Plane Wave Solutions of the Cubic Schrödinger Equation. *Stud. Appl. Math.* **1979**, *60*, 43–58.

79. Tracy, E.R.; Chen, H.H. Nonlinear self-modulation: An exactly solvable model. *Phys. Rev. A* **1988**, *37*, 815–839.

80. Osborne, A. *Nonlinear Ocean Waves and the Inverse Scattering Transform*; International Geophysics Series; Academic Press: Burlington, MA, USA, 2010; Volume 97.

81. Olivier, C.P.; Herbst, B.M.; Molchan, M.A. A numerical study of the large-period limit of a Zakharov–Shabat eigenvalue problem with periodic potentials. *J. Phys. A Math. Theor.* **2012**, *45*, 255205–255211.

82. Wahls, S.; Poor, H.V. Introducing the fast nonlinear Fourier transform. In Proceedings of the 2013 IEEE International Conference on Acoustics, Speech and Signal Processing (ICASSP), Vancouver, BC, Canada, 26–31 May 2013.

83. Wahls, S.; Poor, H.V. Fast numerical nonlinear Fourier transforms. *IEEE Trans. Inf. Theory* **2015**, *61*, 6957–6974.

84. Randoux, S.; Suret, P.; El, G. Inverse scattering transform analysis of rogue waves using local periodization procedure. *Sci. Rep.* **2016**, *6*, 29238.

85. Weerasekara, G.; Tokunaga, A.; Terauchi, H.; Eberhard, M.; Maruta, A. Soliton's eigenvalue based analysis on the generation mechanism of rogue wave phenomenon in optical fibers exhibiting weak third order dispersion. *Opt. Express* **2015**, *23*, 143–153.

86. Weerasekara, G.; Maruta, A. Characterization of optical rogue wave based on solitons' eigenvalues of the integrable higher-order nonlinear Schrödinger equation. *Opt. Commun.* **2017**, *382*, 639–645.

87. Akhmediev, N.; Soto-Crespo, J.M.; Devine, N. Breather turbulence versus soliton turbulence: Rogue waves, probability density functions, and spectral features. *Phys. Rev. E* **2016**, *94*, 022212.

88. Böhm, M.; Mitschke, F. Soliton radiation beat analysis. *Phys. Rev. E* **2006**, *73*, 066615.

89. Böhm, M.; Mitschke, F. Soliton content of arbitrarily shaped light pulses in fibers analysed using a soliton-radiation beat pattern. *Appl. Phys. B* **2007**, *86*, 407–411.

90. Böhm, M.; Mitschke, F. Solitons in lossy fibers. *Phys. Rev. A* **2007**, *76*, 063822.

91. Zajnulina, M.; Böhm, M.; Blow, K.; Rieznik, A.A.; Giannone, D.; Haynes, R.; Roth, M.M. Soliton radiation beat analysis of optical pulses generated from two continuous-wave lasers. *Chaos* **2015**, *25*, 103104.

92. Zajnulina, M.; Böhm, M.; Bodenmüller, D.; Blow, K.; Chavez Boggio, J.M.; Rieznik, A.A.; Roth, M.M. Characteristics and stability of soliton crystals in optical fibres for the purpose of optical frequency comb generation. *Opt. Commun.* **2017**, *393*, 95–102.

93. Hartwig, H.; Böhm, M.; Hause, A.; Mitschke, F. Slow oscillations of dispersion-managed solitons. *Phys. Rev. A* **2010**, *81*, 033810.

94. Mitschke, F.; Halama, I.; Schwache, A. Soliton Gas. *Chaos Solitons Fractals* **1999**, *10*, 913–920.

95. Kruskal, M.D.; Zabusky, N.J. Interaction of "solitons" in a collisionless plasma and the recurrence of initial states. *Phys. Rev. Lett.* **1965**, *15*, 240–243.

![applied sciences logo] *applied sciences*

MDPI

Review

Lévy Statistics and the Glassy Behavior of Light in Random Fiber Lasers

Cid B. de Araújo [1,*]**, Anderson S. L. Gomes** [1] **and Ernesto P. Raposo** [2]

[1]   Departamento de Física, Universidade Federal de Pernambuco, Recife-PE 50670-901, Brazil;
     anderson@df.ufpe.br
[2]   Laboratório de Física Teórica e Computacional, Departamento de Física, Universidade Federal de
     Pernambuco, Recife-PE 50670-901, Brazil; ernesto@df.ufpe.br
*   Correspondence: cid@df.ufpe.br; Tel.: +55-81-2126-7630

Academic Editor: Boris Malomed
Received: 12 May 2017; Accepted: 15 June 2017; Published: 22 June 2017

Abstract: The interest in random fiber lasers (RFLs), first demonstrated one decade ago, is still growing and their basic characteristics have been studied by several authors. RFLs are open systems that present instabilities in the intensity fluctuations due to the energy exchange among their non-orthogonal quasi-modes. In this work, we present a review of the recent investigations on the output characteristics of a continuous-wave erbium-doped RFL, with an emphasis on the statistical behavior of the emitted intensity fluctuations. A progression from the Gaussian to Lévy and back to the Gaussian statistical regime was observed by increasing the excitation laser power from below to above the RFL threshold. By analyzing the RFL output intensity fluctuations, the probability density function of emission intensities was determined, and its correspondence with the experimental results was identified, enabling a clear demonstration of the analogy between the RFL phenomenon and the spin-glass phase transition in disordered magnetic systems. A replica-symmetry-breaking phase above the RFL threshold was characterized and the glassy behavior of the emitted light was established. We also discuss perspectives for future investigations on RFL systems.

Keywords: random fiber laser; Lévy statistics; photonic spin-glass behavior

---

1. Introduction

Proposals for the operation of random lasers (RLs) were made five decades ago by Ambartsumyan and co-workers [1,2], who visualized the possibility of a new kind of laser that does not require the use of optical cavities. Initially, they reported on the operation of a laser in which one of the cavity mirrors was replaced by a piece of paper that scattered the light in such way that a fraction of the backscattered light was enough to provide feedback for the laser operation. Following the original work, the same group published a series of papers studying the line-narrowing [3], frequency stability [4], and the statistical emission properties [5] of lasers with the so-called nonresonant feedback.

Apparently, the initial motivation for these studies was the observation of laser emission from interstellar media [6], and the interest of the group on this subject continued in the subsequent years [7–9].

For about fifteen years, the majority of research on this theme was pursued by groups that concentrated their effort on the operation of RLs based on microcrystals doped by rare-earth ions [10]. However, the first efficient RL system built in a laboratory environment was reported in 1994 by Lawandy and co-workers [11], who demonstrated the operation of an RL based on dye molecules dissolved in alcohol with suspended titanium dioxide particles. That report was followed by a great number of papers from different authors that investigated other physical systems for efficient RL operation. A large variety of materials have been tested in the past years, and recent publications on RLs describe, for example, experiments with dyes dissolved in transparent liquids, gels or liquid crystals with suspended micro or nanoparticles as light scatterers [12–17], powders of semiconductor quantum dots [18,19], dielectric nanocrystals doped with rare-earth ions [20,21], polymers and organic membranes doped by luminescent molecules [22–27], semiconductor and metallic nanowires structures [28–31], and even atomic vapors that present interesting analogies with astrophysical lasers [32].

There is a large literature on RLs motivated by the interest in a deeper understanding of the fundamental properties of RLs [33–36], as well as reports on their possible application in sensing [37], optofluidics [38,39], and imaging [40], among other fields [41]. Moreover, although from a fundamental point of view there are many reports focusing on the basic characteristics of RLs and their operation, the analogies between RLs and other complex systems have only recently been investigated by experiments. For example, in the work by Ghofraniha and co-authors [42], the analogy between RLs and the spin-glass phenomenon typical of highly-disordered magnetic systems was demonstrated for the first time.

In the present work, we review the recent advances on the characteristics of random fiber lasers, which have large potential for applications in various areas, as mentioned below. The article is organized as follows. In Section 2, we describe the Materials and Methods used. In Section 3, the experiments with erbium-doped fibers to characterize the RL behavior and the analysis of the intensity fluctuations are presented, along with the theoretical framework to understand the system behavior. Finally, in Section 4, a summary of the article contents and a discussion on perspectives for future work are presented.

2. Materials and Methods

2.1. Random Fiber Lasers

Random fiber lasers (RFLs) are akin to RLs, being the one-dimensional (1D) or quasi-1D version of the 2D or 3D RLs. They bear the same nonconventional remarkable characteristic: the optical feedback is provided by a scattering medium, rather than by fixed mirrors or fiber Bragg gratings (FBGs), as in conventional fiber lasers. Similarly to conventional fiber lasers, a gain medium is excited by an appropriate optical pump source.

The first demonstrated RFL, by de Mattos and co-workers in 2007 [43], can be seen as a quasi-1D extension of the colloidal-based RL reported by Lawandy and co-workers in 1994 [11]. In [43], the hollow core of a photonic crystal fiber was filled with a colloid al consisting of Rhodamine 6 G and 250 nm rutile (TiO_2) particles suspended in ethylene glicol. By transversely pumping with nanosecond pulses from the second harmonic of a Nd:YAG laser, directional emission was generated axially, and the feedback was due to the TiO_2 scatterers. Shortly after the report of ref. [43], Lizárraga and co-authors [44] and Gagné and Kashyap [45] demonstrated the operation of a continuous-wave (CW) pumped erbium-based RFL (Er-RFL), with random FBGs providing the scattering mechanism. We anticipate that this special type of RFL will be exploited as the photonic platform for all of the work described here, and will be detailed later.

A new breakthrough in the research of RFL systems occurred in 2010, when Turitsyn and co-workers [46] first reported on the operation of an RFL system which exploits Rayleigh scattering as the optical feedback mechanism in rather long (~83 km) conventional single-mode optical fibers. In this pioneer work, the gain mechanism was the stimulated Raman scattering excited in the fiber.

We observe that following this work, the interest in RFL systems and applications has fantastically grown, as reviewed in refs. [47,48]. Indeed, by further exploiting the Rayleigh scattering due to refractive index fluctuations as the mechanism for the multiple light scattering, a myriad of novel types of RFL systems have been demonstrated using a stimulated Raman or Brillouin scattering process.

As most of the works between 2007 and 2014 have been reviewed in refs. [47,48], including polymer-based optical fibers or plasmonically-enhanced RFLs, we highlight here the diversity of works reported over the years 2015 and 2016 (see [49–66] and references therein). As examples, we mention that a Q-switched operation has been reported using Brillouin scattering [56], with pulses as short as 42 ns at 100 kHz being demonstrated. Regarding the fiber length, since the first observation of RFL made using conventional fibers with 83 km [46], RFLs with fiber lengths as short as 120 m have been reported [58], also providing 200 W of output power. Other recent features of RFLs include tunability using graphene-based devices [56] or high-order Raman scattering [57], second harmonic generation [59], polarized emission from disordered polymer optical fibers [60], and photonic turbulence [67].

2.2. Fiber Bragg Grating-Based Random Fiber Lasers

As mentioned above, FBG-based RFLs were first introduced in 2009 [44,45]. Currently, the FBG fabrication methods, characterization, and management constitute a well-developed field, and further information on this subject is deferred to ref. [68].

Typically, a writing setup based, for instance, on CW UV radiation or femtosecond sources at 800 nm, is employed to inscribe an FBG into a conventional or core-doped optical fiber. The case of interest here exploits an active core single-mode fiber, using trivalent erbium ions, Er^{3+}, which can be excited at 980 nm or 1480 nm, and emits in the 1540–1560 nm spectral region. Instead of inscribing the FBGs in an evenly spaced way, thus leading to conventional resonators, the erbium-based FBGs are randomly spaced and play the role of random scatterers, leading to RFL emission. Generally, the fiber is placed on a movable stage which is randomly displaced, thus providing the randomness in the FBG writing process. Several tens to hundreds of gratings can be inscribed along several tens of cm of fiber length.

In the next section, we will describe the fabrication and characterization of the Er-RFL system with a specially-designed FBG [45]. We comment that this system has been lately used as an experimental platform to study complex photonic phenomena, such as the observation of unconventional Lévy-like statistics of output intensity values and the demonstration of the nontrivial replica-symmetry-breaking regime, which marks the signature of the phase transition from a photonic paramagnetic to a photonic spin-glass phase.

3. Results

3.1. Characterization of the FBG-Based Er-RFL Explored as a Statistical-Physics Experimental Platform

As reported by Gagné and Kashyap [45], a unique FBG was produced by writing an exceptionally high number of gratings (>>1000) over a 30 cm length fiber. A polarization-maintaining erbium-doped fiber from CorActive (peak absorption 28 dB/m at 1530 nm, NA 0.25, mode field diameter 5.7 μm) was employed, in which the randomly distributed phase errors grating was written, instead of a random array of gratings as in [44]. It was realized in [45] that during the movement of the translation stage, the friction between the fiber and the mount introduced irregularities in the grating spectrum that could be controlled by managing the air flow from the vacuum. Such irregularities were perceived as small phase errors, randomly but continuously distributed along the grating profile being inscribed. Thanks to this procedure, a high number of modes were observed, which is very important for several applications, as we will see later.

The Er-RFL described in [45] had a very low threshold of 3 mW, with a throughput efficiency of ~4.5% for 100 mW pump power. The number of emitted modes was dependent on the pump power and fiber length (20 or 30 cm fiber lengths were characterized in [45]), and a single or few modes were observed, limited by the system measurement resolution [69].

For the experiments described here, the 30 cm long fiber was employed with the experimental setup shown in Figure 1, reproduced from ref. [68]. The pump source was a semiconductor laser operating in the CW regime at 1480 nm, delivering 150 mW output power at the fiber pigtail. The Er-RFL output was split, through a 1480 nm/1550 nm wavelength-division multiplexer (WDM), with a split ratio of 10/90 for 1480 nm and 1550 nm, respectively, to a power meter and a spectrometer.

The employed fiber splices were lossy, if compared to the original work [45], leading to a higher threshold. The spectrometer (SpectraPro 300i, Acton Research, Acton, MA, USA), coupled to a liquid-N_2 cooled InGaAs CCD camera, had a nominal resolution of 0.1 nm.

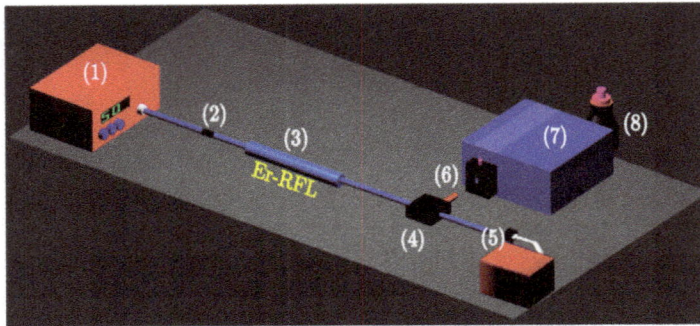

Figure 1. Experimental setup for the erbium-based random fiber laser (Er-RFL) system. (1) Fiber pigtailed semiconductor laser operating in the continuous-wave (CW) regime. (2) Fiber connector. (3) Er-doped RFL. (4) Wavelength-division multiplexer (WDM) 1480–1550. (5) Power meter to measure the output power P_{out} at 1480 nm. (6) RFL emission out to the spectrometer. (7) Spectrometer. (8) Liquid-N_2 cooled InGaAs CCD camera. (Reproduced with permission from ref. [68]).

Figure 2a shows the Er-RFL output spectrum for intensities below and above the RFL threshold, while Figure 2b shows the linewidth narrowing (left y-axis) and emitted Er-RFL intensity (right y-axis) as a function of the pump power P normalized to the threshold power P_{th}. The threshold power was $P_{th} = 16.30 \pm 0.05$ mW [68], which is higher than in the original work of ref. [45] due to the lossy components employed. However, this fact did not affect the experimental studies, and the output was typically around 1–2 mW.

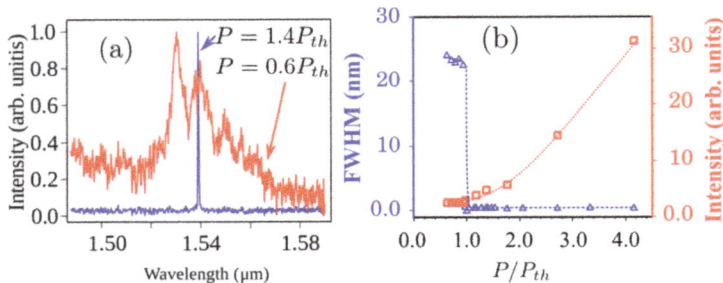

Figure 2. (a) Emitted spectrum of Er-RFL before (red) and after (blue) the laser threshold; (b) Emitted intensity (squares) and FWHM (triangles) of the Er-RFL system as a function of the normalized input power. The measured threshold power was $P_{th} = 16.30 \pm 0.05$ mW. The dotted lines are a guide to the eyes. (Reproduced with permission from ref. [68]).

Besides the routine characterization illustrated in Figure 2, two other analyzes were performed in order to show that the laser is multimode, i.e., with many longitudinal modes, and that the intensity fluctuations do not depend on the pump laser fluctuations.

First, in order to demonstrate the multimode characteristic of the Er-RFL, we employed the technique of speckle contrast [69], following the work of refs. [40,70,71]. To generate the speckle, a scattering medium with dried TiO_2 (250 nm) nanoparticles in water solution on a microscope slide along with a Kohler illumination system was used. Different CCD cameras were employed for the data acquisition, depending upon the source wavelength in the visible or near-infrared. The relation between the speckle contrast C and the number m of longitudinal modes of the source is given by [71] $C = \sigma/\langle I \rangle = 1/\sqrt{m}$, where σ and $\langle I \rangle$ are the standard deviation and the average intensity determined from the speckle, respectively. Figure 3 shows the obtained results for different light sources.

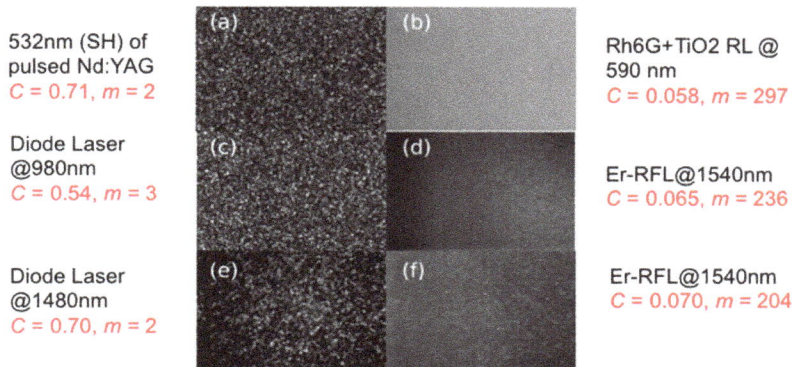

Figure 3. (**a,c,e**) display the measured speckle of the second harmonic of a pulsed Nd:YAG laser (**a**); a CW diode laser operating at 980 nm (**c**); and a similar diode laser, but operating at 1480 nm (**e**); (**b,d,f**) show a colloid random laser (RL) (**b**); the Er-RFL at 1540 nm pumped at 980 nm (**d**); and the same Er-RFL pumped at 1480 nm (**f**). The values of the speckle contrast C and number of longitudinal modes m for each optical source are indicated at the side. (Adapted with permission from ref. [69]).

To confirm the experimental results for the Er-RFL system, we initially characterized a well-known colloidal-based RL consisting of Rhodamine 6 G and TiO_2 nanoparticles. The pump source was the second-harmonic of a pulsed Nd:YAG laser (Ultra, BigSky Laser, Paris, France), operating at 5 Hz and delivering pulses of ~7 ns. Figure 3a shows the speckle image from the pump source, which is a highly coherent source (basically a single mode), and Figure 3b shows the equivalent image from the colloidal RL. Just as in the work of refs. [40,70], the speckle-free RL emission is corroborated, and the calculated values of the parameters C and m for the RL are 0.058 and 297, respectively, which are very much distinct from the respective values obtained for the pump laser. Additionally, the speckle images from Figure 3a,b are strikingly different, as already reported [40].

The same behavior is reproduced for the Er-RFL system, in which the pump laser at 1480 nm presents $m = 2$ longitudinal modes (Figure 3e), whereas the Er-RFL displays the presence of $m = 204$ longitudinal modes (Figure 3f). For completeness, the pump semiconductor laser at 980 nm was also employed, giving similar results, i.e., a 980 nm pump laser was a quasi-single mode ($m = 3$, Figure 3c), whereas the Er-RFL system showed $m = 236$ longitudinal modes, as illustrated in Figure 3d. Actually, being a multimode system is a fundamental requirement for the observation of the spin-glass type of behavior in RLs, as discussed below.

On the other hand, we also remark that the fluctuations of the pump source (less than 5%) were not correlated with the RFL fluctuations, as similarly demonstrated in [42,72]. This important point is corroborated by the results displayed in Figure 4a,b, showing, respectively, the spectral variance of the Er-RFL system and the normalized standard deviation of both the pump laser and Er-RFL. It is thus clear from Figure 4b that the pump laser fluctuations do not affect the Er-RFL fluctuations, particularly because the pump laser was kept working all the time well above the threshold, so that the kind of new physics observed around the threshold in the Er-RFL would not be detected, even if present.

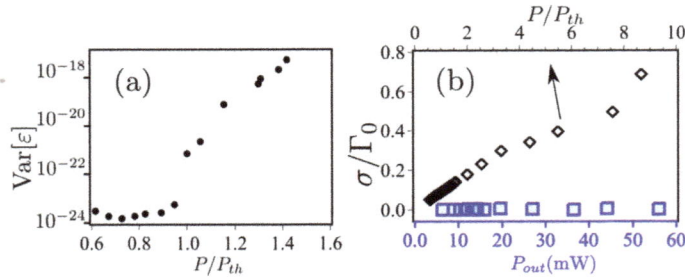

Figure 4. (**a**) Variance of the emitted intensity as a function of the normalized input power P/P_{th}; (**b**) Standard deviation of the maximum intensity (normalized by its average value) of the CW semiconductor pump laser (as function of the output power P_{out}; squares) and the Er-RFL system (as function of the normalized input power P/P_{th}; diamonds). (Reproduced with permission from ref. [68]).

3.2. Theoretical Framework

A great advance in the theoretical understanding of the combined effect of amplification, nonlinearity, and disorder in RL systems was put forward in a series of articles [73–82] published within the last decade. In this subsection, we review the physical mechanisms and some analytical developments underlying the richness of photonic behaviors displayed in the phase diagram of RL systems, as a function of the input excitation power and disorder strength. In fact, in refs. [73–80], a variety of interesting photonic phases emerge, which keep close analogies with some characteristic behaviors of magnetic systems, such as spin glass, paramagnetism, and ferromagnetism.

Moreover, such richness is also present in the diversity of statistical regimes observed in the output intensity emitted by RL systems. Interestingly, as we shall see below, the same theoretical starting point that gives rise to the variety of photonic behaviors also explains the shifts in the statistical properties of the distribution of intensity values. In fact, as discussed in the following, the increasing of the excitation power promotes a sequence of changes in the distribution of intensity values, from a Gaussian regime below the RL threshold to a Lévy-type behavior around the RL transition, and back to a second Gaussian regime well above the threshold. The theoretical basis [81,82] for these statistical aspects of RL emission is also reviewed below.

We start by reviewing the theoretical background [73–80] underlying the diverse photonic behaviors displayed by RLs. In a photonic system with intrinsic disorder, arising either from the inherent optical random noise or from the presence of randomly-located light scatterers (e.g., micro or nanoparticles [12–17], or dielectric nanocrystals doped with rare-earth ions [20,21]), the amplitudes a_k of the electromagnetic modes generally present stochastic dynamics. The Langevin approach is thus a suitable theoretical framework to describe such dynamics through the following set of equations,

$$\frac{da_k}{dt} = -\frac{\partial H}{\partial a_k^*} + F_k, \tag{1}$$

where F_k represents a Gaussian (white) uncorrelated optical noise term, and in the slow-amplitude regime of the modes, the general complex-valued functional H is given by [73–80] (closely following the notation of [78]):

$$H = \sum_{\{k_1 k_2\}'} g_{k_1 k_2}^{(2)} a_{k_1} a_{k_2}^* + \frac{1}{2} \sum_{\{k_1 k_2 k_3 k_4\}'} g_{k_1 k_2 k_3 k_4}^{(4)} a_{k_1} a_{k_2}^* a_{k_3} a_{k_4}^* \tag{2}$$

The symbol $\{\ldots\}'$ implies the frequency-matching conditions $|\omega_{k_1} - \omega_{k_2}| < \gamma$ and $|\omega_{k_1} - \omega_{k_2} + \omega_{k_3} - \omega_{k_4}| < \gamma$ in the quadratic and quartic terms, respectively, with γ denoting the finite linewidth of the modes. The disordered nature of the scattering medium directly affects both the quadratic and quartic interactions among the spatially-overlapping modes in Equation (2). Indeed, the physical origin of the quadratic coupling $g^{(2)}_{k_1 k_2}$ lies in the spatially inhomogeneous refractive index, as well as in the nonuniform distribution of the gain and effective damping contribution due to the cavity leakage. In systems with null or weak leakage, in which the off-diagonal contribution in $g^{(2)}_{k_1 k_2}$ is negligible, the coefficient rates of amplification (γ_k) and radiation loss (α_k) are related to the real part of the diagonal coupling through $g^{(2)R}_{kk} = Re\{g^{(2)}_{kk}\} = \alpha_k - \gamma_k$. On the other hand, the quartic coupling $g^{(4)}_{k_1 k_2 k_3 k_4}$ is associated with the modulation of the nonlinear $\chi^{(3)}$-susceptibility with a random spatial profile [73–80].

The spatial disorder in the scattering medium generally makes the explicit calculation of the quadratic and quartic couplings in Equation (2) rather difficult. Consequently, in [73–80], these couplings were considered as quenched Gaussian variables, with probability distributions independent of the mode combinations $\{k_1 k_2\}'$ and $\{k_1 k_2 k_3 k_4\}'$, respectively. In addition, in the mean-field approach of refs. [73–80], the above frequency-matching constraints were relaxed, implying that all modes interact unrestrictedly. In the case in which the total optical intensity, $I = \sum_k c_k |a_k|^2$, is a constant, with time-independent prefactors c_k, the real part H^R of the Functional (2) is shown in [78,80] to become analogous to the Hamiltonian of the magnetic p-spin model with a spherical constraint [83]. Indeed, the p-spin Hamiltonian also presents a sum of quadratic ($p = 2$) and quartic ($p = 4$) interaction terms, just as in Equation (2), with the couplings drawn from Gaussian distributions [83]. Moreover, in the p-spin model, the sum of the squared spin variables is a constant (spherical constraint), as also happens to the total optical intensity I in the photonic system. This photonic-to-magnetic analogy is indeed relevant, since it allows identifying the amplitudes of the modes with the spin variables, and the excitation (pump) energy in the photonic system with the inverse temperature in the magnetic one. In this sense, we comment below that the typical magnetic phases exhibited by the disordered p-spin model can also find an analogous counterpart in the phase diagram of an RL system.

Once the disordered Hamiltonian H^R, given by the real part of Equation (2), has been built in terms of mode amplitudes that are spin analogues, a repertoire of statistical-physics-based analytical techniques to treat disordered magnetic systems becomes immediately available to the photonic system. In particular, the so-called replica trick [84] can be readily applied to H^R. This approach essentially consists of considering identical copies (i.e., replicas) of the system in order to compute the powers Z^n of the partition function, while calculating the free energy from the limit expression $\ln Z = \lim_{n \to 0} (Z^n - 1)/n$. At the end, the phase diagram of the RL system is obtained as a function of the input pumping rate and disorder strength [77–79]. Remarkably, the physical equivalence (or symmetry) among these replicas can be broken in some circumstances, as discussed below.

Following this photonic-to-magnetic analogy, the photonic phases identified in the RL system maintain some resemblance to the magnetic behaviors of the p-spin model. Indeed, we next describe the main properties of the four photonic regimes obtained in the phase diagram of refs. [77–79], namely: incoherent wave, mode-locking laser, phase-locking wave, and spin-glass RL behavior.

In the incoherent-wave regime, which occurs for low input powers and any disorder strength, the modes oscillate incoherently in an uncorrelated way. In this case, the system operates in a regime with amplified spontaneous emission. According to the analogy above, this phase is similar to the paramagnetic behavior in spin systems at high temperatures and for any degree of disorder, in which the uncorrelated spin directions are random and present fast dynamics. Moreover, just as in the paramagnetic phase, the incoherent-wave solution preserves the symmetry among the photonic replicas (see also below).

The mode-locking laser behavior presents modes oscillating coherently with the same phase, at high input powers, without disorder or for low degrees of disorder. It corresponds to the ferromagnetic phase in spin systems at low temperatures, either without disorder, in which all spins align parallel, or in the presence of low disorder, as in the case of the random bond ferromagnet, with a few clusters of disordered spins in a predominantly ferromagnetic background.

Remarkably, the regime described as the phase-locking wave has no perfect analogous counterpart in disordered spin modes. In the pumping rate versus disorder strength diagram, it occupies an intermediate region between the incoherent-wave and spin-glass RL phases. In this regime, the mode phases are only partially locked.

Finally, in the RL regime, obtained for input powers above the threshold and in the presence of strong disorder, the synchronous oscillation of the modes is frustrated (in contrast with the mode-locking laser behavior), and they acquire phase coherence and nontrivial correlations (differently from the incoherent-wave regime). In this case, the analogue magnetic phase is the spin-glass regime observed in highly disordered magnetic systems at low temperatures, in which the spins point at random directions, while being strongly correlated in time and with rather slow (frozen) dynamics. In the photonic, as well as in the magnetic spin-glass phase, the replicas undergo a nontrivial breaking of symmetry, which is explained as follows.

The concept of replica symmetry breaking (RSB) was introduced by G. Parisi in 1979 in the context of the theory of disordered magnetic systems [84]. In this framework, for sufficiently low temperatures and strong disorder (e.g., in the spin couplings or locations), the free energy landscape breaks into a large number of local minima in the configuration space. Due to the frustrated magnetic interactions in the disordered Hamiltonian, the spins fail to align in a spatially regular configuration, as in the ferromagnetic state. Instead, spins "freeze" along random directions, with rather slow dynamics, in a spin-glass state. As a given spin configuration can be trapped for a long time in a local free energy minimum, metastability and irreversibility effects arise in the spin-glass phase, e.g., magnetic hysteresis. Consequently, identical systems, with the same distribution of spin interactions and prepared under identical conditions (i.e., replicas of the spin system), can reach rather distinct states that lead to different measures of observable quantities and nontrivial correlation patterns. In this case, the system replicas are no longer physically equivalent (or symmetric), and an RSB scenario emerges. Later on, the scope of the concept of RSB was much extended to reach other complex systems [84], including neural networks and structural glasses.

In order to identify a regime with RSB, it becomes necessary to calculate a correlation function that gives a measure of the overlap between two given replicas [84]. In the case of magnetic systems, the replica overlap parameter is a spin-spin correlation function defined by the product of a certain spin occupying the same position in two distinct replicas. In the sequence, the sum of all spins is performed. By considering each pair of replicas, a distribution $P(q)$ of values of such an overlap parameter q is thus obtained. If this distribution is centered around zero, the replicas are considered symmetric, a scenario that is observed in the paramagnetic phase in spin systems. However, if the distribution peaks at non-zero values of the replica overlap parameter, then the symmetry of replicas is broken, and an RSB spin-glass phase can emerge. Therefore, in this sense, a parameter q_{max} can be defined to indicate the locus of the maximum of the distribution $P(q)$, which is considered as the Parisi order parameter [84]. The value of q_{max} thus signalizes a replica-symmetric paramagnetic or an RSB spin-glass phase, respectively, if the maximum of $P(q)$ occurs exclusively at $q_{max} = 0$ (no RSB) or also at values $|q_{max}| \neq 0$ (RSB).

In the photonic context, an analogue correlation function between modes can also be suitably defined [42,79] (see details below). Its distribution of values determines, in a similar way, the presence or absence of the photonic RSB spin-glass phase.

On the experimental side, the very first evidence of photonic RSB glassy behavior in an RL system arose in the 2D functionalized T_5OC_x oligomer amorphous solid-state material [42]. Subsequent demonstrations appeared in 3D functionalized TiO_2 particle-based dye-colloidal [85] and neodymium-doped YBO_3 solid-state [82] RLs. Here, we highlight that the RSB spin-glass phase has also been characterized in the above-mentioned Er-RFL system [68,69]. The emergence of such behavior in Er-RFL is actually justified since this system also presents the disorder and nonlinear ingredients necessary to induce the RSB glassy RL phase, as discussed above in the context of the effective photonic Hamiltonian (2).

We now turn to the discussion on the statistical regimes of output intensities emitted by RL systems.

We first review some analytical developments regarding the distribution of intensity values. Noteworthy, the set of Langevin equations, given by Equation (1), also provides the underlying theoretical basis for such analysis. Indeed, by writing $I_k = c_k |a_k|^2$, a manipulation of Equation (1) yields [82]:

$$\frac{1}{c_{k_2}} \frac{dI_{k_2}}{dt} = -2Re \left\{ \sum_{\{k_1\}'} g^{(2)}_{k_1 k_2} a_{k_1} a^*_{k_2} + \frac{1}{2} \sum_{\{k_1 k_3 k_4\}'} [g^{(4)}_{k_1 k_2 k_3 k_4} + g^{(4)}_{k_1 k_4 k_3 k_2}] a_{k_1} a^*_{k_2} a_{k_3} a^*_{k_4} + a^*_{k_2} F_{k_2} \right\} \quad (3)$$

The restricted sum in the quartic coupling generally involves three classes of mode: combinations [73,86]: $\omega_{k_1} = \omega_{k_2}$ and $\omega_{k_3} = \omega_{k_4}$, $\omega_{k_1} = \omega_{k_4}$ and $\omega_{k_2} = \omega_{k_3}$, and the remaining possibilities satisfying the frequency-matching conditions, which have been usually disregarded [73,86]. We consider the diagonal contribution in the quadratic coupling to dominate over the off-diagonal part. By expressing the optical white noise as the sum of additive and multiplicative statistically independent stochastic processes [87], so that $F_k(t) = F_k^{(0)}(t) + a_k(t) F_k^{(1)}(t)$, and considering slow-amplitude modes $a_k(t)$ (if compared to the rapidly evolving phase dynamics), we obtain the Fokker-Planck equation [81,82,87] for the probability density function (PDF) of the output intensity,

$$\frac{\partial P}{\partial t} = -\frac{\partial}{\partial I_k} [(-d_k I_k - b_k I_k^2 + 2Q I_k) P] + 2Q \frac{\partial^2}{\partial I_k^2} (I_k^2 P) \quad (4)$$

where the parameter Q controls the magnitude of the multiplicative fluctuations through: $[F_k^{(1)R}(t) F_m^{(1)R}(t')] = 2Q \delta_{k,m} \delta(t - t')$, $b_k = g^{(4)R}_{kkkk}/c_k$, and

$$d_k = \sum_{n \neq k} [g^{(4)R}_{kknn} + g^{(4)R}_{knnk} + g^{(4)R}_{nkkn} + g^{(4)R}_{nnkk}] I_n / c_n - 2(\gamma_k - \alpha_k) \quad (5)$$

The steady-state solution of Equation (4) is [81,82,87]:

$$P(I_k) = A_k I_k^{-\mu_k} \exp\left(-b_k I_k / 2Q\right) \quad (6)$$

with $I_k > 0$, A_k as the normalization constant, and $\mu_k = 1 + d_k/2Q$. This PDF presents a power-law decay combined with exponential attenuation. Its second moment can be very large, though still finite, depending on the value of μ_k, mainly if $b_k/2Q \ll 1$. Indeed, the experimental results obtained for the Er-RFL (see below) indicate [68] that the distribution of output intensities displays much larger variance and much stronger fluctuations close to the threshold, if compared with those below and above the threshold. In this sense, as argued below, the PDF of intensities is most properly described by the Lévy α-stable distribution [88] for long time measurements performed in CW pumped RLs, such as the Er-RFL system [68], or during an extensive number of shots in the case of pulsed RLs [89].

As we now turn the focus to the physical discussion on the PDF of intensity values, Equation (6), we initially observe that the analysis is strictly connected with the central limit theorem (CLT) and generalized CLT of statistics. These theorems determine the attraction of the PDF of the sum of a large number of random variables to one of the possible asymptotic stable distributions, namely the Gaussian or the Lévy α-stable family [88]. In the present photonic context, if the stochastic values assumed by the intensity I are identically distributed and uncorrelated over the long sequence of output spectra (or even if they present finite-time correlations), and if the second moment of the PDF $P(I)$ is finite, then the CLT assures [88] that the intensity fluctuations are driven by the Brownian (Gaussian, normal) dynamics. On the other hand, if the second moment of $P(I)$ diverges, the generalized CLT states [88] that the fluctuations are asymptotically governed by the Lévy statistics. The continuous family of Lévy α-stable distribution is described [88] by the Fourier transform of the characteristic function defined in k-space,

$$\overline{P}(k) = \exp\{-|ck|^{\alpha}[1 - i\beta\,\mathrm{sgn}(k)\Phi] + ik\nu\}. \tag{7}$$

The Lévy index $\alpha \in (0,2]$ is the most important parameter, since it drives the magnitude of the intensity fluctuations. Indeed, whereas strong fluctuations with relevant deviations from the Gaussian behavior are associated with values in the range $0 < \alpha < 2$, the Gaussian statistics with relatively weak fluctuations and the result of the CLT are recovered for the boundary value $\alpha = 2$. Therefore, Equation (7) can suitably describe both Gaussian and Lévy statistical regimes, depending only on the value of the single parameter α. In other words, the parameter α, which can be experimentally determined from the direct analysis of the PDF $P(I)$, effectively works as an indicator of the statistical regime (Gaussian or Lévy) of intensity fluctuations. The other independent parameters describe the asymmetry or skewness of the distribution ($\beta \in [-1, 1]$), location ($\nu \in (-\infty, \infty)$), and scale ($c \in (0, \infty)$), along with the function $\Phi(k) = -(2/k)\ln|k|$ if $\alpha = 1$, whereas $\Phi = \tan(\pi\alpha/2)$ if $\alpha \neq 1$.

Though the Lévy PDF, given by the Fourier transform of Equation (7), displays closed analytical form only for a few values of α (e.g., the Cauchy distribution arises for $\alpha = 1$ and $\beta = 0$), its large-I asymptotic behavior is power-law tailed, $P(I) \sim I^{-\mu}$, with exponent $\mu = 1 + \alpha$. Conversely, it is also true that random variables with power-law distribution are governed by the Lévy PDF with $\alpha = \mu - 1$ if $1 < \mu < 3$ (diverging second moment), and by the $\alpha = 2$ Gaussian statistics if $\mu \geq 3$ (finite second moment) [88]. Therefore, if the PDF of intensities presents asymptotic power-law behavior, then the power-law exponent μ also indicates the type of statistical regime (Gaussian or Lévy) of the output intensity values.

At this point, some words of caution are necessary in order to properly interpret the actual experimental data of the Er-RFL system under the statistical framework of the CLT and generalized CLT.

We initially remark that in the present case of intensity fluctuations, as well as in any case of realistic stochastic phenomena, a PDF with a diverging second moment actually represents an unphysical possibility. Nevertheless, it has been demonstrated [90] that a *truncated* power-law PDF, with a large but finite second moment, behaves rather similarly to the Lévy PDF to a considerable extent, defining the so-called Lévy-type (or Lévy-like) statistical behavior. In this case, the crossover to the Gaussian dynamics, predicted by the CLT, is only attained in a very long term [90,91]. In this sense, theoretically justified truncation schemes have been suitably implemented, for example, by restricting the values of the random variable to a finite range [90,91], with $P(I) = 0$ for $I > I_{cutoff}$, or by tempering the power law with an exponential attenuation [81,82,89], $P(I) \sim \exp(-\eta I)/I^{\mu}$, in a form similar to Equation (6). Therefore, the experimental reports of Lévy PDFs of intensities with index $0 < \alpha < 2$ should be properly interpreted as representative of this extensive Lévy-like statistical regime of intensity measurements.

Lastly, in addition to the description above of the truncated power-law with exponential attenuation, which was based on the Langevin dynamics of the amplitudes of the normal modes, we also comment that a PDF of output intensities emitted by RL systems with power-law form, $P(I) \sim I^{-\mu}$, has been derived in [92]. In contrast with the above developments, in this case, the theoretical approach took into account the statistics of the photon trajectories subjected to multiple scatterings within the sample [92]. The power-law exponent was found to be $\mu = 1 + \ell_g/\langle l\rangle$, where ℓ_g and $\langle l\rangle$ denote, respectively, the gain length of the active medium and the average length of the photon paths.

3.3. Lévy Statistics and Glassy Behavior in Er-RFL

In this subsection, we focus on the statistical analysis of the experimental data of the Er-RFL system.

We start by analyzing the intensity spectra. Figure 5a–c display 5000 spectra for each input power, from which the intensity fluctuations can be appreciated in the regimes below (Figure 5a), around (Figure 5b) and above (Figure 5c) the RL threshold [68]. According to the discussion in Section 3.2, the strong intensity fluctuations observed near the threshold suggest that the PDF of the output intensities can be described by the family of Lévy α-stable distributions, including the Lévy statistical regime if $0 < \alpha < 2$ and the Gaussian limit if $\alpha = 2$.

Figure 5d–f portrays the distributions $P(I)$ obtained from the data of Figure 5a–c, as well as the respective best fits to Equation (7) by applying the quantile-based method [93,94]. Best-fit values of the parameters are summarized in Table 1. The values of α are consistent with the Gaussian profiles ($\alpha = 2.0$) shown in Figure 5d,f, respectively, below and above the threshold, and also with the Lévy-like PDF ($\alpha = 1.3$) around the threshold, observed in Figure 5e. The unit value of β, indicating the maximum skewness of the distribution, in all cases reflects the asymmetry related to the positiveness of the intensity. Furthermore, the values of the location parameter v in the Gaussian regimes, $v = 0.858$ for $P/P_{th} = 0.6$ and $v = 0.682$ for $P/P_{th} = 1.8$, agree with the mean values, respectively, observed in Figure 5d,f. Indeed, the actual Gaussian distributions, which are equivalent to the $\alpha = 2$ Lévy PDFs in Figure 5d,f, present the mean v and standard deviation $\sqrt{2}c$, as theoretically predicted [88]. Interestingly, when comparing the Gaussian regimes below and above the threshold, we notice a considerable broadening of the PDF $P(I)$ at $P/P_{th} = 1.8$, leading to a wider spread of intensities (Figure 5f). This result is associated with the larger second moment of the distribution and more intense fluctuations observed above the threshold. Those fluctuations still remain, however, much weaker than the ones measured in the crossover region.

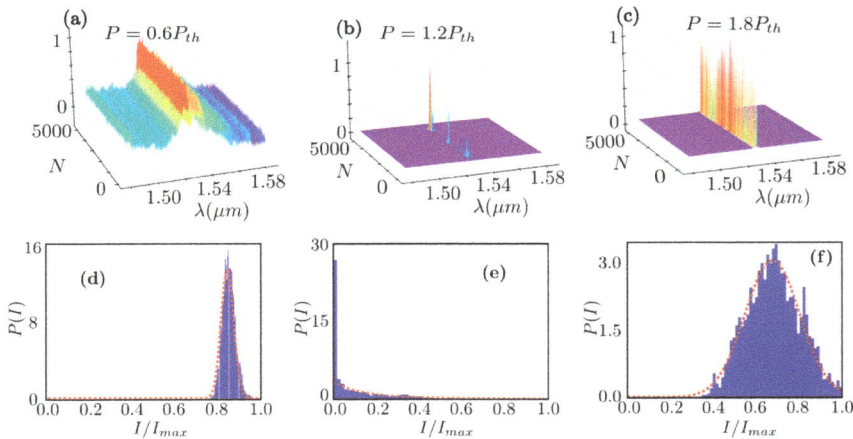

Figure 5. (**a**–**c**) 5000 intensity spectra of the Er-RFL system for each input power (**a**) below; (**b**) around; and (**c**) above the RL threshold; (**d**–**f**) Probability density functions (PDFs) $P(I)$ of maximum intensities obtained from the data shown in (**a**–**c**). The best fits using Equation (7) are depicted in dotted lines and portrait Gaussian profiles (**d**) below and (**f**) above the threshold ($\alpha = 2.0$); and (**e**) a Lévy distribution ($\alpha = 1.3$) around the threshold. (Reproduced with permission from ref. [68]).

Table 1. Summary of best fit parameters to Equation (7) for the intensity distributions of Figure 5d–f [68].

Input Excitation Power	α	β	c	v
$P/P_{th} = 0.6$	1.9	1.0	0.021	0.858
$P/P_{th} = 1.2$	1.3	1.0	0.061	−0.193
$P/P_{th} = 1.8$	2.0	1.0	0.091	0.682

The main result confirming the Lévy behavior of the output intensities of the Er-RFL system is shown in Figure 6. With a basis on ref. [68], we first notice that the variation of the Lévy index α as a function of the normalized input power clearly points to the presence of three distinct statistical regimes. Indeed, as P/P_{th} increases, the statistics of the output intensities progressively shift from the prelasing Gaussian ($\alpha = 2$) to the Lévy ($0 < \alpha < 2$) behavior around the threshold, and to the subsequent Gaussian ($\alpha = 2$) regime deep in the RL phase. Noticeably, this sequence also resembles the statistical behavior of the output intensity of 3D bulk RLs [82,89,93,95,96]. We also notice that, as pointed in ref. [93], the second Gaussian regime is rather distinct from the first one. Indeed, in the second Gaussian phase above the RL threshold, the system is in the regime with self-averaging of the gain [93]. In contrast, in the first Gaussian phase, it is still in the prelasing regime. Moreover, we also comment that the observed independence on the spatial dimensionality of the Lévy character of intensity fluctuations finds support in the theoretical analysis based on Langevin equations, previously discussed in Section 3.2, which is considered to hold, irrespective of the spatial dimension [69,82].

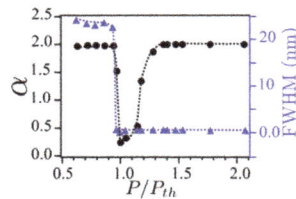

Figure 6. Lévy index α (circles) and FWHM (triangles) as a function of the normalized input pump power P/P_{th}. Three statistical regimes of intensity fluctuations are observed: prelasing Gaussian with amplified spontaneous emission ($\alpha = 2$), Lévy-type RL ($0 < \alpha < 2$) around the threshold, and Gaussian RL ($\alpha = 2$) well above the threshold. The sharp decrease in α at the first Gaussian-to-Lévy transition nicely coincides with the abrupt change in FWHM at the RL threshold. (Reproduced with permission from ref. [68]).

We further notice in Figure 6 that the abrupt decrease in the Lévy index α at the onset of RL behavior is closely related to the sharp linewidth reduction observed in Figure 2. Therefore, our results for the Er-RFL system corroborate the suggestion of [93], that the transition from the Gaussian to the Lévy regime in RLs could be used as a universal identifier of the RL threshold (however, see also the comment below on the results of ref. [97]). In fact, we also included in Figure 6 the FWHM measurement, whose drastic change at the RL threshold nicely coincides with the first Gaussian-to-Lévy statistical transition. For higher values of the input power, after reaching a minimum around the threshold, the index α smoothly rises back to the Gaussian value $\alpha = 2$ achieved above the threshold.

We finally discuss the connection of the above findings with the photonic spin-glass behavior recently reported in the Er-RFL system [68,69].

As mentioned in Section 3.2, the characterization of the photonic RSB glassy phase in the RL regime requires the calculation of a specific two-point correlation function, which in the present context, can be defined either among the mode amplitudes, mode phases, or intensity fluctuations. In the latter case, that can be accessed experimentally, where the replica overlap parameter is defined as [42,79]:

$$q_{\gamma\beta} = \frac{\sum_k \Delta_\gamma(k)\Delta_\beta(k)}{\sqrt{\left[\sum_k \Delta_\gamma^2(k)\right]\left[\sum_k \Delta_\beta^2(k)\right]}},\tag{8}$$

where $\gamma, \beta = 1, 2, \ldots, N_s$ denote the replica labels ($N_s = 5000$ in ref. [68]), the average intensity at the wavelength indexed by k reads $<I>(k) = \sum_{\gamma=1}^{N_s} I_\gamma(k)/N_s$, and the intensity fluctuation is

$\Delta_\gamma(k) = I_\gamma(k) - <I>(k)$. In the photonic context, each output spectrum is considered a replica, i.e., a copy of the Er-RFL system under fairly identical experimental conditions. The PDF P(q) represents the distribution of values $q = q_{\gamma\beta}$ of the mode-mode correlations between intensity fluctuations, Equation (8). In fact, as discussed in the previous subsection, P(q) is analogous to the Parisi order parameter in the RSB spin-glass theory of disordered magnetic systems [84]. In the present context, the distribution P(q) signalizes a photonic replica-symmetric paramagnetic prelasing regime if its maximum occurs exclusively at $q_{max} = 0$ (no RSB), or a RSB spin-glass RL phase if the maximum also assumes values $|q_{max}| \neq 0$ (RSB) (see Section 3.2).

Figure 7 shows a remarkable agreement between the onset of the Lévy statistical regime of intensity fluctuations and the emergence of the RSB glassy RL phase in Er-RFL. Indeed, we observe that both the Lévy and spin-glass behaviors are simultaneously present around the RL threshold. Therefore, besides signaling the Gaussian-to-Lévy shift in the statistical characteristics of intensity fluctuations, which is illustrated by Figure 6, the RL threshold also marks the sharp phase transition from the $q_{max} \cong 0$ replica-symmetric paramagnetic prelasing regime with amplified spontaneous emission to the $q_{max} \cong 1$ spin-glass RL phase with RSB properties. This coinciding behavior, firstly demonstrated in a 3D RL [82], is thus also shared by the Er-RFL system, although a recent report [97] has pointed out that this might not be a universal property of RL systems. Indeed, in ref. [97] it was demonstrated that a rigorous connection between the photonic phases and the statistics of intensity fluctuations is not mandatory, so that there can be circumstances in which, for example, a glassy phase emerges along with a Gaussian statistical regime of fluctuations. Therefore, though a complete theoretical understanding of such a finding is still lacking, it is possible in some instances to trace back the common physical origin of the Lévy and glassy behaviors to the Langevin equations for the amplitudes of the normal modes, which, as discussed, are the basis on which to explain both the statistical regimes of intensity fluctuations and the photonic RSB spin-glass behavior of RL systems.

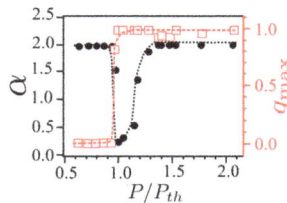

Figure 7. Lévy index α (circles) and overlap parameter $|q| = q_{max}$ at which the PDF P(q) of the photonic replica overlaps with output intensity fluctuations at a maximum (squares) as a function of the normalized input power P/P_{th}. The abrupt decrease in the parameter α at the first Gaussian-to-Lévy transition coincides nicely with the photonic phase transition observed from the $q_{max} \cong 0$ replica-symmetric paramagnetic prelasing behavior, with amplified spontaneous emission, to the $q_{max} \cong 1$ replica-symmetry-breaking (RSB) spin-glass phase of the Er-RFL system. Dotted lines are a guide to the eyes. (Reproduced with permission from ref. [68].)

4. Summary and Discussion

In this work, we described the operation of an RFL based on an erbium-doped fiber imprinted with randomly-spaced Bragg gratings. By exciting the fiber with a CW diode laser, we investigated the statistical fluctuations of the output intensity emitted by the erbium ions in the near infrared. The results allowed us to identify different statistical regimes for excitation powers P below, around, and above the laser threshold, P_{th}. In particular, the Lévy statistics were clearly identified for $P \approx P_{th}$, while the Gaussian statistics were observed for $P < P_{th}$ and $P > P_{th}$. Moreover, we also found that the probability distribution for the emitted intensity around the laser threshold reveals a glassy phase of light that is compatible with an RSB analogue of the spin-glass phase transition.

The studies reported here led to a deeper understanding of the physical processes underlying the RFL operation, and the results were also consistent with recent findings in 2D and 3D RLs.

As illustrated by the references cited in this work, the research on RFLs is still a hot subject after ten years since the first demonstration. The low fabrication cost, small fiber length, and simple operation scheme enable various potential applications of RFLs, for example, in imaging, sensing, and optofluidics. Further research with a basis on the investigation of the intensity fluctuations in RFL systems may include the study of their temporal dynamics, as well as the emergency of extreme events, analogues of rogue waves, and photonic turbulent transitions around the excitation threshold.

Acknowledgments: We thank S.J.M. Carreño, S.I. Fewo, M. Gagné, V. Jerez, R. Kashyap, B.C. Lima, L.J.Q. Maia, A.L. Moura, P.I.R. Pincheira and A.F. Silva for the fruitful collaboration. We also acknowledge the financial support from the Brazilian Agencies: Conselho Nacional de Desenvolvimento Científico e Tecnológico (CNPq) and Fundação de Amparo à Ciência e Tecnologia do Estado de Pernambuco (FACEPE). The work was performed in the framework of the National Institute of Photonics (INCT de Fotônica) and PRONEX-CNPq/FACEPE projects.

Author Contributions: C.B.d.A., A.S.L.G. and E.P.R. contributed equally to this review.

Conflicts of Interest: The authors declare no conflict of interest.

References

1. Ambartsumyan, R.V.; Basov, N.G.; Kryukov, P.G.; Letokhov, V.S. Laser with nonresonant feedback. *JETP Lett.* **1966**, *3*, 167–169.
2. Letokhov, V.S. Generation of light by a scattering medium with negative resonance absorption. *Sov. Phys. JETP* **1968**, *26*, 835–840.
3. Ambartsumyan, R.V.; Kryukov, P.G.; Letokhov, V.S. Dynamics of emission line narrowing for a laser with nonresonant feedback. *Sov. Phys. JETP* **1967**, *24*, 1129–1134.
4. Ambartsumyan, R.V.; Basov, N.G.; Letokhov, V.S. Frequency stability of a HeNe laser with nonresonant feedback. *IEEE Trans. Instrum. Meas.* **1968**, *17*, 338–343.
5. Ambartsumyan, R.V.; Kryukov, P.G.; Letokhov, V.S.; Matveets, Y.A. Emission statistics of a laser with nonresonant feedback. *JETP Lett.* **1967**, *5*, 312–314.
6. Letokhov, V.S. Stimulated radio emission of the interstellar medium. *JETP Lett.* **1966**, *4*, 321–323.
7. Lavrinovich, N.N.; Letokhov, V.S. The possibility of laser effect in stellar atmospheres. *Sov. Phys. JETP* **1975**, *40*, 800–805.
8. Letokhov, V.S. Astrophysical lasers. *Quantum Electron.* **2002**, *32*, 1065–1079. [CrossRef]
9. Johansson, S.; Letokhov, V.S. Astrophysical lasers operating in optical Fe II lines in stellar ejecta of Eta Carinae. *Astron. Astrophys.* **2004**, *428*, 427–509. [CrossRef]
10. Noginov, M.A. *Solid State Random Lasers*; Springer Series in Optical Sciences; Springer: New York, NY, USA, 2005.
11. Lawandy, N.M.; Balachandran, R.M.; Gomes, A.S.L.; Sauvian, E. Laser action in strongly scattering medium. *Nature* **1994**, *368*, 436–438. [CrossRef]
12. Ye, L.; Zhao, C.; Feng, Y.; Gu, B.; Cui, Y.; Lu, Y. Study on the polarization of random lasers from dye-doped nematic liquid crystals. *Nanoscale Res. Lett.* **2017**, *12*, 27. [CrossRef] [PubMed]
13. Abegão, L.M.G.; Manoel, D.S.; Otuka, A.J.G.; Ferreira, P.H.D.; Vollet, D.R.; Donatti, D.A.; De Boni, L.; Mendonça, C.R.; de Vicente, F.S.; Rodrigues, J.J., Jr.; et al. Random laser emission from a rhodamine B-doped GPTS/TEOS-derived organic/sílica monolithic xerogel. *Laser Phys. Lett.* **2017**, *14*, 065801. [CrossRef]
14. Shasti, M.; Coutino, P.; Mukherjee, S.; Varanytsia, A.; Smith, T.; Luchette, A.P.; Sukhomlinova, L.; Kosa, T.; Munoz, A.; Taheri, B. Reverse mode switching of the random laser emission in dye doped liquid crystals under homogeneous and inhomogeneous electric fields. *Photonics Res.* **2016**, *4*, 7–12. [CrossRef]
15. Anderson, B.R.; Gunawidjaja, R.; Eilers, H. Self-healing organic-dye-based random lasers. *Opt. Lett.* **2015**, *40*, 577–580. [CrossRef] [PubMed]
16. Gomes, A.S.L.; Carvalho, M.T.; Dominguez, C.T.; de Araújo, C.B.; Prasad, P.N. Direct three-photon excitation of upconversion random laser emission in a weakly scattering organic colloidal system. *Opt. Express* **2014**, *22*, 14305–14310. [CrossRef] [PubMed]

17. Knitter, S.; Kues, M.; Fallnich, C. Emission polarization of random lasers in organic dye solutions. *Opt. Lett.* **2012**, *37*, 3621–3623. [CrossRef] [PubMed]

18. Chen, Y.; Herrnsdorf, J.; Guilhabert, B.; Zhang, Y.; Watson, I.M.; Gu, E.; Laurand, N.; Dawson, M.D. Colloidal quantum dot random laser. *Opt. Express* **2011**, *19*, 2996–3003. [CrossRef] [PubMed]

19. Dominguez, C.T.; Gomes, M.A.; Macedo, Z.S.; de Araújo, C.B.; Gomes, A.S.L. Multi-photon excited coherent random laser emission in ZnO powders. *Nanoscale* **2015**, *7*, 317–323. [CrossRef] [PubMed]

20. Moura, A.L.; Jerez, V.; Maia, L.J.Q.; Gomes, A.S.L.; de Araújo, C.B. Multi-wavelength emission through self-induced second-order wave-mixing processes from a Nd^{3+} doped crystalline powder random laser. *Sci. Rep.* **2015**, *5*, 13816. [CrossRef] [PubMed]

21. Moura, A.L.; Carreño, S.J.M.; Pincheira, P.I.R.; Fabris, Z.V.; Maia, L.J.Q.; Gomes, A.S.L.; de Araújo, C.B. Tunable ultraviolet and blue light generation from Nd:YAB random laser bolstered by second-order nonlinear processes. *Sci. Rep.* **2016**, *6*, 27107. [CrossRef] [PubMed]

22. Polson, R.C.; Chipouline, A.; Vardeny, Z.V. Random lasing in π-conjugated films and infiltrated opals. *Adv. Mater.* **2001**, *13*, 760–764. [CrossRef]

23. Meng, X.; Fujita, K.; Murai, S.; Tanaka, K. Coherent random lasers in weakly scattering polymer films containing silver nanoparticles. *Phys. Rev. A* **2009**, *79*, 053817. [CrossRef]

24. Costela, A.; Garcia-Moreno, I.; Cerdan, L.; Martin, V.; Garcia, O.; Sastre, R. Dye-doped POSS solutions: Random nanomaterials for laser emission. *Adv. Mater.* **2009**, *21*, 4163–4166. [CrossRef]

25. Tulek, A.; Polson, R.C.; Vardeny, Z.V. Naturally occurring resonators in random lasing of π-conjugated polymer films. *Nat. Phys.* **2010**, *6*, 303–310. [CrossRef]

26. Polson, R.C.; Raikh, M.E.; Vardeny, Z.V. Random lasing from weakly scattering media; Spectrum universality in DOO–PPV polymer films. *Phys. E* **2002**, *13*, 1240–1242. [CrossRef]

27. Dos Santos, M.V.; Dominguez, C.T.; Schiavon, J.V.; Barud, H.S.; de Melo, L.S.A.; Ribeiro, S.J.L.; Gomes, A.S.L.; de Araújo, C.B. Random laser action from flexible biocellulose-based device. *J. Appl. Phys.* **2014**, *115*, 083108. [CrossRef]

28. Yu, S.F.; Yuen, C.; Lau, S.P.; Park, W.I.; Yi, G.-C. Random laser action in ZnO nanorod arrays embedded in ZnO epilayers. *Appl. Phys. Lett.* **2004**, *84*, 3241–3243. [CrossRef]

29. Wang, Z.; Shi, X.; Wei, S.; Sun, Y.; Wang, Y.; Zhou, J.; Shi, J.; Liu, D. Two-threshold silver nanowire-based random laser with different dye concentrations. *Laser Phys. Lett.* **2014**, *11*, 095002. [CrossRef]

30. Gao, F.; Morshed, M.M.; Bashar, S.B.; Zheng, Y.; Shi, Y.; Liu, J. Electrically pumped random lasing based on an Au–ZnO nanowire Schottky junction. *Nanoscale* **2015**, *7*, 9505–9509. [CrossRef] [PubMed]

31. Bashar, S.B.; Suja, M.; Morshed, M.; Gao, F.; Liu, J. An Sb-doped p-type ZnO nanowire based random laser diode. *Nanotechnology* **2016**, *27*, 065204. [CrossRef] [PubMed]

32. Baudouin, Q.; Mercadier, N.; Guarrera, V.; Guerin, W.; Kaiser, R. A cold-atom random laser. *Nat. Phys.* **2013**, *9*, 357–360. [CrossRef]

33. Van der Molen, K.L.; Mosk, A.P.; Lagendijk, A.D. Intrinsic intensity fluctuations in random lasers. *Phys. Rev. A* **2006**, *74*, 053808. [CrossRef]

34. Van der Molen, K.L.; Tjerkstra, R.W.; Mosk, A.P.; Lagendijk, A.D. Spatial extent of random laser modes. *Phys. Rev. Lett.* **2007**, *98*, 143901. [CrossRef] [PubMed]

35. Wiersma, D.S. The physics and applications of random lasers. *Nat. Phys.* **2008**, *4*, 359–367. [CrossRef]

36. Leonetti, M.; Conti, C.; Lopez, C. The mode-locking transition of random lasers. *Nat. Photonics* **2011**, *5*, 615–617. [CrossRef]

37. Ignesti, E.; Tommasi, F.; Fini, L.; Martelli, F.; Azzali, N.; Cavalieri, S. A new class of optical sensors: A random laser based device. *Sci. Rep.* **2016**, *6*, 35225. [CrossRef] [PubMed]

38. Bhaktha, B.N.S.; Bachelard, N.; Noblin, X.; Sebbah, P. Optofluidic random laser. *Appl. Phys. Lett.* **2012**, *101*, 151101. [CrossRef]

39. Bachelard, N.; Gigan, S.; Noblin, X.; Sebbah, P. Adaptive pumping for spectral control of random lasers. *Nat. Phys.* **2014**, *10*, 426–431. [CrossRef]

40. Redding, B.; Choma, M.A.; Cao, H. Speckle-free laser imaging using random laser illumination. *Nat. Photonics* **2012**, *6*, 355–359. [CrossRef] [PubMed]

41. Luan, F.; Gu, B.B.; Gomes, A.S.L.; Yong, K.T.; Wen, S.C.; Prasad, P.N. Lasing in nanocomposite random media. *Nano Today* **2015**, *10*, 168–192. [CrossRef]

42. Ghofraniha, N.; Viola, I.; Di Maria, F.; Barbarella, G.; Gigli, G.; Leuzzi, L.; Conti, C. Experimental evidence of replica symmetry breaking in random lasers. *Nat. Commun.* **2015**, *6*, 6058. [CrossRef] [PubMed]

43. De Mattos, C.J.S.; Menezes, L.S.; Brito-Silva, A.M.; Gámez, M.A.M.; Gomes, A.S.L.; de Araújo, C.B. Random fiber laser. *Phys. Rev. Lett.* **2007**, *99*, 153903. [CrossRef] [PubMed]

44. Lizárraga, N.; Puente, N.P.; Chaikina, E.I.; Leskova, T.A.; Méndez, E.R. Single-mode Er-doped fiber random laser with distributed Bragg grating feedback. *Opt. Express* **2009**, *17*, 395–404. [CrossRef] [PubMed]

45. Gagné, M.; Kashyap, R. Demonstration of a 3 mW threshold Er-doped random fiber laser based on a unique fiber Bragg grating. *Opt. Express* **2009**, *17*, 19067–19074. [CrossRef] [PubMed]

46. Turitsyn, S.K.; Babin, S.A.; El-Taher, A.E.; Harper, P.; Churkin, D.V.; Kablukov, S.I.; Ania-Castanon, J.D.; Karalekas, V.; Podivilov, E.V. Random distributed feedback fibre laser. *Nat. Photonics* **2010**, *4*, 231–235. [CrossRef]

47. Turitsyn, S.K.; Babin, S.A.; Churkin, D.V.; Vatnik, I.D.; Nikulin, M.; Podivilov, E.V. Random distributed feedback fibre lasers. *Phys. Rep.* **2014**, *542*, 133–193. [CrossRef]

48. Churkin, D.V.; Sugavanam, S.; Vatnik, I.D.; Wang, Z.; Podivilov, E.V.; Babin, S.A.; Rao, Y.; Turitsyn, S.K. Recent advances in fundamentals and applications of random fiber lasers. *Adv. Opt. Photonics* **2015**, *7*, 516–569. [CrossRef]

49. Wang, Z.; Wu, H.; Fan, M.; Zhang, L.; Rao, Y.; Zhang, W.; Jia, X. High power random fiber laser with short cavity length: Theoretical and experimental investigations. *IEEE J. Sel. Top. Quantum Electron.* **2015**, *21*, 0900506.

50. Li, S.W.; Ma, R.; Rao, Y.J.; Zhu, Y.Y.; Wang, Z.N.; Jia, X.H.; Li, J. Random distributed feedback fiber laser based on combination of Er-doped fiber and single-mode fiber. *IEEE J. Sel. Top. Quantum Electron.* **2015**, *21*, 0900406.

51. Wu, H.; Wang, Z.; Fan, M.; Zhang, L.; Zhang, W.; Ra, Y. Role of the mirror's reflectivity in forward-pumped random fiber laser. *Opt. Express* **2015**, *23*, 1421–1427. [CrossRef] [PubMed]

52. Zhang, H.; Zhou, P.; Wang, X.; Du, X.; Xiao, H.; Xu, X. Hundred-watt-level high power random distributed feedback Raman fiber laser at 1150 nm and its application in mid-infrared laser generation. *Opt. Express* **2015**, *23*, 17138–17144. [CrossRef] [PubMed]

53. Zhang, W.L.; Ma, R.; Tang, C.H.; Rao, Y.J.; Zeng, X.P.; Yang, Z.J.; Wang, Z.N.; Gong, Y.; Wang, Y.S. All optical mode controllable Er-doped random fiber laser with distributed Bragg gratings. *Opt. Lett.* **2015**, *40*, 3181–3184. [CrossRef] [PubMed]

54. Wang, L.; Dong, X.; Shum, P.P.; Liu, X.; Su, H. Random laser with multiphase-shifted Bragg grating in Er/Yb-codoped fiber. *J. Lightwave Technol.* **2015**, *33*, 95–99. [CrossRef]

55. Du, X.; Zhang, H.; Wang, X.; Zhou, P.; Liu, Z. Investigation on random distributed feedback Raman fiber laser with linear polarized output. *Photonics Res.* **2015**, *3*, 28–31. [CrossRef]

56. Tang, Y.; Xu, J. A random Q-switched fiber laser. *Sci. Rep.* **2015**, *5*, 9338. [CrossRef] [PubMed]

57. Yao, B.C.; Rao, Y.J.; Wang, Z.N.; Wu, Y.; Zhou, J.H.; Wu, H.; Fan, M.Q.; Cao, X.L.; Zhang, W.L.; Chen, Y.F.; et al. Graphene based widely-tunable and singly-polarized pulse generation with random fiber lasers. *Sci. Rep.* **2015**, *5*, 18526. [CrossRef] [PubMed]

58. Zhang, L.; Jiang, H.; Yang, X.; Pan, W.; Feng, Y. Ultra-wide wavelength tuning of a cascaded Raman random fiber laser. *Opt. Lett.* **2016**, *41*, 215–218. [CrossRef] [PubMed]

59. Du, X.; Zhang, H.; Wang, X.; Zhou, P.; Liu, Z. Short cavity-length random fiber laser with record power and ultrahigh efficiency. *Opt. Lett.* **2016**, *41*, 571–574. [CrossRef] [PubMed]

60. Dontsova, E.I.; Kablukov, S.I.; Vatnik, I.D.; Babin, S.A. Frequency doubling of Raman fiber lasers with random distributed feedback. *Opt. Lett.* **2016**, *41*, 1439–1442. [CrossRef] [PubMed]

61. Hu, Z.; Liang, Y.; Qian, X.; Gao, P.; Xie, K.; Jiang, H. Polarized random laser emission from an oriented disorder polymer optical fiber. *Opt. Lett.* **2016**, *41*, 2584–2587. [CrossRef] [PubMed]

62. Ardakani, A.G.; Rafieipour, P. Investigation of one-dimensional Raman random lasers based on the finite-difference-time-domain method: Presence of mode competition and higher-order Stokes and anti-Stokes modes. *Phys. Rev. A* **2016**, *93*, 023833. [CrossRef]

63. Zhang, W.L.; Song, Y.B.; Zeng, X.P.; Ma, R.; Yang, Z.J.; Rao, Y.J. Temperature-controlled mode selection of Er-doped random fiber laser with disordered Bragg gratings. *Photonics Res.* **2016**, *4*, 102–105. [CrossRef]

64. Wu, H.; Wang, Z.; Rao, Y. Tailoring the properties of cw random fiber lasers. In Proceedings of the Advanced Photonics Congress 2016 (IPR, NOMA, Sensors, Networks, SPPCom, SOF), Vancouver, NA, Canada, 18–20 July 2016.

65. Babin, S.A.; Zlobina, E.A.; Kablukov, S.I.; Podivilov, E.V. High-order random Raman lasing in a PM fiber with ultimate e efficiency and narrow bandwidth. *Sci. Rep.* **2016**, *6*, 22625. [CrossRef] [PubMed]

66. Zhang, W.L.; Zheng, M.Y.; Ma, R.; Gong, C.Y.; Yang, Z.J.; Peng, G.D.; Rao, Y.J. Fiber-type random laser based on a cylindrical waveguide with a disordered cladding layer. *Sci. Rep.* **2016**, *6*, 26473. [CrossRef] [PubMed]

67. González, I.R.R.; Lima, B.C.; Pincheira, P.I.R.; Brum, A.A.; Macêdo, A.M.S.; Vasconcelos, G.L.; Menezes, L.S.; Raposo, E.P.; Gomes, A.S.L.; Kashyap, R. Turbulence hierarchy in a random fibre laser. *Nat. Commun.* **2017**, *8*, 15731. [CrossRef] [PubMed]

68. Lima, B.C.; Gomes, A.S.L.; Pincheira, P.I.R.; Moura, A.L.; Gagné, M.; Raposo, E.P.; de Araújo, C.B.; Kashyap, R. Observation of Lévy statistics in one-dimensional erbium-based random fiber laser. *J. Opt. Soc. Am. B* **2017**, *34*, 293–299. [CrossRef]

69. Gomes, A.S.L.; Lima, B.C.; Pincheira, P.I.R.; Moura, A.L.; Gagné, M.; Raposo, E.P.; de Araújo, C.B.; Kashyap, R. Glassy behavior in a one-dimensional continuous-wave erbium-doped random fiber laser. *Phys. Rev. A* **2016**, *94*, 011801. [CrossRef]

70. Hokr, B.H.; Cerjan, A.; Thompson, J.V.; Yuan, L.; Liew, S.F.; Bixler, J.N.; Noojin, G.D.; Thomas, R.J.; Cao, H.; Stone, A.D.; et al. Evidence of Anderson localization effects in random Raman lasing. *SPIE Proc.* **2016**, *9731*, 973110.

71. Goodman, J.W. *Speckle Phenomena in Optics: Theory and Applications*; Roberts & Company: Englewood, CO, USA, 2007.

72. Zhu, G.; Gu, L.; Noginov, M.A. Experimental study of instability in a random laser with immobile scatterers. *Phys. Rev. A* **2012**, *85*, 043801. [CrossRef]

73. Angelani, L.; Conti, C.; Ruocco, G.; Zamponi, F. Glassy behavior of light in random lasers. *Phys. Rev. B* **2006**, *74*, 104207. [CrossRef]

74. Angelani, L.; Conti, C.; Ruocco, G.; Zamponi, F. Glassy behavior of light. *Phys. Rev. Lett.* **2006**, *96*, 065702. [CrossRef] [PubMed]

75. Leuzzi, L.; Conti, C.; Folli, V.; Angelani, L.; Ruocco, G. Phase diagram and complexity of mode-locked lasers: From order to disorder. *Phys. Rev. Lett.* **2009**, *102*, 083901. [CrossRef] [PubMed]

76. Conti, C.; Leuzzi, L. Complexity of waves in nonlinear disordered media. *Phys. Rev. B* **2011**, *83*, 134204. [CrossRef]

77. Antenucci, F.; Conti, C.; Crisanti, A.; Leuzzi, L. General phase diagram of multimodal ordered and disordered lasers in closed and open cavities. *Phys. Rev. Lett.* **2015**, *114*, 043901. [CrossRef] [PubMed]

78. Antenucci, F.; Crisanti, A.; Leuzzi, L. Complex spherical 2 + 4 spin glass: A model for nonlinear optics in random media. *Phys. Rev. A* **2015**, *91*, 053816. [CrossRef]

79. Antenucci, F.; Crisanti, A.; Leuzzi, L. The glassy random laser: Replica symmetry breaking in the intensity fluctuations of emission spectra. *Sci. Rep.* **2015**, *5*, 16792. [CrossRef] [PubMed]

80. Antenucci, F.; Crisanti, A.; Ibáñez-Berganza, M.; Marruzzo, A.; Leuzzi, L. Statistical mechanics models for multimode lasers and random lasers. *Philos. Mag.* **2016**, *96*, 704–731. [CrossRef]

81. Raposo, E.P.; Gomes, A.S.L. Analytical solution for the Lévy-like steady-state distribution of intensities in random lasers. *Phys. Rev. A* **2015**, *91*, 043827. [CrossRef]

82. Gomes, A.S.L.; Raposo, E.P.; Moura, A.L.; Fewo, S.I.; Pincheira, P.I.R.; Jerez, V.; Maia, L.J.Q.; de Araújo, C.B. Observation of Lévy distribution and replica symmetry breaking in random lasers from a single set of measurements. *Sci. Rep.* **2016**, *6*, 27987. [CrossRef] [PubMed]

83. Crisanti, A.; Sommers, H.-J. The spherical p-spin interaction spin glass model: The statics. *Z. Phys. B* **1992**, *87*, 341–354. [CrossRef]

84. Mézard, M.; Parisi, G.; Virasoro, M.A. *Spin Glass Theory and Beyond*; World Scientific: Singapore, 1987.

85. Pincheira, P.I.R.; Silva, A.F.; Carreño, S.J.M.; Moura, A.L.; Fewo, S.I.; Raposo, E.P.; Gomes, A.S.L.; de Araújo, C.B. Observation of photonic to paramagnetic spin-glass transition in specially-designed TiO_2 particles-based dye-colloidal random laser. *Opt. Lett.* **2016**, *41*, 3459–3462. [CrossRef] [PubMed]

86. O'Bryan, I.C.L.; Sargent, I.M. Theory of multimode laser operation. *Phys. Rev. A* **1973**, *8*, 3071–3092. [CrossRef]

87. Schenzle, A.; Brand, H. Multiplicative stochastic processes in statistical physics. *Phys. Rev. A* **1979**, *20*, 1628–1647. [CrossRef]

88. Samorodnitsky, G.; Taqqu, M.S. *Stable Non-Gaussian Random Processes*; Chapman and Hall: London, UK, 1994.

89. Uppu, R.; Mujumdar, S. Exponentially tempered Lévy sums in random lasers. *Phys. Rev. Lett.* **2015**, *114*, 183903. [CrossRef]

90. Mantegna, R.N.; Stanley, H.E. Stochastic process with ultraslow convergence to a Gaussian: The truncated Lévy flight. *Phys. Rev. Lett.* **1994**, *73*, 2946–2949. [CrossRef] [PubMed]

91. Bartumeus, F.; Raposo, E.P.; Viswanathan, G.M.; da Luz, M.G.E. Stochastic optimal foraging: Tuning intensive and extensive dynamics in random searches. *PLoS ONE* **2014**, *9*, e106373. [CrossRef] [PubMed]

92. Lepri, S.; Cavalieri, S.; Oppo, G.-L.; Wiersma, D.S. Statistical regimes of random laser fluctuations. *Phys. Rev. A* **2007**, *75*, 063820. [CrossRef]

93. Uppu, R.; Mujumdar, S. Lévy exponents as universal identifiers of threshold and criticality in random lasers. *Phys. Rev. A* **2014**, *90*, 025801. [CrossRef]

94. McCulloch, J.H. Simple consistent estimators of stable distribution parameters. *Commun. Stat. Simul.* **1986**, *15*, 1109–1136. [CrossRef]

95. Uppu, R.; Tiwari, A.K.; Mujumdar, S. Identification of statistical regimes and crossovers in coherent random laser emission. *Opt. Lett.* **2012**, *37*, 662–664. [CrossRef] [PubMed]

96. Uppu, R.; Mujumdar, S. Dependence of the Gaussian-Lévy transition on the disorder strength in random lasers. *Phys. Rev. A* **2013**, *87*, 013822. [CrossRef]

97. Tommasi, F.; Ignesti, E.; Lepri, S.; Cavalieri, S. Robustness of replica symmetry breaking phenomenology in random laser. *Sci. Rep.* **2016**, *6*, 37113. [CrossRef] [PubMed]

![applied sciences logo]
applied
sciences

MDPI

Article

Modulational Instability in Linearly Coupled Asymmetric Dual-Core Fibers

Arjunan Govindarajan [1,†], **Boris A. Malomed** [2,3,*], **Arumugam Mahalingam** [4] and **Ambikapathy Uthayakumar** [1,*]

[1] Department of Physics, Presidency College, Chennai 600 005, Tamilnadu, India; govind@cnld.bdu.ac.in
[2] Department of Physical Electronics, School of Electrical Engineering, Faculty of Engineering, Tel Aviv University, Tel Aviv 69978, Israel
[3] Laboratory of Nonlinear-Optical Informatics, ITMO University, St. Petersburg 197101, Russia
[4] Department of Physics, Anna University, Chennai 600 025, Tamilnadu, India; mahabs22@gmail.com
* Correspondence: malomed@post.tau.ac.il (B.A.M.); uthayk@yahoo.com (A.U.)
† Current address: Centre for Nonlinear Dynamics, School of Physics, Bharathidasan University, Tiruchirappalli 620 024, Tamilnadu, India.

Academic Editor: Christophe Finot
Received: 25 April 2017; Accepted: 13 June 2017; Published: 22 June 2017

Abstract: We investigate modulational instability (MI) in asymmetric dual-core nonlinear directional couplers incorporating the effects of the differences in effective mode areas and group velocity dispersions, as well as phase- and group-velocity mismatches. Using coupled-mode equations for this system, we identify MI conditions from the linearization with respect to small perturbations. First, we compare the MI spectra of the asymmetric system and its symmetric counterpart in the case of the anomalous group-velocity dispersion (GVD). In particular, it is demonstrated that the increase of the inter-core linear-coupling coefficient leads to a reduction of the MI gain spectrum in the asymmetric coupler. The analysis is extended for the asymmetric system in the normal-GVD regime, where the coupling induces and controls the MI, as well as for the system with opposite GVD signs in the two cores. Following the analytical consideration of the MI, numerical simulations are carried out to explore nonlinear development of the MI, revealing the generation of periodic chains of localized peaks with growing amplitudes, which may transform into arrays of solitons.

Keywords: modulational instability; asymmetric nonlinear fiber couplers; linear stability approach; coupled nonlinear Schrödinger equations

1. Introduction

The modulational instability (MI) is a ubiquitous phenomenon originating from the interplay of linear dispersion or diffraction and the nonlinear self-interaction of wave fields. This effect was first theoretically identified by Benjamin and Feir in 1967 for waves on deep water [1]; hence, MI is often called the Benjamin–Feir instability. Studies of the MI draw steadily growing interest in nonlinear optics [2–4], fluid dynamics [5,6], Bose–Einstein condensates [7–9], plasma physics [10,11] and other fields.

In its standard form, the MI applies to continuous waves (CWs) or quasi-CW states in media featuring cubic (Kerr) self-focusing nonlinearity and anomalous group-velocity dispersion (GVD), giving rise to the instability against infinitesimal perturbations in the form of amplitude and phase modulations, which eventually generates trains of soliton-like pulses [12]. MI can also be observed in the normal-GVD regime in systems incorporating additional ingredients, such as the cross-phase modulation interaction between two components [13], in the case of the co-propagation of optical fields and other effects, in particular the loss dispersion [14] or fourth-order GVD [15]. In all of these cases, destabilizing perturbation may originate from quantum noise or from an additional weak frequency-shifted wave [16]. Based on the nature of the underlying optical propagation, the MI is classified as the temporal (longitudinal) instability [17,18], if the CW is subject to the GVD in fibers, or spatial (transverse) instability [19], if the CW state experiences the action of diffraction in a planar waveguide. More general spatio-temporal MI occurs in bulk optical media when both the GVD and diffraction are essential [20].

The MI has found many important applications, including the creation of pulses with ultra-high repetition rates [21,22], the expansion of the bandwidth of Raman fiber amplifiers [23], the generation of optical supercontinuum [24] and all-optical switching [25]. In the context of nonlinear fiber optics, MI can also drive the four-wave mixing initiated by the interaction of a signal wave with random noise [13]. MI is also often regarded as a precursor to soliton formation, since the same nonlinear Schrödinger equation, which governs the MI, gives rise to stable solitary pulses. Indeed, the breakup of the original CW into soliton arrays may be an eventual outcome of the development of the MI [16].

Starting from the theoretical analysis by Jensen [26], followed by the experimental verification [27], nonlinear directional couplers (NLDC), which are built as dual-core fibers, have been one of the promising elements of integrated photonic circuits for the realization of ultrafast all-optical switches, as well as a subject of intensive fundamental studies [25,28–33]. The operation of the NLDC is governed by the interplay of the Kerr self-focusing, which induces a change in the refractive index in each core, intra-core linear GVD, and linear coupling between the cores. The linear-coupling coefficient determines the critical value of the power, which gives rise to the spontaneous breaking of the symmetry between the two cores [34]. Based on such power-dependent transmission characteristics, many applications of the NLDC have been proposed, such as all-optical switching and power splitting [25], logic operations [35,36], pulse compression [37] and bistability [38].

The MI dynamics in NLDC models was investigated in many works. In particular, in [39], Trillo et al., who first studied soliton switching in NLDC [25], also investigated the MI, considering different combinations of linear and nonlinear effects in a saturable nonlinear medium. In [40], the MI was investigated for antisymmetric and asymmetric CW states in the dual-core fibers, demonstrating that they are subject to the MI even in the normal-GVD regime. In [41], MI was explored by considering the effects of intermodal dispersion, along with higher-order effects, such as the third-order dispersion (TOD) and self-steepening, leading to the conclusion that the intermodal dispersion does not affect the MI growth rate of symmetric or antisymmetric CW states, but can drastically modify the MI of asymmetric CW configurations. Moreover, TOD, as usual, has no influence on the MI gain spectrum in NLDC, while self-steepening can significantly shift the dominant MI band at a sufficiently high input power level. In [42], Li et al. extended the MI to birefringent fiber couplers by including the cross-phase modulation, polarization mode dispersion, and polarization-dependent coupling. Furthermore, in [43], MI was studied under the combined effects of the intermodal dispersion and saturable nonlinear response. In [44], Porsezian et al. carried out analytical and numerical investigation of MI for asymmetric CW states in a dissipative NLDC model, based on cubic-quintic complex Ginzburg–Landau equations. In a similar way, in [45], MI was investigated for asymmetric dissipative fiber couplers, which are used in fiber lasers. In that work, the system was asymmetric, as the bar channel was an active (amplified) one, while the cross channel was a passive lossy core (the same setting was investigated as a nonlinear amplifier [46]).

In all of the works dealing with the MI, except for [45], it was assumed that the NLDC is completely symmetric with respect to the two cores. Extension of the analysis to asymmetric nonlinear couplers is a subject of obvious interest, as new degrees of freedom introduced by the asymmetry may enhance the functionality of NLDC-based devices [47,48]. In a simple way, an asymmetric NLDC (ANLDC) can be manufactured using the difference in diameters of the cores, which tends to produce not only the phase-velocity mismatch between them, but also a change in nonlinearity coefficients. Further, the asymmetry can be imposed by deforming transverse shapes of the cores, while maintaining their areas equal. In such birefringent couplers, one can induce a phase-velocity mismatch without a change in the nonlinearity coefficients. Furthermore, to attain the asymmetry, cores with different GVD coefficients may be used as well. A number of works addressed the switching dynamics [32,49–53], stability of solitons [47,54–57], logic operations [35,36], etc., to elucidate possible advantages of the ANLDC over the symmetric couplers.

In particular, switching of bright solitons has been studied [49] in the model taking into regard the group- and phase-velocity mismatch and differences in the GVD coefficients and effective mode areas of the two cores. Recently, switching dynamics of dark solitons and interaction dynamics of bright solitons have been investigated in [48,58]. However, systematic investigation of the MI dynamics and ensuing generation of pulse arrays in ANLDC has not been reported, as of yet. This is the subject of the present work.

The remainder of the paper is structured as follows. Section 2 introduces the coupled-mode system for the propagation of electromagnetic fields in the asymmetric coupler. Section 3 presents the linear-stability analysis for the MI induced by small perturbations, followed by further analysis in Section 4. Section 5 reports direct simulations of the nonlinear development of the MI. Section 6 concludes the paper.

2. Coupled-Mode Equations

The propagation of optical waves in asymmetric nonlinear couplers is governed by a pair of linearly-coupled nonlinear Schrödinger equations [22,48];

$$i\frac{\partial q_1}{\partial z} + i\beta_{11}\frac{\partial q_1}{\partial t} - \frac{\beta_{21}}{2}\frac{\partial^2 q_1}{\partial t^2} + \gamma_1|q_1|^2 q_1 + cq_2 + \delta_a q_1 = 0, \tag{1}$$

$$i\frac{\partial q_2}{\partial z} + i\beta_{12}\frac{\partial q_2}{\partial t} - \frac{\beta_{22}}{2}\frac{\partial^2 q_2}{\partial t^2} + \gamma_2|q_2|^2 q_2 + cq_1 - \delta_a q_2 = 0, \tag{2}$$

where q_1, q_2 and γ_1, γ_2 are amplitudes of slowly varying envelopes and nonlinearity coefficients in the two cores of the ANLDC, while δ_a accounts for the phase-velocity difference between the cores. Further, $\beta_{1j} \equiv 1/v_{gj}$ and β_{2j} ($j = 1,2$) are the group-velocity and GVD parameters in the j-th core, and c is the coefficient of the linear coupling between the cores.

To derive normalized coupled equations, we perform rescaling,

$$q_j \equiv (\gamma_1 L_D)^{1/2} u_j, \tau \equiv t - \beta_{11}z/T_0, \xi \equiv z/L_D, \tag{3}$$

where $L_D = T_0^2/|\beta_{21}|$ is the dispersion length corresponding to a characteristic pulse width T_0, the result being:

$$i\frac{\partial u_1}{\partial \xi} + \frac{\sigma_1}{2}\frac{\partial^2 u_1}{\partial \tau^2} + |u_1|^2 u_1 + \kappa u_2 + \chi u_1 = 0, \tag{4}$$

$$i\frac{\partial u_2}{\partial \xi} + i\rho\frac{\partial u_2}{\partial \tau} + \alpha\frac{\sigma_1}{2}\frac{\partial^2 u_2}{\partial \tau^2} + \Gamma|u_2|^2 u_2 + \kappa u_1 - \chi u_2 = 0. \tag{5}$$

Here, the normalized coupling coefficient is $\kappa \equiv cL_D$, $\sigma_1 = +1$ and -1 correspond to the anomalous and normal GVD in the first core, while the normalized phase- and group-velocity mismatches and differences in the GVD and effective mode areas are represented, respectively, by:

$$\chi = \delta_a L_D, \rho = (\beta_{12} - \beta_{11})L_D/T_0, \alpha = \beta_{22}/\beta_{21}, \Gamma = \gamma_2/\gamma_1. \tag{6}$$

To design such asymmetric fiber couplers and to calculate the asymmetry coefficients, we adopt physical parameters for the first core, corresponding to standard nonlinear directional couplers, as follows: $\beta_{21} = 0.02$ ps^2/m, $\gamma_1 = 10$ kW^{-1}/m, $T_D = 50$ fs at wavelength $\lambda = 1.5$ µm. Physical parameters for the second core are then determined by normalized coefficients, according to the design outlined above. Furthermore, in terms of this normalized system, we will call "bar" and "cross" the cores corresponding to Equations (4) and (5), respectively.

3. The Linear-Stability Approach

Steady-state CW solutions with common propagation constant Q are looked for as:

$$u_1 = A_1 \exp(iQ\xi), \quad u_2 = A_2 \exp(iQ\xi), \tag{7}$$

where A_1, A_2 are real amplitudes, which determine the total intensity and asymmetry ratio:

$$P = A_1^2 + A_2^2, \ \eta = A_1/A_2. \tag{8}$$

The substitution of Ansatz (7) in Equations (4) and (5) yields an expression for propagation constant Q and a relation between η and the phase velocity mismatch, χ:

$$Q = \frac{P(\Gamma + \eta^2)}{2(1 + \eta^2)} + \frac{\kappa(\eta^2 + 1)}{2\eta}, \tag{9}$$

$$\chi = \frac{P(\Gamma - \eta^2)}{2(1 + \eta^2)} + \frac{\kappa(\eta^2 - 1)}{2\eta}, \tag{10}$$

Next, we add infinitesimal perturbations a_j to the CW solutions, as:

$$\begin{aligned} u_1 &= [A_1 + a_1] \exp(iQ\xi), \\ u_2 &= [A_2 + a_2] \exp(iQ\xi). \end{aligned} \tag{11}$$

Substituting Expression (11) into Equations (4) and (5), we arrive at linearized equations for the complex perturbations:

$$i\frac{\partial a_1}{\partial \xi} + \frac{\sigma_1}{2}\frac{\partial^2 a_1}{\partial \tau^2} + \eta^2 \frac{P}{1 + \eta^2}(a_1 + a_1^*) + \kappa a_2 - \kappa \eta^{-1} a_1 = 0, \tag{12}$$

$$i\frac{\partial a_2}{\partial \xi} + i\rho\frac{\partial a_2}{\partial \tau} + \alpha\frac{\sigma_1}{2}\frac{\partial^2 a_2}{\partial \tau^2} + \Gamma\frac{P}{1 + \eta^2}(a_2 + a_2^*) + \kappa a_1 - \kappa \eta a_2 = 0. \tag{13}$$

Solutions to Equations (12) and (13) are looked for, in the usual form, as:

$$a_1 = F_1 e^{i(K\xi - \Omega\tau)} + G_1 e^{-i(K\xi - \Omega\tau)}, \tag{14}$$

$$a_2 = F_2 e^{i(K\xi - \Omega\tau)} + G_2 e^{-i(K\xi - \Omega\tau)}, \tag{15}$$

where K and Ω are a (generally, complex) wave number and an arbitrary frequency of the perturbation. A set of linear coupled equations for perturbation amplitudes F_j and G_j are derived by substituting Expressions (14) and (15) in Equations (12) and (13):

$$\mathbf{M} \times (F_1, F_2, G_1, G_2)^T = 0, \tag{16}$$

where $\mathbf{M}$ is a 4×4 matrix, whose elements are written in Appendix A. A nontrivial solution exists under condition $\det \mathbf{M} = 0$. Straightforward algebraic manipulations transform the latter condition into a dispersion relation, in the form of a quartic equation for K as a function of Ω:

$$K^4 - aK^3 + bK^2 + cK + d = 0. \tag{17}$$

Rather cumbersome expressions for coefficients (a, b, c, d) are also given in Appendix A. The MI growth rate (gain), defined here for the amplitude of the waves (rather than for the power), is determined by the largest absolute value of the imaginary part of the wave number:

$$G = \{|\mathrm{Im}(K)|\}_{\max}. \tag{18}$$

4. Analysis of the Modulational Instability

4.1. The Anomalous-Dispersion Regime

We start by considering the case of the anomalous GVD in both cores, i.e., $\sigma_1 = 1$ in Equations (4) and (5), as in this case the MI is well known to occur in nonlinear optical fibers. First, in Figure 1, the red line shows the MI gain in the conventional symmetric NLDC ("SNLDC"), with $\alpha = \Gamma = 1$ and $\rho = \chi = 0$. In the same figure, the solid blue line shows the gain for the asymmetric NLDC ("ANLDC") with a particular choice of asymmetry parameters (the reason for choosing these values is explained below), such that the effective mode area of the second core is twice that of the first core, and the GVD of the bar channel is ten-times higher than in the cross one. The figure makes it evident that the MI gain increases by a factor >2 in the ANLDC, and the MI bandwidth is wider by a factor $\simeq 4$. The enhancement of the MI is a new result in the context of the nonlinear directional coupler (similar enhancement was earlier found in the single-core decreasing-GVD fibers with a tapered core [59].

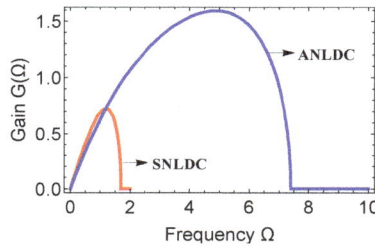

Figure 1. Modulational instability (MI) gain spectra for symmetric nonlinear directional couplers (SNLDC) and asymmetric (ANLDC) couplers in the anomalous group-velocity dispersion (GVD) regime ($\sigma_1 = 1$). Parameters of the symmetric system are $P = \eta = \alpha = \Gamma = \kappa = 1$ and $\rho = \chi = 0$. For the asymmetric ones, the parameters are the same, except for $\alpha = 0.1, \Gamma = 2, \chi = 0.66$ and $\rho = 0.1$.

4.1.1. The Effect of the Input Power on the Instability Spectrum

To elucidate the role of individual effects in the dramatic expansion of the MI region in the asymmetric coupler, we first examine the variation of the MI gain spectrum as a function of the CW power, in both the symmetric and asymmetric systems. Figure 2a clearly demonstrates that the MI gain in the former case increases as in the case of the usual MI [16], i.e., linearly with the power. For the asymmetric system, Figure 2b shows not only the growth of the MI gain with the increase of the power, but also strong expansion of the MI bandwidth.

4.1.2. The Role of the Coupling Coefficient

Figure 3a shows the MI spectrum as a function of the normalized coupling coefficient in the ANLDC, i.e., κ in Equations (12) and (13). The limit case of zero coupling, i.e., the system with decoupled cores, is included too. It is seen that the dependence of the largest gain and MI bandwidth on κ is very weak.

4.1.3. The Impact of Asymmetry Parameters

The influence of the GVD difference, α, on the instability spectra is presented in Figure 3b. The limit case of the coupler with zero GVD in the cross channel, $\alpha = 0$, is included as well. As seen in the figure, the MI bandwidth of MI is infinite in the limit case. Both the gain and bandwidth of the MI monotonically decrease with the increase of α, with the MI vanishing in the limit of $\alpha \to +\infty$. In other words, relatively weak anomalous GVD in the cross channel strongly affects MI bandwidth in the ANLDC.

The influence of the difference in effective mode areas of two cores (Γ) is illustrated by Figure 3c. In this case too, we start with the limit case of an extremely asymmetric coupler, in which the second core is purely dispersive, with zero nonlinearity ($\Gamma = 0$). In this limit, the MI gain vanishes. The MI gain and bandwidth monotonically increase with the growth of Γ. This dependence on Γ is opposite to that on α, which is displayed in Figure 3b. Thus, the MI can be effectively controlled by means of the two asymmetry parameters, Γ and α.

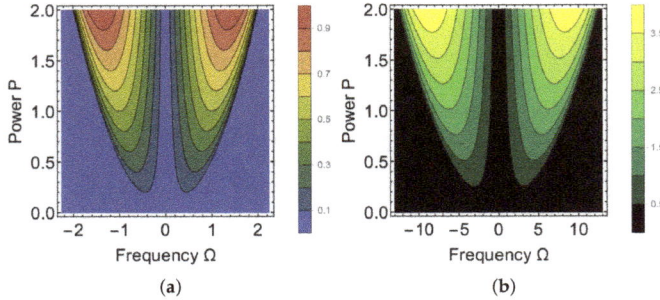

Figure 2. Contour plots showing the dependence of the MI gain on the continuous wave (CW) power, P, and perturbation frequency, Ω, for symmetric and asymmetric couplers in the anomalous-GVD regime ($\sigma_1 = 1$). Parameters of symmetric system (**a**) are $\eta = \alpha = \Gamma = 1$, $\kappa = 2$ and $\rho = \chi = 0$. For the asymmetric system (**b**), $\eta = \alpha = 0.1$, $\Gamma = 2$, $\kappa = 1$, $\rho = 0.1$, and the phase-velocity mismatch is defined in terms of P, in order to produce the largest gain: $\chi = -4.95 + 0.985P$. Note the difference in horizontal scales between (a) and (b).

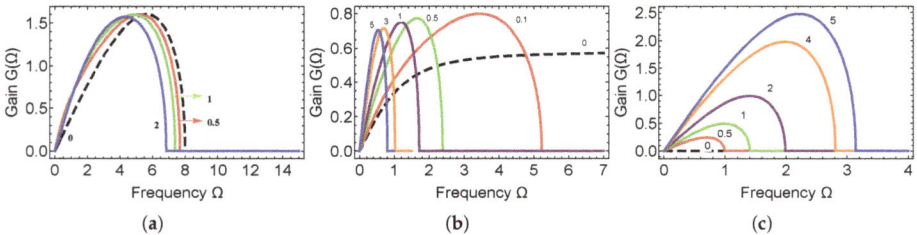

Figure 3. The MI gain spectra in the asymmetric coupler with anomalous GVD ($\sigma_1 = 1$). (**a**) The results for different values of the normalized coupling coefficient, κ. Parameters of the system are $P = 2$, $\eta = 0.5$, $\alpha = 0.1$, $\Gamma = 2$ and $\rho = 0.01$. (**b**) For different values of the ratio of the GVD coefficients in the cross and bar channels, α, indicated near each curve with $P = 1$, $\eta = \kappa = 0.5$, $\Gamma = 1$ and $\rho = 0.01$. (**c**) For different values of the ratio of the nonlinearity coefficients in the two cores, Γ, which are indicated near the curves. Other parameters are $P = 0.5$, $\eta = 0.1$, $\kappa = 0.2$, $\alpha = 1$ and $\rho = 0.1$.

Next, we study the effect of the group-velocity mismatch (walk-off between the cores), ρ. Figure 4a shows the impact of ρ when the asymmetry is represented only by the GVD ratio, $\alpha = 0.1$, while the nonlinear coefficients in both cores are equal. The figure demonstrates that the variation of ρ in the range of $\rho \lesssim 1$ weakly affects the MI spectrum. The effect is much stronger at larger values of the walk-off. In particular, the MI spectral band splits into two at $\rho = 2$. The latter effect seems interesting even if the value of $\rho = 2$ may be difficult to attain in real couplers. On the other hand, the analysis demonstrates that the variation of ρ produces almost no effect on the MI gain in the case when the asymmetry is determined by the difference in the nonlinearity coefficients ($\Gamma \neq 1$), while the GVD coefficients are equal ($\alpha = 1$). The latter result is not shown here in detail, as it does not display noteworthy features.

It is obviously interesting as well to investigate the effect of the CW asymmetry ratio, η (see Equation (8)), on the MI. These results are presented in Figure 4b, which makes it obvious that the gain and bandwidth of the MI quickly decrease with the increase of η from small values ~ 0.1 to $\eta = 2$. With the further increase of the asymmetry ratio to values $\eta > 2$, the largest MI gain slightly increases, while the bandwidth remains practically constant.

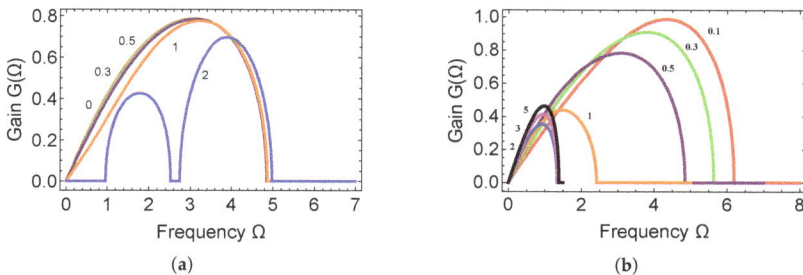

Figure 4. The MI gain spectra in the anomalous-GVD regime ($\sigma_1 = 1$). (**a**) Results for different values of the group-velocity mismatch (ρ in Equation (5)), which are indicated near the curves. Other parameters are $P = \kappa = \Gamma = 1$, $\eta = 0.5$ and $\alpha = 0.1$ (**b**) Results for different values of asymmetry ratio η of the CW state (see Equation (8)), which are indicated near the curves. Other parameters are $P = 0.5, \kappa = 1$, $\alpha = 0.1, \Gamma = 2$ and $\rho = 0.01$.

4.2. The Normal-Dispersion Regime

The combination of the self-focusing Kerr nonlinearity and normal GVD usually supports stable CWs. However, as mentioned in the Introduction, MI may occur under the normal GVD in more complex systems, including couplers. Following the pattern of the MI investigation presented above for the anomalous GVD, we first consider the effect of the CW power, P, on the MI gain. We also compare the instability spectrum of the asymmetric system with that of the symmetric one in Figure 5a,b, respectively. As seen in Figure 5, in both cases, two distinct MI bands determine the instability, and (similar to the anomalous-GVD regime) the MI gain of the asymmetric system linearly grows with P, featuring a broad bandwidth.

To illustrate the essential effect of the coupling coefficient κ, Figure 6a depicts the MI gain spectra for various values of κ. Naturally, no MI takes place in the normal-GVD regime in the absence of the coupling, $\kappa = 0$. It is worthy to note the appearance of two separated MI bands at $\kappa > 1$, the MI gain increasing in both bands, along with their widths, with the growth of κ.

The effect of the relative difference in the magnitude of the normal GVD in the two cores, α, is shown in Figure 6b. Like in the anomalous-GVD regime, here, as well, the MI bandwidth is infinite for $\alpha = 0$ (it also contains a separate finite MI band). The MI spectrum features two separate bands at $\alpha > 0$ and the largest gain at $\alpha = 0.1$. The gain decreases with the subsequent increase of α.

Figure 7a shows the effect of the relative difference in the effective mode areas between the two channels, i.e., the ratio of the nonlinearity coefficient, Γ. It is seen that no MI occurs when the cross channel is linear ($\Gamma = 0$), and two distinct MI bands emerge and expand, featuring a growing largest value of the instability gain, with the increase of Γ.

The influence of the group-velocity mismatch (walk-off between the cores), ρ, is depicted in Figure 7b. Once again, the MI appears in the form of two separated bands. The MI gain and bandwidth nontrivially depend on ρ: at $\rho < 1$ the low-frequency band is narrower, with smaller values of the instability gain, while at $\rho \geq 1$, the situation is inverted.

We have also analyzed the effect of the CW's asymmetry η (see Equation (8)) on the MI in the normal-GVD regime. No MI occurs for small values of η, viz. $\eta < 0.2$. At $\eta > 0.2$ (in particular, at $\eta = 0.5$), there again emerge two separate MI bands, as shown in Figure 8. The MI gain and bandwidth attain their maxima at $\eta = 1$ (equal amplitudes of the CW in the two cores), decreasing with the further increase of η.

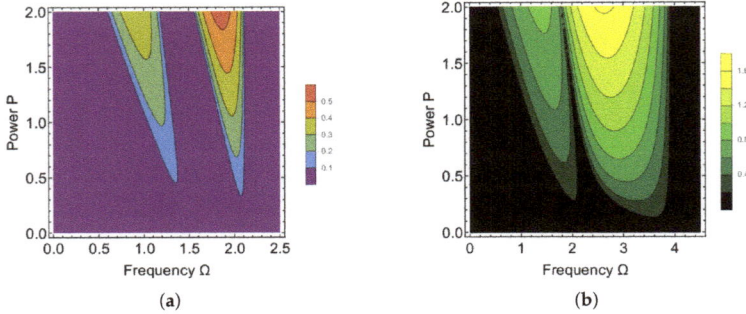

(a) (b)

Figure 5. Contour plots showing the dependence of the MI gain on the CW power, P, in the normal-GVD regime ($\sigma = -1$) in the symmetric and asymmetric systems. Parameters of the symmetric system (a) are $\alpha = \Gamma = 1$, $\eta = 2$, $\kappa = 0.9$ and $\rho = \chi = 0$. For the asymmetric system (b), $\eta = 0.5, \alpha = 0.1, \Gamma = 2, \kappa = 1.1$ and $\rho = 0.01, \chi = -0.825 + 0.7P$.

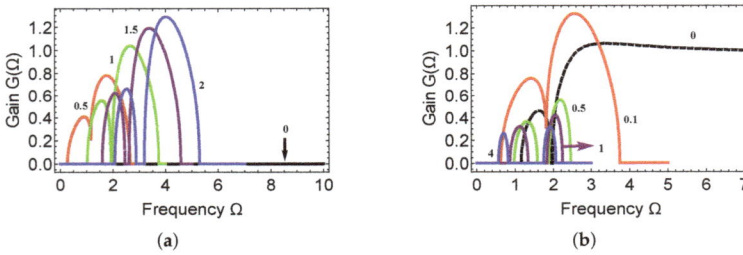

(a) (b)

Figure 6. (a) The MI gain spectra in the normal-GVD regime ($\sigma_1 = -1$) for different values of (a) the coupling coefficient, κ, indicated near the curves. Other parameters are $P = 1, \eta = 0.5, \alpha = 0.1, \Gamma = 2$ and $\rho = 0.01$. (b) The change of values of the difference in normal-GVD coefficients ($\sigma_1 = -1$), α, in the two cores of the nonlinear coupler (the values of α are indicated near the curves). Other parameters are $P = 1.5, \eta = 0.5, \kappa = \Gamma = 1$ and $\rho = 0.1$.

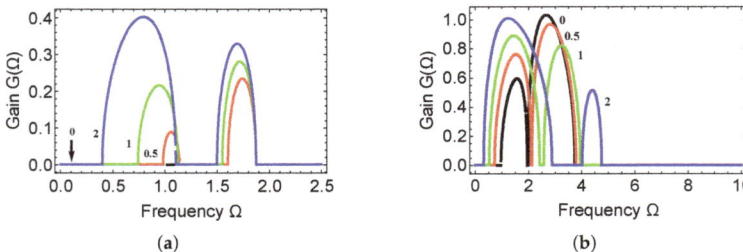

(a) (b)

Figure 7. (a) The MI gain spectra in the normal-GVD regime ($\sigma_1 = -1$); (a) the results for different values of the ratio of the nonlinearity coefficient in the two cores (Γ), indicated near the curves. Other parameters are $P = 1, \eta = 0.5, \alpha = 1, \kappa = 0.7$ and $\rho = 0.1$. (b) The results for different values of the group-velocity mismatch (ρ), indicated near the curves. Other parameters are $P = 1, \eta = 0.5$, $\alpha = 0.1, \Gamma = 2, \kappa = 1$.

Appl. Sci. **2017**, *7*, 645

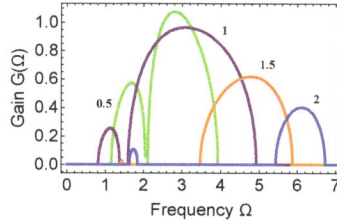

Figure 8. The MI gain spectra in the normal-GVD regime ($\sigma_1 = -1$) at different values of the asymmetry ratio of the CW state (η in Equation (8)). Other parameters are $P = 1, \kappa = 1.1, \alpha = 0.1, \Gamma = 2$ and $\rho = 0.01$.

4.3. The Coupler with Opposite Signs of the Dispersion in the Two Cores

The case of the opposite ("mixed") GVD signs in the two cores of the coupler, which corresponds to $\alpha < 0$ in Equation (5), is obviously interesting, as well [47]. For this purpose, we assume the anomalous and normal GVD in the bar and cross channels, respectively.

Figure 9 shows the effect of the CW power, P, on the MI in the mixed-GVD coupler. The figure demonstrates that the MI gain and bandwidth monotonically increase with the growth of P. It should be noted that the spectra obtained for this case are somewhat different in comparison with the conventional MI spectra, as the gain is stretched over a broad interval of the perturbation frequency when the CW power is low ($P < 1$). In the present case, the effect of the coupling coefficient, κ, on the MI, which is shown in Figure 10, is essentially the same as demonstrated above for the coupler with the normal GVD in both cores; see Figure 6a. Namely, the MI gain and bandwidth increase with the growth of κ.

The effects of the negative value of the ratio of the GVD coefficients, $\alpha < 0$, and the ratio of the nonlinearity coefficients (Γ) in the two cores are shown in Figure 11. Similar to the coupler with the anomalous GVD in each core, cf. Figure 3b, the increase of α (see Figure 11a) leads to shrinkage of the MI band. Like in the coupler with the anomalous GVD in both cores, cf. Figure 3c, the MI gain increases with the growth of Γ, which is depicted in Figure 11b; however, the difference is that, in the present case of the mixed-GVD coupler, the bandwidth is not affected by the variation of Γ.

Figure 12 displays quite nontrivial evolution of the MI spectra with the variation of the group-velocity mismatch (walk-off) between the cores, ρ in Equation (5). The evolution is very different from what is demonstrated above for the coupler with the anomalous GVD in both cores, cf. Figure 4a. Namely, Figure 12 shows that the increase of ρ from zero to one suppresses the MI, which completely vanishes at $\rho = 1$. The system recovers the MI, which features monotonically increasing gain and bandwidth, with the further increase of ρ to values $\rho > 1$.

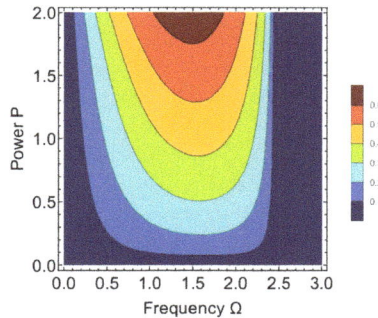

Figure 9. The contour plot showing the dependence of MI gain on the CW total power, P, in the mixed-GVD coupler ($\alpha < 0$). The parameters are $\eta = 0.5, \alpha = -0.1, \Gamma = 2, \kappa = 1$, and $\rho = 0.1$.

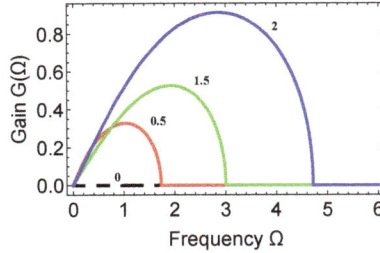

Figure 10. The MI gain spectra for different values of the coupling coefficient, κ (indicated near the curves), in the mixed-GVD coupler ($\alpha < 0$) for $P = 1, \eta = 0.7, \alpha = -0.1, \Gamma = 2$ with $\rho = 0.01$.

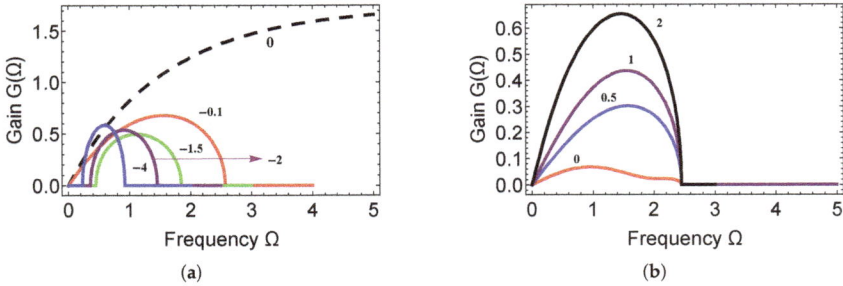

(a)

(b)

Figure 11. The MI gain spectra in the mixed-GVD coupler ($\alpha < 0$). (**a**) The results for different negative values of the ratio of the GVD coefficients in the two cores (α), which are indicated near the curves for $P = 2, \eta = 0.5, \kappa = 1.1, \Gamma = 2$ with $\rho = 0.01$. (**b**) The results for different values of the ratio of the nonlinearity coefficients in the two cores (Γ), which are indicated near the curves. Parameters are same as in (a), except for $\kappa = 1$ and $\alpha = -0.1$.

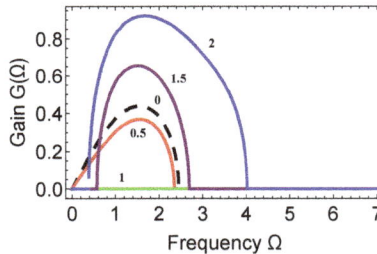

Figure 12. The MI gain spectra for different values of the group-velocity mismatch (walk-off) between the cores in the mixed-GVD coupler ($\alpha < 0$) for $P = \kappa = 1, \eta = 0.5, \alpha = -0.1, \Gamma = 2$.

Next, we consider the impact of the asymmetry parameter η in the CW state; see Equation (8). As show in Figure 13, there is no MI at small values of η, such as $\eta = 0.1$. With the subsequent increase of η up to $\eta = 1$, the MI gain and bandwidth increase, similar to what was observed above in the coupler with anomalous GVD in both cores; see Figure 4b. However, the situation becomes completely different at $\eta > 1$, when the CW amplitude is higher in the bar channel: the MI band splits into two narrower ones, with smaller values of the gain.

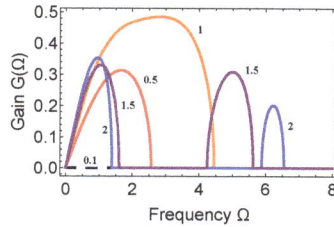

Figure 13. The MI gain spectra for different values of the asymmetry ratio, η, of the CW state (see Equation (8)), which are indicated near the figures, in the mixed-GVD coupler ($\alpha < 0$) for $P = 0.5, \kappa = 1.1, \alpha = -0.1, \Gamma = 2$ with $\rho = 0.01$.

5. Direct Simulations

The analytical results obtained above for the MI have been checked against numerical calculations of the instability spectra. Numerical methods are actually more relevant for direct simulations of the nonlinear evolution of the MI, which was analyzed above in the linear approximation. The simulations were carried out by dint of the well-known split-step Fourier method [48] (using MATLAB). Most results displayed below were obtained using numerical meshes with 512 Fourier points and periodic boundary conditions with respect to variable τ. Simulations performed with denser meshes have produced virtually identical results. Furthermore, results of the nonlinear development of the MI are not sensitive to details of initial small perturbations, which initiate the onset of the MI.

The initial conditions were taken in the form of the CW to which a small periodic perturbation was added:

$$u_j(0, \tau) = A_j + a_0 \cos(\omega_0 \tau), \quad (j = 1, 2), \tag{19}$$

where a_0 is a small amplitude of the perturbation and ω_0 is its frequency.

Various outcomes of the MI development for CW states with different parameters are displayed in Figures 14–23. First, in Figure 14, we show the results for the symmetric coupler in the anomalous-GVD regime when the amplitudes of two CW components are equal ($A_1 = A_2 = 1$). As seen in the figure, a periodic chain of well-shaped soliton-like pulses is produced on top of the nonzero background in both cores. Longer simulations demonstrate regular dynamics of the quasi-soliton arrays. In this work, we do not aim to study the latter in detail, as it is not closely related to the initial MI.

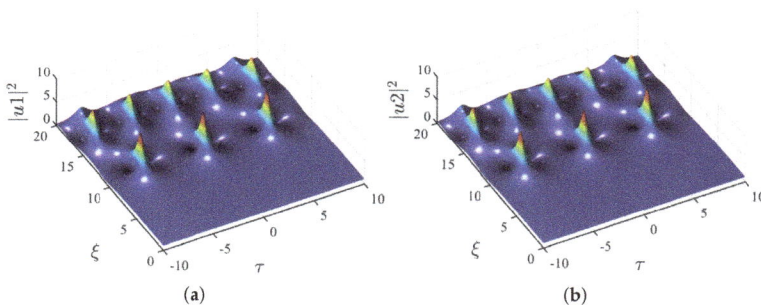

(a)   **(b)**

Figure 14. The evolution of the MI in the symmetric coupler with anomalous GVD ($\sigma = 1$) in the bar (**a**) and cross (**b**) channels for equal amplitudes of the underlying CW state, $A_1 = A_2 = 1$, with perturbation parameters $a_0 = 0.0001$ and $\omega_0 = 1$, in Equation (19). Other parameters are $\alpha = \Gamma = \kappa = 1$ and $\rho = \chi = 0$.

We now turn to simulations of the MI in the asymmetric coupler and the analysis of effects of its different parameters. The impact of the group-velocity mismatch (walk-off) between the cores in the anomalous-GVD regime is presented in Figure 15. As seen in the figure, pulses generated by the MI drift away from their original positions, which implies spontaneous symmetry breaking, as a particular drift direction is selected by the system. We have also investigated the spectral evolution of the MI for different values of the group-velocity mismatch. The results (not shown here in detail) corroborate, in particular, that the group-velocity mismatch has no impact on the instability spectrum, as predicted by the analytical results in Figure 4a. Further, Figure 16 shows the influence of the phase-velocity mismatch on the MI evolution in the anomalous-GVD regime. In this case, the main effects are oscillations of the background and retaining of the power chiefly in the bar channel.

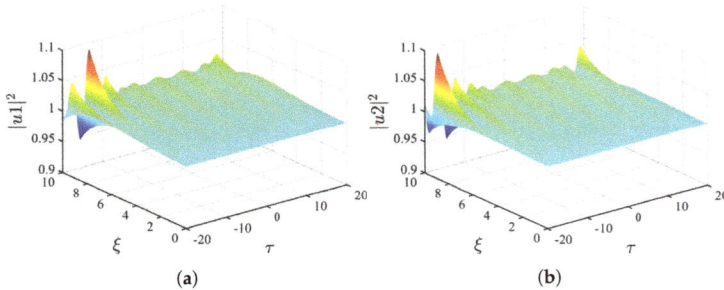

Figure 15. The influence of the group-velocity mismatch, $\rho = 1$, on the evolutions of the MI in the bar (**a**) and cross (**b**) channels in the anomalous-GVD regime. Other system parameters are, $A_1 = A_2 = \omega_0 = \alpha = \Gamma = \kappa = 1, \chi = 0$ and $a_0 = 0.0001$.

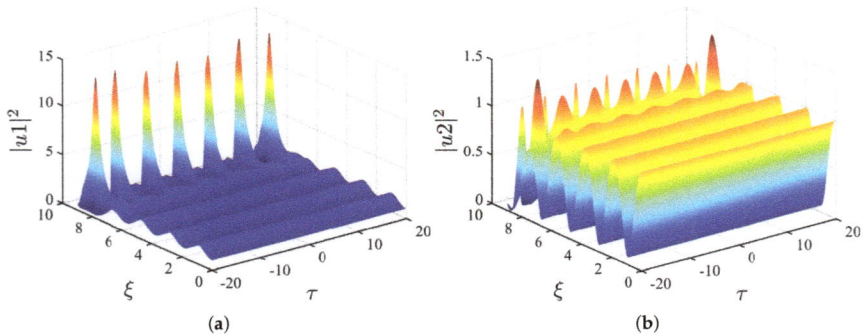

Figure 16. The influence of the phase-velocity mismatch, $\chi = 1$, on the evolution of the MI in the bar (**a**) and cross (**b**) channels in the anomalous-GVD dispersion regime. Other system parameters are $A_1 = A_2 = \omega_0 = \alpha = \Gamma = \kappa = 1, \rho = 0$ and $a_0 = 0.0001$.

The role of the ratio of the GVD coefficients in the two cores, α, is shown in Figures 17 and 18, for the case of the anomalous GVD in both cores. In the case of zero GVD in the cross channel ($\alpha = 0$) (Figure 17) shows that a chain of quasi-solitons with growing amplitudes is generated on top of a nonzero background in the bar channel, while narrow growing peaks emerge at edges of the background in the cross channel. If α increases to $\alpha = 2$, the former picture is essentially reversed, so that a chain of solitons on top of the background appears in the cross channel, and a chain of very narrow solitons is generated in the bar channel. In all of these cases, the soliton chains keep the initial modulation period, $2\pi/\omega_0$.

Figure 19 reveals the impact of the ratio of nonlinearity coefficients between the two cores. In this case, the MI generates a chain of very narrow solitons with a higher amplitude, whose peak powers are growing in the cross channel and growing peaks on an oscillating background with a relatively low amplitude in the bar channel. Next, we plug in all of the parameters, to identify their combined effect on the MI evolution in the anomalous-GVD regime, in Figures 20 and 21. In the former case, it is observed that the MI gives rise to a single soliton in the bar channel, whereas the field in the cross channel decays into radiation. In the latter case, a single soliton is generated too (which is natural for the case of the anomalous GVD), but with components in both cores.

Focusing our attention on the asymmetric coupler in the normal-GVD regime, in Figure 22, we address the case when the amplitudes of the two CW components are equal. In this case as well, a periodic array of peaks with growing amplitudes is generated in both the bar and cross channels. However, its shape is essentially different from the soliton chains displayed above in the anomalous-GVD regime, as in the present case, the array is built of alternating peaks and wells. Lastly, if the amplitudes of the two CW states are widely different, such as in the case of a large amplitude in the bar channel and a relatively small one in the cross channel, the MI evolution leads to a chaotic state, as shown in Figure 23.

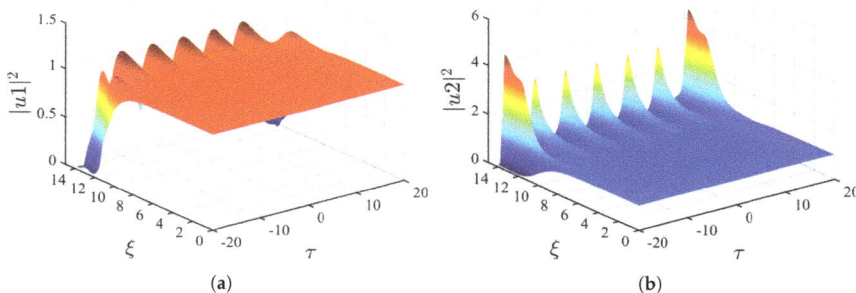

Figure 17. The MI evolution in the bar (**a**) and cross (**b**) cores, in the case of the anomalous GVD in the bar channel, and zero GVD ($\alpha = 0$) in the cross channel. Other parameters are $A_1 = A_2 = \omega_0 = \Gamma = \kappa = 1$ and $a_0 = 0.0001$.

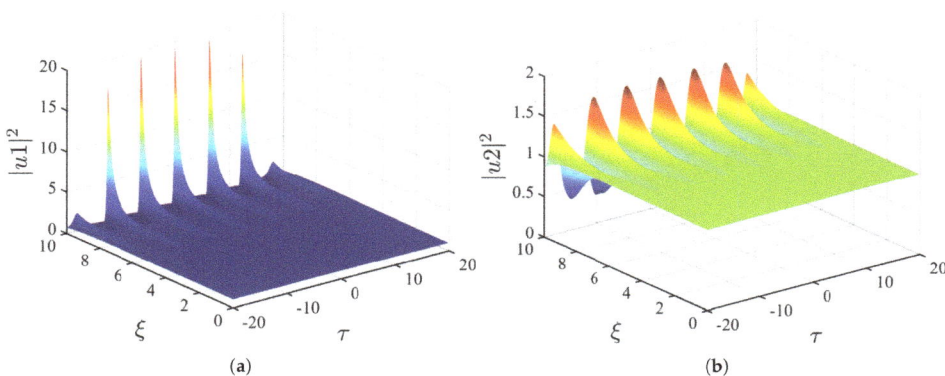

Figure 18. The same as in Figure 17, but when the difference in the GVD coefficients is $\alpha = 2$.

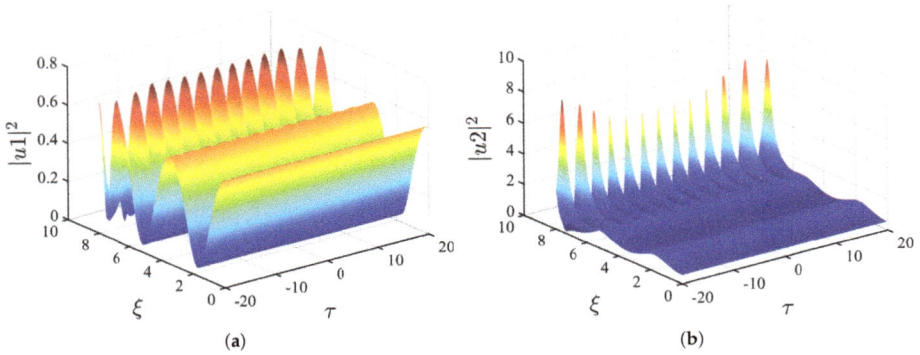

Figure 19. The influence of the ratio of the nonlinearity coefficients in the two cores, $\Gamma = 2$, on the MI evolution in the bar (**a**) and cross (**b**) channels in the anomalous-GVD regime. Other parameters are $A_1 = A_2 = 0.75, \omega_0 = 2, \alpha = \kappa = 1, \rho = \chi = 0$ and $a_0 = 0.0001$.

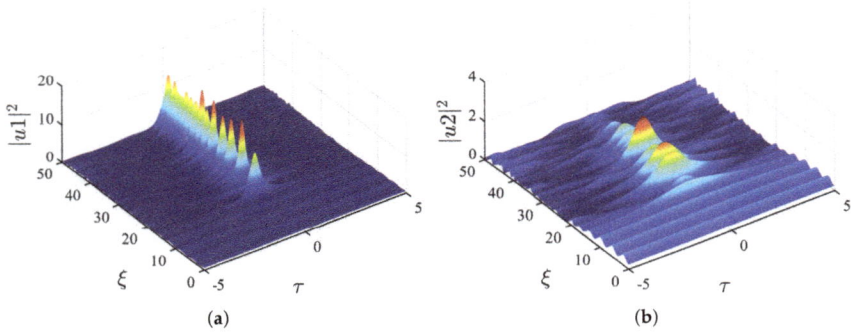

Figure 20. The MI-driven evolution in the bar (**a**) and cross (**b**) channels in the anomalous-GVD regime ($\sigma = 1$) for initial CW amplitudes $A_1 = 0.75$, $A_2 = 0.5$ and perturbation parameters $a_0 = 0.0009, \omega_0 = 1$. Other parameters are $\alpha = 2, \Gamma = 1, \rho = 0.01, \chi = 0.001$ and $\kappa = 1$.

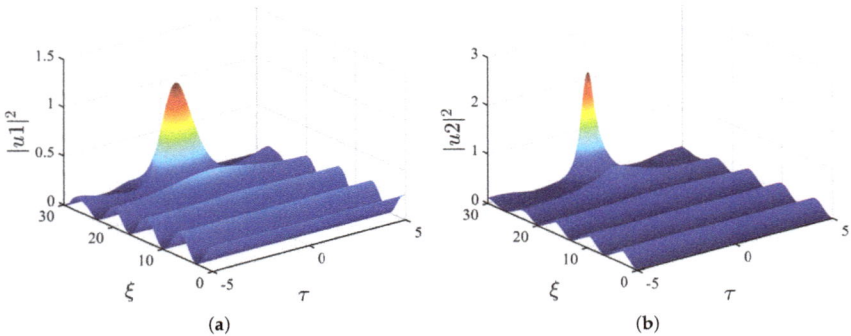

Figure 21. The same as in Figure 20, but for $A_1 = 0.5$, $A_2 = 0.1$, $a_0 = 0.0007$, $\chi = 0.01$ and $\kappa = 0.5$.

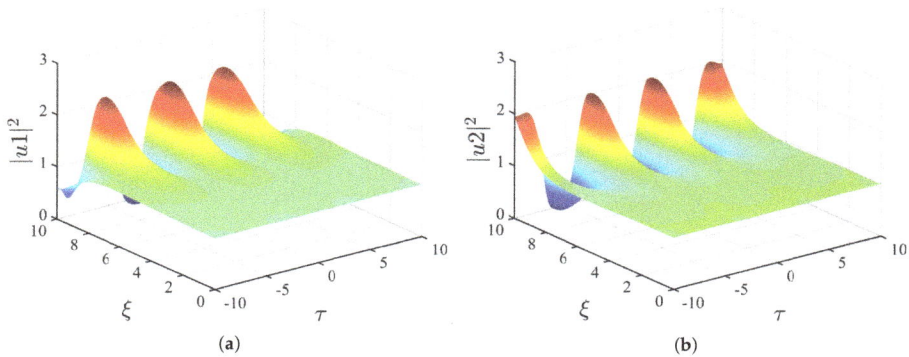

Figure 22. The MI evolution in the bar (**a**) and cross (**b**) channels in the normal-GVD regime ($\sigma = -1$) for the amplitudes of the CW components $A_1 = A_2 = 1$ and perturbation parameters $a_0 = 0.002, \omega_0 = 1$. Other parameters are $\alpha = 2, \Gamma = 1, \rho = 0.01, \chi = 0.001$ and $\kappa = 1$.

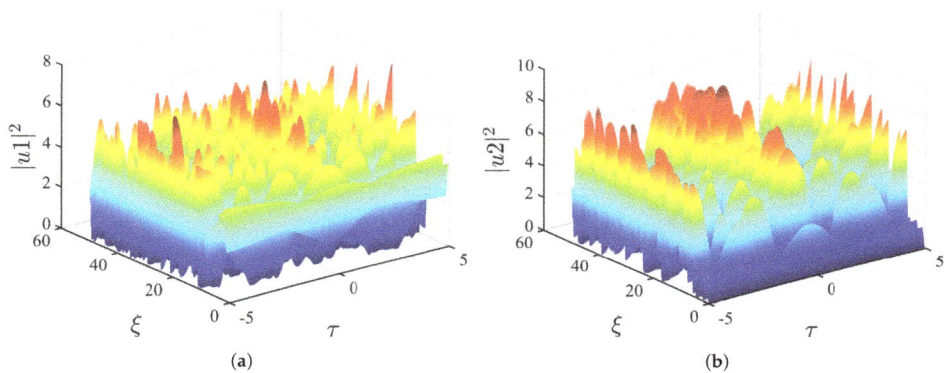

Figure 23. Creation of a chaotic (turbulent) state by the MI in the bar (**a**) and cross (**b**) channels in the normal-GVD dispersion regime ($\sigma = -1$) for CW amplitudes $A_1 = 2, A_2 = 0.001$, with perturbation parameters $a_0 = 0.09, \omega_0 = 1$. Other parameters are $\alpha = 2, \Gamma = 0.5, \rho = 0.01, \chi = 0.02$ and $\kappa = 1$.

6. Conclusions

In this work, we have investigated the MI (modulational instability) in the model of asymmetric dual-core NLDCs (nonlinear directional couplers), based on the system of nonlinear Schrödinger equations, which include differences in the GVD and nonlinearity coefficient in the two cores, as well as the group- and phase-velocity mismatch between them. The MI of symmetric and asymmetric CW states in the NLDC against small perturbations was investigated using the linearized equations for the perturbations. This was followed by direct simulations to investigate the nonlinear development of the MI.

First, we have considered the dependence of the MI gain spectra on the total power of the two-component CW states in the coupler with the anomalous sign of the GVD in both cores. It was found that the MI bands in the asymmetric couplers are broader in comparison with their symmetric counterparts. Then, we focused on the impact of the magnitude of the inter-core coupling coefficient, κ, demonstrating that the increase of κ leads to gradual suppression of the MI. Next, a large GVD coefficient in the bar channel, in comparison with the cross channel, generates very broad MI spectra with large values of the instability gain. If the asymmetry between the cores is introduced only through the difference in the GVD coefficients, high values of the group-velocity mismatch cause splitting of the single MI band into two. The effect of asymmetry between two components of the CW state, η, was identified as well. It was found that the MI gain and bandwidth reduce with the increase of η from small values to one, while the further increase of η leads to shrinkage of the MI band.

Next, the MI was explored in the normal-GVD regime, in which the MI occurs in two separated spectral bands. The increase of the coupling coefficient makes the size of the MI gain in the two bands strongly different. The influence of the difference in the GVD and nonlinearity coefficients was analyzed, as well. The increase of these coefficients leads, respectively, to the decrease and increase of the MI gain in the two bands.

Noteworthy results were produced by the analysis of the MI in the coupler with opposite signs of the GVD in the cores. While the difference in the negative values of the GVD coefficient, and in the nonlinearity coefficients, produce approximately the same effects as in the anomalous-GVD regime, the response to the increase of the coupling coefficient is similar to that in the case of the normal GVD, leading to the increase of the MI gain. A notable effect was observed with the variation of the group-velocity mismatch, ρ, between the cores: the increase of ρ from small values to one suppresses the MI, which disappears at $\rho = 1$. It appears again and enhances at $\rho > 1$. The asymmetry ratio of the two components of the underlying CW state, η, also produces a nontrivial effect: while the MI is absent at small values of η, it appears at $\eta \gtrsim 0.5$ in the form of a single spectral band, which grows up to $\eta = 1$ and then splits into two bands.

Finally, we have also performed systematic simulation of the nonlinear development of the MI in different regimes, which were studied analytically. Typical outcomes feature the generation of periodic chains of growing peaks in the anomalous-GVD regime. In particular, the group-velocity mismatch naturally causes a walk-off effect, while the phase-velocity mismatch and difference in the nonlinearity coefficients produce oscillations on the background, on top of which soliton arrays emerge. The difference in the GVD coefficients facilitates the generation of arrays of very narrow solitary pulses in the bar channel, whereas arrays of regular pulses appear in the cross channel. The formation of a single soliton is possible as well. In the normal-GVD regime, the formation of arrayed peaks with a growing amplitude was observed. The MI of the CW states with widely different amplitudes of its two components may produce a turbulent state.

These results, especially the generation of regular arrays of solitary pulses and of a single pulse, can find applications for the design of signal sources for optical systems. The variations of many parameters that control the dynamics of the asymmetric couplers may be used to optimize these applications.

For further work, it may be relevant to take into regard higher-order terms, such as those accounting for the third-order dispersion and self-steepening, and analyze their effects on the MI in the asymmetric dispersive nonlinear couplers.

Acknowledgments: A.G. is grateful to M. Lakshmanan and National Academy of Sciences, India (NASI), for providing the Senior Research Fellowship (Research Associate) under the NASI Platinum Jubilee Senior Scientist Fellowship project of M.L.

Author Contributions: A.G. conceived of the idea of the present work and has performed the analytical and numerical computations. B.A.M refined the analytical calculations. B.A.M., A.G., A.M. and A.U. conducted the interpretation of the results. A.G. wrote an initial draft. B.A.M. finalized the article.

Conflicts of Interest: The authors declare no conflict of interest.

Appendix A

Elements of matrix **M** in Equation (16) are:

$$
\begin{aligned}
m_{11} &= -K - (\sigma_1 \Omega^2/2) + \eta^2 S - \kappa/\eta, \\
m_{12} &= m_{21} = m_{34} = m_{43} = \kappa, \\
m_{13} &= \eta^2 S, \\
m_{14} &= m_{23} = m_{32} = m_{41} = 0, \\
m_{22} &= -K + \rho\Omega - (\sigma_1 \alpha \Omega^2/2) + \Gamma S - \kappa\eta, \\
m_{24} &= \Gamma S, \\
m_{31} &= \eta^2 S, \\
m_{33} &= K - (\sigma_1 \Omega^2/2) + \eta^2 S - \kappa/\eta, \\
m_{42} &= \Gamma S, \\
m_{44} &= K - \rho\Omega - (\sigma_1 \alpha \Omega^2/2) + \Gamma S - \kappa\eta,
\end{aligned}
\tag{A1}
$$

where $S = P/(1 + \eta^2)$. Coefficients of quartic Equation (17) for K, as functions of Ω, are given by:

$$
a = 2\rho\Omega \tag{A2}
$$

$$
b = 2S\eta\kappa + 2S\Gamma\eta\kappa - 2\kappa^2 - \frac{\kappa^2}{\eta^2} - \eta^2\kappa^2 + \Omega^2\left(\rho^2 + S\eta^2\sigma_1 - \frac{\kappa\sigma_1}{\eta} + S\alpha\Gamma\sigma_1 - \alpha\eta\kappa\sigma_1\right) + \Omega^4\left(-\frac{\sigma_1^2}{4} - \frac{1}{4}\alpha^2\sigma_1^2\right) \tag{A3}
$$

$$
c = \left(-4S\eta\kappa\rho + 2\kappa^2\rho + \frac{2\kappa^2\rho}{\eta^2}\right)\Omega + \Omega^3\left(-2S\eta^2\rho\sigma_1 + \frac{2\kappa\rho\sigma_1}{\eta}\right) + \frac{1}{2}\rho\Omega^5\sigma_1^2 \tag{A4}
$$

$$
\begin{aligned}
d &= \Omega^2\left(2S\eta\kappa\rho^2 - \frac{\kappa^2\rho^2}{\eta^2} + 2S^2\Gamma\eta^3\kappa\sigma_1 - S\Gamma\kappa^2\sigma_1 - S\eta^4\kappa^2\sigma_1 + 2S^2\alpha\Gamma\eta\kappa\sigma_1 - \frac{S\alpha\Gamma\kappa^2\sigma_1}{\eta^2} - S\alpha\eta^2\kappa^2\sigma_1\right) \\
&\quad + \Omega^4\left(S\eta^2\rho^2\sigma_1 - \frac{\kappa\rho^2\sigma_1}{\eta} - \frac{1}{2}S\Gamma\eta\kappa\sigma_1^2 + \frac{1}{4}\eta^2\kappa^2\sigma_1^2 + S^2\alpha\Gamma\eta^2\sigma_1^2 - \frac{S\alpha\Gamma\kappa\sigma_1^2}{\eta} - S\alpha\eta^3\kappa\sigma_1^2\right. \\
&\quad \left. + \frac{1}{2}\alpha\kappa^2\sigma_1^2 - \frac{1}{2}S\alpha^2\eta\kappa\sigma_1^2 + \frac{\alpha^2\kappa^2\sigma_1^2}{4\eta^2}\right) + \Omega^6\left(-\frac{1}{4}\rho^2\sigma_1^2 - \frac{1}{4}S\alpha\Gamma\sigma_1^3 + \frac{1}{4}\alpha\eta\kappa\sigma_1^3 - \frac{1}{4}S\alpha^2\eta^2\sigma_1^3 + \frac{\alpha^2\kappa\sigma_1^3}{4\eta}\right) \\
&\quad + \frac{1}{16}\alpha^2\Omega^8\sigma_1^4
\end{aligned}
\tag{A5}
$$

References

1. Benjamin, T.B.; Feir, J. The disintegration of wave trains on deep water Part 1. Theory. *J. Fluid Mech.* **1967**, *27*, 417–430.
2. Hasegawa, A. Generation of a train of soliton pulses by induced modulational instability in optical fibers. *Opt. Lett.* **1984**, *9*, 288–290.
3. Tai, K.; Hasegawa, A.; Tomita, A. Observation of modulational instability in optical fibers. *Phys. Rev. Lett.* **1986**, *56*, 135.
4. Agrawal, G.P. Modulation instability induced by cross-phase modulation. *Phys. Rev. Lett.* **1987**, *59*, 880.
5. Zakharov, V.E.; Dyachenko, A.; Prokofiev, A. Freak waves as nonlinear stage of Stokes wave modulation instability. *Eur. J. Mech. B/Fluids* **2006**, *25*, 677–692.
6. Melville, W. The instability and breaking of deep-water waves. *J. Fluid Mech.* **1982**, *115*, 165–185.
7. Konotop, V.; Salerno, M. Modulational instability in Bose-Einstein condensates in optical lattices. *Phys. Rev. A* **2002**, *65*, 021602.
8. Li, L.; Li, Z.; Malomed, B.A.; Mihalache, D.; Liu, W. Exact soliton solutions and nonlinear modulation instability in spinor Bose-Einstein condensates. *Phys. Rev. A* **2005**, *72*, 033611.
9. Bhat, I.A.; Mithun, T.; Malomed, B.; Porsezian, K. Modulational instability in binary spin-orbit-coupled Bose-Einstein condensates. *Phys. Rev. A* **2015**, *92*, 063606.
10. Taniuti, T.; Washimi, H. Self-trapping and instability of hydromagnetic waves along the magnetic field in a cold plasma. *Phys. Rev. Lett.* **1968**, *21*, 209.
11. Galeev, A.; Sagdeev, R.; Sigov, Y.S.; Shapiro, V.; Shevchenko, V. Nonlinear theory of the modulation instability of plasma waves. *Sov. J. Plasma Phys.* **1975**, *1*, 5–10.

12. Zakharov, V.; Ostrovsky, L. Modulation instability: The beginning. *Phys. D Nonlinear Phenom.* **2009**, *238*, 540–548.
13. Boggio, J.; Tenenbaum, S.; Fragnito, H. Amplification of broadband noise pumped by two lasers in optical fibers. *J. Opt. Soc. Am. B* **2001**, *18*, 1428–1435.
14. Tanemura, T.; Ozeki, Y.; Kikuchi, K. Modulational instability and parametric amplification induced by loss dispersion in optical fibers. *Phys. Rev. Lett.* **2004**, *93*, 163902.
15. Höök, A.; Karlsson, M. Ultrashort solitons at the minimum-dispersion wavelength: Effects of fourth-order dispersion. *Opt. Lett.* **1993**, *18*, 1388–1390.
16. Agrawal, G. *Nonlinear Fiber Optics*; Optics and Photonics, Academic Press: London, UK, 2006.
17. Hasegawa, A.; Tappert, F. Transmission of stationary nonlinear optical pulses in dispersive dielectric fibers. I. Anomalous dispersion. *Appl. Phys. Lett.* **1973**, *23*, 142–144.
18. Rehberg, I.; Rasenat, S.; Fineberg, J.; De La Torre Juarez, M.; Steinberg, V. Temporal modulation of traveling waves. *Phys. Rev. Lett.* **1988**, *61*, 2449.
19. Malendevich, R.; Jankovic, L.; Stegeman, G.; Aitchison, J.S. Spatial modulation instability in a Kerr slab waveguide. *Opt. Lett.* **2001**, *26*, 1879–1881.
20. Liou, L.; Cao, X.; McKinstrie, C.; Agrawal, G.P. Spatiotemporal instabilities in dispersive nonlinear media. *Phys. Rev. A* **1992**, *46*, 4202.
21. Greer, E.; Patrick, D.; Wigley, P.; Taylor, J. Generation of 2 THz repetition rate pulse trains through induced modulational instability. *Electron. Lett.* **1989**, *25*, 1246–1248.
22. Agrawal, G. *Applications of Nonlinear Fiber Optics*; Academic Press: London, UK, 2001; Chapter 2.
23. Ellingham, T.; Ania-Castañón, J.; Turitsyn, S.; Pustovskikh, A.; Kobtsev, S.; Fedoruk, M. Dual-pump Raman amplification with increased flatness using modulation instability. *Opt. Express* **2005**, *13*, 1079–1084.
24. Dudley, J.M.; Genty, G.; Dias, F.; Kibler, B.; Akhmediev, N. Modulation instability, Akhmediev Breathers and continuous wave supercontinuum generation. *Opt. Express* **2009**, *17*, 21497–21508.
25. Trillo, S.; Wabnitz, S.; Wright, E.; Stegeman, G. Soliton switching in fiber nonlinear directional couplers. *Opt. Lett.* **1988**, *13*, 672–674.
26. Jensen, S. The nonlinear coherent coupler. *IEEE Trans. Microw. Theory Tech.* **1982**, *30*, 1568–1571.
27. Maier, A. Optical transistors and bistable elements on the basis of non-linear transmission of light by the systems with unidirectional coupled waves. *Kvantovaya Elektron.* **1982**, *9*, 2296–2302.
28. Kivshar, Y.S. Switching dynamics of solitons in fiber directional couplers. *Opt. Lett.* **1993**, *18*, 7–9.
29. Friberg, S.; Weiner, A.; Silberberg, Y.; Sfez, B.; Smith, P. Femotosecond switching in a dual-core-fiber nonlinear coupler. *Opt. Lett.* **1988**, *13*, 904–906.
30. Malomed, B.A.; Skinner, I.; Chu, P.; Peng, G. Symmetric and asymmetric solitons in twin-core nonlinear optical fibers. *Phys. Rev. E* **1996**, *53*, 4084–4091.
31. Chiang, K.S. Intermodal dispersion in two-core optical fibers. *Opt. Lett.* **1995**, *20*, 997–999.
32. Chen, Y.; Snyder, A.W.; Payne, D.N. Twin core nonlinear couplers with gain and loss. *IEEE J. Quantum Electron.* **1992**, *28*, 239–245.
33. Govindaraji, A.; Mahalingam, A.; Uthayakumar, A. Femtosecond pulse switching in a fiber coupler with third order dispersion and self-steepening effects. *Optik Int. J. Light Electron Opt.* **2014**, *125*, 4135–4139.
34. Snyder, A.W.; Mitchell, D.; Poladian, L.; Rowland, D.R.; Chen, Y. Physics of nonlinear fiber couplers. *J. Opt. Soc. Am. B* **1991**, *8*, 2102–2118.
35. Yang, C.C. All-optical ultrafast logic gates that use asymmetric nonlinear directional couplers. *Opt. Lett.* **1991**, *16*, 1641–1643.
36. Yang, C.C.; Wang, A. Asymmetric nonlinear coupling and its applications to logic functions. *IEEE J. Quantum Electron.* **1992**, *28*, 479–487.
37. Kitayama, K.I.; Wang, S. Optical pulse compression by nonlinear coupling. *Appl. Phys. Lett.* **1983**, *43*, 17–19.
38. Thirstrup, C. Optical bistability in a nonlinear directional coupler. *IEEE J. Quantum Electron.* **1995**, *31*, 2101–2106.
39. Trillo, S.; Stegeman, G.; Wright, E.; Wabnitz, S. Parametric amplification and modulational instabilities in dispersive nonlinear directional couplers with relaxing nonlinearity. *J. Opt. Soc. Am. B* **1989**, *6*, 889–900.
40. Tasgal, R.S.; Malomed, B.A. Modulational instabilities in the dual-core nonlinear optical fiber. *Phys. Scr.* **1999**, *60*, 418.

41. Li, J.H.; Chiang, K.S.; Chow, K.W. Modulation instabilities in two-core optical fibers. *J. Opt. Soc. Am. B* **2011**, *28*, 1693–1701.

42. Li, J.H.; Chiang, K.S.; Malomed, B.A.; Chow, K.W. Modulation instabilities in birefringent two-core optical fibres. *J. Phys. B At. Mol. Opt. Phys.* **2012**, *45*, 165404.

43. Nithyanandan, K.; Raja, R.V.J.; Porsezian, K. Modulational instability in a twin-core fiber with the effect of saturable nonlinear response and coupling coefficient dispersion. *Phys. Rev. A* **2013**, *87*, 043805.

44. Porsezian, K.; Murali, R.; Malomed, B.A.; Ganapathy, R. Modulational instability in linearly coupled complex cubic–quintic Ginzburg–Landau equations. *Chaos Solitons Fractals* **2009**, *40*, 1907–1913.

45. Ganapathy, R.; Malomed, B.A.; Porsezian, K. Modulational instability and generation of pulse trains in asymmetric dual-core nonlinear optical fibers. *Phys. Lett. A* **2006**, *354*, 366–372.

46. Malomed, B.A.; Peng, G.; Chu, P. Nonlinear-optical amplifier based on a dual-core fiber. *Opt. Lett.* **1996**, *21*, 330–332.

47. Kaup, D.J.; Malomed, B.A. Gap solitons in asymmetric dual-core nonlinear optical fibers. *J. Opt. Soc. Am. B* **1998**, *15*, 2838–2846.

48. Govindaraji, A.; Mahalingam, A.; Uthayakumar, A. Numerical investigation of dark soliton switching in asymmetric nonlinear fiber couplers. *Appl. Phys. B* **2015**, *120*, 341–348.

49. Li, Q.; Zhang, A.; Hua, X. Numerical simulation of solitons switching and propagating in asymmetric directional couplers. *Opt. Commun.* **2012**, *285*, 118–123.

50. Govindaraji, A.; Mahalingam, A.; Uthayakumar, A. Dark soliton switching in nonlinear fiber couplers with gain. *Opt. Laser Technol.* **2014**, *60*, 18–21.

51. He, X.; Xie, K.; Xiang, A. Optical solitons switching in asymmetric dual-core nonlinear fiber couplers. *Optik Int. J. Light Electron Opt.* **2011**, *122*, 1222–1224.

52. Shum, P.; Liu, M. Effects of intermodal dispersion on two-nonidentical-core coupler with different radii. *IEEE Photonics Technol. Lett.* **2002**, *14*, 1106–1108.

53. Nóbrega, K.; da Silva, M.; Sombra, A. Multistable all-optical switching behavior of the asymmetric nonlinear directional coupler. *Opt. Commun.* **2000**, *173*, 413–421.

54. Kaup, D.J.; Lakoba, T.I.; Malomed, B.A. Asymmetric solitons in mismatched dual-core optical fibers. *J. Opt. Soc. Am. B* **1997**, *14*, 1199–1206.

55. Atai, J.; Malomed, B.A. Stability and interactions of solitons in asymmetric dual-core optical waveguides. *Opt. Commun.* **2003**, *221*, 55–62.

56. Atai, J.; Malomed, B.A. Spatial solitons in a medium composed of self-focusing and self-defocusing layers. *Phys. Lett. A* **2002**, *298*, 140–148.

57. Zafrany, A.; Malomed, B.A.; Merhasin, I.M. Solitons in a linearly coupled system with separated dispersion and nonlinearity. *Chaos Interdiscip. J. Nonlinear Sci.* **2005**, *15*, 037108.

58. Govindaraji, A.; Mahalingam, A.; Uthayakumar, A. Interaction dynamics of bright solitons in linearly coupled asymmetric systems. *Opt. Quantum Electron.* **2016**, *48*, 563.

59. Xu, W.C.; Zhang, S.M.; Chen, W.C.; Luo, A.P.; Liu, S.H. Modulation instability of femtosecond pulses in dispersion-decreasing fibers. *Opt. Commun.* **2001**, *199*, 355–360.

MDPI AG

St. Alban-Anlage 66

4052 Basel, Switzerland

Tel. +41 61 683 77 34

Fax +41 61 302 89 18

http://www.mdpi.com

Applied Sciences Editorial Office

E-mail: applsci@mdpi.com

http://www.mdpi.com/journal/applsci